Sustaining Power Resources through Energy Optimization and Engineering

Pandian Vasant
Universiti Teknologi PETRONAS, Malaysia

Nikolai Voropai
Energy Systems Institute SB RAS, Russia

A volume in the Advances in Computer and
Electrical Engineering (ACEE) Book Series

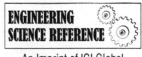

An Imprint of IGI Global

Published in the United States of America by
    Engineering Science Reference (an imprint of IGI Global)
    701 E. Chocolate Avenue
    Hershey PA, USA 17033
    Tel: 717-533-8845
    Fax: 717-533-8661
    E-mail: cust@igi-global.com
    Web site: http://www.igi-global.com

Copyright © 2016 by IGI Global. All rights reserved. No part of this publication may be reproduced, stored or distributed in any form or by any means, electronic or mechanical, including photocopying, without written permission from the publisher. Product or company names used in this set are for identification purposes only. Inclusion of the names of the products or companies does not indicate a claim of ownership by IGI Global of the trademark or registered trademark.
                      Library of Congress Cataloging-in-Publication Data

Names: Vasant, Pandian, editor. | Voropa?i, N. I. (Nikola?i Ivanovich),
  editor.
Title: Sustaining power resources through energy optimization and engineering
  / P. Vasant and Nikolai Voropai, editors.
Description: Hershey PA : Engineering Science Reference, [2016] | Includes
  bibliographical references and index.
Identifiers: LCCN 2015042055| ISBN 9781466697553 (hardcover) | ISBN
  9781466697560 (ebook)
Subjects: LCSH: Electric power systems--Mathematical models. | Energy
  transfer--Mathematical models. | Energy conservation--Mathematical models.
  | Renewable energy sources. | Sustainable development.
Classification: LCC TK1001 .S87 2016 | DDC 621.042--dc23 LC record available at http://lccn.loc.gov/2015042055

This book is published in the IGI Global book series Advances in Computer and Electrical Engineering (ACEE) (ISSN: 2327-039X; eISSN: 2327-0403)

British Cataloguing in Publication Data
A Cataloguing in Publication record for this book is available from the British Library.

All work contributed to this book is new, previously-unpublished material. The views expressed in this book are those of the authors, but not necessarily of the publisher.

For electronic access to this publication, please contact: eresources@igi-global.com.

Advances in Computer and Electrical Engineering (ACEE) Book Series

Srikanta Patnaik
SOA University, India

ISSN: 2327-039X
EISSN: 2327-0403

Mission

The fields of computer engineering and electrical engineering encompass a broad range of interdisciplinary topics allowing for expansive research developments across multiple fields. Research in these areas continues to develop and become increasingly important as computer and electrical systems have become an integral part of everyday life.

The **Advances in Computer and Electrical Engineering (ACEE) Book Series** aims to publish research on diverse topics pertaining to computer engineering and electrical engineering. **ACEE** encourages scholarly discourse on the latest applications, tools, and methodologies being implemented in the field for the design and development of computer and electrical systems.

Coverage

- Optical Electronics
- Analog Electronics
- Algorithms
- Programming
- Microprocessor Design
- Sensor Technologies
- Circuit Analysis
- Computer Hardware
- VLSI Design
- Digital Electronics

IGI Global is currently accepting manuscripts for publication within this series. To submit a proposal for a volume in this series, please contact our Acquisition Editors at Acquisitions@igi-global.com or visit: http://www.igi-global.com/publish/.

The Advances in Computer and Electrical Engineering (ACEE) Book Series (ISSN 2327-039X) is published by IGI Global, 701 E. Chocolate Avenue, Hershey, PA 17033-1240, USA, www.igi-global.com. This series is composed of titles available for purchase individually; each title is edited to be contextually exclusive from any other title within the series. For pricing and ordering information please visit http://www.igi-global.com/book-series/advances-computer-electrical-engineering/73675. Postmaster: Send all address changes to above address. Copyright © 2016 IGI Global. All rights, including translation in other languages reserved by the publisher. No part of this series may be reproduced or used in any form or by any means – graphics, electronic, or mechanical, including photocopying, recording, taping, or information and retrieval systems – without written permission from the publisher, except for non commercial, educational use, including classroom teaching purposes. The views expressed in this series are those of the authors, but not necessarily of IGI Global.

Titles in this Series

For a list of additional titles in this series, please visit: www.igi-global.com

Operation, Construction, and Functionality of Direct Current Machines
Muhammad Amin (COMSATS Institute of Information Technology, Wah Cantt, Pakistan) and Mubashir Husain Rehmani (COMSATS Institute of Information Technology, Wah Cantt, Pakistan)
Engineering Science Reference • copyright 2015 • 290pp • H/C (ISBN: 9781466684416) • US $200.00 (our price)

Performance Optimization Techniques in Analog, Mixed-Signal, and Radio-Frequency Circuit Design
Mourad Fakhfakh (University of Sfax, Tunisia) Esteban Tlelo-Cuautle (INAOE, Mexico) and Maria Helena Fino (New University of Lisbon, Portugal)
Engineering Science Reference • copyright 2015 • 464pp • H/C (ISBN: 9781466666276) • US $235.00 (our price)

Agile and Lean Service-Oriented Development Foundations, Theory, and Practice
Xiaofeng Wang (Free University of Bozen/Bolzano, Italy) Nour Ali (Lero- The Irish Software Engineering Research Centre, University of Limerick, Ireland) Isidro Ramos (Valencia University of Technology) and Richard Vidgen (Hull University Business School, UK)
Information Science Reference • copyright 2013 • 312pp • H/C (ISBN: 9781466625037) • US $195.00 (our price)

Electromagnetic Transients in Transformer and Rotating Machine Windings
Charles Q. Su (Charling Technology, Australia)
Engineering Science Reference • copyright 2013 • 586pp • H/C (ISBN: 9781466619210) • US $195.00 (our price)

Design and Test Technology for Dependable Systems-on-Chip
Raimund Ubar (Tallinn University of Technology, Estonia) Jaan Raik (Tallinn University of Technology, Estonia) and Heinrich Theodor Vierhaus (Brandenburg University of Technology Cottbus, Germany)
Information Science Reference • copyright 2011 • 578pp • H/C (ISBN: 9781609602123) • US $180.00 (our price)

Kansei Engineering and Soft Computing Theory and Practice
Ying Dai (Iwate Pref. University, Japan) Basabi Chakraborty (Iwate Prefectural University, Japan) and Minghui Shi (Xiamen University, China)
Engineering Science Reference • copyright 2011 • 436pp • H/C (ISBN: 9781616927974) • US $180.00 (our price)

Model Driven Architecture for Reverse Engineering Technologies Strategic Directions and System Evolution
Liliana Favre (Universidad Nacional de Centro de la Proviencia de Buenos Aires, Argentina)
Engineering Science Reference • copyright 2010 • 461pp • H/C (ISBN: 9781615206490) • US $180.00 (our price)

www.igi-global.com

701 E. Chocolate Ave., Hershey, PA 17033
Order online at www.igi-global.com or call 717-533-8845 x100
To place a standing order for titles released in this series, contact: cust@igi-global.com
Mon-Fri 8:00 am - 5:00 pm (est) or fax 24 hours a day 717-533-8661

Editorial Advisory Board

Mbohwa Charles, *University of Johannesburg, South Africa*
Vassili Kolokoltsov, *University of Warwick, UK*
Igor Litvinchev, *Nuevo Leon State University, Mexico*
Igor Tyukhov, *Moscow State University of Mechanical Engineering, Russia*
Dieu Ngoc Vo, *Ho Chi Minh University of Technology, Vietnam*
Gerhard Wilhelm Weber, *METU, Turkey*

List of Reviewers

Alibek Issakhov, *Al-Farabi Kazakh National University, Kazakhstan*
Goran Klepac, *Raiffeisen Bank Austria d. d., Croatia*
Gerardo Mendez, *Instituto Technologico de Nuevo Leon, Mexico*
Provas Kumar Roy, *Dr. B. C. Roy Engineering College, India*
Sergey V. Zharkov, *Siberian Branch of the Russian Academy of Sciences, Russia*

Table of Contents

Foreword *by Valeriy V. Kharchenko* xvii

Foreword *by Vassili N Kolokoltsov* xviii

Foreword *by Gerhard-Wilhelm Weber & N. Serhan Aydın & Erik Kropat* xix

Preface xxi

Acknowledgment xxv

Chapter 1
Assessment and Enhancement of the Energy Supply System Efficiency with Emphasis on the Cogeneration and Renewable as Main Directions for Fuel Saving 1
    Sergey Zharkov, Energy Systems Institute, Siberian Branch of the Russian Academy of Sciences, Russia

Chapter 2
Problems of Modeling and Optimization of Heat Supply Systems: Methods to Comprehensively Solve the Problem of Heat Supply System Expansion and Reconstruction 26
    Valery Stennikov, Energy Systems Institute, Siberian Branch of the Russian Academy of Sciences, Russia
    Tamara Oshchepkova, Energy Systems Institute, Siberian Branch of the Russian Academy of Sciences, Russia
    Nikolay Stennikov, Energy Systems Institute, Siberian Branch of the Russian Academy of Sciences, Russia

Chapter 3
Problems of Modeling and Optimization of Heat Supply Systems: Bi-Level Optimization of the Competitive Heat Energy Market 54
    Valery Stennikov, Energy Systems Institute, Siberian Branch of the Russian Academy of Sciences, Russia
    Andrey Penkovskii, Energy Systems Institute, Siberian Branch of the Russian Academy of Sciences, Russia
    Oleg Khamisov, Energy Systems Institute, Siberian Branch of the Russian Academy of Sciences, Russia

Chapter 4
Problems of Modeling and Optimization of Heat Supply Systems: New Methods and Software for Optimization of Heat Supply System Parameters 76

    Valery Stennikov, Energy Systems Institute, Siberian Branch of the Russian Academy of Sciences, Russia

    Evgeny Barakhtenko, Energy Systems Institute, Siberian Branch of the Russian Academy of Sciences, Russia

    Dmitry Sokolov, Energy Systems Institute, Siberian Branch of the Russian Academy of Sciences, Russia

    Tamara Oshchepkova, Energy Systems Institute, Siberian Branch of the Russian Academy of Sciences, Russia

Chapter 5
Problems of Modeling and Optimization of Heat Supply Systems: Methodological Support for a Comprehensive Analysis of Fuel and Heat Supply Reliability 102

    Valery A. Stennikov, Energy Systems Institute, Siberian Branch of the Russian Academy of Sciences, Russia

    Ivan V. Postnikov, Energy Systems Institute, Siberian Branch of the Russian Academy of Sciences, Russia

Chapter 6
Fuzzy Random Regression-Based Modeling in Uncertain Environment 127

    Nureize Arbaiy, University Tun Hussein Onn Malaysia, Malaysia

    Junzo Watada, Waseda University, Japan

    Pei-Chun Lin, Waseda University, Japan

Chapter 7
A Novel Optimization Algorithm for Transient Stability Constrained Optimal Power Flow 147

    Sourav Paul, Dr. B. C. Roy Engineering College, India

    Provas Kumar Roy, Jalpaiguri Government Engineering College, India

Chapter 8
Improved Pseudo-Gradient Search Particle Swarm Optimization for Optimal Power Flow Problem 177

    Jirawadee Polprasert, Asian Institute of Technology, Thailand

    Weerakorn Ongsakul, Asian Institute of Technology, Thailand

    Vo Ngoc Dieu, Ho Chi Minh City University of Technology, Vietnam

Chapter 9
Engineering QoS and Energy Saving in the Delivery of ICT Services 208

    Alessandra Pieroni, Guglielmo Marconi University of Study, Italy

    Giuseppe Iazeolla, Guglielmo Marconi University of Study, Italy

Chapter 10
Mathematical Modelling of the Thermal Process in the Aquatic Environment with Considering the Hydrometeorological Condition at the Reservoir-Cooler by Using Parallel Technologies 227

    Alibek Issakhov, al-Farabi Kazakh National University, Kazakhstan

Chapter 11
A Novel Evolutionary Optimization Technique for Solving Optimal Reactive Power Dispatch Problems 244
    Provas Kumar Roy, Jalpaiguri Government Engineering College, India

Chapter 12
Application of Adaptive Tabu Search Algorithm in Hybrid Power Filter and Shunt Active Power Filters: Application of ATS Algorithm in HPF and APF 276
    Saifullah Khalid, IEEE, India

Chapter 13
Recent Techniques to Identify the Stator Fault Diagnosis in Three Phase Induction Motor 309
    K. Vinoth Kumar, Karunya Institute of Technology and Sciences University, India
    S. Suresh Kumar, Dr. N. G. P. Institute of Technology, India
    A. Immanuel Selvakumar, Karunya Institute of Technology and Sciences University, India
    R. Saravana Kumar, Vellore Institute of Technology University, India

Chapter 14
Optimal Reactive Power Dispatch Incorporating TCSC-TCPS Devices Using Different Evolutionary Optimization Techniques 326
    Provas Kumar Roy, Jalpaiguri Government Engineering College, India
    Susanta Dutta, Dr. B. C. Roy Engineering College, India
    Debashis Nandi, National Institute of Technology, India

Chapter 15
Scope of Biogeography-Based Optimization for Economic Load Dispatch and Multi-Objective Unit Commitment Problem 360
    Vikram Kumar Kamboj, I. K. Gujral Punjab Technical University, India
    S. K. Bath, Giani Zail Singh Campus College of Engineering & Technology, Bathinda, India

Chapter 16
Modern Optimization Algorithms and Applications in Solar Photovoltaic Engineering 390
    Igor Tyukhov, All-Russian Research Institute for Electrification of Agriculture (VIESH), Russia
    Hegazy Rezk, Minia University, Egypt
    Pandian Vasant, Universiti Teknologi PETRONAS, Malaysia

Compilation of References 446

About the Contributors 486

Index 492

Detailed Table of Contents

Foreword *by Valeriy V. Kharchenko* xvii

Foreword *by Vassili N Kolokoltsov* xviii

Foreword *by Gerhard-Wilhelm Weber & N. Serhan Aydın & Erik Kropat* xix

Preface xxi

Acknowledgment xxv

Chapter 1
Assessment and Enhancement of the Energy Supply System Efficiency with Emphasis on the
Cogeneration and Renewable as Main Directions for Fuel Saving 1
    Sergey Zharkov, Energy Systems Institute, Siberian Branch of the Russian Academy of
        Sciences, Russia

The paper presents methods for assessing economic, resource and environmental efficiency of energy supply systems and ways of its improvement, the main of which are the development of cogeneration and renewable energy sources (RES). The problem of allocating fuel and financial costs in the case of the combined production is solved. The methods allow determining specific indicators of supplied products which makes it possible to compare the efficiency of energy supply systems of different companies and countries, and to define their future target indicators. The technology of introducing RES-based power plants to the energy supply systems by means of using unstabilized RES-based power for direct fuel substitution at thermal power plants. The paper can be interesting to power engineering specialists, businessmen and economists, and also participants of the upcoming UN Climate Change Conference aimed at achieving a universal agreement on climate, which will be held in 2015 in Paris.

Chapter 2
Problems of Modeling and Optimization of Heat Supply Systems: Methods to Comprehensively Solve the Problem of Heat Supply System Expansion and Reconstruction 26

    Valery Stennikov, Energy Systems Institute, Siberian Branch of the Russian Academy of Sciences, Russia

    Tamara Oshchepkova, Energy Systems Institute, Siberian Branch of the Russian Academy of Sciences, Russia

    Nikolay Stennikov, Energy Systems Institute, Siberian Branch of the Russian Academy of Sciences, Russia

The paper addresses the issue of optimal expansion and reconstruction of heat supply systems, which includes a set of general and specific problems. Therefore, a comprehensive approach to their solving is required to obtain a technically admissible and economically sound result. Solving the problem suggests search for effective directions in expansion of a system in terms of allocation of new heat sources, their type, output; construction of new heat networks, their schemes and parameters; detection of "bottlenecks" in the system and ways of their elimination (expansion, dismantling, replacement of heat pipeline sections, construction of pumping stations). The authors present a mathematical statement of the problem, its decomposition into separate subproblems and an integrated technique to solve it. Consideration is given to a real problem solved for a real heat supply system. A set of arising problems is presented. The application of developed methodological and computational tools is shown.

Chapter 3
Problems of Modeling and Optimization of Heat Supply Systems: Bi-Level Optimization of the Competitive Heat Energy Market 54

    Valery Stennikov, Energy Systems Institute, Siberian Branch of the Russian Academy of Sciences, Russia

    Andrey Penkovskii, Energy Systems Institute, Siberian Branch of the Russian Academy of Sciences, Russia

    Oleg Khamisov, Energy Systems Institute, Siberian Branch of the Russian Academy of Sciences, Russia

In the chapter, one of widespread models of the organization of heat supply of consumers presented in the "Single buyer" format is considered. The scientific and methodical base for its description and research offers to accept the fundamental principles of the theory of hydraulic circuits, bi-level programming, principles of economics in the energy sector. Distinctive feature of the developed mathematical model is that it, along with traditionally solved tasks within the bilateral relations heat sources – consumers of heat, considers a network component with physics and technology properties of a heat network inherent in it, and also the economic factors connected with costs of production and transport of heat energy. This approach gives the chance to define the optimum levels of load of heat sources, providing the set demand for heat energy from consumers taking into account receiving with heat sources of the greatest possible profit and performance thus of conditions of formation of the minimum costs in heat networks during the considered time period. Practical realization of the developed mathematical model is considered by the example of the heat supply system with two heat sources, working for single heat networks. With the help of the developed model are carried plotting optimal load heat sources during the year, as well as a nomogram for determining the average tariffs for heat energy heat sources.

Chapter 4
Problems of Modeling and Optimization of Heat Supply Systems: New Methods and Software for
Optimization of Heat Supply System Parameters 76
> *Valery Stennikov, Energy Systems Institute, Siberian Branch of the Russian Academy of Sciences, Russia*
> *Evgeny Barakhtenko, Energy Systems Institute, Siberian Branch of the Russian Academy of Sciences, Russia*
> *Dmitry Sokolov, Energy Systems Institute, Siberian Branch of the Russian Academy of Sciences, Russia*
> *Tamara Oshchepkova, Energy Systems Institute, Siberian Branch of the Russian Academy of Sciences, Russia*

This chapter presents new methods and software system SOSNA intended for the parameter optimization of heat supply systems. They make it possible to calculate large-scale systems which have a complex structure with any set of nodes, sections, and circuits. A new methodological approach to solving the problem of the parameter optimization of the heat supply systems is developed. The approach is based on the multi-level decomposition of the network model, which allows us to proceed from the initial problem to less complex sub-problems of a smaller dimension. New algorithms are developed to numerically solve the parameter optimization problems of heat supply systems: an effective algorithm based on the multi-loop optimization method, which allows us to consider hierarchical creation of the network model in the course of problem solving; a parallel high-speed algorithm based on the dynamic programming method. The new methods and algorithms were used in the software system SOSNA.

Chapter 5
Problems of Modeling and Optimization of Heat Supply Systems: Methodological Support for a
Comprehensive Analysis of Fuel and Heat Supply Reliability 102
> *Valery A. Stennikov, Energy Systems Institute, Siberian Branch of the Russian Academy of Sciences, Russia*
> *Ivan V. Postnikov, Energy Systems Institute, Siberian Branch of the Russian Academy of Sciences, Russia*

This chapter deals with the problem of comprehensive analysis of heat supply reliability for consumers. It implies a quantitative assessment of the impact of all stages of heat energy production and distribution on heat supply reliability for each consumer of the heat supply system. A short review of existing methods for the analysis of fuel and heat supply reliability is presented that substantiates the key approaches to solving the problem of comprehensive analysis of heat supply reliability. A methodological approach is suggested, in which mathematical models and methods for nodal evaluation of heat supply reliability for consumers are developed and the studies on the impact of different elements of fuel and heat supply systems on its level are described. Mathematical modeling is based on the Markov random processes, models of flow distribution in a heat network, deterministic dependences of thermal processes of heat energy consumption and some other models.

Chapter 6
Fuzzy Random Regression-Based Modeling in Uncertain Environment 127
    Nureize Arbaiy, University Tun Hussein Onn Malaysia, Malaysia
    Junzo Watada, Waseda University, Japan
    Pei-Chun Lin, Waseda University, Japan

The parameter value determination is important to avoid the developed mathematical model is troublesome and may yield inappropriate results. However, estimating the weights of the parameter or objective functions in the mathematical model is sometimes not easy in real situations, especially when the values are unavailable or difficult to decide. Additionally, various uncertainties include in the statistical data makes common mathematical analysis is not competent to deal with. Hence, this paper presents the Fuzzy Random Regression approach to determine the coefficient whereby statistical data used contain uncertainties namely, fuzziness and randomness. The proposed methods are able to provide coefficient information in the model setting and consideration of uncertainties in the evaluation process. The assessment of coefficient value is given by Weight Absolute Percentage Error of Fuzzy Decision. It clarifies the results between fuzzy decision and non-fuzzy decision that shows the distance of different between both approaches. Finally, a real-life application of production planning models is provided to illustrate the applicability of the proposed algorithms to a practical case study.

Chapter 7
A Novel Optimization Algorithm for Transient Stability Constrained Optimal Power Flow 147
    Sourav Paul, Dr. B. C. Roy Engineering College, India
    Provas Kumar Roy, Jalpaiguri Government Engineering College, India

Optimal power flow with transient stability constraints (TSCOPF) becomes an effective tool of many problems in power systems since it simultaneously considers economy and dynamic stability of power system. TSC-OPF is a non-linear optimization problem which is not easy to deal directly because of its huge dimension. This paper presents a novel and efficient optimisation approach named the teaching learning based optimisation (TLBO) for solving the TSCOPF problem. The quality and usefulness of the proposed algorithm is demonstrated through its application to four standard test systems namely, IEEE 30-bus system, IEEE 118-bus system, WSCC 3-generator 9-bus system and New England 10-generator 39-bus system. To demonstrate the applicability and validity of the proposed method, the results obtained from the proposed algorithm are compared with those obtained from other algorithms available in the literature. The experimental results show that the proposed TLBO approach is comparatively capable of obtaining higher quality solution and faster computational time.

Chapter 8
Improved Pseudo-Gradient Search Particle Swarm Optimization for Optimal Power Flow Problem 177
    Jirawadee Polprasert, Asian Institute of Technology, Thailand
    Weerakorn Ongsakul, Asian Institute of Technology, Thailand
    Vo Ngoc Dieu, Ho Chi Minh City University of Technology, Vietnam

This paper proposes an improved pseudo-gradient search particle swarm optimization (IPG-PSO) for solving optimal power flow (OPF) with non-convex generator fuel cost functions. The objective of OPF problem is to minimize generator fuel cost considering valve point loading, voltage deviation and voltage stability index subject to power balance constraints and generator operating constraints, transformer tap setting constraints, shunt VAR compensator constraints, load bus voltage and line flow constraints. The proposed IPG-PSO method is an improved PSO by chaotic weight factor and guided by pseudo-gradient search for particle's movement in an appropriate direction. Test results on the IEEE 30-bus and 118-bus systems indicate that IPG-PSO method is superior to other methods in terms of lower generator fuel cost, smaller voltage deviation, and lower voltage stability index.

Chapter 9
Engineering QoS and Energy Saving in the Delivery of ICT Services 208
    Alessandra Pieroni, Guglielmo Marconi University of Study, Italy
    Giuseppe Iazeolla, Guglielmo Marconi University of Study, Italy

ICT service-providers are to daily face the problem of delivering ICT services (data processing (Dp) and/ or telecommunication (Tlc) services) assuring the best compromise between Quality of Service (QoS) and Energy Optimization. Indeed, any operation of saving energy involves waste in the QoS. This holds both for Dp and for Tlc services. This paper introduces models the providers may use to support their decisions in the delivery of ICT services. Dp systems totalize millions of servers all over the world that need to be electrically powered. Dp systems are also used in the government of Tlc systems, which also require Tlc-specific power, both for mobile networks and for wired networks. Research is thus expected to investigate into methods to reduce ICT power consumption. This paper investigates ICT power management strategies that look at compromises between energy saving and QoS. Various optimizing ICT power management policies are studied that optimize the ICT power consumption (minimum absorbed Watts), the ICT performance (minimum response-time), and the ICT performance-per-Watt.

Chapter 10
Mathematical Modelling of the Thermal Process in the Aquatic Environment with Considering
the Hydrometeorological Condition at the Reservoir-Cooler by Using Parallel Technologies 227
    Alibek Issakhov, al-Farabi Kazakh National University, Kazakhstan

This paper presents the mathematical model of the thermal power plant in reservoir under different hydrometeorological conditions, which is solved by three dimensional Navier - Stokes and temperature equations for an incompressible fluid in a stratified medium. A numerical method based on the projection method, which divides the problem into four stages. At the first stage it is assumed that the transfer of momentum occurs only by convection and diffusion. Intermediate velocity field is solved by fractional steps method. At the second stage, three-dimensional Poisson equation is solved by the Fourier method in combination with tridiagonal matrix method (Thomas algorithm). At the third stage it is expected that the transfer is only due to the pressure gradient. Finally stage equation for temperature solved like momentum equation with fractional step method. To increase the order of approximation compact scheme was used. Then qualitatively and quantitatively approximate the basic laws of the hydrothermal processes depending on different hydrometeorological conditions are determined.

Chapter 11
A Novel Evolutionary Optimization Technique for Solving Optimal Reactive Power Dispatch
Problems 244
    Provas Kumar Roy, Jalpaiguri Government Engineering College, India

Biogeography based optimization (BBO) is an efficient and powerful stochastic search technique for solving optimization problems over continuous space. Due to excellent exploration and exploitation property, BBO has become a popular optimization technique to solve the complex multi-modal optimization problem. However, in some cases, the basic BBO algorithm shows slow convergence rate and may stick to local optimal solution. To overcome this, quasi-oppositional biogeography based-optimization (QOBBO) for optimal reactive power dispatch (ORPD) is presented in this study. In the proposed QOBBO algorithm, oppositional based learning (OBL) concept is integrated with BBO algorithm to improve the search space of the algorithm. For validation purpose, the results obtained by the proposed QOBBO approach are compared with those obtained by BBO and other algorithms available in the literature. The simulation results show that the proposed QOBBO approach outperforms the other listed algorithms.

Chapter 12
Application of Adaptive Tabu Search Algorithm in Hybrid Power Filter and Shunt Active Power
Filters: Application of ATS Algorithm in HPF and APF 276
    Saifullah Khalid, IEEE, India

A novel hybrid series active power filter to eliminate harmonics and compensate reactive power is presented and analyzed. The proposed active compensation technique is based on a hybrid series active filter using ATS algorithm in the conventional Sinusoidal Fryze voltage (SFV) control technique. This chapter discusses the comparative performances of conventional Sinusoidal Fryze voltage control strategy and ATS-optimized controllers. ATS algorithm has been used to obtain the optimum value of Kp and Ki. Analysis of the hybrid series active power filter system under non-linear load condition and its impact on the performance of the controllers is evaluated. MATLAB/Simulink results and Total harmonic distortion (THD) shows the practical viability of the controller for hybrid series active power filter to provide harmonic isolation of non-linear loads and to comply with IEEE 519 recommended harmonic standards. The ATS-optimized controller has been attempted for shunt active power filter too, and its performance has also been discussed in brief.

Chapter 13
Recent Techniques to Identify the Stator Fault Diagnosis in Three Phase Induction Motor 309
    K. Vinoth Kumar, Karunya Institute of Technology and Sciences University, India
    S. Suresh Kumar, Dr. N. G. P. Institute of Technology, India
    A. Immanuel Selvakumar, Karunya Institute of Technology and Sciences University, India
    R. Saravana Kumar, Vellore Institute of Technology University, India

Induction motors have gained its popularity as most suitable industrial workhorse, due to its ruggedness and reliability. With the passage of time, these workhorses are susceptible to faults, some are incipient and some are major. Such fault can be catastrophic, if unattended and may develop serious problem that may lead to shut down the machine causing production and financial losses. To avoid such breakdown, an early stage prognosis can help in preparing the maintenance schedule, which will lead to improve its life span. Scientist and engineers worked with different scheme to diagnose the machine faults. In this paper, the authors diagnose the turn-to-turn faults condition of the stator through symmetrical component analysis. The results of the analysis also verified through Power Decomposition Technique (PDT) in Matlab /SIMULINK. The results are compatible with the published results for known faults.

Chapter 14
Optimal Reactive Power Dispatch Incorporating TCSC-TCPS Devices Using Different Evolutionary Optimization Techniques 326
    Provas Kumar Roy, Jalpaiguri Government Engineering College, India
    Susanta Dutta, Dr. B. C. Roy Engineering College, India
    Debashis Nandi, National Institute of Technology, India

The chapter presents two effective evolutionary methods, namely, artificial bee colony optimization (ABC) and biogeography based optimization (BBO) for solving optimal reactive power dispatch (ORPD) problem using flexible AC transmission systems (FACTS) devices. The idea is to allocate two types of FACTS devices such as thyristor-controlled series capacitor (TCSC) and thyristor-controlled phase shifter (TCPS) in such a manner that the cost of operation is minimized. In this paper, IEEE 30-bus test system with multiple TCSC and TCPS devices is considered for investigations and the results clearly show that the proposed ABC and BBO methods are very competent in solving ORPD problem in comparison with other existing methods.

Chapter 15
Scope of Biogeography-Based Optimization for Economic Load Dispatch and Multi-Objective Unit Commitment Problem 360
    Vikram Kumar Kamboj, I. K. Gujral Punjab Technical University, India
    S. K. Bath, Giani Zail Singh Campus College of Engineering & Technology, Bathinda, India

Biogeography Based Optimization (BBO) algorithm is a population-based algorithm based on biogeography concept, which uses the idea of the migration strategy of animals or other spices for solving optimization problems. Biogeography Based Optimization algorithm has a simple procedure to find the optimal solution for the non-smooth and non-convex problems through the steps of migration and mutation. This research chapter presents the solution to Economic Load Dispatch Problem for IEEE 3, 4, 6 and 10-unit generating model using Biogeography Based Optimization algorithm. It also presents the mathematical formulation of scalar and multi-objective unit commitment problem, which is a further extension of economic load dispatch problem.

Chapter 16
Modern Optimization Algorithms and Applications in Solar Photovoltaic Engineering 390
    Igor Tyukhov, All-Russian Research Institute for Electrification of Agriculture (VIESH),
      Russia
    Hegazy Rezk, Minia University, Egypt
    Pandian Vasant, Universiti Teknologi PETRONAS, Malaysia

This chapter is devoted to main tendencies of optimization in photovoltaic (PV) engineering showing the main trends in modern energy transition - the changes in the composition (structure) of primary energy supply, the gradual shift from a traditional (mainly based on fossil fuels) energy to a new stage based on renewable energy systems from history to current stage and to future. The concrete examples (case studies) of optimization PV systems in different concepts of using from power electronics (particularly maximum power point tracking optimization) to implementing geographic information system (GIS) are considered. The chapter shows the gradual shifting optimization from specific quite narrow areas to the new stages of optimization of the very complex energy systems (actually smart grids) based on photovoltaics and also other renewable energy sources and GIS.

Compilation of References 446

About the Contributors 486

Index 492

Foreword

Nowadays the sustainable development is probably the most important problem in view of mankind surviving. It is complex, multidisciplinary concept for which the standard, finally ascertained definition is still absent. But it is absolutely clear that among various factors defining character of development of the countries, regions and territories is energy. According one of existing points of view sustainable energy is any kind of energy that can sufficiently cater to the energy needs of the present generation without endangering the supply of energy to future generations. Sometimes the term is defined to include any kind of substitute form of energy that is used as the current generation tries to come up with technological innovations that will lead to the discovery of sustainable energy. That's why the book *Sustaining Power Resources through Energy Optimization and Engineering* is very timely and very and important. The purpose of the book is to gather together chapters dedicated to different aspects of effective energy resources generation and consuming which cover a huge topics dealing in very different degree with sustainable consumption of energy such as energy management policy, economic Load Dispatch, Biogeography Based Optimization, Heat Supply Complex, Markov Random Process, Multi-loop Optimization Method, three Induction motor, bi-level modeling, competitive market, thermal power plant, heat network, teaching learning based optimization, optimal power flow, fuzzy decision, hybrid power filter, evolutionary optimization, artificial bee colony, greenhouse gas emissions, valve-point loading effect and ground source energy.

Valuable emphasis has been put on two the most important tendency in the modern energy such as rational use of energy (RES) and renewable energy sources (RES) implementations.

Actually the book is in some way the encyclopedia of efficiency in an energy sector. In spite of the fact that in the book such important aspects as photo-voltaic conversion of solar energy, electricity generation at wind and small power stations as well as distributed energy aspects are absent it certainly is useful to a wide range of readers, as for energy experts, as well for specialists of different profile dealing with consumption of various energy resources.

There is a wish to express deep gratitude to all authors who were taking part in preparation of the manuscript of the book for their efforts, for their job, and generosity with which they impart experience and knowledge with readers. The special gratitude is deserved publishing house of IGI Global in particular editorial team, consisting of Prof. Dr. Pandian Vasant (Universiti Teknologi PETRONAS, Malaysia) and Prof. Nikolai Voropai (Irkutsk Technical University, Russia) and the Editorial Advisory Board, for having provided the opportunity for authors to publish their newest results.

Thanks to all who take part in preparing, editing and publishing the premium book of a high academic standards.

Valeriy V. Kharchenko
All-Russian Scientific Research Institute for Electrification of Agriculture (VIESH), Russia

Foreword

Energy optimization and engineering are the key problems for the modern society. The efforts of many research groups and the whole Institutes all over the globe are devoted to a proper storage, distribution and production of power resources. This book "Sustaining Power Resources through Energy Optimization and Engineering" is aimed at giving the state-of-the-art of various directions of research in this direction. The wideness of the cutting edge contribution is quite impressive. In the 16 chapters of the book the reader will follow the development of cogeneration and renewable energy sources, the discussion of optimal expansion and reconstruction of heat supply systems, with related questions of reliability for consumers, specific software systems (e. g. SOSNA) for the parameter optimization of such systems, exploitation of renewable energy sources dealt with by various optimization models and statistical tools, related problems from biogeography. The expositions combine clear laymen-term explanations, arguments and suggestions with deeper mathematical developments. The book is highly recommended for all interested in this vibrant and very timely topic of sustaining power resources.

Vassili N Kolokoltsov
University of Warwick, UK

Vassili Kolokoltsov *is a professor at the Department of Statistics of the University of Warwick. He receives his PhD in 1985 from the Moscow State University and the Full Doctor of Science in Mathematics and Physics from the Steklov Mathematics Institute of the Russia Academy of Science in 1993. His research is in various areas of applied mathematics, mostly in stochastic analysis, control theory and games, and mathematical physics. He published 8 monographs and over 100 papers in peer reviewed journals. He presented invited lectures on numerous international conferences in US, Mexico, Japan, Russia and Europe. In 2011 he received (jointly with O. A. Malafeyev) the award of St.Petersburg University for the Research in Game Theory.*

Foreword

The main purpose of the project of this book "Sustaining Power Resources through Energy Optimization and Engineering" has been to introduce advanced results in the areas of sustainable development and operation of energy systems. In fact, these have become very important problems from the perspectives of an effective energy supply to consumers with required quality and of the reliability of supply, both intra- and inter-generationally. An effective operation of the economy and of the life of the people strongly depends on a sustainable development and an efficient and reliable operation of energy systems. Herewith, through its collection and display of the state-of-the-art in those areas and of emerging trends, this book project has become very worthwhile. Soon it could mean a precious and very much needed service for the academic sector, for the practice of engineering and of our economies and, eventually, for developing policies and innovative solutions to improve the living conditions of the people on earth.

In the course of the recent decades, the toolbox of engineering, economics, computer science, statistics and applied mathematics, informatics, bio and life sciences, has gained the interest of so many researchers and practitioners from all around the world, in emerging analytics, algorithms and information technologies, giving a strong impact to all areas of engineering and information technologies, but also in economy, finance and social sciences. A core role in this context is played by Optimization, Optimal Control and Probability Theory, in theory, methods and practice. This book benefits from that vast growth as much as it is applied on "Sustaining Power Resources through Energy Optimization and Engineering."

The authors of the chapters of this book are experienced and enthusiastic researchers and scholars from all over the world, who refine, associate and employ the less model-based but more data-driven techniques of engineering and computer science and the deep model-based techniques methods from mathematics. The first ones are also called smart or intelligent algorithms; they have their roots in the engineering disciplines, in computer science, informatics, in bio- and nature-inspired traditions of reasoning. Indeed, amazing challenges exist in all fields of the modern live, in high technology and economy, in the sectors of development, the improvement of living conditions, of gaining future chance and perspectives for us humans. In this context, without any doubt, the areas studied in this book, namely, power, energy and electricity, as well as the related fields of climate change, economics, technology and environmental sciences, play a huge role which can hardly be overestimated.

Special attention paid in this work is the presence of *uncertainties* of multiple kinds, the high complexity of analysing and interpreting large data sets and of entire problems that we nowadays find almost everywhere. In the present book, these two main academic cultures and traditions, namely, the more model-free and data-driven one from engineering and the more model-based one from mathematics, are not considered as separated or disjoint from each other, but from their potential of *common* chances, of synergies and of a promise to humankind.

This valuable compendium addresses the following research topics and techniques, and much more, in relation with "Sustaining Power Resources through Energy Optimization and Engineering":

- *Energy Management Policy,*
- *Economic Load Dispatch,*
- *Biogeography Based Optimization,*
- *Heat Supply Complex,*
- *Markov Random Process,*
- *Multi-loop Optimization Method,*
- *Three Induction Motor,*
- *Bi-level Modeling,*
- *Competitive Market,*
- *Thermal Power Plant,*
- *Heat Network,*
- *Teaching Learning Based Optimization,*
- *Optimal Power Flow,*
- *Fuzzy Decision,*
- *Hybrid Power Filter,*
- *Evolutionary Optimization,*
- *Artificial Bee Colony,*
- *Greenhouse Gas Emissions,*
- *Valve-Point Loading Effect,*
- *Ground source Energy.*

This means a huge variety and prosperity indeed, unfolded in sixteen chapters.

To every one of the authors of these valuable *chapter* contributions, we extend our sincere appreciation and our gratitude for having shared their devotion, expertise and new insights with the whole academic family and with mankind. We are very thankful to the Publishing House IGI Global, to the editorial team which consists of the editors Prof. Dr. Pandian Vasant and Prof. Dr. Nikolai Voropai, and to the Editorial Advisory Board Members, for having offered and maintained the chance and the stage for experts to publish their emerging advances and suggestions. We convey to them all our heartily thanks for having made possible a premium book on a high academic level, of an intellectual, a real-world and a human importance.

Now, we wish you and each one of us a great pleasure and gain when reading this new IGI Global work, a great benefit through it in all personal, professional and social directions.

Sincerely yours,

Gerhard-Wilhelm Weber
METU, Institute of Applied Mathematics, Turkey

N. Serhan Aydın
METU, Institute of Applied Mathematics, Turkey

Erik Kropat
University of the Bundeswehr Munich, Germany
July 22, 2015

Preface

Sustainable development and operation of energy systems are very important problems from the viewpoint of effective energy supply to consumers with required quality and reliability of supply. These problems are the most important for infrastructural energy systems which have developed transmission and distribution network structures. The effective operation of the economy and the life of people strongly depend on sustainable development and operation of energy systems. The objective of this book is to introduce advanced results in the foregoing area.

The following research topics are well covered in this book:

- Energy Management Policy
- Economic Load Dispatch
- Biogeography-Based Optimization
- Heat Supply Complex
- Markov Random Process
- Multi-Loop Optimization Method
- Three Induction motor
- Bi-Level Modeling
- Competitive Market
- Thermal Power Plant
- Heat Network
- Teaching Learning-Based Optimization
- Optimal Power Flow
- Fuzzy Decision
- Hybrid Power Filter
- Evolutionary Optimization
- Artificial Bee Colony
- Greenhouse Gas Emissions
- Valve-Point Loading Effect
- Ground Source Energy

The book is organized into 16 chapters. A brief description of each of the chapters is as follows:

Chapter 1: The chapter presents methods for assessing economic, resource and environmental efficiency of energy supply systems and ways of its improvement, the main of which are the development of cogeneration and renewable energy sources (RES). The problem of allocating fuel and financial costs in the case of the combined production is solved. The methods allow determining specific indicators of supplied products which makes it possible to compare the efficiency of energy supply systems of different companies and countries, and to define their future target indicators. The technology of introducing RES-based power plants to the energy supply systems by means of using unstabilized RES-based power for direct fuel substitution at thermal power plants is discussed.

Chapter 2: The chapter addresses the issue of optimal expansion and reconstruction of heat supply systems, which includes a set of general and specific problems. Therefore, a comprehensive approach to their solving is required to obtain a technically admissible and economically sound result. Solving the problem suggests search for effective directions in expansion of a system in terms of allocation of new heat sources, their type, output; construction of new heat networks, their schemes and parameters; detection of "bottlenecks" in the system and ways of their elimination (expansion, dismantling, replacement of heat pipeline sections, construction of pumping stations). The authors present a mathematical statement of the problem, its decomposition into separate subproblems and an integrated technique to solve it. Consideration is given to a real problem solved for a real heat supply system.

Chapter 3: In this chapter one of widespread models of the organization of heat supply of consumers presented in the Single buyer format is considered. The scientific and methodical base for its description and research offers to accept the fundamental principles of the theory of hydraulic circuits, bi-level programming, and principles of economics in the energy sector. Distinctive feature of the developed mathematical model is that it, along with traditionally solved tasks within the bilateral relations heat sources – consumers of heat, considers a network component with physics and technology properties of a heat network inherent in it, and also the economic factors connected with costs of production and transport of heat energy.

Chapter 4: This chapter presents new methods and software system SOSNA intended for the parameter optimization of heat supply systems. They make it possible to calculate large-scale systems which have a complex structure with any set of nodes, sections, and circuits. A new methodological approach to solving the problem of the parameter optimization of the heat supply systems is developed. The approach is based on the multi-level decomposition of the network model, which allows us to proceed from the initial problem to less complex sub-problems of a smaller dimension. New algorithms are developed to numerically solve the parameter optimization problems of heat supply systems: an effective algorithm based on the multi-loop optimization method, which allows us to consider hierarchical creation of the network model in the course of problem solving; a parallel high-speed algorithm based on the dynamic programming method.

Chapter 5: This chapter deals with the problem of comprehensive analysis of heat supply reliability for consumers. It implies a quantitative assessment of the impact of all stages of heat energy production and distribution on heat supply reliability for each consumer of the heat supply system. A methodological approach is suggested, in which mathematical models and methods for nodal evaluation of heat supply reliability for consumers are developed and the studies on the impact of different elements of fuel and heat supply systems on its level are described. Mathematical modeling is based on the Markov random processes, models of flow distribution in a heat network, deterministic dependences of thermal processes of heat energy consumption and some other models.

Chapter 6: In the recent attempts to stimulate alternative energy sources for heating and cooling of buildings, emphasise has been put on utilisation of the ambient energy from ground source heat pump systems (GSHPs) and other renewable energy sources. Exploitation of renewable energy sources and particularly ground heat in buildings can significantly contribute towards reducing dependency on fossil

Preface

fuels. Energy Research Institute (ERI), between July 2011 and November 2011. This chapter highlights the potential energy saving that could be achieved through use of ground energy source.

Chapter 7: This chapter presents a Fuzzy Random Regression-based model. The proposed method demonstrates the ability to provide coefficient information and consideration of hybrid uncertainties in the evaluation process. The schemes discussed in this work satisfy decision-maker intentions, which address some limitation to determine the coefficient and deals with fuzzy random uncertainties in the multi-objective problem formulation using satisfaction-based optimization.

Chapter 8: Optimal power flow with transient stability constraints becomes an effective tool of many problems in power systems since it simultaneously considers economy and dynamic stability of power system. TSC-OPF is a non-linear optimization problem which is not easy to deal directly because of its huge dimension. This chapter presents a novel and efficient optimisation approach named the teaching learning based optimisation (TLBO) for solving the TSCO-PF problem. The quality and usefulness of the proposed algorithm is demonstrated through its application to four standard test systems namely, IEEE 30-bus system, IEEE 118-bus system, WSCC 3-generator 9-bus system and New England 10-generator 39-bus system. To demonstrate the applicability and validity of the proposed method, the results obtained from the proposed algorithm are compared with those obtained from other algorithms available in the literature.

Chapter 9: This chapter proposes an improved pseudo-gradient search particle swarm optimization (IPG-PSO) for solving optimal power flow (OPF) with non-convex generator fuel cost functions. The objective of OPF problem is to minimize generator fuel cost considering valve point loading, voltage deviation and voltage stability index subject to power balance constraints and generator operating constraints, transformer tap setting constraints, shunt VAR compensator constraints, load bus voltage and line flow constraints. The proposed IPG-PSO method is an improved PSO by chaotic weight factor and guided by pseudo-gradient search for particle's movement in an appropriate direction.

Chapter 10: ICT service-providers are to daily face the problem of delivering ICT services (data processing (Dp) and telecommunication (Tlc) services) assuring the best compromise between Quality of Service (QoS) and Energy Optimization. Indeed, any operation of saving energy involves waste in the QoS. This holds both for Dp and for Tlc services. This chapter introduces models the providers may use to support their decisions in the delivery of ICT services. Dp systems totalize millions of servers all over the world that need to be electrically powered. Dp systems are also used in the government of Tlc systems, which also require Tlc-specific power, both for mobile networks and for wired networks. Research is thus expected to investigate into methods to reduce ICT power consumption. This chapter investigates ICT power management strategies that look at compromises between energy saving and QoS.

Chapter 11: This chapter presents the mathematical model of the thermal power plant in reservoir under different hydrometeorological conditions, which is solved by three dimensional Navier - Stokes and temperature equations for an incompressible fluid in a stratified medium. A numerical method based on the projection method, which divides the problem into four stages. Then qualitatively and quantitatively approximate the basic laws of the hydrothermal processes depending on different hydrometeorological conditions are determined.

Chapter 12: Biogeography based optimization (BBO) is an efficient and powerful stochastic search technique for solving optimization problems over continuous space. Due to excellent exploration and exploitation property, BBO has become a popular optimization technique to solve the complex multimodal optimization problem. However, in some cases, the basic BBO algorithm shows slow convergence rate and may stick to local optimal solution. To overcome this, quasi-oppositional biogeography based-optimization (QOBBO) for optimal reactive power dispatch (ORPD) is presented in this chapter.

Chapter 13: Biogeography Based Optimization (BBO) algorithm is a population-based algorithm based on biogeography concept, which uses the idea of the migration strategy of animals or other spices for solving optimization problems. Biogeography Based Optimization algorithm has a simple procedure to find the optimal solution for the non-smooth and non-convex problems through the steps of migration and mutation. This research chapter presents the solution to Economic Load Dispatch Problem for IEEE 3, 4, 6 and 10-unit generating model using Biogeography Based Optimization algorithm. It also presents the mathematical formulation of scalar and multi-objective unit commitment problem, which is a further extension of economic load dispatch problem.

Chapter 14: A novel hybrid series active power filter to eliminate harmonics and compensate reactive power is presented and analyzed. The proposed active compensation technique is based on a hybrid series active filter using ATS algorithm in the conventional Sinusoidal Fryze voltage (SFV) control technique. This chapter discusses the comparative performances of conventional Sinusoidal Fryze voltage control strategy and ATS-optimized controllers. The ATS-optimized controller has been attempted for shunt active power filter too, and its performance has also been discussed in brief.

Chapter 15: Induction motors have gained its popularity as most suitable industrial workhorse, due to its ruggedness and reliability. With the passage of time, these workhorses are susceptible to faults, some are incipient and some are major. Such fault can be catastrophic, if unattended and may develop serious problem that may lead to shut down the machine causing production and financial losses. To avoid such breakdown, an early stage prognosis can help in preparing the maintenance schedule, which will lead to improve its life span. Scientist and engineers worked with different scheme to diagnose the machine faults. In this chapter, the authors diagnose the turn-to-turn faults condition of the stator through symmetrical component analysis.

Chapter 16: This chapter is devoted to main tendencies of optimization in photovoltaic (PV) engineering showing the main trends in modern energy transition - the changes in the composition (structure) of primary energy supply, the gradual shift from a traditional (mainly based on fossil fuels) energy to a new stage based on renewable energy systems from history to current stage and to future. The chapter shows the gradual shifting optimization from specific quite narrow areas to the new stages of optimization of the very complex energy systems (actually smart grids) based on photovoltaics and also other renewable energy sources and GIS.

The book editors are very grateful to the editorial team and entire staff of IGI Global, for their confidence, interest, continuous guidance and support at all levels of preparation of this book's preparation.

We editors wish all the readers a pleasant and enjoyable, insightful and inspiring lecture of the contributions of this IGI Global book. In fact, we cordially hope that this special issue will present and value IGI Global as a premium publishing house in science, engineering, economics, and finance, which strongly fosters very much needed intellectual advances and their contributions to humanity and mankind in all over the world.

Pandian Vasant
Universiti Teknologi PETRONAS, Malaysia

Nikolai Voropai
Energy Systems Institute SB RAS, Russia

Acknowledgment

The editors would like to acknowledge the help of all the people involved in this project and, more specifically, to the authors and reviewers that took part in the review process. Without their support, this book would not have become a reality.

First, the editors would like to thank each one of the authors for their contributions. Our sincere gratitude goes to the chapter's authors who contributed their time and expertise to this book.

Second, the editors wish to acknowledge the valuable contributions of the reviewers regarding the improvement of quality, coherence, and content presentation of chapters. Most of the authors also served as referees; we highly appreciate their double task.

Finally, the editors sincerely the authorities of Universiti Teknologi PETRONAS and Irkutsk Technical University for their strong encouragement, motivation and support in making this book very successful publication in the global stage.

Pandian Vasant
Universiti Teknologi PETRONAS, Malaysia

Nikolai Voropai
Energy Systems Institute SB RAS, Russia

Chapter 1
Assessment and Enhancement of the Energy Supply System Efficiency with Emphasis on the Cogeneration and Renewable as Main Directions for Fuel Saving

Sergey Zharkov
Energy Systems Institute, Siberian Branch of the Russian Academy of Sciences, Russia

ABSTRACT

The paper presents methods for assessing economic, resource and environmental efficiency of energy supply systems and ways of its improvement, the main of which are the development of cogeneration and renewable energy sources (RES). The problem of allocating fuel and financial costs in the case of the combined production is solved. The methods allow determining specific indicators of supplied products which makes it possible to compare the efficiency of energy supply systems of different companies and countries, and to define their future target indicators. The technology of introducing RES-based power plants to the energy supply systems by means of using unstabilized RES-based power for direct fuel substitution at thermal power plants. The paper can be interesting to power engineering specialists, businessmen and economists, and also participants of the upcoming UN Climate Change Conference aimed at achieving a universal agreement on climate, which will be held in 2015 in Paris.

INTRODUCTION

The nation that leads the world in creating new sources of clean energy will be the nation that leads the 21st century global economy -Barack Obama

We must leave future generations with the world energy which will help to avoid conflicts and unreasonable rivalry for energy security. -V. V. Putin

DOI: 10.4018/978-1-4666-9755-3.ch001

The recent report of the UN predicts catastrophic deterioration of a climatic situation on the planet. Nobel Prize laureate, economist Michael Spence calls for immediate actions (the article "Growth in the New Climate Economy" at Project Syndicate), otherwise future expenses on neutralization of ecological damage will be "very heavy" for global economy. Recently published report of the Global Commission on the Economy and Climate "The New Climate Economy: Better Growth, Better Climate", proves that the actions for fight against climate change not only do not slow down the economic growth, but even can accelerate it considerably - and in the near future. Therefore, the measures to reduce environmental pollution, save fuel and other resources, develop cogeneration, use renewable energy resources on a large scale, and assess their efficiency are very important (Jovanovic, Turanjanin, Bakic, Pezo & Vucicevic B, 2011; Pezzini, Gomis-Bellmunt & Sudrià-Andreu, 2011; Melentiev, 1987; Melentiev, 1993; Twidell & Weir 1990). Besides, cogeneration and wind power industry are the most significant technologies for fuel saving and reducing greenhouse gas emissions.

Russia and other countries apply methods for separation of fuel and financial costs in the energy sector mainly at cogeneration (combined heat and power production) plants (CPs) in order to assess their efficiency and fix heat and electricity tariffs (Alexanov, 1995; Arakelyan, Kozhevnikov & Kuznetsov, 2006; Gitelman & Ratnikov, 2008; Kharaim, 2003; Kuznetsov, 2006; Malafeev, Smirnov, Kharaim, Khrilev & Livshits, 2003; Melentiev, 1987; Melentiev, 1993; Padalko & Zaborovsky, 2006; Popyrin, Denisov & Svetlov, 1989; Semenov, 2002). The problem of interrelated pricing of electric and heat energy is especially urgent for the countries that consume a lot of fuel for heat supply and have a high share of CPs in their energy systems. However, in this case it appears to be important to pay more attention to the assessment of the overall efficiency of energy supply systems.

The issue of separating total costs of cogeneration is highly topical not only for the energy sector, but also for other branches of industry producing several types of products. Moreover, a great number of methods for separating costs are applied in the energy sector and industry.

Development of methods for separating costs at CPs leads in particular to the following conclusion (Gitelman & Ratnikov, 2008; Kharaim, 2003; Malafeev, Smirnov, Kharaim, Khrilev & Livshits, 2003; Semenov, 2002): Allocation of CPs costs requires that the principle of equivalence between electricity and heat markets (Gitelman & Ratnikov, 2008) (equivalent equilibrium (Malafeev, Smirnov, Kharaim, Khrilev & Livshits, 2003)) be satisfied, since the use of combined generation technology involves no individual heat and power generation businesses. Here it is possible to speak just about the combined generation business proper, because the inefficiency of one decreases the efficiency of the other, so they can be only equally efficient or inefficient. And only a temporary imbalance is acceptable.

In author's opinion the efficiency of energy supply systems is characterized by three aspects – economic, resource (including fuel) and environmental. Therefore, new methods are suggested by the author (Zharkov, 2009) for the overall assessment of economic, resource and environmental efficiency of CPs and the energy supply systems as a whole and the ways to enhance it with emphasis on the cogeneration and renewables as the main directions for fuel economy.

METHODS OF EXPENSES ALLOCATION

I. Cogeneration (Combined Heat and Power Generation at Cogeneration Power Plants (CP))

Let us consider the economic efficiency of CPs. Let

$X = W_{CP}^{\Sigma} s_E$ – cost of electricity from CP,

where

W_{CP}^{Σ} – amount of electricity supplied by CP to consumers,
s_E – cost of electricity in the market,

$Y = Q_{CP}^{\Sigma} s_H$ – cost of heat from CP,

where

Q_{CP}^{Σ} – amount of heat supplied by CP to consumers,
s_H – heat cost in the market.

If there are no heat and electricity markets, it is necessary to calculate several variants of separate production and determine values s_E and s_H.

S_{CP}^{Σ} – total costs (expenditures) of CP,

x – share of electricity in S_{CP}^{Σ},

y – share of heat in S_{CP}^{Σ}, i.e. $x + y = 1$ (1)

Proportional cost allocation (according to the principle of equivalence between electricity and heat markets) requires that the following condition be met

$$\frac{x}{X} = \frac{y}{Y} \qquad (2)$$

By solving the system of equations (1) and (2), we find:

$$x = \frac{X}{X+Y} = \frac{W_{CP}^{\Sigma} s_E}{W_{CP}^{\Sigma} s_E + Q_{CP}^{\Sigma} s_H} \qquad (3)$$

$$y = \frac{Y}{X+Y} = \frac{Q_{CP}^{\Sigma} s_H}{W_{CP}^{\Sigma} s_E + Q_{CP}^{\Sigma} s_H} \qquad (4)$$

Hence:
The electricity cost price is

$$s_E^{CP} = \frac{xS_{CP}^{\Sigma}}{W_{CP}^{\Sigma}} = \frac{S_{CP}^{\Sigma}s_E}{W_{CP}^{\Sigma}s_E + Q_{CP}^{\Sigma}s_H} \qquad (5)$$

The heat cost price is

$$s_H^{CP} = \frac{yS_{CP}^{\Sigma}}{Q_{CP}^{\Sigma}} = \frac{S_{CP}^{\Sigma}s_H}{W_{CP}^{\Sigma}s_E + Q_{CP}^{\Sigma}s_H} \qquad (6)$$

If the cost prices of electricity and heat are higher than the market prices, the system is inefficient (Zharkov, 2009). Knowing the energy cost price, it is possible to set tariffs for it (Malafeev, Smirnov, Kharaim, Khrilev & Livshits, 2003).

II. Polygeneration (P)

In the general case, the problem of allocating costs at the combined production in accordance with the criterion of equal profitability of supplied energy products is solved as follows:

$$\sum_{i=1}^{n} x_i = 1 \qquad (7)$$

$$\frac{x_1}{A_1^P s_1^A} = \frac{x_2}{A_2^P s_2^A} = \frac{x_i}{A_i^P s_i^A} = \frac{x_n}{A_n^P s_n^A} \qquad (8)$$

where A_i^P - amount of supplied energy product i, s_i^A - alternative (market) cost of product i, n - number of products supplied, x_i – share of product i in S_P^{Σ}, S_P^{Σ} - total costs of the combined production or supply.
Then:

$$s_i^P = \frac{S_P^{\Sigma}s_i^A}{\sum_{i=1}^{n}\left[A_i^P s_i^A\right]} = s_i^A k_P^F \qquad (9)$$

$$k_P^F = \frac{S_P^{\Sigma}}{\sum_{i=1}^{n}\left[A_i^P s_i^A\right]} \qquad (10)$$

where k_P^F - factor of financial (economic, commercial) effect of the combined production or supply, s_i^P - cost price of energy product i. If $k_P^F > 1$, the system is inefficient.

Assessment and Enhancement of the Energy Supply System Efficiency

The formulae are true for the producers (suppliers) of electricity, heat, process steam, cold for conditioning, gasoline, diesel fuel, coke, hydrogen, etc. produced from oil, natural gas, coal, etc.

The optimal structure of products from the viewpoint of the maximum profit can be determined by using variant calculations: $k_P^F \to \min$ (Zharkov, 2009).

III. Combined Output of Products

In the more general case expressions (9, 10) has the form.

The cost price of the product is:

$$s_i^P = s_i^A k_P^F \qquad (11)$$

$$k_P^F = \frac{S_P^\Sigma}{\sum_{i=1}^{n}\sum_{j=1}^{m}\left[A_{ij}^P s_i^A\right]} \qquad (12)$$

where: k_P^F – coefficient of cost saving, S_P^Σ – full production costs, s_i^A – alternative (competitive, market) cost of product i, A_{ij}^P – amount of product i supplied to consumer j; n – number of product types; m – number of consumers. $(1-k_P^F)100\%$ - achieved percent of cost saving with respect to the market environment.

If an enterprise is oriented to several markets, we can determine the averaged value of s_i^A (proportionally to the planned supplies of product to each market), and then the average cost price of product i. Besides each market (consumer) j can be represented by own, A_{ij}^P and s_{ij}^A, then the cost price of one and the same product intended for supply to different markets can be different: s_{ij}^P (Zharkov, 2014):

$$s_{ij}^P = s_{ij}^A k_P^F \qquad (13)$$

$$k_P^F = \frac{S_P^\Sigma}{\sum_{i=1}^{n}\sum_{j=1}^{m}\left[A_{ij}^P s_{ij}^A\right]} \qquad (14)$$

S_P^Σ includes delivery costs of A_{ij}^P. The optimal structure of products and the structure of supplies to the considered markets from the viewpoint of the maximum profitability can be calculated on the basis of variants: $k_P^F \to \min$.

Index k_P^F characterizes the financial (economic, commercial) efficiency of the enterprise relative to the formed current prices of supplied products and allows the choice of the most effective measures to improve the economic efficiency (Zharkov, 2009, 2014).

IV. Energy Supply Systems (ESS) of a Region (Province, Territory or City)

Here $s_i^A = b_i^A$ is the specific fuel consumption to supply consumer i by the alternative method, $S_P^\Sigma = B_{ESS}^\Sigma$ is the total fuel consumption in the energy supply system which is expressed in thermal units.

$$b_i^{ESS} = \frac{B_{ESS}^\Sigma b_i^A}{\sum_{i=1}^{n}\left[A_i^{ESS} b_i^A\right]} = b_i^A k_{ESS}^{Fl} \qquad (15)$$

$$k_{ESS}^{Fl} = \frac{B_{ESS}^\Sigma}{\sum_{i=1}^{n}\left[A_i^{ESS} b_i^A\right]} \qquad (16)$$

where: k_{ESS}^{Fl} - coefficient of the fuel saving effect of combined energy supply, A_i^{ESS} - amount of supplied product i, b_i^{ESS} - specific fuel consumption to produce and supply the i-th kind of energy.

In the general case the specific fuel consumption to produce and supply the i-th energy kind for the energy supply system as a whole is determined by the formula:

$$b_i^{ESS} = b_i^N k_{ESS}^{Fl} \qquad (17)$$

$$k_{ESS}^{Fl} = \frac{B_{ESS}^\Sigma}{\sum_{i=1}^{n}\sum_{j=1}^{m}\left[A_{ij}^{ESS} b_i^N\right]} \qquad (18)$$

where: k_{ESS}^{Fl} – coefficient of fuel saving, B_{ESS}^Σ – total fuel consumption in the energy supply system which is expressed in thermal units, e.g. in tce; b_i^N – normative specific fuel consumption for supply of the i-th energy product; A_{ij}^{ESS} – amount of the product i supplied to consumer j; n – number of energy product kinds; m – number of consumers.

$(1 - k_{ESS}^{Fl})100\%$ - percentage of fuel economy achieved with respect to the standard fuel consumption for energy supply. The value $k_{ESS}^{Fl} > 1$ indicates an excessive fuel consumption in the energy supply system with respect to the standard.

In this case, if the energy (for example, electricity) is imported, a virtual (fictitious) energy source is added to the energy supply system, and the parameters of this source are determined by the amount of fuel consumed to produce energy and supply it to this source. When the system exports energy, the fuel consumption for energy production is excluded from the balance of energy supply to the territory, but is taken into account in calculations as a virtual consumer.

For example, when calculating the net indices of energy supply each fuel unit consumed in the energy supply system, besides its energy content, should have an additional parameter represented by its own energy intensity, in the same way as its cost includes expenditure on production and supply. Thus,

Assessment and Enhancement of the Energy Supply System Efficiency

we can determine the optimal set of equipment at the plant or in the energy system, capacity and operating conditions of the equipment, in particular the optimal degree of cogeneration centralization: $k_{ESS}^{Fl} \to \min$; $k_{ESS}^{Fl} = f(S, P, M)$, S – a set of equipment, P - equipment capacity, M - operating conditions of the equipment (Zharkov, 2009, 2014).

V. An Example for Cogeneration Plant (Combined Heat and Power Production)

According to Academician L.A. Melentiev, who laid the scientific foundation for cogeneration in Russia, "the cogeneration efficiency should be determined with regard to the separate electricity and heat production" (Melentiev, 1948). Therefore, the alternative power and heat sources for cogeneration plants are represented by condensing power plants (CPPs) and boiler plants (BP), for which $b_1^A = b_{CPP}$ and $b_2^A = b_{BP}$, where b_{CPP} and b_{BP} are the specific (standard) fuel consumption for electricity and heat production at CPPs and boiler plants, respectively. As a result the coefficient of the fuel consumption reduction through the combined energy production is (Popyrin, Denisov & Svetlov, 1989):

$$k_{CP}^{Fl} = \frac{B_{CP}^{\Sigma}}{W_{CP}^{\Sigma} b_{CPP} + Q_{CP}^{\Sigma} b_{BP}} \quad (19)$$

W_{CP}^{Σ} and Q_{CP}^{Σ} – electricity and heat production at CP, B_{CP}^{Σ} - total fuel consumption at CP.

Specific fuel consumption for electricity and heat production at CP (Popyrin, Denisov & Svetlov, 1989):

$$b_{CP}^{E} = b_{CPP} k_{CP}^{Fl} \quad (20)$$

$$b_{CP}^{H} = b_{BP} k_{CP}^{Fl} \quad (21)$$

Then $(1 - k_{CP}^{Fl}) 100\%$ – percentage of fuel economy achieved as compared to the separate electricity and heat production (Zharkov, 2009, 2014). Or:

The cogeneration plant efficiency in electricity production

$$\eta_{CP}^{E} = \frac{W_{CP}^{\Sigma}}{B_{CP}^{\Sigma}} \quad (22)$$

The cogeneration plant efficiency in heat production

$$\eta_{CP}^{H} = \frac{Q_{CP}^{\Sigma}}{B_{CP}^{\Sigma}} \quad (23)$$

$$k_{CP}^{Ef} = 1 / k_{CP}^{Fl} = \frac{\eta_{CP}^{E}}{\eta_{CPP}} + \frac{\eta_{CP}^{H}}{\eta_{BP}} \quad (24)$$

η_{CPP} - thermal efficiency of CPP, η_{BP} - thermal efficiency of BP.

k_{CP}^{Ef} – coefficient of efficiency of using the heat of fuel combustion.

$$^{ext}\eta_{CP}^{E} = \eta_{CPP} k_{CP}^{Ef} \qquad (25)$$

$$^{ext}\eta_{CP}^{H} = \eta_{BP} k_{CP}^{Ef} \qquad (26)$$

$^{ext}\eta_{CP}^{E}$ and $^{ext}\eta_{CP}^{H}$ – external (as compared to other energy sources) indices of fuel utilization efficiency (efficiency coefficient) for power and heat generation at CPs.

Excessive fuel consumption for the case of separate production is:

$$(k_{CP}^{Ef} - 1) 100\% \qquad (27)$$

The above indices can be calculated both for individual cogeneration plants and energy companies, territories, settlements, countries, etc. Such parameters characterize the energy supply system efficiency and are integral, i.e. taking into account the losses in energy production and transportation, operating conditions of power plants and level of using renewable (secondary) energy resources.

At present the actively developed smart grid technology does not take into account the heat supply. Therefore, in the course of its development in the cold countries it is necessary to include the criterion of total fuel use efficiency which takes into consideration the whole energy supply system. Thus, a purely electric smart grid technology is transformed into the technology of energy supply system development and functioning, which can be called the "smart energy supply system" (SESS) technology that includes cold supply to stimulate the development of trigeneration and energy supply for preparing food to stimulate the switch from gas ovens to electric ones. The suggested integrated assessment of energy supply shows that in most of the cases it is more profitable (in terms of financial and fuel costs) to burn natural gas at an efficient CP and provide consumers with electricity than to supply gas to houses for preparing food, let alone the system simplification and safety improvement. Fuel saving should become a chief goal of SESS operation ($k_{SESS}^{Fl} \to \min$).

In the calculations of power plant parameters (territory) based on the above formulae each consumer can be represented by its own indices A_{ij} and, s_{ij}^{A}; by several, if it consumes different energy products (electric power, heat, hot water, gas for cooking, cold for conditioners, etc.) or by one, but at different levels of the load curve.

Thus, here it is possible to apply an individual approach to each energy consumer and producer. And the indicated analytic expressions make it possible to quantitatively estimate the level of the overall price and fuel competitiveness of CPs (ESS) and fix tariffs for heat and electricity generated by them (or ESS) (Zharkov, 2009, 2014).

VI. Use of Resources (R) in the System

See the following equation pertaining to the use of resources in the system:

Assessment and Enhancement of the Energy Supply System Efficiency

$$^S k_i^R = \frac{R_i^\Sigma}{\sum_{j=1}^{n}\left[A_j^S r_j^i\right]}, \quad ^S r_j^i = {}^S k_i^R r_j^i \tag{28}$$

where R_i^Σ - consumption of resource i in the system (S), n - number of products obtained, $^S r_j^i$ - specific consumption of resource i required to produce product j; r_j^i - standard (target) specific consumption of resource i required to produce product j. The values of indices of the best companies, world average, average national, standard, theoretically achievable or the indices of separate production (if there is any) can be represented by r_j^i. Probable r_j^i can include electricity, heat, labor force, raw materials, equipment, water, fuel, etc.

The total coefficient of resource use efficiency is:

$$^S k_\Sigma^R = \sum_{i=1}^{m}\left[^S k_i^R \phi_i\right] \tag{29}$$

where m - number of resources used, ϕ_i - significance (weight) of resource i in the total consumption of the company. For example, ϕ_i is the share of resource i in the total cost of resources consumed (Zharkov, 2009), i.e.:

$$\sum_{i=1}^{m} \phi_i = 1 \tag{30}$$

VII. CO$_2$ Emissions

See the following equation pertaining to CO$_2$ emissions:

$$k_S^{CO_2} = \frac{C_S^\Sigma}{\sum_{i=1}^{n}\left[A_i^S c_i^N\right]}, \quad c_i^S = c_i^N k_S^{CO_2} \tag{31}$$

where c_i^S - specific emissions of CO$_2$ during production and supply of energy product i; C_S^Σ - total CO$_2$ emissions from the plant or system (S); A_i^S - amount of energy product i produced or supplied to consumers, n - number of products produced or supplied to consumers; c_i^N - standard specific emissions, $k_S^{CO_2}$ - coefficient of energy production/supply efficiency from the viewpoint of emission minimization (Zharkov, 2009).

VIII. Environmental Pollution

See the following equation pertaining to environmental pollution:

$$^Sk_i^V = \frac{V_i^\Sigma}{\sum_{j=1}^{n}\left[A_j^V \, ^Nv_j^i\right]} \qquad (32)$$

$$^Sv_j^i = \, ^Sk_i^V v_j^i \qquad (33)$$

where i - type of pollution, V_i^Σ - total i emissions, n - number of products obtained, $^Sv_j^i$ - specific pollution with i during generation of product j, $^Nv_j^i$ - normative specific pollution with i during generation of product j.

The total pollution coefficient is:

$$^Sk_\Sigma^V = \sum_{i=1}^{m}\left[^Sk_i^V \phi_i\right] \qquad (34)$$

where m - number of pollution types, ϕ_i - danger of pollution with i is fixed based on the expert judgment (Zharkov, 2009).

IX. Total Efficiency

The coefficient of total system efficiency is:

$$k_S^\Sigma = k_S^F \phi_F + \, ^Sk_\Sigma^R \phi_R + \, ^Sk_\Sigma^V \phi_V \qquad (35)$$

where ϕ_F, ϕ_R, ϕ_V are the relative importance of the activity aspects (Zharkov, 2009).

X. Efficiency of Investment

1. By reduction of the current costs

$$E_I^F = (1 - \, ^{ex}k^F / \, ^0k^F)/K, \; k^F = (k_{CP}^F, \; k_P^F, \; k_{ESS}^F, \; k_S^F) \qquad (36)$$

$$\text{or } E_I^F = \Delta k^F / K, \; \Delta k^F = (\Delta k_{CP}^F, \; \Delta k_P^F, \; \Delta k_{ESS}^F, \; \Delta k_S^F) \qquad (37)$$

2. By fuel saving

$$E_I^{Fl} = (1 - \, ^{ex}k^{Fl} / \, ^0k^{Fl})/K \qquad (38)$$

$$\text{or } E_I^{Fl} = \Delta k^{Fl} / K \qquad (39)$$

3. By resource saving

$$E_I^R = \Delta k_\Sigma^R / K \tag{40}$$

4. By CO_2 emission reduction

$$E_I^{CO_2} = \Delta k^{CO_2} / K \tag{41}$$

5. By environmental pollution reduction

$$E_I^V = \Delta k_\Sigma^V / K \tag{42}$$

6. Total efficiency

$$E_I^\Sigma = \Delta k^\Sigma / K \tag{43}$$

Δk^F, Δk^{Fl}, Δk_Σ^R, Δk^{CO_2}, Δk_Σ^V, Δk^Σ - expected changes in the coefficients as a result of taking measures. For example,

$$\Delta k^F = {}^0k^F - {}^{ex}k^F \tag{44}$$

${}^{ex}k^F$, ${}^{ex}k^{Fl}$, ${}^{ex}k^R$, ${}^{ex}k^{CO_2}$, ${}^{ex}k^V$, ${}^{ex}k^\Sigma$ - expected (forecast) values of the coefficients as a result of taking measures, K - investment required, E_I - relative investment efficiency. For example, ${}^0k^F$ and ${}^0k^{Fl}$ are initial values of the parameter. Achieved percentage of reduction in specific fuel consumption for energy supply is:

$$(1 - {}^{ex}k^{Fl}/{}^0k^{Fl})\ 100\% \tag{45}$$

Fixed parameters s_i^A, s_{ij}^A, b_i^N (η_{CPP}, η_{BP}), r_j^i, c_i^N, ${}^N v_j^i$ determine a level (benchmark), relative to which the efficiency of energy supply systems, energy technologies and productions, and modernization actions is estimated (compared).

The above mentioned is true for non-energy productions of territories and countries as well. Possible areas for the modernization are: reduction in losses and costs, optimization of the structure of products made, production of new products from waste, etc., for example, fish farming and cultivation of vegetables with the use of waste heat.

Thus, we can rank the planned measures by the efficiency and the priority of implementation. The tables, ranked by decreasing E_I, may be the basis for the company's investment plans. In fact, the suggested method enables us to determine the most efficient ways to develop the energy sector and economy of the territory (company) at a certain point of time.

Possible measures to be taken in the energy sector are: improvement in fuel use efficiency (including steam turbine cogeneration plants with condensing boilers: fuel economy reaches more than 40%, $^{ext}\eta_{CP}^{E}$ =0.745, $^{ext}\eta_{CP}^{H}$ =1.22; Stennikov V.A. & Zharkov S. V., 2013, 1, Teemu Tolvo, 2015), reduction in energy losses, and replacement of fuel by renewable energy sources (RES). Practical application of the suggested efficiency criteria will contribute to transition to the "green" energy and economy along the most effective trajectory of their modernization (Zharkov, 2009, 2014).

UNSTABILIZED ELECTRICITY OF WIND POWER PLANTS AS A SUBSTITUTE OF HIGH-QUALITY FUEL AT THERMAL POWER PLANTS

The use of RES is the next most effective fuel saving measure after cogeneration in terms of efficiency. Among all RESs we can single out the wind as a widely spread energy resource with relatively high density (as compared to insolation). However, the sea wave energy has higher density (Zharkov S. V., 2013, 1).

In recent years the world wind power industry has experienced an exponential increase in the installed capacity of power plants. By the end of 2014 it had reached 370 GW. However, the high cost of wind turbines and many unsolved problems in the operation of wind power plants (WPPs) hinder the large-scale use of wind energy in the most frequently applied type of direct connection of a wind power plant to the power grid. The main problems caused by the wind flow instability are (Figueiredo, 2012; Perez-Arriaga & Batlle, 2012; Renzo, 1982; Slootweg & Kling, 2003; Twidell & Weir, 1990; Yuan Zhao, Li-Sha Hao, & Yu-Ping Wang, 2009):

- Difficulties in providing acceptable quality of power supplied to the power system under unsteady wind wheel torque of wind power installations and transient processes during their connection to /disconnection from the grid;
- An essential increase in the cost and decrease in the WPP reliability due to the complex systems used to maintain the rotational frequency of generators or application of inverters, switch them on, synchronize them with the grid; switch them off (at the wind speed below the permissible minimum or above the maximum), restart, active and reactive power distribution, etc.;
- A need to monitor the WPP capacity and compensate for it, if necessary, by a mobile reserve (including the spinning one used to meet the demand), which complicates the power system operation proportionally to the share of WPPs in the power system.

Besides, the use of WPPs as a fuel-saving energy source has the following disadvantages:

1. Parallel operation of a WPP capacity covering the base electric load curves in power systems requires a highly maneuverable back-up capacity of the same magnitude, which is usually uneconomical in terms of fuel consumption and covers the load for most (about 65-70%) of the year. This, in its turn, casts doubt on the fuel saving expected from the use of WPPs in the power system as a whole (in particular, when there are no free maneuverable capacities of hydro power plants) and limits their potential share in the power system to the value of the available maneuverable capacity;
2. There appears a contradiction between the trends towards fuel saving in power generation that are actively developing in the world: on the one hand, WPPs are the most economically efficient in

the areas of decentralized power supply with expensive fuel and its high specific consumption for power generation; on the other hand, in such areas it is necessary to develop cogeneration. This is explained by the fact that cogeneration plants have, firstly, low specific fuel consumption for power generation, secondly, low maneuverability, and, thirdly, their technical and economic parameters drastically decrease at electric load drop of the power plant. Therefore, their parallel operation with WPPs is irrational;

3. Trends in the development of WPPs and CPs are quite opposite. Following the traditions of the previous century, WPPs in the grid become larger and larger (up to several hundreds of MW) to reduce specific costs, at present CPs tend towards the reduction in unit capacity (to several tens of kW) following the strategy of using distributed energy sources.

Thus, it is necessary to create a technology that would make it possible: a) to reserve (back-up) the WPP capacity by using highly efficient base power plants, ideally – cogeneration plants, including small ones; b) to soften the requirements for the quality of power generated by WPPs.

Therefore, it seems worthwhile to unite WPP and the back-up power plant in one plant taking into account the specific features of power generation by WPPs. As opposed to the traditionally applied scheme, it is suggested by the author (Zharkov, 2004) not connecting WPPs to the power grid directly, but using the WPP power for direct fuel substitution in thermal cycles of gas-turbine (combined cycle) plants – GTPs (CCPs): by means of an electric heater (electric coil) installed in the gas turbine duct right in front of the combustion chamber to heat the air entering the combustion chamber (Figure 1).

Thus, the fuel (perhaps, including hydrogen) consumption regulated to maintain the given temperature of gases at the turbine inlet depending on the varying electric heater capacity will decrease. As a highly-maneuverable heat source (with no thermal inertia unlike the boiler of a steam-turbine plant) combustion chamber can accurately keep the required temperature. For smoothing out the greatly uneven amounts of electricity coming to electric heaters in an electric network it is possible to install before them an electric power accumulator of capacitor type (on the basis of condensers). Fuel economy has some additional benefits:

- The power system receives high-quality power from a GTP (CCP) generator irrespective of the wind force fluctuations. Here there is no power quality deterioration in the grid, even due to the transient processes during WPP connection/disconnection;
- There is no need for parallel maneuverable capacity, covering the base electric load, and associated fuel overconsumption that exceeds fuel saving expected from the WPP use. GTPs (CCPs)+WPPs used as a complex can be base-oriented and, hence, highly economical in terms of fuel consumption;
- There appears an opportunity to use wind power generation together with combined power and heat production at a cogeneration plant. It seems doubtful to save fuel in the area of power supply, as well as to solve the above problems of using wind power generation and subsequent burning of fuel at boiler plants, since heat supply is usually more fuel-consuming, than power generation. The coefficient of heat utilization at CPs reaches 0.8-0.9, i.e. all the heat energy, entering the cycle as the WPP electric power and fuel, is actually utilized. Thus, depending on the used GTP, the share of heat, entering the cycle and converted into electric power with stable parameters, can reach up to 25-40% at a gas-turbine cogeneration plant and up to 40-55% at a combined-cycle cogeneration plant. Wind is the second most important climatic parameter (after outdoor temperature), which

Figure 1. The use of unstabilized WPP power in the GTP cycle: 1 – wind turbine; 2 - electric heater; 3 - compressor; 4 - combustion chamber; 5 - gas turbine; 6 - system water heater or steam boiler; 7 – boiler plant.

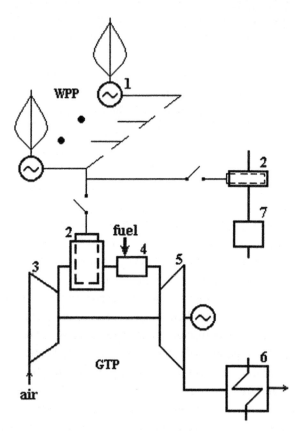

determines the heat demand. The use of WPPs will help compensate for increased heat losses by providing peak heat supply for heating exactly when it is needed, i.e. during windy periods;
- Energy from many WPPs can be concentrated at one or several GTPs (CCPs) by a local grid with a gradual increase in the WPP capacity and in the number of unstabilized power consumers. Both major consumers (CCPs-CPs, energy storage systems), and small ones (GTPs-CPs, boiler plants), each with its own connection priority, can be connected to the local grid. The priorities can be changed depending on the seasonal and daily changes in the energy consumption structure;
- The switch to a variable frequency of wind wheel rotation (without applying inverters), which will allow an increase in the coefficient of wind energy utilization by 20-35%, thus, decreasing the specific cost of WPPs, expansion of the effective wind speed range with a rise in CF_{WPP}, and a decrease in mechanical stress on WPP blades and shafts;
- The maximum simplification of the electric circuit, control system and WPP design, as in this case, the generators operate to cover active load, and the requirements for power quality are extremely low; an appropriate decrease in cost and increase in WPP reliability, which is especially important for the remote areas with severe climatic conditions, where there are no qualified experts, who can perform technical works;

- The switch from horizontal-axial WPPs to cheaper and more reliable vertical-axial ones, which are characterized by the higher unevenness of torque and, consequently, their direct connection to the power system is complicated. However, the inclined axis wind turbines (including a floating design) are more promising for some (coastal) areas (Zharkov, 2013, 2).

The GTP (CCP)-CP+WPP scheme allows one to combine the achievements of the conventional energy sector (gas-turbine and combined-cycle technologies, cogeneration) and non-conventional (wind power plants), rather than oppose them, which is often observed and makes WPPs compete with GTPs and CCPs that are more economical in terms of fuel consumption (while the efficiency of GTP (CCP)+WPP just increases). The 50% replacement of fuel by the WPP power is even more beneficial from the energy standpoint, than a two-fold increase of the GTP efficiency, since with equal fuel consumption the thermal capacity of GTP(CCP)-CP and electric capacity of CCP-CP are higher. Moreover, during the whole useful life, GTP(CCP)+WPP will not become obsolete in terms of continuously updated fuel GTP(CCP), since, in this case, the specific fuel consumption for power generation will always be lower. To reduce fuel consumption it is possible to install an additional up-to-date WPP, if necessary.

The proposed scheme is most promising in the isolated energy supply systems. Therefore, GTP(CCP)-CP+WPP systems become more attractive in the context of wider use of distributed energy sources on the basis of small cogeneration plants and local renewables. The heat from CPs can be used for heating, hot water supply, and air-conditioning by the absorption of heat converters.

High-quality fuel at its current cost of about 300-500 $/thous. m^3 (or 9-15 $/GJ) for the European consumers can be replaced by the power generated by WPPs at the cost of 3-5 cent/kWh. The power produced by the present-day WPPs costs 3-6 cent/kWh, therefore, taking into account the above factors of potential decrease in the cost of WPPs, we can say that even now for some consumers it is economically efficient to use wind energy according to the GTP(CCP)+WPP scheme, and the efficiency will enhance with the fuel price rise, and improvement of WPP technology and increase in the scale of WPP use.

The technology can be used at steam-turbine plants (STPs). There is a proposal (the Siemens patent) to raise the upper steam temperature of the steam-turbine cycle by hydrogen-oxygen boiler superheaters. Here it is worthwhile to replace part of the hydrogen burnt with the WPP unstabilized power by using the electric heater to heat the steam, entering the superheater (Figure 2).

In this case there is a decrease in the fuel (and correspondingly oxidizer) consumption that can be adjusted to maintain the given steam temperature at the turbine inlet depending on the varying electric heater capacity (due to a variable wind force). The average annual saving of expensive pure hydrogen and oxygen to be consumed may reach 30-50% with a WPP installed capacity equal to the boiler superheater thermal capacity, and may be higher with the electric heater powered from the local unstabilized power grid on the basis of a large WPP. Hydrogen and oxygen can also be produced by using electric power from the local grid - in the water electrolyzer.

STP can also operate on the basis of the "wind force change": at a sliding steam pressure without boiler superheater. Such a scheme is the simplest and, therefore, the most attractive for application at the initial stage of the technology development. For example, the electric heaters capacity of STP in CCP (Figure 3) makes up only a few per cent of the electric heater capacity of GTP. Therefore, the electric heater of STP can expect practically 100-per cent power supply from a large WPP during a year, which is intended for combined operation with CPs and boiler plants. The additional (reserve) source of the electric power 4 (for example, diesel generator or hydrogen fuel cells) provides with the electric power the electric heaters of STP during the windless periods of time (Stennikov V.A. & Zharkov S. V., 2013, 2).

Figure 2. The use of unstabilized WPP power in the STP cycle with hydrogen-oxygen superheater: 1 – wind turbine; 2 - electric heater; 3 - hydrogen-oxygen superheater; 4 - steam boiler; 5 - steam turbine; 6 – electrolyzer; 7 - condenser; 8 - pump.

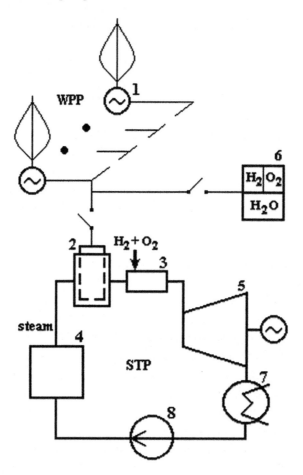

It seems expedient to run such STP+WPP complexes along with improving the technology of high-temperature STP before the introduction of STP with hydrogen-oxygen boiler superheaters, since the STP+WPP scheme takes an intermediate position both in cost, and in the implementation complexity between the traditional and new (boiler superheater) methods of improving the STP efficiency.

The approach of the age of hydrogen energy is basically associated with a wide use of fuel cells. However, according to some experts, the fuel-cell-based hydrogen power plants will be the most efficient at the capacities below 10 MW. At the capacities over 10 MW the steam-turbine cycle power plants with hydrogen-oxygen contact type steam generators are more efficient. The STP+WPP scheme is applicable in this case as well: both for undercritical steam parameters (Figure 4), and supercritical steam parameters (the scheme in Figure 2 is updated: boiler 4 is excluded, and devices 2 and 3 are used as components of the contact type direct-flow boiler).

The potential hydrogen and oxygen saving is the same as in the previous case. The STP+WPP scheme can be used as a prototype and a component of the plants based on the high-temperature STPs with hydrogen-oxygen superheaters and steam generators. The STP+WPP complexes can promote the increase in fuel utilization efficiency, expansion of wind energy use and smooth transition to hydrogen energy.

Figure 3. The use of unstabilized WPP power in the CCP cycle: 1 – wind turbine; 2 - electric heater; 3 - steam waste heat boiler; 4 – additional (reserve) source of the electric power.

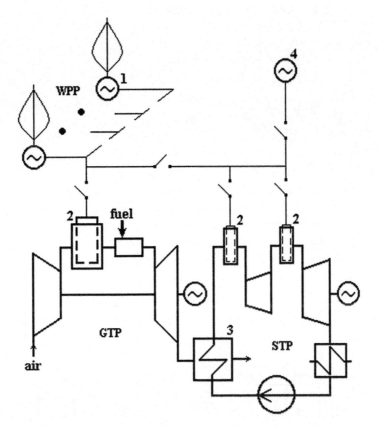

The application of the STP+WPP technology is reasonable, first of all, at thermal power plants (TPPs), located on the coast, since there is a necessary potential of wind energy resource. The offshore location of TPP is also more preferable because it gives the opportunity to use cold sea water for condenser cooling, which improves the efficiency of plants, and provides cheap fuel delivery by sea.

Usually the inclusion of large WPPs in the power system increases the unevenness of operation of the existing thermal power plants and decreases their efficiency, especially at CPs, i.e. the total efficiency of investment in decreasing fuel consumption (emissions) is low (Smith, J., Nayak, D. R., & Smith, P., 2014). The use of the suggested technology is more effective (YuanZhao, Li-ShaHao & Yu-PingWang, 2009).

The use of an offshore plant in combination with heat-consuming industries (for example, production of liquid fuel from natural gas, desalination, extraction of gold, uranium etc. from sea water) is especially effective. Such a plant can help decrease the unevenness of power and imported natural gas consumption and reduce the consumption of imported natural gas, produce hydrogen and oxygen from water, as well as liquid fuel from hydrogen and carbon dioxide (Zharkov, 2004).

Here we can use the experience in designing a solar power plant on the basis of GTP (Buck & Schwarzbözl, 2009). It is effective to use solar (including photovoltaic) and wind energy jointly within a single plant, since the variations in the intensity usually occur during the antiphase. Such a technology removes the technological constraints on the use of wind and solar energy in the energy supply systems of regions and countries. Correspondingly it becomes possible to use renewable energy sources on a large scale

Figure 4. The use of unstabilized WPP power in the STP cycle with hydrogen-oxygen superheater: 1 – wind turbine; 2 – electric heater; 3 - hydrogen-oxygen vapor superheater; 4 - hydrogen-oxygen steam generator; 5 – system water heater or reheat generator.

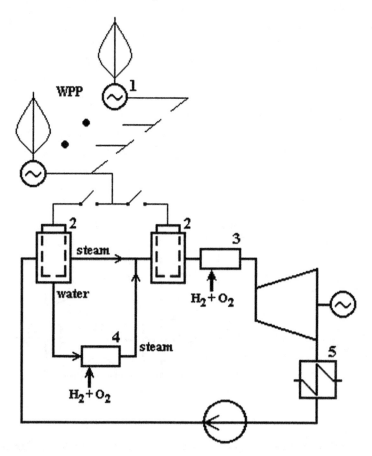

without application of expensive electric power storage and, hence, to reduce CO_2 emissions without expensive technologies of its extraction and burial. Besides it is more important that the hydrocarbons are saved. The suggested technology is the only possibility of the large-scale development of wind and solar power industries. In so doing the grids that connect WPPs, power systems and consumers are used more effectively, since the TPP+WPP complex generates a stable power flow. Besides, the European super-grid (O'Connor, 2009) becomes unnecessary for WPPs.

For example, in author's opinion supply of Europe with environmentally clean power is more valid by using wind energy of the north-european coast of Russia and the Siberia tundra with hydrogen storage in the exhausted gas fields than from the planned solar power plants in Sahara. The similar project also can be used for power supply of Japan from wind power stations of the Sakhalin Island.

Thus:

1. Connection of WPP to an electric grid through GTPs (STPs,CCPs) allows us to avoid solving the problems of maintaining power quality and operating reserve of the WPP capacity in the power system and also to use wind energy at the plants of combined heat and high-quality electric power

production, small ones included. In this case technological restrictions on the use of WPPs (Hirth L., 2015) are removed: the WPP capacity can surpass the total capacity of the power plants and boiler plants in the energy supply systems. Thus, the wind energy development becomes power system-independent, and the WPP owners and operators of electric networks do not have points of contact and the commercial WPP efficiency does not depend on constraints, tariffs and operating conditions of power systems and hence, no special bills regulating interrelations between WPP and the power system are needed.

2. Owing to local effect of WPPs the application of GTP(STP,CCP)+WPP makes it possible to exactly estimate real energy and economic benefits (or losses) of WPP use, while at direct WPP connection to the power grid, which is usually in use, all the problems of WPP utilization are shifted to the power system and the effect of WPPs on the economic efficiency of the power system as a whole can hardly be assessed and as a rule they are not considered.

3. The use of the technology is the top priority in the isolated energy supply systems based on the local grids of unstabilized power that combine WPPs, GTP(STP,CCP)-CPs and boiler plants. There is an opportunity for maximum substitution of fossil fuel with wind energy. Also WPPs can participate in a covering of peak power loadings (Zharkov, 2006).

4. The GTP(CCP)-CP+WPP complexes can foster the extensive use of WPPs, GTPs, CCPs, CPs and distributed energy plants in energy supply systems, since the wind becomes an energy resource to be used at GTPs(STPs,CCPs).

An Example of Using the Technique

Criteria E_I gives the opportunity to determine the line of the most effective development of an energy company and energy supply system of the territory from the fuel-saving viewpoint.

Graphic illustration of the technique is presented in Figure 5. The sequence of measures is $A_0 \rightarrow A_1 \rightarrow A_2 \rightarrow A_i$ subject to $\alpha_1 > \alpha_2 > \alpha_3 > \alpha_i$, where $\alpha_i = \operatorname{arctg}(\Delta k_i / k_i)$, where Δk_i is the expected (forecasted) changes in the coefficient value as a result of taking the i-th measure.

The tables of ranking by decreasing E_i can underlie the investment plans of the companies.

First of all the "choice" will be made of the measures with small capital investment (required investment) K_i, since today the energy companies and existing energy supply systems have considerable reserves for increasing the production efficiency.

For instance, we have an energy supply system (ESS) with the electric loads N_{ESS}=100 MW and the thermal loads Q_{ESS}=350 MW(th), which are covered by a condensing power plant (CPP) with the efficiency η_{CPP} =0.4 and a boiler plant with the efficiency η_{BP}=0.9, respectively. Instantaneous fuel consumption is B_{ESS0}=100/0.4+350/0.9=639 MW(th).

The following variants of energy supply system modernization are considered:

Variant 1. Construction of a WPP with the capacity N_{WPP}=100 MW, which substitutes some amount of electricity of CPP, in the energy supply system. It is assumed that the specific cost of WPP is 1000 $/kW, CF_{WPP}=0.35. Uneven operation of CPP decreases its average annual efficiency to η_{CPP}=0.33.

Fuel consumption in the energy supply system is B_{ESS1}=100·(1-0.35)/0.33+350/0.9=586 MW(th).

Variant 2. Construction of a CP (including a new heat network) with the total efficiency $\eta_{CP}^{\Sigma} = \eta_{CP}^{E} + \eta_{CP}^{H}$ =0.9.

Figure 5. Graphic illustration of the technique for determining the line of the most effective development of the energy company or energy supply system

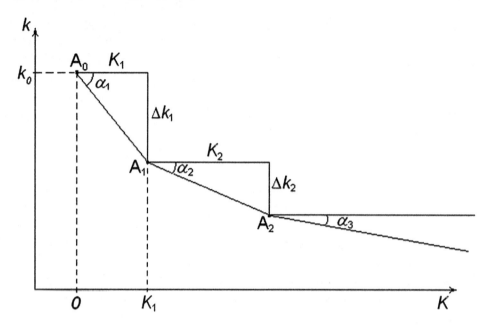

The required electric efficiency of the CP (η_{CP}^E) at known loadings and the total efficiency of CP (η_{CP}^Σ) is defined as follows.

The condition to be met is:

$$\frac{N_{ESS}}{Q_{ESS}} = \frac{\eta_{CP}^E}{\eta_{CP}^H} = \frac{\eta_{CP}^E}{\eta_{CP}^\Sigma - \eta_{CP}^E} \tag{46}$$

Hence:

$$N_{ESS}(\eta_{CP}^\Sigma - \eta_{CP}^E) = \eta_{CP}^E Q_{ESS} \tag{47}$$

$$\eta_{CP}^E N_{ESS} + \eta_{CP}^E Q_{ESS} = N_{ESS}\eta_{CP}^\Sigma \tag{48}$$

$$\eta_{CP}^E = \frac{N_{ESS}\eta_{CP}^\Sigma}{N_{ESS} + Q_{ESS}} = \eta_{CP}^\Sigma \frac{N_{ESS}}{N_{ESS} + Q_{ESS}} \tag{49}$$

Therefore, in this case:

$$\eta_{CP}^E = \eta_{CP}^\Sigma \frac{N_{ESS}}{N_{ESS} + Q_{ESS}} = 0.9\frac{100}{100 + 350} = 0.2 \tag{50}$$

Electric capacity of CP is N_{CP}=100 MW, and thermal capacity is Q_{CP}=350 MW(th). The CP cost is 2000 $/kW(e).

Fuel consumption in the energy supply system is B_{ESS2}=100/0.2=500 MW(th).

Variant 3. Construction of a WPP in the energy supply system, which substitutes some amount of fuel at CP constructed in Variant 2 (Figure 1). The WPP capacity equals the fuel consumption capacity at CP: N_{WPP}=(100+350)/0.9=500 MW. Due to the simplification of the electric circuit of WPP and increase in its operation efficiency, its specific cost decreases to 750 $/kW, and CF_{WPP} increases up to 0.4.

Fuel consumption in the energy supply system is B_{ESS3}=100(1-0.4)/0.2=100/0.2-0.4·500=300 MW(th).

Results of calculations are presented in Table 1.

Table 1. Comparison of the variants on energy supply system modernization

	Modernization Variants				
	0	1	2	3	2+3
Fuel consumption, MW(th)	639	586	500	300	300
Fuel saving $\Delta B_{ESS}=B_{ESS(i-1)}-B_{ESSi}$, MW(th)	-	53	-	200	-
Fuel saving $\Delta B_{ESS}=B_{ESS0}-B_{ESSi}$, MW(th)	-	-	139	-	339
Fuel saving, %	-	8	22	40	53
Investment, million $	-	100	200	375	575
k_{ESS}^{Fl}	1	0.92	0.78	0.47	0.47
$^i k^{Fl}/^0 k^{Fl}$	-	0.92	0.78	0.6	0.47
$1 - {^i k^{Fl}}/{^0 k^{Fl}}$	-	0.08	0.22	0.4	0.53
Investment efficiency $E_I^{Fl} = (1 - {^i k^{Fl}} / {^0 k^{Fl}}) / K$, 1/billion $	-	0.8	1.1	1.07	0.92
Δk_i^{Fl}	-	0.08	0.22	0.28	0.53
Investment efficiency $E_I^{Fl} = \Delta k_i^{Fl} / K$, 1/billion $	-	0.8	1.1	0.75	0.92
Investment efficiency $E_I^{Fl} = \Delta B_{ESS} / K$, MW(th)/million $	-	0.53	0.695	0.53	0.59
Specific fuel consumption for electric energy, GJ/MWh	9	8.28	7.02	4.23	4.23
Specific fuel consumption for heat energy, GJ/MWh(th)	4	3.68	3.12	1.88	1.88
Decrease in specific fuel consumption for electric and heat energy, %	-	8	22	40	53
$^{ext}\eta_{ESS}^E$	0.4	0.435	0.513	0.851	0.851
$^{ext}\eta_{ESS}^H$	0.9	0.978	1.154	1.915	1.915

Thus, under the given conditions the strategy to effectively develop the considered energy supply system consists in taking successive measures:

1. Switch to combined energy production;
2. Introduction of WPPs by means of the existing CP.
3. It is also possible to combine these two measures in one.
4. Cogeneration saves more than 20% of fuel, and its combination with wind power station – more than 50%.
5. For comparison of the efficiency of different measures on the energy supply system modernization it is expedient to apply the formulae $E_I^{Fl} = \Delta k_i^{Fl} / K$, and for calculation of the attainable fuel saving – $E_I^{Fl} = (1 - {}^i k^{Fl} / {}^0 k^{Fl}) / K$. Specific fuel consumption decreases in proportion to a reduction in the total fuel consumption.

CONCLUSION

1. The author developed methods for assessing economic, resource and environmental efficiency of cogeneration plants (CPs) and energy supply systems as a whole and ways of its improvement.
2. The problem of allocating fuel and financial costs at the combined production in accordance with the criterion of equal profitability of supplied products (including energy) is solved. The methods allow determining specific indicators of supplied energy products which makes it possible to compare the efficiency of energy supply systems of different companies and countries, and to define their future target indicators. The suggested technique offers the opportunity to determine the line of the most effective development of energy supply system and economy as a whole of the territory (the company). Thus, we can rank the planned measures by the efficiency and the priority of implementation.
3. The technique is suitable for the energy supply systems with any set of equipment. Practical application of the suggested efficiency criteria will contribute to the transition to the "green" energy and economy along the most effective trajectory of their modernization.
4. The technology of introducing RES-based power plants to the energy supply systems by means of using unstabilized RES-based power for direct fuel substitution in thermal cycles of gas-turbine (combined cycle) and steam-turbine plants (the wind is viewed as the most promising type of RES). Connection of wind power plants to an electric grid through thermal power plants allows us to avoid solving the problems of maintaining power quality and operating reserve of the wind power plants capacity in the power system and also to use wind energy at the plants of combined heat and high-quality electric power production, small ones included. The technology is the only possibility of the large-scale development of wind and solar power industries and it can promote a smooth transition to hydrogen energy.
5. The suggested technology will provide a sustainable development of the world's energy and economy on the basis of balance between the scarce conventional fuels rising in price and RES technologies becoming cheaper in combination with an increase in the environmental friendliness and availability of energy, which is necessary for the future world without conflicts and frictions.

6. The paper can be interesting to power engineering specialists, businessmen and economists, and also organizers of the upcoming UN Climate Change Conference to definition of obligations for the future specific indicators of countries greenhouse gases emissions on the made energy.

REFERENCES

Alexanov, A. P. (1995). Distribution of fuel costs of energy supplied by CPs. *Energetik, 1,* 7–8.

Arakelyan, E. K., Kozhevnikov, N. N., & Kuznetsov, A. M. (2006). Tariffs for electricity and heat. *Teploenergetika, 11,* 60–64.

Buck, R., & Schwarzbözl, P. (2009). Solarized gas turbine power systems. *Gas Turbo Technology, 2,* 17–21.

Figueiredo, E. F. (2012). Study Committee A1 (Rotating electrical machines). SC Annual report 2011. *Electra, 260,* 20–26.

Gitelman, L. D., & Ratnikov, B. E. (2008). *Energy business: textbook* (3rd ed.). Moscow: Publishing House "Delo" ANKh.

Hirth, L. (2015). The Optimal Share of Variable Renewables: How the Variability of Wind and Solar Power affects their Welfare-optimal Deployment. *The Energy Journal (Cambridge, Mass.), 36*(1), 149–184. doi:10.5547/01956574.36.1.6

Jovanovic, M., Turanjanin, V., Bakic, V., Pezo, M., & Vucicevic, B. (2011). Sustainability estimation of energy system options that use gas and renewable resources for domestic hot water production. *Energy, 36*(4), 2169–2175. doi:10.1016/j.energy.2010.08.042

Kharaim, A. A. (2003). How to calculate the tariffs for heat and electricity generated at CPs without separating the fuel? *Novosti Teplosnabzheniya, 11,* 3-7.

Kuznetsov, A. M. (2006). Comparison of results of separation of fuel consumption for electricity and for heat supplied by CP by different methods. *Energetik, 7,* 21.

Malafeev, V. A., Smirnov, I. A., Kharaim, A. A., Khrilev, L. S., & Livshits, I. M. (2003). Formation of tariffs at CP in the market environment. *Teploenergetika, 4,* 55–63.

Melentiev, L. A. (1948). *Cogeneration. Part 2.* Moscow: Publishing House of the USSR Academy of Sciences.

Melentiev, L. A. (1987). *Outlines of Russian energy sector history.* Moscow: Nauka.

Melentiev, L. A. (1993). *Selected works. Scientific fundamentals of cogeneration and energy supply of cities and industrial enterprises.* Moscow: Nauka.

O'Connor, E. (2009). Going Mainstream. *Modern Power Systems, 11,* 17–18.

Padalko, L. P., & Zaborovsky, A. M. (2006). On the principles of mutually agreed distribution of energy system costs between heat and electric energy. *Energiya i Menedzhment, 1,* 8-12.

Perez-Arriaga, I. J., & Batlle, C. (2012). Impacts of Intermittent Renewables on Electricity Generation System Operation. *Economics of Energy & Environmental Policy, 1*(2), 3–16. doi:10.5547/2160-5890.1.2.1

Pezzini, P., Gomis-Bellmunt, O., & Sudrià-Andreu, A. (2011). Optimization techniques to improve energy efficiency in power systems. *Renewable & Sustainable Energy Reviews, 15*(4), 2028–2041. doi:10.1016/j.rser.2011.01.009

Popyrin, L. S., Denisov, V. I., & Svetlov, K. S. (1989). On the methods for distributing CP costs. *Elektricheskie Stantsii, 11*, 20-25.

Renzo de, D. J. (1982). Wind power: Recent developments. Moscow: Energoatomizdat.

Semenov, V. G. (2002). Analysis of possibility for CP operation in the electricity market. *Novosti Teplosnabzheniya, 12*, 45-47.

Slootweg, J. G., & Kling, W. L. (2003). Is the answer blowing in the wind? *IEEE Power&Energy, 1*(6), 26–33.

Smith, J., Nayak, D. R., & Smith, P. (2014). Wind farms on undegraded peatlands are unlikely to reduce future carbon emission. *Energy Policy, 66*, 585–591. doi:10.1016/j.enpol.2013.10.066

Stennikov, V. A., & Zharkov, S. V. (2013a). *The gas-burning CPs of the increased fuel efficiency*. First International Forum "Renewable Energy: Towards Raising Energy and Economic Efficiencies" - REENFOR-2013, Moscow.

Stennikov, V. A., & Zharkov, S. V. (2013b). *Problems of WF use in the isolated power systems and possible ways of their solving*. First International Forum "Renewable Energy: Towards Raising Energy and Economic Efficiencies" - REENFOR-2013, Moscow.

Tolvo, T. (2015). Flue gas condensing and scrubbing: a winning combination. *Modern Power Systems, 35*(3), 18-20.

Twidell, J., & Weir, A. (1990). *Renewable energy resources: Transl. from the Engl*. Moscow: Energoatomizdat.

Zhao, Hao, & Wang. (2009). Development strategies for wind power industry in Jiangsu Province, China: Based on the evaluation of resource capacity. *Energy Policy, 37*, 1736–1744.

Zharkov, S. (2004). Wind use at thermal power plants. *RE-GEN. Wind, 3*, 13–15.

Zharkov S. V. (2004). Wind energy use at gas-turbine and steam-turbine plants. *EW, 11*, 58-61.

Zharkov, S. V. (2006). Use of wind power plants in power systems. *Энергия: экономика, техника, экология, 10*, 38–39.

Zharkov, S. V. (2009). How to estimate efficiency of energy supply systems. *Gas Turbo Technology, 4*, 7–10.

Zharkov, S. V. (2013a). *Wave power stations as possible export subject of shipyards.* First International Forum "Renewable Energy: Towards Raising Energy and Economic Efficiencies" - REENFOR-2013, Moscow.

Zharkov, S. V. (2013b). *Wind power installation with an inclined axis – the perspective direction of WPIs development for conditions of the RF.* First International Forum "Renewable Energy: Towards Raising Energy and Economic Efficiencies" - REENFOR-2013, Moscow.

Zharkov, S. V. (2014). About methods of an assessment of energy supply efficiency and decrease stimulation energy consumption of economy of the RF. *Energetik, 3*, 34–40.

Chapter 2
Problems of Modeling and Optimization of Heat Supply Systems:
Methods to Comprehensively Solve the Problem of Heat Supply System Expansion and Reconstruction

Valery Stennikov
Energy Systems Institute, Siberian Branch of the Russian Academy of Sciences, Russia

Tamara Oshchepkova
Energy Systems Institute, Siberian Branch of the Russian Academy of Sciences, Russia

Nikolay Stennikov
Energy Systems Institute, Siberian Branch of the Russian Academy of Sciences, Russia

ABSTRACT

The paper addresses the issue of optimal expansion and reconstruction of heat supply systems, which includes a set of general and specific problems. Therefore, a comprehensive approach to their solving is required to obtain a technically admissible and economically sound result. Solving the problem suggests search for effective directions in expansion of a system in terms of allocation of new heat sources, their type, output; construction of new heat networks, their schemes and parameters; detection of "bottlenecks" in the system and ways of their elimination (expansion, dismantling, replacement of heat pipeline sections, construction of pumping stations). The authors present a mathematical statement of the problem, its decomposition into separate subproblems and an integrated technique to solve it. Consideration is given to a real problem solved for a real heat supply system. A set of arising problems is presented. The application of developed methodological and computational tools is shown.

DOI: 10.4018/978-1-4666-9755-3.ch002

INTRODUCTION

The problem of heat supply system (HSS) construction and expansion is hot for both new and most of the existing systems. It is brought about by the technical transformations, emergence of and increase in heat loads which should be met. Solution to the problem should be oriented to the use of cutting-edge technological capabilities, enhancement in reliability and efficiency of heat supply. This should be accompanied by the improvement in the quality of the rendered services, and the creation of more comfortable conditions for consumers. The use of energy resources, material and financial expenditure should be kept at minimum.

To cope with these tasks it is necessary to consider a set of variants for the route of the heat network, form an optimal configuration of the system, choose a set and types of heat sources and address other issues. It is necessary to detect the "weak" (congested) places in the existing heat supply systems and find the methods of their elimination. The methodological complexity of this problem lies in the fact that it is necessary to take into account the existing part of the system, a diversity of alternative options for its expansion and a large dimension of the problems to be solved. The set of conditions and constraints is added by a system of inequalities that take into consideration the requirements of heat supply reliability, and by the logical conditions that determine the potential ways to reconstruct the heat supply system components and improve its reliability. The obtained variants of the system configuration and parameters should be tested for their feasibility and operability.

The considered problem of the heat supply system expansion has two important aspects. On the one hand, it is connected to the creation of scientific and methodological basis for the informed decisions on the rational structure of the systems, their expansion and operation. The other aspect is a set of technical issues related to the identification and development of technical directions in the construction of the systems, their equipment and technologies for their operation. The most challenging of the technical issues are methodological issues of modeling, calculating and optimizing complex heat supply systems. Here we can highlight a number of subproblems. The main of them are: selection of an optimal configuration, determination of optimal parameters, and calculation of flow distribution in the heat supply system.

The expansion optimization problem of multi-loop heat supply systems cannot be completely formalized and solved using a single universal method. However, the decomposition of the general problem into several subproblems has made it possible to distribute the solving process in time and adjust solutions, taking into account individual characteristics of the systems.

In the paper we present a new technique for solving the problem and results of its practical application to the expansion and reconstruction of a heat supply system in one of Russia's city.

BACKGROUND

The following methods are applied to determine an optimal structure of heat supply system:

1. Comparison of variants specified by an engineer (Robineau, Fazlollahi, Fournier, Berthalon & Verdier, 2014; Shifrinson, 1940; Vasant, 2013; Zanfirov, 1933).
2. Search for an optimal configuration of the system (Courant & Robbins, 1941; Fazlollahi, 2014; Weber, 2008).

3. Solving a transport problem (Ford & Fulkerson, 1962).
4. Application of the technique of "redundant" design schemes (Merenkova, Sennova & Stennikov, 1982; Nekrasova & Khasilev, 1970; Stennikov, Oshchepkova & Stennikov, 2013).

The second and third methods do not take account of such features of HSSs as their distribution on the territory, obstacles to their construction, etc. They have limited practical application, whereas the method for comparison of variants and the technique of redundant design schemes are applied more widely to design heat supply systems.

In their research paper (Robineau, Fazlollah, Fournier, Berthelon. & Verdier, 2014), the authors present a methodology for multi-objective optimization of heat networks. A computer tool using this methodology allows decision makers to economically determine a number of optimal configurations of the heat network. The main advantage of this methodology is the fact that the optimal solutions can be selected from a large set of variants (combinations of technical solutions). Furthermore, this methodology ensures the selection of optimal variants, taking into account local features. This tool can be used to obtain several potential variants of the hetwen network expansion, which are further analyzed in more detail with corresponding modeling tools.

In the papers (Fazlollahi, Mandel, Becker & Maréchal, 2012; Haikarainen, Pettersson, & Saxén, 2014; Weber, 2008) the authors suggest the heat supply system design and optimization approaches. The approaches make it possible to select the configuration of a heat supply system, taking into account the sites of energy sources and potential routes for laying the heating pipes of the heat network.

In the study (Dorfner & Hamacher, 2014) the optimization problem of heat network configuration is formulated as a mixed-integer problem on a graph. The heat network model employs an idealized graph for the representation of potential heating pipe locations. Such a graph is developed on the basis of publicly available data. Demand for heat is calculated separately based on the points (consumers) specified in advance on the graph. The network configuration optimization is based on a concave investment cost function of the heat networks. The validity of the obtained results is demonstrated on a case study in Munich.

Recently meta-heuristic algorithms (MHA) have gained noteworthy attention for their abilities to solve difficult optimization problems in engineering, economies etc (Vasant, 2013).

It should be noted that the attention paid to the issues of selection of the optimal heat supply system configuration is insufficient. One of the reasons is complexity of this problem. Another reason is that researchers are not interested in solving this problem. This can probably be explained by the fact that the route of the heat network in any case is determined by politicians and city planners, and no quantitative evaluations are involved. Therefore, there is the assurance that it is no use including the heat network configuration development and some other technical and economic optimization problems in the projects. At the same time many experts are convinced that the computational tools make it possible to obtain quantitative estimates which will facilitate making the best decision.

In this paper, we develop a methodology of "redundant" design schemes suggested in (Merenkova, Sennova & Stennikov, 1982; Nekrasova & Khasilev, 1970; Stennikov, Oshchepkova & Stennikov, 2013). The methodology takes into consideration the specific features of heat supply systems to the maximum extent possible, allows us to consider all possible design variants of the heat supply system scheme in a single initial graph, and obtain optimal solutions.

Optimization problem of HSS parameters is described by a set of balance equations (flow rates and heads) and by technical and economic relationships. Optimization criterion is the minimum estimated (discounted) costs.

Traditionally heat supply systems are reconstructed by determining pipeline diameters by the linear programming methods (Dantzig, 1963) or by the dynamic programming (DP) method for branched systems (Appleyard, 1978; Garbai & Molnar, 1974). However, for Russian large and multi-loop HSSs with a heat capacity above 4500 MW this leads to non-optimal solutions.

Researchers often combine the operation optimization problems and the parameter optimization problems of heat networks.

In their research (Robineau, Fazlollah, Fournier, Berthelon & Verdier, 2014), the authors present a flexible methodology which can be applied to a great number of problems including the optimization of existing and new systems, the calculation of operating conditions and selection of parameters, the assessment of efficiency, and others. The methodology is intended for obtaining preliminary solutions and aggregate estimates.

The methods for parameter optimization are considered in the studies (Fazlollahi, 2014; Phetteplace, 1995; Valdimarsson, 2001). They make it possible to solve the problems on the determination of equipment parameters of energy sources and pipelines of heat networks, and can take into consideration potential constraints on their selection. In these studies the researchers use the linear and nonlinear programming methods which can be applied to the aggregate estimation of the obtained solutions.

To optimize the parameters of branched (without loops) heat supply systems, we apply the DP method which is based on the Bellman's principle of optimality (Bellman, 1957). The method is actively developed in the studies (Barakhtenko, Oshchepkova, Sokolov & Stennikov, 2013; Merenkov (Ed.), 1992; Sokolov, Stennikov, Oshchepkova & Barakhtenko, 2012). The parameter optimization of multi-loop heat supply systems is a multi-extremal problem of nonlinear programming with discrete variables and complex constraints. It can be solved by the *method of multi-loop optimization* which was suggested in (Barakhtenko, Sokolov & Stennikov 2014; Stennikov, Sennova & Oshchepkova, 2006; Sumarokov, 1976). This method combines the methods for the calculation of flow distribution and parameter optimization of branched heat supply systems. To date, no methods similar to the multi-loop optimization method have appeared in the literature.

The flow distribution model includes a set of balance equations (analogues of Kirchhoff laws) (Cross, 1936; Kirchhoff, 1847; Maxwell, 1873) and closing relationships (Equations 11, 12, 13, and 14) that describe the relationship between pressure losses and flow rate in a heat network section.

The problem of flow distribution in heat networks was and has been studied by a great number of researchers (Merenkov & Khasilev, 1985; Novitsky, 2014; Robineau, 2014; Sennova & Sidler, 1987; Vasant, Barsoum &Webb, 2012; Vasant, 2014; Vasant, 2015;). The paper (Ancona, Bianchi, Branchini & Melino, 2014) presents new software to design and analyze the district heating networks. This software is developed by researchers from the University of Bologna on the basis of the Todini-Pilati algorithm generalized by the use of Darcy-Weisbach equation. The developed software can model the operating conditions of the district heating network under specified conditions.

The studies discussed in this paper are based on the flow distribution models (Novitsky, 2014; Sennova & Sidler, 1987) and information-computation system developed at the Energy Systems Institute SB RAS (Alekseev, Novitsky, Tokarev & Shalaginova, 2007).

PROBLEM STATEMENT

Modern HSSs have become complex engineering structures. They represent an integrated system of heat generation plants and numerous nodes for connection of various consumers. They are characterized by large scales and distribution throughout a vast territory. The structural complexity of HSSs manifest itself in a great number of loops and their spatial schemes, presence of active devices and regulators, inhomogeneity and nonlinearity of individual technical and economic characteristics of the network components. The large number of loops in the HSSs schemes is determined by the presence of two-line heat networks (HNs) that provide closed circulation of heat carrier and by the process of expansion, when individual main lines are connected by links, thus forming multi-loop systems of supply and return pipelines. Considerable length of HSSs, rugged relief and their irregular operating conditions require a large number of active components (pumping and throttling stations, regulators of flow rate, pressure, temperature, etc.).

Continuous complication and development of the existing HSSs call for expansion of old and construction of new heat sources (HSs), reconstruction of operating heat pipelines and pumping stations as well as construction of new heat generation and network facilities. The entire range of subproblems that constitute this problem can be solved both at the stage of planning the current measures on reconstruction of HSSs and at the stage of their expansion. The subproblems may need to be solved not only during expansion of the existing HSSs but also during construction of new ones. However, these subproblems are to the greatest extent typical of the existing heat networks. Only by solving the issues of HSS expansion and reconstruction in a timely and substantiated manner it will be possible to provide reliable heat supply to consumers, enhance economic efficiency of capital investments, decrease metal intensity of the system and improve other technical and economic indices.

The problem of optimal expansion and reconstruction of HSSs includes a variety of general and relatively specific subproblems.

The general problem can be easily decomposed into the following subproblems:

1. Optimization of HSS structure.
2. Optimization of heat network parameters.
3. Calculation of flow distribution (hydraulic calculation).
4. Calculation and estimation of heat supply reliability indices.

Thus, integrating and solving the subproblems in a system is very important to obtain technically acceptable and economically feasible result. A quasi-dynamic approach is suggested to solve this problem. The approach is based on the organization of a continuous process with adjustment of changing information at each stage.

The problem can be formulated as follows. For a HSS operating with one or several heat sources it is necessary to determine the measures on its reconstruction and identify the parameters of new components capable of supplying the required amount of heat. This should be done taking into account the existing part of the system and its operating conditions.

For solving the problem posed we set the structure of sources, their spatial location, mix and sizes of equipment, heat network scheme with existing and new parts highlighted, a discrete set of pipe diameters, sizes of pumping stations and other equipment.

The objective is to find an optimal distribution of heat load among the sources; detect bottlenecks in the heat networks and identify rational methods for their reconstruction; and optimize parameters of new system components. It is also necessary to determine the effective directions in HSS reconstruction and construction in order to create optimal service areas for heat sources, to lay new heat networks, to determine their schemes and parameters, and connection types of heat consumers.

The solutions to be obtained should meet the imposed technical requirements and involve minimum expansion and operation costs.

MATHEMATICAL MODEL

The problem is solved by finding the minimum objective function which represents a sum of costs of heat sources (Z_{HS}), costs of heat networks (Z_{HN}) and costs of electricity (Z_{CE}) that are reduced (discounted) to one year (hereinafter the values of costs are given in the National currency unit (NCU):

$$Z_{HS} + Z_{HN} + Z_{CE} \to \min. \tag{1}$$

The objective function (Equation 1) is calculated taking into account a payback period of capital received as a credit to implement the solutions proposed:

$$Z = aC + U, \tag{2}$$

where C – capital investments; U – annual costs; a – a discount factor, being, in fact, an annual rate of return.

The value of factor a is calculated considering capital investment dynamics by discounting:

$$a = \frac{1}{(1+r)^{-1} + (1+r)^{-2} + \ldots + (1+r)^{-k}}, \tag{3}$$

where r – a discount interest rate that can be represented by a real interest on principal; k – a payback period.

This method, also called the least cost method or the method of annual discounted costs, is widely used in the market economy to estimate various projects and compare them with one another.

The costs of heat sources are made up of a discounted capital component, semi-fixed costs of heat source operation and maintenance, and fuel costs:

$$Z_{HS} = a \sum_{j \in J_2} k_{HSj}(g_j) Q_j + \sum_{j \in J_2} U_j + \sum_{t=1}^{T} \sum_{j \in J_2} z_{Bjt} B_{jt} h_p, \tag{4}$$

where $k_{HSj}(g_j)$ – capital investments in the j-th heat source per unit of capacity that are determined depending on type and size of the source (g_j); Q_j - capacity of the j-th heat source (GJ/h); $U_j = \varepsilon_{HS} k_{HSj}(g_j) Q_j$ – semi-fixed annual costs of the j-th heat source operation and maintenance; ε_{HS} – a standard coefficient of annual deductions; z_{Bjt} – cost of 1 ton of coal equivalent (tce) (29307.6 kJ/kg) for the j-th heat source; $B_{jt} = f(Q_{jt})$ – hourly fuel consumption by the j-th heat source during time period t; Q_{jt} – capacity of the j-th heat source at period t (GJ/h); h_p - duration of period t (hour), T – annual number of hours.

Hereinafter J_1, J_2 – subsets of demand and supply nodes that belong to the set of all nodes J; I– a set of sections; I_1, I_2 – subsets of existing and newly designed network sections, respectively; I_3, I_4 – subsets of sections with existing and new pumping stations ((PSs).

The costs of heat networks Z_{HN} include the costs of heat network construction, reconstruction, operation and maintenance ($\sum_{i \in I} Z_{HNi}$), the costs of PS construction and operation ($\sum_{i \in I} Z_{PSi}$), the costs of electricity ($\sum_{i \in I} Z_{Eit}$) which is used to pump a heat carrier and costs of compensation for heat losses ($\sum_{i \in I} Z_{HLit}$).

$$Z_{HN} = \sum_{i \in I} Z_{HNi} + \sum_{i \in I} Z_{PSi} + \sum_{t=1}^{T} (\sum_{i \in I} Z_{Eit} + \sum_{i \in I} Z_{HLit}) h_p. \qquad (5)$$

The constant constituent of annual costs of the heat networks is determined by the expression:

$$\sum_{i \in I_1 \cup I_2} Z_{HNi} = a \sum_{i \in I_1 \cup I_2} k_{HNi}(d_i) l_i + \sum_{i \in I_1 \cup I_2} U_{HNi}, \qquad (6)$$

where $k_{HNi}(d_i)$ – capital investments in construction and reconstruction of heat networks per unit of length depending on the pipeline diameter in the i-th heat network section; l_i – length of the i-th heat network section (m); $U_{HNi} = \varepsilon_{HN} k_{HNi}(d_i) l_i$ – semi-fixed annual operating costs of heat network; ε_{HN} – a standard coefficient of annual deductions.

The constant constituent of the heat network costs is equal to:

$$\sum_{i \in I_3 \cup I_4} Z_{PSi} = a \sum_{i \in I_3 \cup I_4} c_1 k_{PSi}(\varphi_i) (\frac{x_i H_i}{\eta_i}) + \sum_{i \in I_3 \cup I_4} U_{PSi}, \qquad (7)$$

where c_1 – a coefficient depending on dimension of values; $k_{PSi}(\varphi_i)$ – capital investments in pumping station construction and reconstruction per unit of capacity in the i-th section depending on the station type and size φ_i; $U_{PSi} = \varepsilon_{PS} k_{PSi}(\varphi_i) c_1 (\frac{x_i H_i}{\eta_i})$ – semi-fixed annual operating costs of pumping station; ε_{PS} – a standard coefficient of annual deductions.

Costs of electricity consumed to pump a heat carrier in the system at each period t are calculated by the following relationship:

$$\sum_{i \in I_3 \cup I_4} Z_{Eit} = c_1 z_{Et} \sum_{i \in I_3 \cup I_4} \frac{x_{it} H_{it} B_{it}}{\eta_{it}}, \quad t = \overline{1, T}, \qquad (8)$$

where z_{Et} – electricity cost (tariff); x_{it} – an hourly flow rate of heating-system water in the i-th section at period t; H_{it} – operating head at pumping station in the i-th section at period t (mwc); T_{it} – number of pumping station operation hours in the i-th section at period t (hour at period t); η_{it} – efficiency of pumping station in the i-th section at period t.

Costs of heat losses ($\sum_{i \in I} Z_{HLit}$) for each section are determined by the expression:

$$\sum_{i \in I} Z_{HLit} = z_{HLt} \sum_{i \in I} q_{it}(d_i) l_i , \ t = \overline{1,T} , \tag{9}$$

where z_{HLt} – heat cost (price); $q_{it}(d_i)$ – heat losses per unit of the i-th section length that correspond to the type of pipeline insulation and pipeline construction method.

Electricity costs (Z_{CEt}) represent the electricity cost that provides equal electricity supply in all variants. They are equal to the product of maximum electricity tariff (z_{CEt}) and the difference in electricity supplies (W_{jt}, MW) in different variants:

$$Z_{CE} = \sum_{t=1}^{T} z_{CEt} \sum_{j \in J_2} W_{jt} h_p . \tag{10}$$

Correct results cannot be obtained without considering certain conditions and constraints which include:

- Equations of material and energy balances (analogues of Kirchhoff's laws) as well as closing relationships which describe a network part of the heat network and have the following form:

for supply pipeline

$$\begin{cases} A x_t = G_{1t} + G_{2t} \\ A^T P_t = y_t \\ y_t + H_t = h_t \end{cases} , \ t = \overline{1,T}, \tag{11}$$

for return pipeline

$$\begin{cases} A x'_t = G_{1t} \\ A^T P'_t = y'_t \\ y'_t + H'_t = h'_t \end{cases} , \ t = \overline{1,T}, \tag{12}$$

where $A = \{a_{ij}\}$ – the $(m-1) \times n$ matrix of connections for linearly independent nodes; $x_t = \begin{bmatrix} x_{1t} \\ \vdots \\ x_{nt} \end{bmatrix}$,

$x'_t = \begin{bmatrix} x'_{1t} \\ \vdots \\ x'_{nt} \end{bmatrix}$ – vectors of heat carrier flow rates in network sections in supply and return pipelines at period t (t/h); $G_{1t} = \begin{bmatrix} G_{11t} \\ \vdots \\ G_{1mt} \end{bmatrix}$ – vector of heat carrier flow rates at nodes for heating and ventilation at period t (t/h); G_{2t} – vector of heat carrier flow rates at nodes for hot water supply at period t (t/h);

$P_t = \begin{bmatrix} P_{1t} \\ \vdots \\ P_{mt} \end{bmatrix}$, $P'_t = \begin{bmatrix} P'_{1t} \\ \vdots \\ P'_{mt} \end{bmatrix}$ – vectors of heat carrier pressures at heat network nodes in supply and return pipelines at period t (mwc); $y_t = \begin{bmatrix} y_{1t} \\ \vdots \\ y_{nt} \end{bmatrix}$, $y'_t = \begin{bmatrix} y'_{1t} \\ \vdots \\ y'_{nt} \end{bmatrix}$ – vectors of difference in heat carrier pressures in a branch of supply and return pipelines at period t (mwc); $y_t = P_{jt} - P_{j+1\,t}$, $y'_t = P'_{jt} - P'_{j+1\,t}$; $H_t = \begin{bmatrix} H_{1t} \\ \vdots \\ H_{nt} \end{bmatrix}$,

$H'_t = \begin{bmatrix} H'_t \\ \vdots \\ H'_{nt} \end{bmatrix}$ – vectors of pumping station operating heads in a branch of supply and return pipelines at period t (mwc); $h_t = \begin{bmatrix} h_{1t} \\ \vdots \\ h_{nt} \end{bmatrix}$, $h'_t = \begin{bmatrix} h'_t \\ \vdots \\ h'_{nt} \end{bmatrix}$ – vectors of heat carrier pressure losses in a branch of supply and return pipelines at period t (mwc).

- Relationship between pressure losses in the i-th branch and flow rate has the form:

$$h_{it} = \phi_i \frac{x_{it}^2 l_i (1 + \alpha_i)}{d_i^{5,25}}, \quad t = \overline{1,T}, \, i \in I \tag{13}$$

$$h'_{it} = \phi_i \frac{(x'_{it})^2 l_i (1 + \alpha_i)}{d_i^{5,25}}, \quad t = \overline{1,T}, \, i \in I, \tag{14}$$

where ϕ_i - a coefficient depending on equivalent pipe roughness of the *i*-th branch; α_i - a coefficient of local losses in the *i*-th branch.

- Condition for choice and discreteness of:

pipe diameters for new sections, and reconstruction method for existing heat network sections:

$$d_i = \begin{cases} \xi_{i1}\hat{d}_1 + \xi_{i2}\hat{d}_2 + \ldots + \xi_{i\chi_{HN}}\hat{d}_{\chi_{HN}}, & \xi_{i\chi_{HN}} = 1 \vee 0, \sum_{r_{HN}=1}^{\chi_{HN}} \xi_{ir_{HN}} = 1, i \in I_2, \\ \xi_{i1}\hat{d}_1 + \xi_{i2}\hat{d}_2 + \ldots + \xi_{i\chi'_{HN}}\hat{d}'_{\chi_{HN}}, & \xi_{i\chi'_{HN}} = 1 \vee 0, \sum_{r'_{HN}=1}^{\chi'_{HN}} \xi_{ir'_{HN}} = 1, i \in I_1, \end{cases} \quad (15)$$

where χ_{HN} – a discrete set of pipe diameters for construction of new sections; χ'_{HN} – a discrete set of reconstruction methods for existing pipeline sections.

types and sizes of new pumping stations and methods for reconstruction of existing pumping stations:

$$\varphi_i = \begin{cases} \psi_{i1}\hat{\varphi}_1 + \psi_{i2}\hat{\varphi}_2 + \ldots + \psi_{i\chi_{PS}}\hat{\varphi}_{\chi_{PS}}, & \psi_{i\chi_{PS}} = 1 \vee 0, \sum_{r_{PS}=1}^{\chi_{PS}} \psi_{ir_{PS}} = 1, i \in I_4, \\ \psi_{i1}\hat{\varphi}_1 + \psi_{i2}\hat{\varphi}_2 + \ldots + \psi_{i\chi'_{PS}}\hat{\varphi}'_{\chi_{PS}}, & \psi_{i\chi'_{PS}} = 1 \vee 0, \sum_{r'_{PS}=1}^{\chi'_{PS}} \psi_{ir'_{PS}} = 1, i \in I_3, \end{cases} \quad (16)$$

where χ_{PS} – a discrete set of pumping station types and sizes for the case of new construction; χ'_{PS} – a discrete set of possible methods for reconstruction of existing pumping stations.

types and sizes of heat sources:

$$g_j = \begin{cases} \delta_{j1}\hat{g}_1 + \delta_{j2}\hat{g}_2 + \ldots + \delta_{j\chi_{HS}}\hat{g}_{\chi_{HS}}, & \delta_{j\chi_{HS}} = 1 \vee 0, \sum_{r_{HS}=1}^{\chi_{HS}} \delta_{jr_{HS}} = 1, j \in J_2, \\ \delta_{j1}\hat{g}_1 + \delta_{j2}\hat{g}_2 + \ldots + \delta_{j\chi'_{HS}}\hat{g}'_{\chi_{HS}}, & \delta_{j\chi'_{HS}} = 1 \vee 0, \sum_{r'_{HS}=1}^{\chi'_{HS}} \delta_{jr'_{HS}} = 1, j \in J_2, \end{cases} \quad (17)$$

where χ_{HS} – a discrete set of types and sizes of newly constructed heat sources; χ'_{HS} – a discrete set of possible methods for reconstruction of existing heat sources.

- Specific capital investments in:

construction of new heat network sections:

$$k_{HNi}(d_i) = \xi_{i1}k_{HN1}(\hat{d}_1) + \xi_{i2}k_{HN2}(\hat{d}_2) + \ldots + \xi_{i\chi_{HN}}k_{HN\chi_{HN}}(\hat{d}_{\chi_{HN}}),$$

$$\xi_{i\chi_{HN}} = 1 \vee 0, \sum_{r_{HN}=1}^{\chi_{HN}} \xi_{ir_{HN}} = 1, \ i \in I_2; \qquad (18)$$

reconstruction of existing heat network sections:

$$k_{HNi}(d_i) = \xi_{i1} k_{HN1}(\hat{d}_1) + \xi_{i2} k_{HN2}(\hat{d}_2) + \ldots + \xi_{i\chi'_{HN}} k_{HN\chi'_{HN}}(\hat{d}_{\chi'_{HN}}),$$

$$\xi_{i\chi'_{HN}} = 1 \vee 0, \sum_{r'_{HN}=1}^{\chi'_{HN}} \xi_{ir'_{HN}} = 1, \ i \in I_1; \qquad (19)$$

construction of new pumping stations:

$$k_{PSi}(\varphi_i) = \psi_{i1} k_{PS1}(\hat{\varphi}_1) + \psi_{i2} k_{PS2}(\hat{\varphi}_2) + \ldots + \psi_{i\chi_{PS}} k_{PS\chi_{PS}}(\hat{\varphi}_{\chi_{PS}}),$$

$$\psi_{i\chi_{PS}} = 1 \vee 0, \sum_{r_{PS}=1}^{\chi_{PS}} \psi_{ir_{PS}} = 1, \ i \in I_4; \qquad (20)$$

reconstruction of existing pumping stations:

$$k_{PSi}(\varphi_i) = \psi_{i1} k_{PS1}(\hat{\varphi}_1) + \psi_{i2} k_{PS2}(\hat{\varphi}_2) + \ldots + \psi_{i\chi'_{PS}} k_{PS\chi'_{PS}}(\hat{\varphi}_{\chi'_{PS}}),$$

$$\psi_{i\chi'_{PS}} = 1 \vee 0, \sum_{r'_{PS}=1}^{\chi'_{PS}} \psi_{ir'_{PS}} = 1, \ i \in I_3; \qquad (21)$$

construction of new heat sources:

$$k_{HSj}(g_j) = \delta_{j1} k_{HS1}(\hat{g}_1) + \delta_{j2} k_{HS2}(\hat{g}_2) + \ldots + \delta_{j\chi_{HS}} k_{HSe}(\hat{g}_e),$$

$$\delta_{j\chi_{HS}} = 1 \vee 0, \sum_{r_{HS}=1}^{\chi_{HS}} \delta_{jr_{HS}} = 1, \ j \in J_2; \qquad (22)$$

reconstruction of existing heat sources:

$$k_{HSj}(g_j) = \delta_{j1} k_{HS1}(\hat{g}_1) + \delta_{j2} k_{HS2}(\hat{g}_2) + \ldots + \delta_{j\chi'_{HS}} k_{HS\chi'_{HS}}(\hat{g}_{\chi'_{HS}}),$$

$$\delta_{jx'_{HS}} = 1 \vee 0, \sum_{r'_{HS}=1}^{x'_{HS}} \delta_{jr'_{HS}} = 1, j \in J_2. \tag{23}$$

- Specifications and constraints:

    on heat carrier pressures at nodes:
    for supply pipeline

$$\underline{P}_j \leq P_{jt} \leq \overline{P}_j, \ j \in J, \ t = \overline{1,T}, \tag{24}$$

    for return pipeline

$$\underline{P}'_j \leq P'_{jt} \leq \overline{P}'_j, \ j \in J, \ t = \overline{1,T}. \tag{25}$$

    on heat carrier velocity in a branch (m/s):
    for supply pipeline

$$\underline{w}_i \leq w_{it} \leq \overline{w}_i, \ i \in I, \ t = \overline{1,T}, \tag{26}$$

    for return pipeline

$$\underline{w}'_i \leq w'_{it} \leq \overline{w}'_i, \ i \in I, \ t = \overline{1,T}, \tag{27}$$

where the heat carrier velocity in the pipeline is determined by the relationship:
    for supply pipeline

$$w_{it} = \frac{x_{it}}{0,7854 d_i^2 \rho}, \ i \in I, \tag{28}$$

    for return pipeline

$$w'_{it} = \frac{x'_{it}}{0,7854 d_i^2 \rho'}, \ i \in I. \tag{29}$$

    on heat carrier pressure difference at consumers (mwc):

$$y_{jt} \geq \underline{y}_j, \ j \in J_1, \ t = \overline{1,T}, \tag{30}$$

$$y_{jt} = P_{jt} - P'_{jt}, \, j \in J_1, \, t = \overline{1,T}. \tag{31}$$

on difference in heating-system water temperature at consumers (C°):

$$\Delta \tau_{jt} \geq \Delta \underline{\tau}_j, \, j \in J_1, \, t = \overline{1,T}, \tag{32}$$

$$\Delta \tau_{jt} = \tau_{jt} - \tau'_{jt}, \, j \in J_1, \, t = \overline{1,T}. \tag{33}$$

where τ_j and τ'_j - temperature of heating-system water in supply and return pipelines (C°).

on capacity of heat sources (GJ/h):

$$\underline{Q}_j \leq Q_{jt} \leq \overline{Q}_j, \, j \in J_2, \, t = \overline{1,T}, \tag{34}$$

where $Q_{jt} = \Delta \tau_{jt} G_{jt} c_{jt}$ – capacity of the j-th heat source at period t (GJ/h); c_{jt} – heating capacity of heat carrier (kJ/(kg · C°)); $\underline{Q}_j$, $\overline{Q}_j$ – low and upper constraints on capacity of the j-th heat source, that depend on type and size of heat source g_j (GJ/h).

on nodal indices of heat supply reliability that are determined by standard values for all demand nodes: availability factor

$$K_j^{(1)} \geq K_H^{(1)}, \, j \in J_1, \tag{35}$$

probability of failure-free operation

$$R_j^{(2)} \geq R_H^{(2)}, \, j \in J_1. \tag{36}$$

Temperature of heat carrier in supply pipeline (τ_{jt}, C°) depends on the ambient air temperature (τ_{TAt}, C°), the method applied to regulate heat supply and consumption (χ) and a scheme of connecting consumers to heat networks (ζ_j):

$$\tau_{jt} = f(\tau_{TAt}, \chi, \zeta_j), \, j \in J_1, \, t = \overline{1,T}. \tag{37}$$

Temperature of heat carrier in return pipeline (τ'_{jt}, C°) depends on the ambient air temperature (τ_{TAt}, C°), heating-system water temperature in supply pipeline (τ_{jt}), the method applied to regulate heat supply and consumption (χ) and the scheme of connecting consumers to heat networks (ζ_j):

$$\tau'_{jt} = f(\tau_{jt}, \tau_{TAt}, \chi, \zeta_j), \, j \in J_1, \, t = \overline{1,T}. \tag{38}$$

The condition for choosing a scheme of connecting consumers has the following form:

$$\zeta_j = \mu_{j1}\hat{\zeta}_1 + \mu_{j2}\hat{\zeta}_2 + \ldots + \mu_{j\chi_{SC}}\hat{\zeta}_{\chi_{SC}},$$

$$\mu_{j\chi_{SC}} = 1 \vee 0, \sum_{r_{SC}=1}^{\chi_{SC}} \mu_{jr_{SC}} = 1, j \in J_1. \tag{39}$$

The condition for choosing the method for regulation of heat supply and consumption is written in the following form:

$$\Omega_j = \theta_{j1}\hat{\Omega}_1 + \theta_{j2}\hat{\Omega}_2 + \ldots + \theta_{j\chi_{reg}}\hat{\Omega}_{\chi_{reg}},$$

$$\theta_{j\chi_{reg}} = 1 \vee 0, \sum_{r_{reg}=1}^{\chi_{reg}} \theta_{jr_{reg}} = 1, j \in J_1, \tag{40}$$

where χ_{SC} – a set of schemes of connecting consumers to heat network; χ_{reg} – a set of methods for regulation of heat supply and consumption.

Thus, the considered problem can be mathematically formulated as minimization of the function of total costs (Equation 1) subject to (Equations 11, 12, 13, 14, 15, 16, 17, 18, 19, 20, 21, 22, 23, 24, 25, 26, 27, 28, 29, 30, 31, 32, 33, 34, 35, 36, 37, 38, 39, and 40).

PROPERTIES OF THE MATHEMATICAL MODEL

The need to take into account the current state of heat network essentially complicates the problem stated. The schemes get more complicated turning from branched (dead-end) into the multi-loop (ring) ones. The requirements for controllability of systems, and shift from qualitative to quantitative regulation of heat supply and consumption generate the necessity to introduce structural and parametric redundancy in them. Continuous objective function becomes discrete with additional local minimums (Figure1, points K, L, I). Hydraulic and economic parameters of system components have a nonlinear character. Certain complexity of solving the considered problem is caused by a large spatial distribution of the systems, structural inhomogeneity, uneven distribution of heat sources and consumers, and continuous HSS expansion in time.

The above formulated general mathematical problem of nonlinear discrete programming contains some unformalized propositions and procedures. Analysis and study of the problem posed made it possible to replace it by a set of relatively independent subproblems which are then integrated in a single computational process.

Figure 1. Graphical interpretation of optimization problem of heat supply system parameters

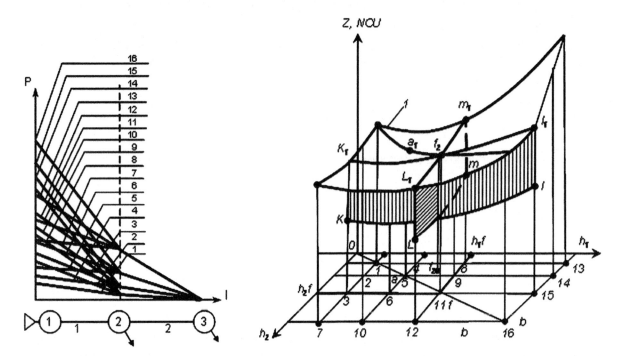

A SET OF PROBLEMS TO BE SOIVED

Solving this set of problems has become possible owing to the methods and algorithms that were developed on the basis of the theory of hydraulic circuits at the Energy Systems Institute (Merenkov (Ed.), 1992; Merenkov & Khasilev, 1985; Sennova & Sidler, 1987). These methods and algorithms were practically tested and proved to be effective.

The optimization problem of HSS structure implies the choice of the number, type, sites and capacity of sources, including determination of whether it is expedient to expand or exclude the existing sources from the scheme and distribute heat load among the sources; and the search for the optimal HSS scheme with its division into reserved and unreserved parts (Merenkova, Sennova & Stennikov, 1982; Nekrasova & Khasilev, 1970).

Despite relatively simple conceptual formulation, the problems of HSS scheme and structure optimization are complex multi-extremal network problems of nonconvex minimization. They have a large dimension, an elaborate system of constraints and present difficulty for application of mathematical methods, particularly when it comes to their computational and practical efficiency. This is first of all due to the fact that the function of discounted costs in these problems, even when new systems are designed, is multi-extremal, since it is convex with respect to pressure losses in the network sections and concave with respect to flow rates (Sennova & Sidler, 1987; Stennikov, Sennova & Oshchepkova, 2006). At the same time they are very important because determine not only the structure of systems, their type, scales but also their parameters. The economic benefit from the appropriate choice of system structure is, as a rule, larger than from the optimization of its parameters.

The mathematical statement of the problem includes an objective function represented by expression (Equation 1), a system of conditions, equality and inequality constraints and logical conditions: (Equations 11, 12, 13, and 14), (Equation 16), (Equation 17), (Equation 24), (Equation 25), (Equation 34). The problem stated is reduced to the problem of the most beneficial flow distribution in a given redundant (virtual) network scheme and implies minimization of nonlinear function (Equation 1) under the foregoing constraints. The optimal solution lies on the boundaries of the feasible region defined by the set of balance (Equations 11 and 12) and corresponds to one of the trees that connect demand nodes with heat sources. The optimal solution is found by the method of purposeful limited enumeration of variants of the redundant scheme trees that has various algorithmic modifications (Merenkova, Sennova & Stennikov, 1982).

The redundant scheme represents a calculated scheme of a system. It reflects demand nodes, potential nodes for heat source construction and all potential (realizable) lines among them. The scheme is preliminarily outlined by a researcher as an aggregate of all admissible options of network configuration and distribution of sources with account taken of constraints on construction of pipelines and location of sources. The redundant scheme marks out the existing sources, network sections, pumping and throttling stations. Potential sites for new sources and consumers as well as pipelines between them are also plotted. It can adequately show the dynamics of system development and stages of putting its components into operation. The scheme is constructed considering the constraints related to natural and artificial obstacles on the territory and the sites chosen for placement of sources.

In the problems of reconstructing and expanding HSSs with numerous sources the redundant scheme (Figure 2) is complemented with a fictitious node and a subset of fictitious sections connecting it with the nodes, in which construction of a heat source is allowed or a heat source is under reconstruction. The number of fictitious sections for each supply node depends on how many alternative types of sources are considered at this node. For example, in Figure 2 for the node HS1 their number equals four while for the node HS2 – two. Capacity of sources Q_j at nodes with fictitious sections are not given but included in the number of decision variables as unknown values of flow rates x_i in these branches. The fictitious sections are assigned to the costs of the sources located at the incident nodes. The costs describing nonlinear dependence on the capacity value can be represented by both function $x = f(Q)$ and table.

Figure 2. A redundant scheme of heat supply system

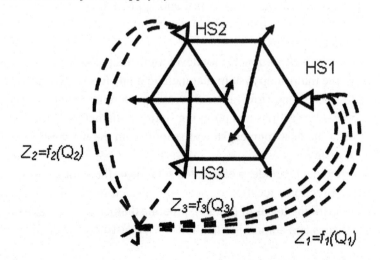

In the most general case an individual (decentralized) source can be considered at each demand node. This will make it possible to solve a structural subproblem and simultaneously determine an optimal level of heat supply centralization (decentralization), a rational concentration of source capacities and an economically efficient distance of heat transportation (an economic radius of heat supply).

In the formalized description of the redundant scheme a set of nodes J is increased by adding a fictitious node $j = m + 1$ to it. In the subset of supply nodes (J_2) we highlight J_4 - the subset of nodes with fixed inflows and J_3 - the subset of nodes without load. The resulting set of nodes has the form $J = J_1 \cup J_2 \cup J_3 \cup J_4 \cup j_{m+1}$.

Unlike the conventional calculated scheme the set of sections I of the redundant scheme is supplemented by the subset of fictitious lines I_{m+1} that connect supply nodes with the fictitious node. The set of sections takes the form $I = I_1 \cup I_2 \cup I_3 \cup I_4 \cup I_{m+1}$.

New parameters of the redundant scheme expanded by the fictitious node will be: the number of nodes – $M = m + 1$; the number of sections – $N = n + n_f$; the number of loops – $C = c + n_f - 1(n_f)$ – the number of fictitious sections in the scheme). Here m, n, c are the number of nodes, sections and loops in the redundant scheme before its expansion.

The obtained redundant scheme is a multiloop network that connects a set of heat sources with consumers. The subproblem of choosing an optimal system structure and the most advantageous configuration of the heat network can be reduced to determination of the optimal (in terms of an assumed criterion) flow distribution. The process of solving the subproblem consists in rejection of unnecessary sections and determination of the most advantageous configuration of the network or several networks in the redundant scheme. The sections with nonzero flow rates remain in the solution sought. The zero flow rate in the fictitious branch means that at the node incident to it construction of a new source is inefficient or removal of the existing source from the scheme is rational. As a result the optimal subnetwork is determined in the initial scheme as a tree or a forest of trees that connects sources with consumers.

Three different models are applied to solve the considered subproblem. The models implement continuous and discrete statements and differ in the accuracy of consideration of the existing part, the discreetness of equipment parameters, the number and type of constraints.

Choice of a specific model depends on the consideration level of systems and the stage of their designing.

The technique of redundant schemes is universal and applied to solve many subproblems of calculation and optimization of complex objects within a network structure. It provides solutions for both centralized and decentralized systems.

The optimization subproblem of parameters for developing multi-loop HSSs consists in determination of optimal parameters of the multi-loop system that include diameters of pipelines of newly constructed sections; placement sites, heads and flow rates of pumping stations; methods of reconstruction (from those permitted) of the existing sections and pumping stations; diameters of the reconstructed sections; available pressure heads at the heat sources; flow distribution in the network, load distribution among sources, and pressure at system nodes.

In this case the specified data include: a calculated scheme of the multi-loop system with indication of existing and newly laid sections; location of heat sources and their calculated parameters (thermal power, water flow rates, operating heads of pumps); pipeline diameters of the existing network sections, their hydraulic characteristics and admissible methods of reconstruction; location of existing pumping stations and their performance; potential sites for new pumping stations; a set of standard pipe diameters, a list of pumping stations used for optimization and also specific investments in them; and other technical and economic indices and coefficients.

The optimization criterion is the minimum discounted costs that are expressed by the function $Z_{HN}(d, x, H, P)$ and include construction and operation costs for heat pipelines and pumping stations, costs of heat losses, costs of electricity used to pump heat carrier. A system of conditions and constraints is represented by subsystems (Equations 11, 12, 13, 14, 15, 16, 17, 18, 19, 20, 21, 22, 23, 24, 25, 26, 27, 28, 29, 30, 31, 32, 33, 34, 35, 36, 37, 38, 39, and 40) that determine the region of feasible solutions, physicotechnical and other requirements that describe specific features of heat supply systems and their individual components.

While to solve the optimization problem of developing multi-loop system parameters the method of multi-loop optimization (MMO) was suggested by researchers of the Energy Systems Institute (Sumarokov, 1976), the optimization problem for treelike heat networks is mostly solved by the DP method (Bellman, 1957). The MMO is based on the principle of decomposition of the stated problem into subproblems of parametric optimization of a branched heat network (without loops) and the subproblem of flow distribution of the multi-loop network that is less complicated and can be solved by available strict formalized methods.

The DP method applied to optimize parameters of the branched HSS takes into consideration individual features of the system and its components, current state, technical constraints and logic conditions, when choice of parameters is made (Appleyard, 1978; Garbai & Molnar, 1974; Merenkov & Khasilev, 1985; Sennova & Sidler, 1987). Graphical interpretation of the method is presented in Figure3. Calculation of flow distribution provides implementation of decisions made and serviceability of system in operation.

Calculations of real HSSs and good results of their practical implementation that were confirmed during operation indicate not only high computational efficiency of the MMO, but practical utility of the results obtained on its base.

The subproblem of flow distribution calculation (hydraulic calculation) is an independent subproblem of calculation, analysis and check of whether the system operating conditions are realizable. It is solved in a multistage process of operation optimization and is included in description of the region of feasible solutions in optimization models (Novitskyy, 2014). Distribution of flow rates and pressures of the heat carrier in the HSS is described by a set of linear (analogues of the first and second Kirchhoff's laws for electric circuit) and nonlinear (the relation between flow rates and pressure losses in the network sections) Equations 11, 12.

The experience gained in solving such subproblems was used as a base for construction of an effective model of flow distribution. The model implements the nodal pressure method, takes into consideration matrix sparsity, and automatically chooses the step at each iteration of the computational process (Alekseev, Novitsky, Tokarev & Shalaginova, 2007). Consideration of the network property of flow distribution subproblems decreases dimension of the systems of equations manifold and provides high speed of the applied models.

At Energy Systems Institute these subproblems are solved using the software tool ANGARA that has a developed user interface (Alekseev, Novitsky, Tokarev & Shalaginova, 2007). It is characterized by some specific features that distinguish it from the other flow distribution models, in particular: possibility of hydraulic calculation of large-dimensional networks of any complex configuration and structure with the arbitrary number and location of sources, consumers, pumping stations, regulators of pressure and flow

Figure 3. Graphical interpretation of DP method for optimization of heat supply system parameters

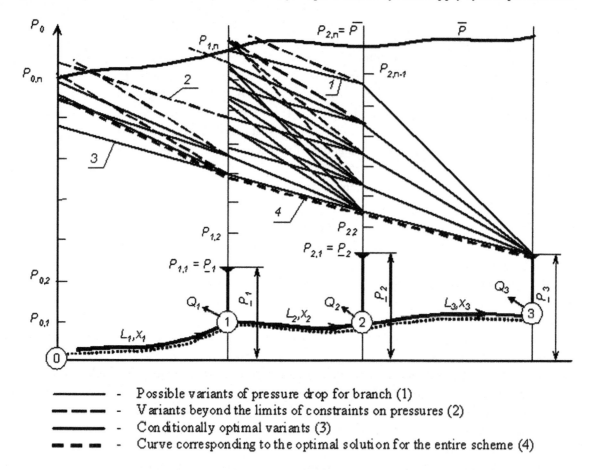

———— - Possible variants of pressure drop for branch (1)
— — — - Variants beyond the limits of constraints on pressures (2)
▬▬▬▬ - Conditionally optimal variants (3)
■ ■ ■ - Curve corresponding to the optimal solution for the entire scheme (4)

rate; high speed, theoretically guaranteed convergence of the computational process; higher reliability; compact storage and accumulation of data on different schemes; possibility of graphical representation and interpretation of initial information and calculation results.

The subproblem of calculation and estimation of heat supply reliability indices supposes consideration of two levels of heat supply to consumers – calculated (Equation 35) and emergency (Equation 36) (Merenkova, Sennova & Stennikov, 1982).

Reliability of the calculated (first) level for the consumer at node j is determined by the availability factor $K_j^{(1)}$ (Takaishvili & Khasilev, 1972). It is a mean value of the heating season part, during which the consumer has a calculated value of indoor temperature (t_{IN}). Instead of factor $K_j^{(1)}$ we can use the value of mean total (during the heating season) time of decrease in the indoor temperature below the calculated value – the time of unavailability ($Z_j^{(1)}$, h), which corresponds to the value of the factor.

Reliability of the emergency (second) level is estimated by the probability of failure-free operation $R_j^{(2)}$. It represents the probability that during the heating season the air temperature for the j–th consumer will not be lower than some boundary value $t_{b.v.}^{(2)}$. This index can also be applied in the form of failure rate $a_j^{(2)}$ (1 per year or 1 in n years).

To solve the problem of heat supply system expansion the reliability indices for heat supply to consumers are calculated by the following relations:

$$K_j^{(1)} = (T_{HP} - \sum_{i \in I_j} Z_i) / T_{HP} ; \tag{41}$$

$$Z_j^1 = T_{HP} (1 - K_j^{(1)}) ; \tag{42}$$

$$R_j^{(2)} = \prod_{n=1}^{N} \exp\left[-\sum_{i \in I_j} \omega_i (\Delta t_n - t_{pj}^{(2)}) \exp(-t_{pj}^{(2)} / \overline{\tau}_{2i})\right]; \tag{43}$$

$$a_j^{(2)} = -\ln R_j^{(2)}, \tag{44}$$

where ω_i, $\overline{\tau}_{2i}$ – a failure flow (1/h) and a mean time to restore the i-th HN element (Z); I_j – a set of elements in the branch to the j-th consumer and in the ring part of the network hydraulically connected with the consumer; Δt_n – duration of the n-th period, into which the heating season is divided $\Delta t_n = \dfrac{T_{HP}}{N}$; $t_{pj}^{(2)}$ – a value of the slack time of the j-th consumer with respect to the lower reliability level for HN elements, $T_{нp}$ – duration of heating period.

The probabilistic estimates of the considered levels of heat supply to consumers make it possible to distinguish reserved and unreserved parts of HN. The parameters of the reserved network part are chosen so that during failure of any section in this network part the consumers received some, as a rule, lower amount of heat. This is the norm of standby heat supply ϕ_j, $j \in J_1$.

A system of reliability standards is suggested to estimate heat supply reliability. It is oriented towards the vector reliability with respect to every consumption node and based on the above indicated indices (Sennova (Ed.), 2000).

A PROBLEM SOLVING TECHNIQUE

The technique for solving the complex optimization problem of HSS expansion is based on successively solving the above enumerated subproblems. In general it can be represented as follows:

1. Researcher prepares a redundant scheme of HSS with information required for calculation.
2. The redundant scheme (if necessary) is automatically supplemented with a fictitious node and expanded with fictitious sections.
3. An optimal system structure and the HN scheme are chosen without considering reliability requirements.

4. A mix of equipment at heat sources is determined, heat load is distributed among them and their technical and economic indices are calculated.
5. The heat network parameters are determined and its technical and economic indices are calculated.
6. The nodal reliability indices of heat supply to consumers are calculated and they are estimated in terms of standard values.
7. If the reliability indices satisfy standard requirements, the problem is solved and point 10 is performed.

When both reliability indices are violated, the capacities of heat sources (radiuses of heat supply) have to be reduced. It is necessary to solve the subproblem of scheme and structure optimization (return to step 3).

If the value $R_j^{(2)}$ does not correspond to the standard, the following stage is performed.

8. The reserved and unreserved system parts are determined based on the value $R_H^{(2)}$.
9. The reserved system part is supplemented by the sections included in the redundant scheme, provided the condition of the bilateral supply of "unreliable" scheme nodes with the heat carrier is satisfied.
10. The reserved network parameters (d, H) are determined.
11. Emergency situations in the ring network are calculated. The parameters (d, H) of sections are adjusted to provide emergency heat supply, when needed.
12. Correspondence of reliability indices to the standard requirements is checked. In case of their violation the system structure is modified (return to step 3).
13. Serviceability of the system is checked and its operating conditions are analyzed.
14. The criteria of economic efficiency and technical characteristics of the system are calculated and the financial possibility of implementing the obtained solutions is analyzed.

The algorithm devised on the basis of this technique is realized in the software tool (the dialog computing system). It includes five software packages and has several levels arranged by the hierarchical principle. When moving from one level to another the subproblems are defined in more detail and their list is extended. Figure 4 presents the software tool structure.

The suggested methodological and computational tool can be applied to calculate HSSs of different complexity. This is very important because there is no ready standard set of recommendations on expansion and reconstruction of heat supply systems. They are determined individually for a concrete HSS depending on its specific features.

SOLUTIONS AND RECOMMENDATIONS

The suggested methodology and software make it possible to prepare the recommendations for making decisions on the expansion of heat supply systems.

The process of generating such recommendations is demonstrated below.

The project on reconstruction of the considered real heat supply system in a city of Russia aims to enhance efficiency, reliability and quality of heat supply to consumers in terms of future increase in the heat loads.

Figure 4. Software for optimization of heat supply system expansion

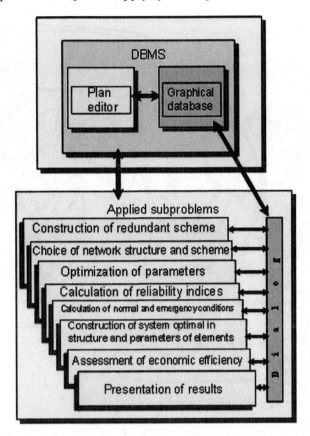

The considered area is supplied with heat from 6 cogeneration plants, 8 large and above 140 medium and small boiler plants. The total heat load here will account for above 7685 GJ/h by the year 2025.

All heat sources operate separately, each for its zone. At the same time they are connected by double-pipe heat networks that are arranged in a multi-loop scheme with numerous pumping stations.

The measures on reconstruction of HSS are chosen for the end of the studied period. Such an approach is suitable because, as the research shows, the successive replacement of heat network pipelines by those with larger diameters is neither economically nor technologically feasible. The sequence of reconstruction or construction of the heat network sections chosen for the calculated period is specified with a minimum lead time depending on a real growth of heat loads.

All of the numerous potential system expansion options are represented by a redundant scheme (Figure 5). It is supplemented with a fictitious node and its fictitious sections (shown by the dashed lines) connected to supply nodes. Each fictitious section is limited by the capacity in accordance with the output of heat sources and described by the function of their construction and operation costs.

The optimization calculations made using the described methodological and computational tools for the obtained redundant scheme enabled us to choose the preferable optimal variant of heat supply system reconstruction, and obtain system recommendations on its improvement. The suggested measures on system reconstruction and the zones, in which heat sources operate in rated conditions, are shown with thick lines in Figure 5. The obtained recommendations are universal and can be useful for other

Figure 5. Redundant scheme of an urban heat supply system and recommendations on its reconstruction

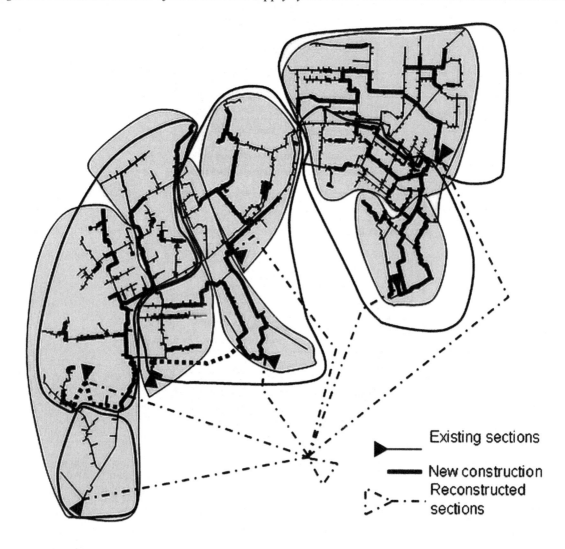

systems. According to these recommendations efficiency, reliability and controllability of heat network of the considered heat supply system can be enhanced, provided the concrete measures are implemented. Among them are the following:

1. Rational choice of system structure including the number, type and capacity of sources, configuration of network scheme, schemes of connecting sources and consumers to the heat network. Construction of a new cogeneration plant and further operation of 3 existing cogeneration plants, at the same time two most remote heat sources start heat supply to meet heat demand of the adjacent areas, one cogeneration plant is dismantled, more than 40 uneconomical boiler plants are removed from service. These recommendations provide an optimal level of heat supply centralization, concentration of heat source capacities, rational distance of heat energy transportation from the standpoint of reliability and economic efficiency.

2. Implementation of the suggested measures on heat network reconstruction with minimum redundancy that would provide reservation and control of thermal and hydraulic conditions in normal and emergency situations and arrange joint work of heat sources. This makes it possible to guarantee qualitative heat supply to consumers in normal conditions and the minimum decrease in the level of comfort in emergency situations.
3. Equipping of HSSs with computerized control nodes and remote terminal units that are intended to improve system controllability and heat carrier distribution in accordance with demand.
4. Joint operation of heat sources for common heat networks, that provides reduction in costs, enhancement of system operation efficiency, reliability and quality of heat supply.
5. Introduction of qualitative and quantitative regulation of heat delivery with independent scheme of connecting heating and ventilation loads, and a closed system of hot water supply as an effective measure on energy conservation and fulfillment of technical requirements to operation.
6. Arrangement of the effective thermal and hydraulic conditions of system operation that take into consideration multilevel regulation and implement innovative technologies of HSS operation.

Solving the stated problem became possible largely owing to application of the suggested methodological approach and the new software tools. As a result an optimal option for system expansion and reconstruction was obtained. The option combines the best of suggested solutions, takes into account all technical and other conditions and constraints. This option is more economical than that obtained by the conventional method of comparing several earlier planned options. It allows a decrease in capital investments by 45% and discounted costs – by 18%.

The proposed approach made it possible to solve the problem of choosing the optimal mix of heat sources, reconstruction and expansion of HSS.

FUTURE RESEARCH DIRECTIONS

Further studies should be aimed at the development of mathematical methods of searching for the most cost-effective configuration of multi-loop heat networks and optimizing their parameters. The discreteness of the set of equipment, the existing heat supply systems, and the changes in their development should be taken into account. The obtained solutions should ensure joined operation of the heat sources for the unified heat networks, including the distributed heat generation. Together, these problems form the problem of optimal synthesis of the heat supply systems. The solutions to these problems should facilitate the creation of controllable heat supply systems on an intelligent basis, which involve consumers' active participation.

CONCLUSION

The problem of heat supply system expansion and reconstruction includes a set of subproblems presented in the paper. The subproblems cannot be solved by one or even several universal methods. For this purpose a combination of dialog procedures and formalized methods is required. Besides, it is difficult to work out uniform approaches and find uniform solutions, since every system is individual in configuration and has its specific features. In this context the paper proposes the methodological and computational tool to take into account different types of heat supply systems, their distribution over a vast territory with broken relief, topological complexity of heat networks, different schedules of heat delivery regulation, etc. The tool allows the solution of most subproblems arising in the process of HSS expansion. Its serviceability, computational capabilities and practical applicability are illustrated on a real system. The recommendations obtained for the system have been either implemented or are at the stage of implementation.

REFERENCES

Alekseev, A. V., Novitsky, N. N., Tokarev, V. V., & Shalaginova, Z. I. (2007). *Principles of development and software support of the information-computing environment for computer simulation technique of pipeline and hydraulic systems. In Truboprovodnye sistemy energetic: Metody matematicheskogo modelirovania i optimizatsii* (pp. 221–229). Novosibirsk: Nauka. (in Russian)

Ancona, M. A., Bianchi, M., Branchini, L., & Melino, F. (2014). District heating network design and analysis. *Energy Procedia*, *45*, 1225–1234. doi:10.1016/j.egypro.2014.01.128

Appleyard, J. R. (1978). Optimal design of distribution networks. *The Building Services Engineer*, *45*(11), 191-204.

Barakhtenko, E. A., Oshchepkova, T. B., Sokolov, D. V., & Stennikov, V. A. (2013). New Results in Development of Methods for Optimization of Heat Supply System Parameters and Their Software Implementation. *International Journal of Energy Optimization and Engineering*, *2*(4), 80–99. doi:10.4018/ijeoe.2013100105

Barakhtenko, E. A., Sokolov, D. V., & Stennikov, V. A. (2014). New Results in Developing of Algorithms Intended for Parameters Optimization of Heat Supply Systems. In *Proceeding of International Conference Mathematical and Informational Technologies, MIT-2013* (pp. 58-66). Beograd, Serbia.

Bellman, R. (1957). *Dynamic Programming*. Princeton, NJ: Princeton University Press.

Courant, R., & Robbins, H. (1941). *What is Mathematics?* Oxford University Press.

Cross, H. (1936). *Analysis of flow in networks of conduits or conductors (Bulletin No 286)*. Urbana, IL: University of Illinois.

Dantzig, G. B. (1963). *Linear Programming and Extensions*. Princeton, NJ: Princeton University press.

Dorfner, J., & Hamacher, T. (2014). Large-scale district heating network optimization. *IEEE Transactions on Smart Grid*, *5*(4), 1884–1891. doi:10.1109/TSG.2013.2295856

Fazlollahi, S. (2014). *Decomposition optimization strategy for the design and operation of district energy system*. (PhD Thesis).

Fazlollahi, S., Mandel, P., Becker, G., & Maréchal, F. (2012). Methods for multi-objective investment and operating optimization of complex energy systems. *Energy*, *45*(1), 12–22. doi:10.1016/j.energy.2012.02.046

Ford, L. R., & Fulkerson, D. R. (1962). *Flows in Networks*. Princeton University Press.

Garbai, L., & Molnar, L. (1974). Optimization of urban public utility networks by discrete dynamic programming. In Colloquia mathematical society János Bolyai. 12: Progress in operations research (pp. 373-390). Eger, Hungary.

Haikarainen, C., Pettersson, F., & Saxén, H. (2014). A model for structural and operational optimization of distributed energy systems. *Applied Thermal Engineering*, *70*(1), 211–218. doi:10.1016/j.applthermaleng.2014.04.049

Kirchhoff, G. (1847). Ueber die Auflösung der Gleichungen, auf welche man bei der Untersuchung der linearen Vertheilung galvanische Ströme geführt wird. *Annalen der Physik und Chemie*, *72*(12), 497–508. doi:10.1002/andp.18471481202

Maxwell, J. C. (1873). *A treatise of electricity and magnetism* (Vol. 1). Oxford.

Merenkov, A. P. (Ed.). (1992). Mathematical modeling and optimization of heat, water, oil and gas supply systems. Novosibirsk: VO "Nauka", Sibirskaya izdatelskaya firma. (in Russian)

Merenkov, A. P., & Khasilev, V. Ya. (1985). *Theory of hydraulic circuits*. Nauka. (in Russian)

Merenkova, N. N., Sennova, E. V., & Stennikov, V. A. (1982). Optimization of schemes and structure of district heating systems. [in Russian]. *Elektronnoye Modelirovaniye*, *6*, 76–82.

Nekrasova, O. A., & Khasilev, V. Ya. (1970). Optimal tree of the pipeline system. [in Russian]. *Ekonomika i Matematicheskiye Metody*, *4*(3), 427–432.

Novitskyy, N. (2014). Calculation of the Flow Distribution in Hydraulic Circuits Based от Their Linearization by Nodal Model of Secants and Chords. *Thermal Engineering*, *60*(14), 1051–1060. doi:10.1134/S004060151314005X

Phetteplace, G. (1995). *Optimal design of piping systems for district heating. U.S. Army Corps of Engineers*. Cold Regions Research & Engineering Laboratory.

Robineau, J., Fazlollahi, S., Fournier, J., Berthalon, A., & Verdier, I. (2014). Multi-objective optimization of the design and operating strategy of a district heating network - application to a case study. In *Proceedings from the 14th International Symposium on District Heating and Cooling* (pp. 79-87). Stockholm, Sweden.

Sennova, E. V. (Ed.). (2000). *Reliability of heat supply systems* (Vol. 4). Novosibirsk: Nauka. (in Russian)

Sennova, E. V., & Sidler, V. G. (1987). *Mathematical modeling and optimization of developing heat supply systems*. Novosibirsk: Nauka. (in Russian)

Shifrinson, B.L. (1940). *Main calculation of heat networks*. Gosenergoizdat. (in Russian).

Sokolov, D. V., Stennikov, V. A., Oshchepkova, T. B., & Barakhtenko, Ye. A. (2012). The New Generation of the Software System Used for the Schematic–Parametric Optimization of Multiple Circuit Heat Supply Systems. *Thermal Engineering*, *59*(4), 337–343. doi:10.1134/S0040601512040106

Stennikov, V. A., Oshchepkova, T. B., & Stennikov, N. V. (2013). Optimal Expansion and Reconstruction of Heat Supply Systems: Methodology and Practice. *International Journal of Energy Optimization and Engineering*, *2*(4), 59–79. doi:10.4018/ijeoe.2013100104

Stennikov, V. A., Sennova, E. V., & Oshchepkova, T. B. (2006). Methods of complex optimization of heat supply systems development. *Izvestiya RAN. Energetika, 3*, 44–54.

Sumarokov, S.V. (1976). A method of solving the multiextremal network problem. *Ekonomika i Matematicheskiye Metody, 12*(5), 1016-1018. (in Russian).

Takaishvili, M. K., & Khasilev, V. Ya. (1972). On the basic concepts of the technique for reliability calculation and reservation of heat networks. *Teploenergetika, 4*, 14–19.

Valdimarsson, P. (2001). Pipe network diameter optimization by graph theory. In B. Ulanicki (Ed.), Water Software Systems: theory and applications (pp. 1-13). Baldock: Research Studies Press.

Vasant, P., Barsoum, N., & Webb, J. (2012). *Innovation in Power, Control, and Optimization: Emerging Energy Technolgies*. Hershey, PA: IGI Global; doi:10.4018/978-1-61350-138-2

Vasant, P. M. (2013). *Meta-Heuristics Optimization Algorithms in Engineering, Business, Economics and Finans*. Hershey, PA: IGI Global; doi:10.4018/978-1-4666-2086-5

Vasant, P. M. (2014). *Handbook of Research on Novel Solf Computing Intelligent Algorithms: Theory and Practical Applications*. Hershey, PA: IGI Global; doi:10.4018/978-1-4666-4450-2

Vasant, P. M. (2015). *Handbook of Research on Artificial Intelligence Techniques and Algorithms*. Hershey, PA: IGI Global; doi:10.4018/978-1-4666-7258-1

Weber, C. I. (2008). *Multi-objective design and optimization of district energy systems including polygeneration energy conversion technologies*. (PhD Thesis).

Zanfirov, A.M. (1933). Technical economic calculation of water and heat networks. *Teplo i Sila, 11*, 4-10. (in Russian)

KEY TERMS AND DEFINITIONS

Discreteness: A characteristic opposed to the continuity. Discrete set is a set consisting of isolated points.

Heat Supply System: A system supplying heat to buildings to provide thermal comfort for people staying in these buildings or to be able to meet technological standards. The heat supply system consists of the following functional parts: heat sources (boiler, cogeneration power plants), heat networks and heat consumers.

Mathematical Modeling: Research into an object or a system of objects through the construction and study of their mathematical models.

Method of Multi-Loop Optimization: A method devised by researcher of the Energy Systems Institute S.V. Sumarokov. The method makes it possible to simultaneously determine the optimal diameters of pipelines in a multi-loop system, identify the sites for the installation and parameters of pumping and throttle stations, find optimal flow distribution in the system, and solve the problem of its existing part reconstruction.

Pipeline: Is an engineering structure intended for the transportation of gas and liquid substances as well as solid fuel under the effect of pressure difference in the cross sections of a pipe.

"Redundant" Design Schemes: Are the schemes, including virtually all potential tie lines between heat sources and consumers.

Reliability: Is an ability of an object to maintain the values of all parameters that characterize the capability to perform the required functions. Intuitively, the reliability of objects is related to inadmissibility of failures in operation.

Chapter 3
Problems of Modeling and Optimization of Heat Supply Systems:
Bi-Level Optimization of the Competitive Heat Energy Market

Valery Stennikov
Energy Systems Institute, Siberian Branch of the Russian Academy of Sciences, Russia

Andrey Penkovskii
Energy Systems Institute, Siberian Branch of the Russian Academy of Sciences, Russia

Oleg Khamisov
Energy Systems Institute, Siberian Branch of the Russian Academy of Sciences, Russia

ABSTRACT

In the chapter, one of widespread models of the organization of heat supply of consumers presented in the "Single buyer" format is considered. The scientific and methodical base for its description and research offers to accept the fundamental principles of the theory of hydraulic circuits, bi-level programming, principles of economics in the energy sector. Distinctive feature of the developed mathematical model is that it, along with traditionally solved tasks within the bilateral relations heat sources – consumers of heat, considers a network component with physics and technology properties of a heat network inherent in it, and also the economic factors connected with costs of production and transport of heat energy. This approach gives the chance to define the optimum levels of load of heat sources, providing the set demand for heat energy from consumers taking into account receiving with heat sources of the greatest possible profit and performance thus of conditions of formation of the minimum costs in heat networks during the considered time period. Practical realization of the developed mathematical model is considered by the example of the heat supply system with two heat sources, working for single heat networks. With the help of the developed model are carried plotting optimal load heat sources during the year, as well as a nomogram for determining the average tariffs for heat energy heat sources.

DOI: 10.4018/978-1-4666-9755-3.ch003

INTRODUCTION

Russia's centralized heat supply system (CHSS) represents the main type of heat energy supply to consumers. Increase in the number of owners in this economic sector that is connected with liberalization of the energy sector of the early 1990s has led to the formation of new economic relations between heat energy producers and consumers, who are inseparably linked to each other by the shared technologies, but are in different economic conditions.

In CHSS that produces 72% of all the heat in the country, the share of boilers plants is above 49%, and that of thermal power plants – about 45%. The rest of the heat is produced by other sources (nuclear power plants, electric boiler plants, etc.).

The key heat consumers in public CHSS are the population and the main part of the social sphere in the country that accounts for 73%. Their share in the final heat consumption is 53%. Thus, heat supply is a socially important segment of Russia's energy sector.

At present the majority of heat sources in CHSS belong to local joint stock companies (i.e. JSC-energo), departmental enterprises and municipalities (usually boiler plants). Main heat networks belong chiefly to heat producers, and distribution networks and internal heat consumption systems are owned by municipal public utilities. However, the interests of the above mentioned enterprises participating in the process of heat supply are different and often opposite.

The present-day business mechanism for management of heat supply systems in municipal organizations combines the elements of market relations and remaining administrative command methods of management. Lack of an effective management system has plunged the heat and power industry into the deepest, almost system crisis, as evidenced by the overall aging of fixed assets, avalanche-like increase in the equipment failure rate, and lack of financial resources for modernization of heat supply systems, which already threatens the quality of people's lives in many cities and towns.

Technical degradation of CHSS, weak regulatory and legal framework, and formation of heat energy prices which is poorly controlled by regulatory authorities have dramatically reduced the CHSS competitiveness. This, in turn, explains why a great number of consumers have switched from CHSS to their own heat sources.

Repeated attempts to reform the sectors of heat and power industry have not produced the expected results, since there is no comprehensive approach to the restructuring of this industry that would balance the interests of all the participants of the heat supply process.

These conditions and the need to take into account so many interests make the problem of reforming CHSS especially urgent for increasing its efficiency, reliability, and competitiveness (including the efficient use of fuel, energy saving, improvement of environmental situation, and highly qualified social services under the conditions of market economy).

The main problems here are: construction of effective decision making procedures for the case where the interests of all CHSS participants are in conflict; integration of several heat sources that belong to different owners for their operation in one heat network and search for the optimal redistribution of heat load among them in the conditions of competitive heat energy market; development of a transparent methodology for forming heat energy tariffs on the basis of the marginal pricing method.

In this context there appears a new scientific aspect related to the development of a methodological basis that allows one to state problems and find their solutions, which are connected with the CHSS reformation and provide its efficient and competitive operation under the conditions of market economy.

BACKGROUND

Today not only Russia but other countries as well pay a lot of attention to the problems of heat supply system optimization, which is eventually aimed at optimizing the scheme of systems and flows, and determining the main design solutions (standard sizes of pipelines, pumping stations, etc.) and right moments to put the facilities into operation. Under the conditions of energy sector liberalization the optimization problem becomes significantly complicated, and its structuring starts to consider by variants depending on the way the industry is organized. The Russian scientific and practical experience in organizing and reforming the heat power industry under the conditions of market economy has been poorly studied and not systematized, and the foreign experience, as practice shows, can not be fully adapted to the conditions of Russia. The following issues are also insufficiently studied:

- Methods and criteria for comprehensive estimation of the efficiency of urban heat supply systems in a market environment;
- Priority lines in improvement of the business mechanism for heat supply system management;
- Improvement of price formation and system of state regulation of heat energy tariffs in new economic conditions;
- Creation of a competitive environment in heat markets;
- Management of demand for heat energy as a new type of activity of heat supply companies;
- Information and mathematical models of heat energy market optimization;
- Methods of inspection of heat supply systems to create a database for monitoring the efficiency of heat supply systems, etc.

A comprehensive resolution of the above mentioned issues, which underlie the reform of the business mechanism for management of urban heat supply systems in the market economy, is only becoming the subject of scientific studies.

A great number of published papers on the problems of heat and power industry and public utilities are mainly devoted to solving certain problems, i.e. organizational, technical, operating and methodological.

The most studied problems and methods in the field of heat supply are:

- Methods for calculation and optimization of heat supply systems;
- Generation and formation of technical directions for expansion of heat supply systems, substantiation of their rational scale and structure, and modernization and reform;
- Determination of quantity and location of heat sources and selection of a heat network scheme;
- Determination of heat network parameters (pipeline diameters, location and parameters of pumping and choke stations, etc.);
- Determination of normative hydraulic conditions of the whole system.

The statement of these problems and methods for their solving have a long history both in Russia and abroad (Basset & Winterbone & Pearson, 2001; Cross, 1936; Khasilev, 1964; Merenkov & Khasilev, 1985; Shifrinson, 1940; Sokolov, 1999; Wood & Reddy & Funk, 1993; Zanfirov, 1934).

Problems of Modeling and Optimization of Heat Supply Systems

Among the papers on optimization of heat network parameters the works (Khasilev, 1985; Merenkov & Khasilev; Shifrinson 1985, 1940; Sokolov, 1999; Zanfirov, 1934). deserve special consideration. The aggregate technical and economic parameters obtained in these papers for the calculation of heat networks are still used in the design of heat supply systems.

As it turned out, the most effective method for solving the problems of choosing the optimal parameters is the method of dynamic programming suggested and applied to the heat supply systems by the author of (Merenkov & Khasilev; Shifrinson 1985). The dynamic programming method allows one to find the global minimum and take into account different physical and technical constraints, discreteness of equipment parameters and other specific features of the considered systems. This method was further developed at Energy Systems Institute SB RAS.

The method of multiloop optimization based on sequential improvement in solutions was proposed to optimize multiloop systems. It is a combination of the method for calculating flow distribution and the method of dynamic programming. Its mathematical description and computational capabilities are presented in paper (Merenkov & Khasilev; Shifrinson 1985).

The integrated direction involving a broad range of issues of mathematical modeling, calculation and optimization of different pipeline systems was generalized in the hydraulic circuit theory, which was founded by V.Ya. Khasilev (Khasilev, 1964), further studied by A.P. Merenkov (Merenkov & Khasilev, 1985) and is now being successfully developed at Energy Systems Institute SB RAS. The hydraulic circuit theory includes the mathematical description of hydraulic circuits, methods of hydraulic calculation, and problems and methods of optimizing hydraulic circuit structure and parameters.

The works by E.V. Sennova (Sennova & Sidler, 1987) made a significant contribution to the development of methods for calculating and optimizing large centralized systems. An integrated approach to the quantitative estimation of heat supply system reliability which is being developed within the hydraulic circuit theory thoroughly considers the network structure of the systems, specific features of their emergency redundancy and different operating conditions.

The problem of development and modernization of heat supply systems relates to the technical and economic problems of structural optimization that are being studied in the hydraulic circuit theory. This problem arises in the process of heat load increase and change in other conditions of system operation. It consists in determination of the rational ways to change structural principles of the system, in detection of bottlenecks and their elimination by means of replacing or strengthening some of the existing components (heat sources, heat network sections, pumping stations), and in choice of concrete structures and parameters of the system. As was mentioned above, to solve these problems, the researchers of the Department of Pipeline and Hydraulic Systems at Energy Systems Institute SB RAS developed the method of multiloop optimization, which was implemented in the computational module SOSNA.

Under new economic conditions the structural integrity of the centralized heat supply system has been destroyed due to privatization of its individual components. This has provoked conflicts between the heat energy producers, heat sales organizations and consumers. Each of these participants of the heat supply has his own interests that do not coincide with the interests of other participants. For instance, the owners of heat sources are interested to make maximum profit on heat energy production, while consumers are interested in minimum tariffs.

Therefore, to solve the problems of operation, development and modernization of heat supply systems under new economic conditions, it is necessary to apply not only the existing optimization methods, but also the game models that are intended for solving the problems in some specific uncertain situations characterized by different, often conflicting interests of the parties. The current situation makes it necessary to find such optimal conditions, under which each of the participants will be interested in the development of the whole heat supply system.

Many problems of operations research and mathematical economy are connected with finding a compromise or equilibrium solution. These are the problems of finding Nash, Cournot and Stackelberg equilibria and problems of hierarchical control (e.g. two level control).

The monograph written by John von Neumann and Oskar Morgenstern (Neumann & Morgenstern, 1970) offered much hope of extension of the game theory to a universal mathematical theory of behavior under the conditions of conflicting interests. Today the game theory has found its rightful place in the studies on social and economic phenomena. Poor understanding of such concepts as Pareto optimality, Nash equilibrium, C-core and hierarchical solution makes the mathematical study of situations with several participants impossible.

Among all decision-making conditions, the conflict conditions deserve special attention. This is explained firstly by the practical importance of conflicts in social life and development, and secondly by the special complexity of a conflict as a phenomenon, depending on which one has to make decisions. In case of a conflict the decision maker has to take into account not only his own goals, but the goals of his partners as well. Besides, he has to take into consideration not only already known objective circumstances of the conflict, but also the decisions taken by his rivals and unknown to him (Moulin, 1985; Moulin, 1991; Owen, 1971).

At present the mathematical modeling of conflict situations has been thoroughly studied for competitive markets of different products. It is aimed at searching for an equilibrium product price, at which the producers make decisions on the production without any accurate initial information of their competitors. Taking into account the behavior of his rivals, each of the competitors develops his own strategy in such a way that each player would makes maximum profit.

The current economic theory describes the agents (consumers and producers) of economic systems using extreme problems. One of the methods for analyzing these systems is based on constructing demand and supply functions in accordance with the extreme problems and matching them with each other afterwards. Problems of constructing demand and supply functions relate to the problems of parametric programming, whose most studied class is the linear optimization problems. Thus, the given problem is reduced to determination of the equilibrium point, i.e. the point of intersection of the demand and supply functions.

Walras model of the competitive economy functioning is central in the economic theory. Fundamental studies devoted to this topic are the studies by Kenneth Arrow and Jerard Debreu, who laid the basis for modern mathematical models of the closed decentralized economic system (Arrow & Debreu, 1954).

The key notion of the up-to-day economic and mathematical studies is the idea of equilibrium. For example, paper (Razumikhin, 1986) considers the interconnection between economic models and a number of physical models and methods of the equilibrium theory.

There are different methods for finding equilibrium (compromise) solutions. However, most of them are efficient only in a certain economic and mathematical model. One of the rapidly developing research areas connected with the Arrow-Debreu model is the equilibrium programming. The theory of equilibrium programming is aimed at working out a single mathematical approach to finding compromise or equilibrium solutions. The central idea of this theory is the fact that, from a mathematical viewpoint, different concepts of equilibrium in economic and mathematical models are mathematically the fixed points of corresponding extreme mappings, which allows one to develop a single methodological approach to finding equilibrium. Moreover, a new broader view of the notion of equilibrium makes it possible, on the one hand, to ease the requirements that guarantee the convergence of existing methods to the equilibrium point, and, on the other hand, to develop new requirements more sensibly. In quite a short period of time this field of knowledge has experienced considerable progress in the theory and methods of finding solutions to equilibrium problems of different types (variational inequalities, complementarity problems, saddle point problems and game equilibrium problems).

To model such economic processes as the market behavior of producers and consumers based on the interaction between two or more parties, the authors of formulate mathematical decision-making in the form of a two level programming problem called Stackelberg non-cooperative game (Bard, 1998; Dempe, 2002). This model considers such interaction between the participants of the decision-making process, where they are combined into a hierarchical system consisting of two levels. The lower level of this model is represented by the producers, acting in the interests of their profit maximization. The upper level is the state, which tries to interest the producers in producing more or less products than they already do. In most of the cases the state pursues social, humanitarian and ecological aims, for instance, to reduce the emissions from some plants, to maintain the normal level of prices of consumer goods, etc. An important feature of this game is that the participants take decisions subsequently, i.e. when the player of the lower level makes a decision, he already knows the decision of the upper level and uses it as an example in his objective function and constraints. One more feature is that the player of the upper level can not make the other player take a decision favorable to the upper level, but he can influence his choice.

Problems of forecasting and developing different economic systems are especially important and characterized first of all by the uncertainty of initial data and complex calculations. The decisions made are often assessed on the basis of many parameters. To take a decision, the player usually considers several scenarios of the system development and uses formal and informal methods of analysis, among which he selects one for implementation. An analogous approach is applied to designing complex production and technological systems. This is a multivariant analysis, which now represents one of the ways to help make decisions. Here the main task consists in choosing the number of variants of the system development that should be finally analyzed in accordance with the chosen parameters.

Economic and mathematical models describe the behavior of the market subjects by extreme problems. However, in reality by virtue of many reasons only few subjects solve these problems. The final consumer is described by the utility function. It is hard to definitely determine, whether the consumer really maximizes the utility under budget constraints, or minimizes the costs of achieving a certain utility level. This makes it necessary to find not only optimal solutions, but also the solutions close to optimal. One of the methods of searching for solutions close to optimal is the discrete programming. The key studies in this research area were conducted by R. Bellman, S. Dreyfus et al. (Bellman & Dreyfus, 1965).

In the past decade a lot of studies have been conducted on the application of game and economic and mathematical models to different energy systems (Green, 1999; Stoft, 2006; Vasant, Barsoum, Webb, 2012; Vasant, 2013; Vasant, 2014; Vasant, 2015).

Issues concerning the construction of effective decision-making schemes and forms of organization of urban heat supply under new economic conditions are described by V.G. Semenov (Semenov, 2003). He considers possible schemes of organizing relationships between the participants of the heat energy market and describes their properties and consequences.

From the above analysis of publications it follows that:

1. The existing methods of optimizing centralized heat supply systems are based on the general economic criterion, which makes it impossible to take into account the interests of different participants in the heat energy market.
2. There is no methodological decision-making basis for designing competitive heat energy markets.
3. There are no mathematical and optimization models of heat energy markets for the liberalized economy which would allow an adequate estimation of market design and operation.

ANALYSIS OF HEAT MARKETS IN RUSSIA

Currently more than 50 000 heat markets are operating in Russia. Every market has its features and specificity. This is determined by both a variety of heat sources taking part in the market and by unique configuration of heat networks. In simplified form the markets can be divided into four main categories:

- Extra-large markets - in 15 cities with production and consumption of above 40 million GJ/year;
- Large markets - in 44 cities with consumption from 8 to 40 million GJ/year;
- Medium-sized markets - in hundreds of cities with consumption from 2 to 8 million GJ/year;
- Small markets - in more than 40 000 settlements that consume less than 2 million GJ/year of heat produced by centralized sources.

Normally heat markets are created on the basis of existing heat supply systems. The model form depends on a strategic goal stated. As in many other areas of economic activity, one of the possible forms of organization of the market in district heating can be gradually focus on the creation of a competitive relationship. This model, in general, corresponds to the situation prevailing in the cities where there are several heat sources belonging to different owners, while heating networks are separated from the generation of heat and should be merged into a single Heating Network Company. From the perspective of microeconomics, this form of organization of the market relates to oligopoly (a limited number of manufacturers, a lot of consumers on the market in question) (Busygin, Zhelobodko, Zyplakov, 2003).

Competition between sources of heat occurs in cases of redundancy of their total capacity, compared with the total demand (competition offers), presence of a competitive threat to the construction of new cost-effective heat sources (competitive threat), or the possibility of withdrawal of consumers heat supply system (competition demand) to an alternative method of heat supply (decentralized source) (Belyaev, 2009).

The competition in the heat market as a whole is an important element of the market economy, in view of the fact that it contributes to the efficiency of heat energy production, improving its quality, and as a consequence reduce its prices, which can positively affect the development of district heating.

Enlargement, the implementation of the model of the competitive market of heat energy can be represented in the following scheme (Figure 1) (Gitel'man, Ratnikov, 2006). Prerequisites necessary for the formation of the model should include the following basic principles:

- the presence of two or more heat sources belonging to different owners;
- association of thermal networks (backbone, distribution, etc.). various forms of ownership (private, municipal, etc.) into a single Heating Network Company;
- presence of excess heat power sources and reserves of hydraulic regimes of heat networks.

Figure 1. Model of a competitive heat market

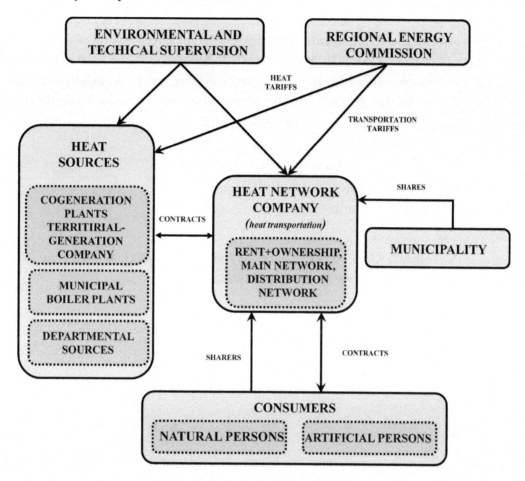

The heat network company purchases heat from heat sources at the lowest tariffs and then sells it to consumers. This model is an organizational and economic mechanism that provides a competitive selection of heat suppliers for a local district heating system based on the technical capabilities of heat sources and its network infrastructure.

The model of the kind is often called a «Single buyer» model (Gitel'man, Ratnikov, 2006; Belyaev, 2009). In Russia this model has also found wide application and is often used to model electricity (Podkovalnikov, Khamisov, 2012) and heating systems (Stennikov, Khamisov, Pen`kovskii, 2011) in the process of their restructuring.

In terms of heat supply organization the competitive heat market can be represented by a bi-level hierarchical model. The upper level of the model is represented by a heat network company that purchases heat at the second level represented by heat sources. Based on the tariff situation in the heat market heat sources generate the amount of heat that maximizes their profit. In doing so the network company tends to minimize network costs, considering physical and technical constraints of the heat network.

DESCRIPTION OF MODELS

Consideration is given to a hydraulic circuit represented by the calculated scheme of a real heat supply system consisting of m nodes and n branches (Merenkov & Khasilev, 1985; Sennova & Sidler, 1987). The hydraulic circuit structure is described by a complete incidence matrix $\overline{A}$, in which the number of rows coincides with the number of nodes and the number of columns coincides with the number of branches. The elements a_{ij} of matrix $\overline{A}$ are determined as follows:

$$a_{ij} = \begin{cases} 0, & \text{if branch } i \text{ has no connection with node } j; \\ 1, & \text{if flow in branch } i \text{ goes from node } j; \\ -1, & \text{if flow in branch } i \text{ enters node } j. \end{cases}$$

According to (Merenkov & Khasilev, 1985; Sennova & Sidler, 1987), the flow distribution in hydraulic circuit can be represented by the following set of equations:

$$Ax = Q \tag{1}$$

$$\overline{A}^T P = h - H \tag{2}$$

$$h = Sx|x| \tag{3}$$

where A is an $(m-1) \times n$ incidence matrix for linearly independent nodes, which is obtained on the basis of complete matrix $\overline{A}$ by deleting any of its rows; $\overline{A}^T$ is a transposed complete matrix of node and branch connections; P is a vector of nodal pressures (mwc); x is a vector of flow rates in the network sections, t/h; Q is a vector of flow rates at nodes, t/h; h, H are vectors of losses and operating heads, mwc; S is a diagonal matrix of hydraulic resistance coefficients s_i, mh²/t², $|x|$ is the module of vector of flow rates in the network sections, t/h.

The first subsystems (Equations 1) reflect the first law of Kirchhoff, the second subsystems (Equations 2) – the second law of Kirchhoff, and the third subsystems (Equations 3) – relationship between the head loss in branches and their hydraulic resistances.

Part of nodes in the hydraulic circuit is supposed to represent the locations of heat sources that belong to various owners. This set of sources is denoted by J_{HS}. The maximum and minimum level of heat production at heat sources is expressed through Q_{j_min} and Q_{j_max}, and is written in the form of system constraints:

$$Q_{j\_min} \leq Q_j \leq Q_{j\_max} \tag{4}$$

The nodes other than heat sources represent either distribution nodes (branchings) or consumers. Denote by J_j a set of consumer nodes, for which the demand for heat Q_j^l is determined by the heat load duration curve Figure 2. Configuration of the curve is well described by the Rossander equation. According to the equation heat load at each time instant τ^* is determined by the expression:

$$Q_j^l = \left[1 - (1-r)\left(\frac{\tau^*}{\tau}\right)^{\frac{g-r}{1-g}}\right]Q_j^{cl} + Q_j^{hws}, \qquad (5)$$

$$r = (1-v)\frac{t_4 - t_2}{t_4 - t_1}, \qquad (6)$$

$$g = (1-v)\frac{t_4 - t_3}{t_4 - t_1}. \qquad (7)$$

where Q_j^{hws} is a calculated heat load of hot water supply, GJ/h; r is a load curve irregularity factor; Q_j^{cl} is a calculated heating load, GJ/h; v is a share of hot water supply load; t_4 is a calculated indoor temperature, °C; t_1, t_2, t_3 are outdoor temperatures: the calculated one, the one that corresponds to the beginning of heating period (+8 °C) and the average one for a heating period °C; τ is heating period duration, h.

The annual volume of heat consumption equals the area S* under the curve of heat load variation during a year and is determined by integration of expression (Equations 5) with respect to time τ^*:

Figure 2. Heat load duration curve

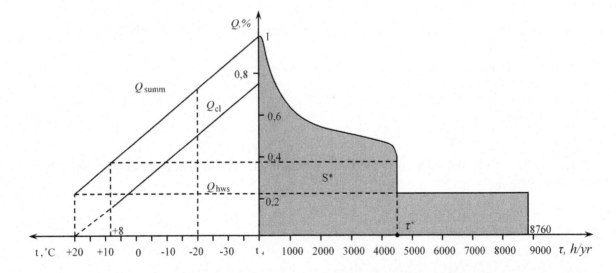

$$Q_C^{year} = \int_0^\tau \left[\left[1-(1-r)\left(\frac{\tau^*}{\tau}\right)^{\frac{g-r}{1-g}}\right]Q_j^{Cl} + Q_j^{hws}\right]d\tau + \int_\tau^{8760} Q_j^{hws} d\tau, \qquad (8)$$

or

$$Q_C^{year} = \left[\tau \sum_j Q_j^{Cl} - \frac{(1-r)\tau^{*\frac{g-r}{1-g}+1}}{\tau^{\frac{g-r}{1-g}}(\frac{g-r}{1-g}+1)}\sum_j Q_j^{Cl} + \tau\sum_j Q_j^{hws}\right]_0^\tau + \left[\tau\sum_j Q_j^{hws}\right]_\tau^{8760}. \qquad (9)$$

Each supply node sells heat to a network company at tariff c_j. Denote by $f_j(Q_j)$ the costs of heat production by the *j-th* node. In the competitive market behavior of heat generation node is dictated by its intention to maximize profit $F_j(Q_j)$. Let c_j, $j \in J_{HS}$ be the price per heat unit, then the heat amount produced by source *j* is found by solving the problem:

$$F_j(Q_j) = c_j Q_j - f_j(Q_j) \to \max, \quad j \in J_{HS}, \qquad (10)$$

$$Q_{j\_min} \leq Q_j \leq Q_{j\_max}, \quad j \in J_{HS}. \qquad (11)$$

The costs of heat production by a heat source have the form of a squared relationship. They were obtained by approximating actual heat supply data depending on the heat source costs by the method of least squares:

$$f_j^q(Q_j) = \alpha_j Q_j^2 + \beta_j Q_j + \gamma_j, \quad \alpha_j > 0 \qquad (12)$$

Thus, the problem corresponding to expressions (Equations 10,11) will have the form:

$$-\alpha_j Q_j^2 + (c_j - \beta_j)Q_j - \gamma_j \to \max, \quad j \in J_{HS}, \qquad (13)$$

$$Q_{j\_min} \leq Q_j \leq Q_{j\_max}, \quad j \in J_{HS}. \qquad (14)$$

Dependence of optimal solution to problem (Equations 13,14) on heat price is determined by the following relations:

$$Q_j^* = \begin{cases} \underline{Q}_j, & c_j < 2\alpha_j \underline{Q}_j + \beta_j, \\ \dfrac{c_j - \beta_j}{2\alpha_j}, & 2\alpha_j \underline{Q}_j + \beta_j \le c_j \le 2\alpha_j \overline{Q}_j + \beta_j, \\ \overline{Q}_j, & c_j > 2\alpha_j \overline{Q}_j + \beta_j. \end{cases} \quad (15)$$

From the set of equations (15) it follows that if the heat prices vary within the values corresponding to the expression

$$2\alpha_j \underline{Q}_j + \beta_j \le c_j \le 2\alpha_j \overline{Q}_j + \beta_j,$$

the amount of the produced heat providing the maximal profit depends on the price linearly:

$$Q_j = \frac{c_j - \beta_j}{2\alpha_j}, \; j \in J_{HS}. \quad (16)$$

A network company purchases heat energy from a heat source on a competitive basis with technical and physical constraints of the network (Equations 1, 2 and 3) and calculates the resulting weighted average tariff w^{HS} at time instant τ^*:

$$w^{HS} = \frac{\sum_j c_j Q_j}{\sum_j Q_j}, \; j \in J_{HS}. \quad (17)$$

In the bi-level model, the network company, as was noted above, being at the top level, tends to minimize the costs of the heat network. The function of annual costs of the heat network is determined by the known relation (Merenkov & Khasilev, 1985; Sennova & Sidler, 1987):

$$Z_i(d_i) = f \sum_i k_i(d_i) l_i \to \min, \quad i \in I \quad (18)$$

where I is a set of heat network sections i, $f = 0.075$ is a share of conditionally constant and operating costs for the heat network, $k_i(d_i)$ is specific capital investment in the i-th section of the heat network according to the standard of its diameter d_i, as well as the pipe laying type and requirements, NCU/m (National Currency Unit), l_i is a length of network section i, m.

The calculated specific capital costs of the existing network sections are determined by the analytic dependence (Merenkov & Khasilev, 1985; Sennova & Sidler, 1987):

$$k_i(d_i) = a_i + b_i d_i^{u_i}, i \in I \quad (19)$$

where a_i, b_i, u_i are approximation coefficients of numerical values for unit cost of laying pipelines of different diameters.

Change in the network pipeline diameter with the change in water flow rate and pressure head is calculated by the Shifrinson formula (Shifrinson, 1940) and has the form:

$$d_i = \chi_i \frac{x_i^{0.38}}{h_i^{0.19}} l_i^{0.19} = \chi_i x_i^{0.38} h_i^{-0.19} l_i^{0.19}, \quad i \in I \tag{20}$$

where d_i is a pipeline diameter, mm; χ_i is a coefficient depending on the pipe inner surface roughness; x_i is a mass heat carrier flow rate in a heat network section, t/h; h_i is friction pressure drop in a heat network section, Pa (kgf/m²); l_i is a heat network section length, m.

Considering the foregoing, the dependence of the estimated costs of the existing network sections, including the costs of electric power used to overcome friction, is of the form:

$$Z_i(x_i, h_i) = f \sum_i [a_i + b_i \chi_i^{u_i} x_i^{0.38 u_i} h_i^{-0.19 u_i} l_i^{0.19 u_i}] l_i + \frac{C \sum_i x_i h_i}{362.7 \eta}, \quad i \in I \tag{21}$$

Let us rewrite expression (Equations 21) in terms of dependence (Equations 3) and we obtain it in the following form:

$$Z_i(x_i) = B + \frac{C \sum_i x_i^2 |x_i| S_i}{362.7 \eta}, \quad i \in I \tag{22}$$

where $B = f \sum_i [a_i + b_i \chi_i^{u_i} S_i^{-0.19 u_i} l_i^{0.19 u_i}] l_i = const$ are conditionally constant annual costs for the heat network; C is a unit cost of electricity, NCU/kWh; η is pumping station efficiency, %.

The network tariff for heat energy transportation at time instant τ^* is calculated by the relation:

$$w^* = \frac{Z_i(x_i)}{\sum_j Q_j}, i \in I, j \in J_{HS} \tag{23}$$

Based on the above said, the problem of load distribution among heat sources in the competitive heat market will be written as follows. Find:

$$Z_i(x_i) = B + \frac{C \sum_i x_i^2 |x_i| S_i}{362.7 \eta} \to \min \ i \in I, \tag{24}$$

subject to conditions for heat network:

$$Ax = Q, \tag{25}$$

$$\overline{A}^T P = h - H, \tag{26}$$

$$h = Sx|x|, \tag{27}$$

subject to conditions and restrictions for heat sources:

$$Q_j = \frac{c_j - \beta_j}{2\alpha_j}, \; j \in J_{HS}, \tag{28}$$

$$\underline{Q_j} \leq Q_j \leq \overline{Q_j}, \; j \in J_{HS}, \tag{29}$$

subject to conditions for consumers:

$$Q_j^C = \left[1 - (1-r)\left(\frac{\tau^*}{\tau}\right)^{\frac{g-r}{1-g}}\right] Q_j^{cl} + Q_j^{hws}, \; j \in J_l, \tag{30}$$

$$r = (1-v)\frac{t_4 - t_2}{t_4 - t_1}, \tag{31}$$

$$g = (1-v)\frac{t_4 - t_3}{t_4 - t_1}. \tag{32}$$

The resulting tariff for heat consumers was determined for any time instant τ^* after solving problem (24) - (32):

$$w = w^{HS} + w^*. \tag{33}$$

The integral annual technical and economic indices of heat supply systems are determined by their summation over time $\tau^*=1..8760$ hours and the average annual tariff for heat energy production and transportation to consumers is calculated.

SOLUTIONS AND RECOMMENDATIONS

Let us consider the calculated scheme of a heat supply system presented in Figure 3. Nodes 1 and 2 correspond to the nodes of heat energy supply, which work in the competitive market, nodes 3-19 and 21-29 correspond to the aggregated consumers of heat energy.

The following initial data are used in the calculation are presented in the Appendix.

The calculation results for the conditions stated above are presented in Table 1 and Table 3.

Table 1 contains basic technical and economic indices and annual average cost of heat energy production and weighted average tariff for heat energy output by heat sources. The consumer tariff depends on the sum of annual weighted average weighted tariff by heat source and annual average tariff of heat transportation in the heat network in Table 2. As per Table 1, the heat tariff of the first and second heat sources is 4% and 15% higher, respectively.

Figure 4 illustrates graphically solutions for heat sources and heat networks for the heat market represented by the bi-level model.

In Figure 4 possible variants of annual heat load distribution between sources at the given demand for heat energy during a year are presented on the abscissa, and the total annual costs for heat production and transportation are shown on the ordinate.

The curves of costs for heat network and heat sources are plotted on the basis of multivariant calculations for different levels of heat source loading with consideration of their associated optimal flow distribution in the network. The curve of total costs is obtained by summation of the values of costs for heat sources and heat networks for different combinations of loads of heat sources within a year.

Figure 3. The scheme of heat supply to consumers.

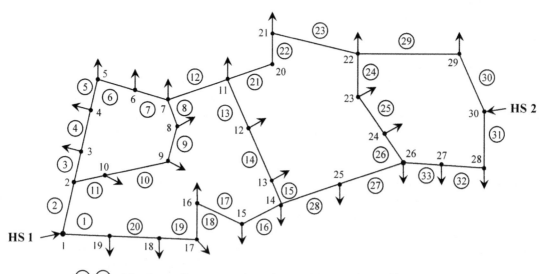

①-㉝ – Numbers of heat network sections, 1-30 – numbers of heat network nodes,
HS 1 – Heat Sources 1, **HS 2** – Heat Sources 2,

Table 1. Calculated indices for heat sources

Calculated Indices	Source 1	Source 2
Heat energy production, 10^3 GJ/yr	21292	16945
Heat energy production costs, 10^3 NCU/yr	1796400	1591514
Profit, 10^3 NCU/yr	72215.57	24476,53
Heat energy sales proceeds, 10^3 NCU/yr	1868661.5	1836280
Production cost price, NCU /GJ	84.3	93.9
Average annual tariff, NCU/GJ	87.7	108.3
Weighted average tariff, NCU / GJ	96.85	

Table 2. Calculated indices for a heat network company

Calculated Indices	Heat Network Company
Network costs, 10^3 NCU/yr	1379230
Average annual tariff for heat carrier pumping, NCU /GJ	36.07
Weighted average tariff, NCU/ GJ	96.85
Average annual tariff for consumer, NCU /GJ	132.92

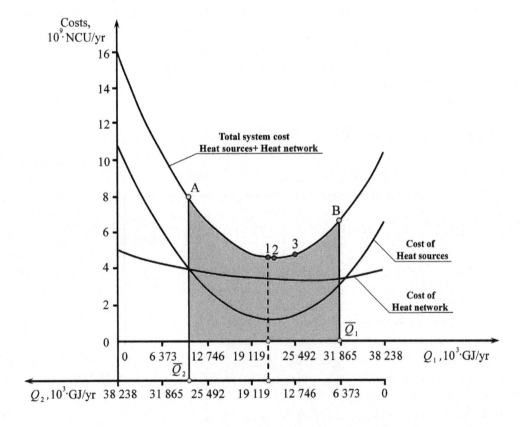

Figure 4. Graphical analysis of the results obtained

Verticals $A\overline{Q2}$ and $B\overline{Q1}$ correspond to the upper-bound constraints on capacity of the first and second heat sources $\overline{Q2}$ and $\overline{Q1}$. The region of feasible solutions to the studied problem for the determined demand for heat energy lies between them. It is presented as the AB section on the curve of total annual costs.

The optimal solution, obtained by using the criterion of minimum reduced costs, corresponds to point 2 on the AB curve in Figure 4. Solution, obtained on the bi-level model, corresponds to point 1 on the AB curve in the same Figure. This solution is close to the solution based on the criterion of minimum reduced costs. A heating system was calculated for comparative analysis by the criterion of minimum weighted average tariff for a heat source without optimization of heat networks. In this statement solution for heat sources was obtained first, then the optimal flow distribution in heat networks and their technical and economic indices were calculated for the determined heat source capacity. This solution corresponds to point 3 on the AB curve in the Figure. Comparison of the results obtained has showed that optimization of heat sources in the competitive market without optimization of heat networks (point 3) may lead to increase in the whole system costs and hence consumer tariff. In particular, in this case they are more than 12% higher than the costs calculated by the bi-level model.

Graphical solution for the heat energy sources is shown in Figure 5.

In Figure 5 the ordinate presents possible combinations of annual volumes of heat energy production by the first and second heat sources at the given annual heat energy demand by consumers, and the abscissa – average annual prices, which correspond to these combinations.

The KN and LM lines are the response functions of heat energy production volume versus its price for the second and first heat sources, respectively. The ABCD rectangle obtained by superposition on the plot of constraints on the maximum capacity of heat sources and their associated optimal constraints on the average annual heat tariffs is a region of feasible solutions for determination of the optimal load distribution between heat sources. The $Q1^* Q2^*$ line is the best loading level of heat sources with heat

Figure 5. Graphical representation of solution for heat sources.

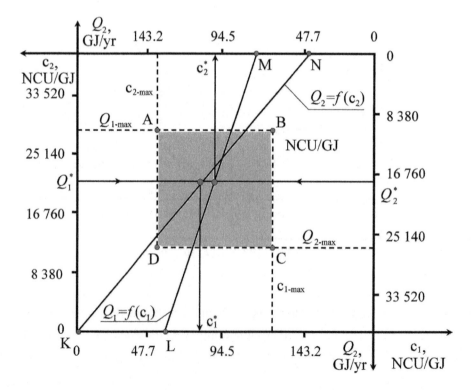

network optimization that is obtained on the basis of the suggested model. It is important to note that the $Q1^*Q2^*$ line departure from the optimal level to the AB section leads to increasing network costs by 34%, to the CD section – by 36%.

FUTURE RESEARCH DIRECTIONS

Further investigation heat markets involve the development of mathematical models of the monopoly market of heat energy in the form of organizational models "United heat supply organization". In contrast to the mathematical model of the competitive market, the mathematical model of a monopoly market should be able to accommodate a lot of interests of consumers of heat energy which will be described in the corresponding demand functions, as well as take into account the different methods of adjusting final tariffs.

CONCLUSION

1. The paper describes a method of heat network optimization in terms of the most beneficial load distribution among heat sources. It is based on the bi-level model of heat supply system representation, when a heat network company, integrating all heat networks, is at the top level and tends to minimize network costs, and heat generating companies (sources), being at the lower level, optimize their profit.
2. The study and its graphical analysis based on various methodological approaches, have showed that heat networks determine, in many respects, the economic and energy benefits of heat sources and the whole heat supply system operation. Optimization of heat sources without regard for heat networks can result in cost overrun by 12% and more.
3. The proposed mathematical bi-level model intended for optimization of heat networks in terms of the optimal heat load distribution among sources allows the best arrangement of heat supply. It exhibits best the conditions being formed in the Russian competitive heat markets. The given model takes into account the established "behavior rules" of the heat market participants (producers – heat network company – consumers), as well as economic, physical and technical constraints of the system under consideration.

REFERENCES

Arrow, K., & Debreu, G. (1954). Existence of equilibrium for a competitive economy. *Econometrica*, *22*(3), 265–290. doi:10.2307/1907353

Bard, J. (1998). *Practical Bi-level Optimization*. Dordrecht, Netherlands: Kluwer Academic Publishers; doi:10.1007/978-1-4757-2836-1

Basset, M., Winterbone, D., & Pearson, R. (2001). Calculation of Steady Flow Pressure Loss Coefficients for Pipe Junctions. *Proceedings - Institution of Mechanical Engineers*, (215): 861–881.

Bellman, R., & Dreyfus, S. (1965). *Applied problems of dynamic programming*. Moscow, USSR: Nauka.

Belyaev, L. S. (2009). *Problems of the electricity market*. Novosibirsk: Nauka.

Cross, H. (1936). *Analysis of flow in network of conduits or conductors*. Urbana, IL: Eng. Exg. Station of University of Illinois.

Dempe, S. (2002). *Foundations of Bi-level Programming*. Dordrecht, Netherlands: Kluwer Academic Publishers.

Gitel'man, L., Ratnikov, B. (2006). *Energy business*. Moskow: Delo.

Green, R. (1999). The electricity contract market in England and Wales. *The Journal of Industrial Economics, 47*(1), 107–124. doi:10.1111/1467-6451.00092

Halseth, A. (1998). Market power in the Nordic electricity market. *Utilities Policy, 7*(4), 259–268. doi:10.1016/S0957-1787(99)00003-X

Khasilev, V. Ya. (1964). Elements of hydraulic circuit theory. *Izv. AN SSSR. Energetika i transport*, (1), 69-88.

Merenkov, A., & Khasilev, V. (1985). *The theory of hydraulic circuits*. Moscow: Nauka.

Moulin, H. (1985). *Game theory with examples from mathematical economics*. Moscow: Mir.

Moulin, H. (1991). *Cooperative decision making: Axioms and models*. Moscow: Mir.

Neumann, J., & Morgenstern, O. (1970). *Theory of game and economic behavior: Tranls. from Engl*. Moscow: Nauka.

Owen, G. (1971). *Game theory*. Moscow: Mir.

Podkovalnikov, S. V., & Khamisov, O. V. (2012). Imperfect electricity markets: Modeling and study of the development of generating capacities, *Izvestiya RAN. Energy, 2*, 57–76.

Razumkhin, B. S. (1986). *Physical models and methods of equilibrium theory in programming and economics*. Moscow: Nauka.

Semenov, V. (2003). Office of heat supply. *Novosti Teplosnabgenia, 2*(18), 31-39.

Sennova, E., & Sidler, V. (1987). *Mathematical modeling and optimization of heat supply systems*. Novosibirsk: Nauka. Sib. Otdeleniye.

Shifrinson, B. L. (1940). *Basic calculation of heat networks. Theory and computational methods*. Moscow: Gosenergoizdat.

Sokolov, E. Ya. (1999). *Combined heat and power production and heat networks*. Moscow: Izdatelstvo MEI.

Stennikov, V. A., Khamisov, O. V., & Pen`kovskii, A. V. (2011). Optimizing the Heat Market on the Basis of a Two-Level Approach. *Thermal Engineering, 12*(58), 1043–1048. doi:10.1134/S0040601511120111

Stoft, S. (2006). *Power system economics: Designing markets for electricity*. Moscow: Mir.

Vasant, P. (2015). *Handbook of Research on Artificial Intelligence Techniques and Algorithms*. Hershey, PA: IGI Global. doi:10.4018/978-1-4666-7258-1

Vasant, P., Barsoum, N., & Webb, J. (2012). *Innovation in Power, Control, and Optimization: Emerging Energy Technologies*. Hershey, PA: IGI Global. doi:10.4018/978-1-61350-138-2

Vasant, P. M. (2013). *Meta-Heuristics Optimization Algorithms in Engineering, Business, Economics, and Finance*. Hershey, PA: IGI Global. doi:10.4018/978-1-4666-2086-5

Vasant, P. M. (2014). *Handbook of Research on Novel Soft Computing Intelligent Algorithms: Theory and Practical Applications*. Hershey, PA: IGI Global. doi:10.4018/978-1-4666-4450-2

Wood, D., Reddy, L., & Funk, J. (1993). Modeling Pipe Networks Dominated by Junctions. *Journal of Hydraulic Engineering, 119*(8), 949–958. doi:10.1061/(ASCE)0733-9429(1993)119:8(949)

Zanfirov, A.M. (1934). About economic calculation of heat networks. *Teplo i Sila*, (12), 38-39.

KEY TERMS AND DEFINITIONS

Bilevel Program: Section of mathematical programming, designed to optimize the different technical, economic and other systems with a hierarchical management structure and consists of two levels (upper and lower).

Graph Rossander: Graph duration of the total heat load (heating, ventilation, hot water, technology), depending on the number of hours of standing during the year, outdoor temperatures.

Heat Energy Market: The scope of a particular product - of heat energy (power, heat carrier) with the participation of district heating and consumers.

Heat Supply System: A system for distributing heat generated in a centralized location for residential and commercial heating requirements such as space heating and water heating.

Oligapoliya: Market structure of imperfect competition, which is dominated by a very small number of producers of goods.

"Single Buyer" Model: Organizational and economic mechanism for competitive selection of suppliers of heat energy for district heating system-wide, taking into account the technical capabilities of heat sources and network infrastructure.

Theory of Hydraulic Circuits: Scientific and technical discipline providing the unique language and methodology for solving the problems of modeling, calculation, identification, and optimization of pipeline and hydraulic systems of different type and purposes.

APPENDIX

Input data:

1. Cost function for the first and second source:

$$f_1(Q_1) = 127033 + 987Q_1 + 0.42Q_1^2, \; f_2(Q_2) = 138630 + 798Q_2 + 1.05Q_2^2;$$

2. Duration of heating period: 5760 h/year;
3. Calculated indoor air temperature: +20°C;
4. Outdoor air temperature: -36 °C;
5. Loads (see Table 3):

Table 3. Consumer heat loads

Consumer Number	Heating Load, GJ	Hot Water Load, GJ	Consumer Number	Heating Load, GJ	Hot Water Load, GJ
3	268.1	67.1	16	335.2	83.8
4	201.1	50.3	17	402.2	100.6
5	167.6	41.9	18	167.6	41.9
6	201.1	50.3	19	268.1	67.1
7	301.6	75.4	21	301.6	75.4
8	234.6	58.7	22	335.2	83.8
9	335.2	83.8	23	234.6	58.7
10	402.2	100.6	24	301.6	75.4
11	301.6	75.4	25	402.2	100.6
12	234.6	58.7	26	335.2	83.8
13	378.7	94.7	27	234.6	58.7
14	536.3	134.1	28	167.6	41.9
15	804.4	201.1	29	201.1	50.3

7. Temperature at the beginning of the heating period: +8 °C;
8. Average temperature of the heating period -8.9 °C;
9. Share of hot water supply: 0.2;
10. Constraints on the thermal capacity of the sources: $0 \leq Q_1 \leq 8380$ GJ/h, $0 \leq Q_2 \leq 7542$ GJ/h;
11. Coefficients of the function of capital investment in the network: $a=6835$, $b=63827$, $u=1.45$;
12. Specific electricity cost: $C = 0,6$ rub./kWh;
13. Efficiency coefficient of the pump-motor unit: $\eta = 0.7$.
14. Characteristics of the heat network (see Table 4):

Table 4. Resistance of the sections

Section Number	Length, m	Resistance, mh²/t²	Section Number	Length, m	Resistance, mh²/t²
1	600	0.00011	18	350	0.000166
2	400	0.000117	19	350	0.0002
3	200	0.000077	20	400	0.00017
4	150	0.000157	21	400	0.00019
5	600	0.000257	22	300	0.0002
6	300	0.000236	23	550	0.00025
7	300	0.000256	24	400	0.00014
8	200	0.000356	25	500	0.000203
9	400	0.000156	26	300	0.00015
10	450	0.000296	27	450	0.0001
11	200	0.000206	28	400	0.000356
12	400	0.000156	29	1000	0.000156
13	500	0.000216	30	700	0.00021
14	550	0.000276	31	700	0.00037
15	350	0.000356	32	400	0.00012
16	400	0.000456	33	400	0.0001
17	400	0.000156	-	-	-

15. For all the participants the coefficient depending on the pipe roughness is taken as:

$\chi = 0.01277 \ (k = 0.5 \, mm)$.

Chapter 4
Problems of Modeling and Optimization of Heat Supply Systems:
New Methods and Software for Optimization of Heat Supply System Parameters

Valery Stennikov
Energy Systems Institute, Siberian Branch of the Russian Academy of Sciences, Russia

Dmitry Sokolov
Energy Systems Institute, Siberian Branch of the Russian Academy of Sciences, Russia

Evgeny Barakhtenko
Energy Systems Institute, Siberian Branch of the Russian Academy of Sciences, Russia

Tamara Oshchepkova
Energy Systems Institute, Siberian Branch of the Russian Academy of Sciences, Russia

ABSTRACT

This chapter presents new methods and software system SOSNA intended for the parameter optimization of heat supply systems. They make it possible to calculate large-scale systems which have a complex structure with any set of nodes, sections, and circuits. A new methodological approach to solving the problem of the parameter optimization of the heat supply systems is developed. The approach is based on the multi-level decomposition of the network model, which allows us to proceed from the initial problem to less complex sub-problems of a smaller dimension. New algorithms are developed to numerically solve the parameter optimization problems of heat supply systems: an effective algorithm based on the multi-loop optimization method, which allows us to consider hierarchical creation of the network model in the course of problem solving; a parallel high-speed algorithm based on the dynamic programming method. The new methods and algorithms were used in the software system SOSNA.

DOI: 10.4018/978-1-4666-9755-3.ch004

INTRODUCTION

Designing heat supply systems (HSSs) of modern cities and industrial centers, we have to solve the problem of choosing optimal HSS parameters which implies determining diameters of new heat pipelines, pressure heads of pumping stations and their sites, as well as the ways to reconstruct existing network sections. The development of market for heat pipelines, equipment and technologies used in HSS construction significantly increases the opportunities for the implementation of technical solutions. This, in turn, requires that the opportunities be taken into account in HSS modeling and optimization, since each type of the equipment has its own characteristics and is described by its set of mathematical models that characterize its parameters and techno-economic relationships.

Modern conditions that require us to have different technological equipment and be capable to flexibly use it in the calculation of various types of pipeline systems make it impossible to implement software within the programming paradigms applied previously. Thus, it becomes necessary to develop and apply new adaptive approaches to adjust the software to the calculation of any types of systems with a wide range of equipment. Moreover, when developing and testing new methods and algorithms of HSS optimization and mathematical modeling, we should be guided by the software expandable architecture, which will make it possible to add new components to the existing calculation scheme.

The methodological approaches traditionally applied in the construction of software have no clear division into the methods (algorithms, techniques) for solving the applied problems and mathematical models of the HSS components. Therefore, the software modules implementing the algorithms become oriented to the specific classes of problems and a set of equipment, which complicates their adjustment to a certain problem and their multiple use when building different software systems. It is impossible for a researcher to create their models of components and integrate them into the software while doing scientific or engineering calculations. A multiple doubling of one and the same model of the HSS component occurs in different computational modules. Therefore, when correction is needed the changes should be made in all the software modules.

To overcome the enumerated difficulties it is necessary to develop a new methodological approach to the construction of the software, in order to automatically create complex software systems aimed at solving the applied problems, considering individual features of the HSSs to be modeled. In the framework of this approach the knowledge from the application domain should be formalized in the form of ontologies (Guarino, 1997; Gavrilova & Khoroshevsky, 2000). This will enable us to use them many times to automatically construct the applied software systems. It is necessary to develop a single software platform to automatically form a software system oriented to a specific applied problem and do complex engineering calculations to find the optimal ways of transforming real HSSs to improve their efficiency and reliability of operation.

BACKGROUND

Analytical methods were the first theoretically substantiated methods for calculating optimal diameters of the pipeline network. The most well-known among them is the method suggested by Grashof in 1875 for a water main with point loads. He solved the problem of search for a constrained minimum of the water main cost function in terms of the given total pressure loss and on the basis of the Lagrange multiplier method.

The first studies in the field of heat supply to focus on the considered problem appeared in the 1930s (Davidson, 1934; Yakimov, 1931). Moreover, their authors aimed at calculating a main with point loads and tried to apply the Grashof method to calculate branched heat networks. The papers by A.M. Zanfirov (Zanfirov, 1933, 1934) present an analytical solution to the problems of optimal diameter selection for the branched network with random topology and an available head in the sources.

The development of analytical methods for technical and economic HSS calculation was significantly affected by the studies conducted by B.L. Shifrinson and V.Ya. Khasilev (Khasilev, 1957; Shifrinson, 1940). B.L. Shifrinson reformulated the problems of techno-economic HSS calculation by assuming not pipeline diameters but pressure losses in the network section to be independent variables.

This made it possible to state the problems for randomly configured branched heat networks (but without loops and with one heat source), and significantly simplify the way of solving them. B.L. Shifrinson obtained a number of key results, which largely underlie the theory and practice of HSS calculation. V.Ya. Khasilev (Khasilev, 1957) summarized and studied the properties of continuous models of HSS parameter optimization and discovered the flatness of the economic criterion (calculated costs) in optimal solutions.

Analytical methods for solving the problems of parameter optimization of branched HSSs had been widely spread in the international design practice before the discrete optimization methods appeared (Heindrun, 1968). The analytical methods for calculation of optimal pipeline diameters in multiloop heat networks did not evolve because they were hard to apply.

Today the most developed and popular methods are the methods of discrete optimization. Their major competitors are the methods of linear and dynamic programming (Bellman, 1957; Dantzig, 1963). Years of experience in solving network problems have proved that the dynamic programming method is the most effective, since it allows the global minimum to be found taking into account the diversity of physical and technical constraints, discreteness of equipment parameters and location, and other specific features of the system and its components. This method can be successfully applied to optimize newly created systems and solve the reconstruction problems of the expanding ones. This method was first implemented by A.P. Merenkov in 1963 to optimize the parameters of branched heat networks (Merenkov & Khasilev, 1985). Later on, these problems were called the parameter optimization problems, and the dynamic programming method was repeatedly implemented at Energy Systems Institute in the form of standard programs for different types of computers (Merenkov & Khasilev, 1985).

Solving the problems of optimizing branched network parameters, researchers often use the method of dynamic programming (Appleyard, 1978; Barreau & Moret-Bailly, 1977; Caizergues & Franenberg, 1977; Garbai & Molnar, 1974). In their studies French authors (Barreau & Moret-Bailly, 1977; Caizergues & Franenberg, 1977) compared the analytical methods (including the method based on Lagrange multipliers) and the dynamic programming method, and came to the conclusion that the latter is more effective.

To optimize loop networks S.V. Sumarokov developed the method of multiloop optimization (Sumarokov, 1976), which makes it possible to simultaneously determine optimal diameters of pipelines in a multiloop system and the sites and parameters of pumping and throttling stations, find optimal flow distribution in the system, and solve the problems of reconstructing its existing part.

Today methods based on linear programming continue to develop. They can take into account the constraints on the use of standard pipelines of a discrete size without rounding off the obtained diameter values. Morgan and Goulter (Morgan & Goulter, 1985) changed the Kally procedure (Kally, 1972) to relate the Hardy-Cross method to the linear optimization model. Their model makes it possible to find optimal parameters both for new and existing systems to be reconstructed. Application of this method to design problems proved its effectiveness, since it reduces the costs of network construction and operation. The main disadvantage of this method is the fact that two possible standard pipeline sizes can be found for some network sections, which makes it difficult to use the method in design practice.

Currently optimal parameters for pipeline systems and energy systems are determined on the basis of different heuristic software methods (Vasant, Barsoum & Webb, 2012; Vasant, 2013) using genetic algorithms (Savic & Walters, 1997; Wenzhong & Jiayou, 2009) and simulated annealing algorithms (Cunha & Sousa, 1999; Khanh, Vasant, Elamvazuthi & Dieu, 2014; Castillo-Villar & Smith, 2014). The advantages of these methods are that they make it possible to take into account the complexity of hydraulic operating conditions and discreteness of equipment parameters.

One of the useful methodologies to solve optimization problems is soft computing, which conducts collaboration, association, and complementariness of the different techniques that integrates them. The applications and developments of soft computing can be found in the field of renewable energy and energy efficiency where soft computing techniques are showing a great potential to solve the problems that arise in this area (Cai, 2015).

The discrete methods developed at the Energy Systems Institute SB RAS make it possible to obtain technically and economically sound solutions allowing solving the following subproblems in a coordinated manner: calculation of optimal diameters for new sections of a heat network; determination of necessary types of reconstruction of the existing sections of a heat network; calculations of heads and sites for the installation of pumping stations; calculation of head values for the heat sources.

An important feature of these methods is that they make it possible to fully consider many characteristics of heat network: a discrete set of the equipment applied, nonlinearity of hydraulic relationships, a wide variety and complexity of mathematical models. The method of dynamic programming, unlike the existing similar methods, allow us to find a solution to the optimization problem, considering the specific features of the existing equipment.

The said specific features of the discrete methods developed at the Energy Systems Institute make it possible to conclude that they are in demand in the research and engineering field.

The methods and algorithms for solving the optimization problem of HSS parameters have been developed at the Energy Systems Institute SB RAS for decades. They are well tested and adjusted. The experience gained in the process of their numerous practical applications was used for their further improvement. At the same time, there are some drawbacks of the existing versions of the algorithms:

- Low-speed operation in solving the optimization problems for the networks with a great number of components and complex configuration;
- Impossibility of applying a multilevel modeling technique developed on the basis of the theory of hydraulic circuits which was devised at the Energy Systems Institute SB RAS.

The considered properties of the algorithms allow us to conclude that it is necessary to develop the improved versions of the algorithms which are free from the mentioned disadvantages.

STATEMENT OF THE OPTIMIZATION PROBLEM OF HEAT SUPPLY SYSTEM PARAMETERS

The statement of the parameter optimization problem of the branched and multi-loop HSSs is presented in the following form.

The specified characteristics include the following:

1. Scheme of a network consisting of m nodes and n sections, represented by a directed graph $G_{hss} = (J, I)$, where J – a set of vertices (nodes), I – a set of edges (sections); $J = J_1 \cup J_2 \cup J_3 = J^f \cup J^b$, where J_1, J_2 and J_3 – sets of consumers, sources and branch points in the network, respectively, J^f and J^b – sets of nodes in the supplying and return mains, respectively; $I = I^p \cup I^c \cup I^s$, where $I^p = I^f \cup I^b = I_1 \cup I_2$ – a set of pipeline sections, which consists of the sections of the supply I^f and return I^b mains, respectively, I_1 and I_2 – sets of existing and new sections, respectively; I^c and I^s – sets of consumer-sections and source-sections; $I_3 \subset I^p$ – sections, in which pumping stations are either installed or are allowed to be installed;
2. Lengths of pipeline sections $L_i, i \in I^p$;
3. A set of standard pipeline diameters $D = \left\{ d_1^s, \ldots, d_k^s \right\}$;
4. Constraints on the ways of reconstructing the existing pipeline sections $U_i, i \in I_1$ from a set of potential ways U;
5. Constraints on the pressure at nodes and speed in sections;
6. Tabular coefficients and technical and economic indices;
7. Loads of consumers and productivity of sources;
8. A set of types and pump parameters that can be installed in the system $T = \left\{ T_1^s, \ldots, T_r^s \right\}$.

It is necessary to determine:

1. "Bottlenecks" in the system – overloaded pipeline sections;
2. Vector of flow rates in pipes x;
3. Vector of pressures at nodes P;
4. Diameters of new pipeline sections $d_i, i \in I_2$;
5. Pipeline sections that need reconstructing, new diameters of the pipelines $d_i, i \in I_1$ and ways of reconstructing $u_i \in U_i, i \in I_1$;
6. Heads H_i and locations for the pumps for pipelines $i \in I_3$;
7. Parameters ensuring the minimum difference between the heads at nodes of the supplying and return pipelines at consumers ΔH_i, $i \in I^c$.

Mathematical problem statement. We have to find a solution to provide the minimum total estimated costs throughout the entire system, i.e. it is necessary to minimize the function

$$Z(d,H,x,h) = \sum_{i \in I^{\mathrm{p}}} Z_i^{\mathrm{p}}(d_i) + \sum_{i \in I_3} Z_i^{\mathrm{ps}}(H_i, x_i) + \sum_{i \in I} Z_i^{\mathrm{e}}(x_i, h_i) + \sum_{i \in I^{\mathrm{p}}} Z_i^{\mathrm{t}}(d_i) \to \min, \tag{1}$$

where Z_i^{p} – costs of laying pipeline sections; Z_i^{ps} – costs of constructing pumping stations; Z_i^{e} – costs of electricity used to pump heat carrier along the network sections; Z_i^{t} – cost of heat losses; $d = (d_1, \ldots, d_n)^{\mathrm{T}}$, $H = (H_1, \ldots, H_n)^{\mathrm{T}}$, $x = (x_1, \ldots, x_n)^{\mathrm{T}}$ and $h = (h_1, \ldots, h_n)^{\mathrm{T}}$ – vectors of diameters, heads, flow rates and head losses in the sections, respectively.

The losses of head h_i are determined by the expression

$$h_i = s_i x_i^2, i \in I^{\mathrm{p}}, \tag{2}$$

where s_i – coefficient of hydraulic resistance of a branch, which is determined by the equation

$$s_i = \phi_i \frac{(1+\alpha_i)L_i}{d_i^{5.25}}, i \in I^{\mathrm{p}}, \tag{3}$$

where ϕ_i – coefficient that depends on the pipe roughness, α_i – coefficient of local losses that depends on a diameter.

The nodal model of flow distribution (Merenkov & Khasilev, 1985) is:

$$A^s x - Q = 0, \tag{4}$$

$$y - A^T P = 0, \tag{5}$$

$$y + H - SXx = 0, \tag{6}$$

where A and A^s – complete $m \times n$-incidence matrix and $(m-1) \times n$-incidence matrix of linearly independent nodes and sections of the calculated scheme, respectively; $y = (y_1, \ldots, y_n)^{\mathrm{T}}$ – vector of differences between the pressures at branch nodes; $S = \mathrm{diag}(s_1, \ldots, s_n)$ and $X = \mathrm{diag}(|x_1|, \ldots, |x_n|)$ – diagonal matrices of hydraulic resistances and flow rates, respectively; $Q = (Q_1, \ldots, Q_{m-1})^{\mathrm{T}}$ – vector of nodal sources or sinks; $P = (P_1, \ldots, P_m)^{\mathrm{T}}$ – vector of pressures at nodes.

The system of conditions and constraints includes:

- Constraints on pressure at nodes P_j

$$P_j^{\min} \le P_j \le P_j^{\max}, j \in J, \tag{7}$$

where $P_j^{\min}$ and $P_j^{\max}$ – lower and upper constraints, respectively;

- Constraints on the heat carrier flow speed along the sections v_i

$$v_i^{\min} \leq v_i \leq v_i^{\max}, i \in I^{\mathrm{p}}, \tag{8}$$

where $v_i^{\min}$ and $v_i^{\max}$ – lower and upper constraints, respectively;

- Constraints on the difference between heads at nodes in consumers ΔH_i

$$\Delta H_i = P_{j_\mathrm{f}} - P_{j_\mathrm{b}} \geq \Delta H_i^{\min}, i \in I^{\mathrm{c}}, j_\mathrm{f} \in J^{\mathrm{f}}, j_\mathrm{b} \in J^{\mathrm{b}}, \tag{9}$$

where j_f and j_b – numbers of consumer nodes of supply and return pipelines, respectively, $\Delta H_i^{\min}$ – the minimum difference between the heads;

- The conditions for discreteness of pipeline diameters (Equation 10), types of reconstruction (Equation 11) and types of equipment for pumping stations (Equation 12):

$$d_i \in D_i \in D, i \in I^{\mathrm{p}}, \tag{10}$$

$$u_i \in U_i \subset U, i \in I^{\mathrm{p}}, \tag{11}$$

$$H_i \in T_i \in T, i \in I_3, \tag{12}$$

where D_i and T_i – admissible standard pipeline diameters and heads of pumping stations on section i.

By solving the problem of minimizing function (Equation 1) subject to (Equations 2, 3, 4, 5, 6, 7, 8, 9, 10, 11, and 12) it is necessary to determine the optimal parameters of the HSS.

METHODS FOR SOLVING THE PROBLEM OF HSS PARAMETER OPTIMIZATION

To solve the formulated problem, researchers from the Energy Systems Institute developed an optimization algorithm based on *dynamic programming* (Bellman, 1957; Merenkov & Khasilev, 1985). The additive objective functional (Equation 1) makes it possible to apply a dynamic programming method to solve this problem. The idea of this method lies in a multi-step process of determining the parameters of the network components (sections and nodes) by successive fitting them in the direction from consumers to the source (Merenkov & Khasilev, 1985; Merenkov, 1992). The region of search for a solution (Figure 1), which is formed between the upper ($P_j^{\max}, j \in J$) and lower ($P_j^{\min}, j \in J$) constraints on pressure is divided into n intervals (according to the number of sections) which are split into μ cells (which determine the accuracy of the solution). The problem is solved in three stages: 1) forward, dur-

Figure 1. Illustration of the problem solving process by the dynamic programming method

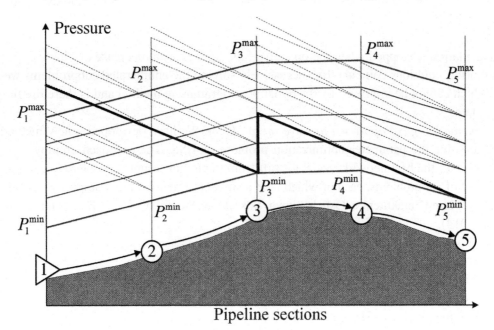

ing which all the cells are successively filled and a set of conditionally optimal variants is formed; 2) selection of the cheapest variant for the whole network; 3) backward, which suggests the selection of parameters and cost components corresponding to the selected variant.

Mathematically, the process of identifying the conditionally-optimal variant in cell z in step i is defined by the recurrent equation:

$$Z_i^*(\tilde{P}_{jz}) = \min_{\substack{d_i^u \in D_i \\ k=1,\ldots,\mu}} \left[Z_i^{pl}(d_i^u) + Z_i^{ps}(P_{jk}^{ap} - P_{jz}^{bp}) + Z_{i-1}^*(P_{j+1,k}) \right], z = 1,\ldots,\mu, \quad (13)$$

where Z_i^{pl} – costs of the pipeline with a diameter d_i^u, including the costs of construction, operation, electricity and heat losses, P_{jz}^{bp} and P_{jk}^{ap} – pressure before the pumping station and after it in cells z and k, respectively; Z_i^{ps} – costs of the pumping station with a head $P_{jk}^{ap} - P_{jz}^{bp} = H_{iz}$; $Z_{i-1}^*(P_{j+1,k})$ – the sum of costs in step $i-1$ in cell k; Z_i^* – a sum of costs for the selected conditionally – optimal variant.

Optimization of Multi-Loop Network Parameters

The optimal parameters of the multi-loop HSS are selected by the multi-loop optimization method, which is based on a methodological principle of successive improvement in solutions, during which the following subproblems are solved (Sumarokov, 1976; Merenkov, 1992):

1. Calculation of flow distribution in the network (flow rates x and pressures P) at fixed diameters d and heads H;

2. Optimization of parameters (diameters d, heads H and pressures P) by the dynamic programming method for the fixed flow rates x.

The calculation is stopped when the objective function ceases to decrease or pipeline diameters in the neighboring iterations coincide. By successively solving the enumerated subproblems we obtain a solution which cannot be improved by the dynamic programming method and is therefore the solution to the problem.

A new methodological approach to solve the problem. We have developed a new methodological approach to solve the parameter optimization problem of HSS. The approach employs a multilevel decomposition of the heat network model to shift from the initial complex problem to a hierarchically-connected set of subproblems, each of which is of a smaller dimension and complexity as compared to the initial problem (Barakhtenko, Oshchepkova, Sokolov & Stennikov, 2013; Barakhtenko, Sokolov & Stennikov, 2014).

The methodological approach includes:

- The principles of a multilevel decomposition of the heat network model;
- The principles of "coordinating the solutions" among the hierarchical levels of the heat network model in the process of HSS parameter optimization;
- An algorithm of dynamic decomposition of the network model into two parts: loop and dead-end branches;
- An algorithm of an upper level for the parameter optimization of the HSS, considering the available heads in consumers;
- Algorithms of the multi-loop optimization and dynamic programming methods intended for solving the problem with respect to the multilevel decomposition of the heat network model.

The methodological approach suggests the following levels of the hierarchical heat network model (Figure 2):

Figure 2. Levels of a hierarchical heat supply network model

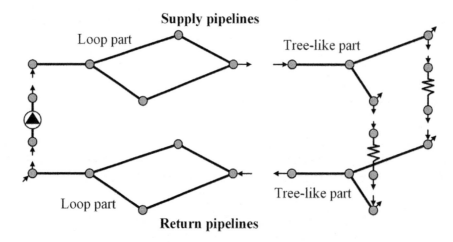

1. The entire heat network;
2. Supply and return mains;
3. Loop and tree-like parts (dead-end branches) of the supply and return main pipelines;
4. Individual components of the heat network (sections, pumping stations, etc.)

The suggested methodological approach supposes the implementation of the following problem solving steps:

Step 1: Calculation of nodal flow rates according to the specified loads, operating conditions and temperature curve.
Step 2: Multilevel decomposition of the heat network model.
Step 3: Calculation of optimal parameters of the return main pipeline by the multi-loop optimization method.
Step 4: Adjustment of constraints on the pressure in consumers at nodes of the supply main pipeline, considering the obtained pressures at the nodes of the return main pipeline to ensure the required available heads in consumers.
Step 5: Calculation of optimal parameters of the supply main pipeline by the multi-loop optimization method.
Step 6: Determination of the total costs and capital investment in the HSS according to the parameters obtained during the optimization.
Step 7: Testing of a solution obtained for the system for the capability to provide different operating conditions of the HSS.
Step 8: In the case the operating conditions are not provided, the initial data are adjusted, and transition to step 1 is made.

Development of algorithms for parameter optimization of HSSs. Software package SOSNA implements a new variant of the algorithm for HSS parameter optimization. The algorithm makes it possible to efficiently calculate branched and multiloop schemes with any number of components and structure of any complexity. The suggested algorithm is based on the depth-first search method (DFS) (Levitin, 2006). Considering the network from its source, the algorithm uses the incidence matrix to determine the nodes (end or already examined nodes), which then serve as the starting points for the construction of trajectories of conditionally optimal piezometers. Then at the branch point the algorithm creates a "junction" between different conditionally optimal trajectories.

One of the main advantages of the new algorithm is that it operates in the networks that split into subnetworks. This makes it possible to calculate optimal parameters for the selected subnetworks simultaneously. The previous algorithm did not provide such opportunities. Besides, it had a disadvantage as it suggested that there were no closed circulation loops in the network (Sennova & Sidler, 1987). An example of a closed circulation loop is presented in Figure 3. However, these loops can appear in the network in the course of iterative calculations. The merit of the new algorithm is that it detects closed circulation loops, opens them on its own, calculates the network correctly, and then finds a solution that does not contain a closed circulation loop.

Today researchers solve the problems of HSS parameter optimization by presenting them in a multilevel form. For instance, data processing environment ANGARA developed at the Energy Systems

Figure 3. Scheme of a heat supply system with a closed circulation loop

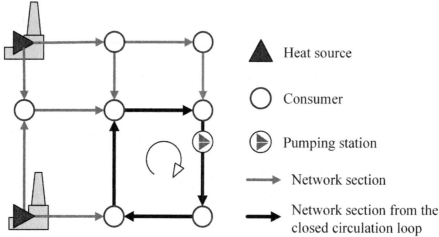

Institute allows one to work with such systems. The main specific feature of the new parameter optimization algorithm unlike the previously used one is that it makes it possible to model HSSs at many levels.

Flow distribution calculation. The traditional problem of calculating flow distribution in hydraulic circuits with lumped parameters (without regulators) is reduced to solving a closed system of equations. For the problem dealing with the calculation of flow distribution, which is described by the nodal model (Equations 4, 5, and 6), there are different methods of solving the system of equations. One of the most effective methods of solving this system of equations is the global gradient algorithm. The algorithm was developed by the professor E. Todini and his colleagues (Todini & Pilati, 1988) from the University of Bologna.

The loop model of flow distribution (Merenkov & Khasilev, 1985) is:

$$\begin{cases} A^s x - Q = 0 \\ By = 0 \\ f(x) = y + H \end{cases} \quad (14)$$

where B – loop matrix, $f(x)$ – vector-function of hydraulic characteristics of network sections which relates technical (hydraulic resistance s) and hydraulic parameters (heat carrier flow rate in section x).

We use one of the most effective ways to solve the system of equations (Equation 14), i.e. the Newton method. The Newton method is sensitive to the initial approximation x^0, which should lie in the proximity to the sought solution. Otherwise, the convergence is not guaranteed. This disadvantage can be overcome by adjusting the step when the process

$$x^{k+1} = x^k + \lambda^k \Delta x^k,$$

is applied and the scalar parameter λ^k, called by the step length, is selected based on the one-dimensional minimization of some norm of a vector of residual

$$\lambda^k = \arg\min_\lambda \left\| w(x^k + \lambda \Delta x^k) \right\|,$$

where w – is the vector of residuals of heads.

In the problems aimed at calculating the flow distribution in the large networks, matrices have large dimensions and contain a great number of zero elements, which makes it possible to use the technologies of sparse matrices for mathematical operations with them in the computer calculations. In the publications there are effective algorithms allowing fast implementation of the above method in software and considerable reduction in the required size of the random access memory.

The use of sparse matrix techniques. We suggest using the sparse matrix technology for storage and processing of matrices with a great number of zero elements (Tewarson, 1973; George & Liu, 1981; Pissanetzky 1984). There are different methods of compact storage of matrix elements in the memory. The most widely used sparse matrix storage scheme, which is called compressed row storage (CRS).

According to the CRS scheme three one-dimensional arrays are required to store the matrices:

1. An array of nonzero matrix elements in which they are enumerated by row from the first to the last one(designate the array as *values*);
2. An array of column numbers for respective elements of the array "*values*" (designate this array as *cols*);
3. An array of pointers which are used to describe each row (designate the array as *pointer*). The description of a row of number k is stored in the positions from *pointer*[k] to *pointer*[k+1]-1 in the arrays *values* and *cols*.

This scheme imposes the minimum requirements for the memory and at the same time proves very convenient for several important operations with sparse matrices: summation, multiplication, permutation of rows and columns, transposition, solving linear systems with sparse matrices of coefficients by both direct and iteration methods, etc.

THE USE OF CONCURRENT COMPUTING

Software implementations of the algorithm for solving the problem of HSS parameter optimization relate to resource-intensive calculations that require powerful computers and considerable time for an engineer to solve the applied problem. To overcome these drawbacks, the researchers developed approaches to optimization of the computational process that help reduce problem-solving time and are based on the concurrent computing techniques, i.e. methods for organizing computer calculations, according to which the programs are developed as a set of interacting computational threads or processes that work simultaneously (concurrently).

We suggest the following ways of applying concurrent computing to the organization of computational process in the problem of HSS parameter optimization:

- Simultaneous calculation of parameters for supply and return main pipelines.
- Simultaneous calculation of several subnetworks, which make up the HSS.

- Organization of the computational process so that the algorithm of dynamic programming can calculate some sections simultaneously.

The first method of organizing the computational process is based on a specific feature of the HSS that makes it different from the other pipeline systems – the pipeline has two lines: the network consists of supply and return mains. According to the suggested way of computational process organization the optimization problems for both supply and return mains are solved simultaneously. Moreover, optimal solutions can be obtained for each pipeline, regardless of each other. The main advantage of the suggested approach is that it helps reduce the calculation time of the entire network (i.e. the time of calculating both mains on a modern computer, which supports concurrent computing, does not exceed the time of calculating either of the mains). The disadvantage of this method for computational process organization is the independence of parameters obtained for each of the mains. For example, the diameters obtained for each network section can be different for supply and return mains, which is inacceptable in some cases, since this does not comply with the standards of HSS design practice.

The second method of organizing the computational process involves simultaneous calculation of several subnetworks that make up the HSS. The advantage of this method is that it reduces the time of calculating the parameters to the time necessary to calculate the parameters of the largest (most complex) HSS subnetwork. The disadvantage of this approach is that it can be used only in the case the network splits into subnetworks.

The developed technique for organizing the computational process implies the modification of the previously suggested algorithm for HSS parameter optimization by adding concurrent computing. This technique suggests the use of multithreaded execution, which begins automatically, when the launched calculation program starts solving the problem of HSS parameter optimization. The main idea of this technique is that in the course of the problem-solving there appear several optimizers, each of which operates in a separate thread of execution (Figure 4). Each optimizer is developed so that it can solve the optimization problem of one separate network section. Numbers of the sections, whose parameters can be optimized, are put in the queue. Optimizer of the section parameters extracts the number of the next section from the queue and calculates its optimal parameters. Here we follow the principles of con-

Figure 4. Scheme of the computational process organized on the basis of concurrent computing

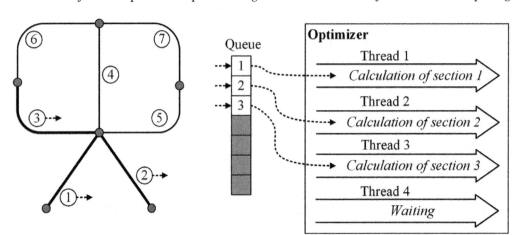

structing the algorithm of dynamic programming: 1) computational process is organized in the direction from consumers to the source (sources); 2) the section is calculated, if all the sections outgoing from its end node are calculated.

The suggested version of the algorithm for solving the problem of HSS parameter optimization on the basis of concurrent computing differs from the algorithm described above in one very important feature – while in this algorithm all the sections of the calculated network are put in the queue before the "forward procedure" of the algorithm, in the concurrent version of the algorithm the sections are put in the queue during the "forward procedure" (this happens automatically, as soon as one of the dynamic programming conditions is met for some section: all the sections outgoing from the end node are already calculated). Before the forward procedure starts in the concurrent version of the algorithm only the sections that end at the end consumer nodes join the queue. Thus, for the HSS scheme presented in Figure 5 such sections are S-1 and S-2.

The suggested algorithm makes it possible to overcome the main difficulty in the design of concurrent programs, i.e. to provide a correct sequence of interactions between different computational threads and coordinate the resources distributed among the threads.

THE USE OF MODERN INFORMATION TECHNOLOGIES

The construction principles and implementation scheme of the software package SOSNA are based on the componential approach, which allows one to make a hierarchical decomposition of the software system into parts and present its logical architecture by components interacting with one another. A distinctive feature of this approach is that a component that has already been implemented (e.g. an implementation method for solving systems of equations, tree construction algorithm, etc.) can be repeatedly used to solve different problems. This property of components allows the implementation of a methodology for solving the problems of optimal HSS expansion and reconstruction which is based on successive stage-by-stage problem-solving and mutual coordination of results obtained at each stage. It becomes possible to use ready-made components and extend the software functionality without restructuring the source code.

Figure 5. HSS sections chosen at the beginning of the calculation

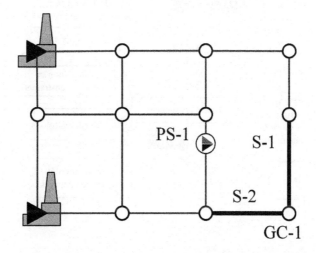

Software components can be divided into groups, depending on the functions to be performed. Each group has a developed interface of interaction between the components, including the rules of their calling, transfer of parameters and results of calculation. We single out the following groups of components:

- Computation components that include software implementations of the algorithms for HSS mathematical modeling and optimization;
- System components that prepare and examine the data (interaction with the database, validation of initial data, etc.);
- Control components intended for organizing the sequence of calls of computation and system components.

The main programming language used by the developers of the software package is Java, which provides the software portability and offers them the opportunity to apply object-oriented programming (Eckel, 2006). Some computation components, whose software implementation is required to be very fast (implementation of the algorithm for determining pipeline diameters in the HSS parameter optimizer, solving the systems of equations in a component for hydraulic calculation), are written in languages C, C++ and Fortran and connected to the software package through the standard interface JNI (Java Native Interface) of the Java platform (Horstmann & Cornell, 2004). The software allows one to organize the interaction with databases oriented towards the operation under the control of different database management systems (DBMSs). To this end, the developers used JDBC (Java DataBase Connectivity) technology, i.e. a platform-independent industry standard of interaction between the Java applications and different DBMSs. The suggested software package architecture is presented in Figure 6.

The software package is based on the principles of software development paradigm controlled by the models (Model-Driven Engineering, MDE). The main idea of MDE is that the software is automatically constructed on the basis of models for solving a concrete problem (Brambilla, Cabot &Wimmer, 2012; Völter, Stahl, Bettin, Haase & Helsen, 2006; Mellor, Scott, Uhl & Weise, 2004). Models, which describe

Figure 6. Software package architecture

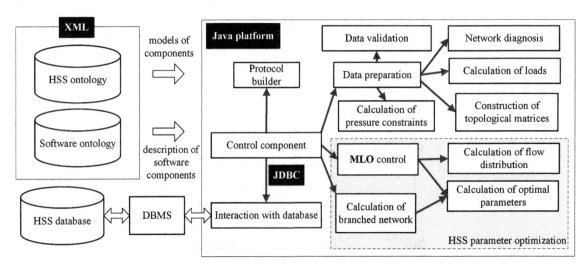

the objects of the current problem and relations between them, are formalized by special languages intended for the description of object domains. The scheme of the computational process is formed on the basis of these models, as is shown in Figure 7.

Models underlying the software package are formalized in the form of ontologies (Guarino, 1997; Gavrilova & Khoroshevsky, 2000). The developers use the following types of ontologies:

- **Ontology of Heat Supply Systems (HSS Ontology):** Includes the description of properties of HSS subsystems; types and parameters of equipment and pipelines; hierarchy of the network components, their properties, relations between components, and their mathematical description (e.g. closing relationships, formulas for calculation of resistance, etc.).
- **Ontology of Problems:** Includes the description of the problem hierarchy, and a set of parameters represented by the initial data, and parameters obtained as a result of solving the problem.
- **Software Ontology:** Includes the description of software components and their properties; input and output parameters; data formats; and technologies for accessing the system components.
- **Ontology of Databases:** Describes the structure of different databases (tables and relations between them, concepts, stored procedures, etc.) that operate under the control of different DBMSs.

The HSS ontology is intended for formation of the model components, which integrate with computational algorithms of the software components during the calculations. The obtained models make it possible to adjust the software to the specific features of equipment used to design a certain HSS. A fragment of the HSS ontology is presented in Figure 8.

The ontology of problems is intended to organize a step-by-step problem solving (call of the software modules and subprograms). It reflects the interaction between the subproblems that help solve the current problem. The ontology of problems is organized in the form of a graph, whose nodes correspond to the steps of problem solving, and edges – to the relations between the steps. The process of solving a concrete problem represents a traversal of the graph constructed for this problem.

Figure 7. Scheme of the computational process on the basis of models

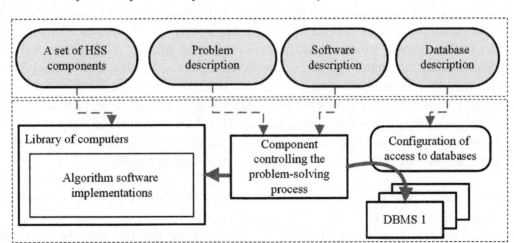

Figure 8. Fragment of the heat supply system ontology

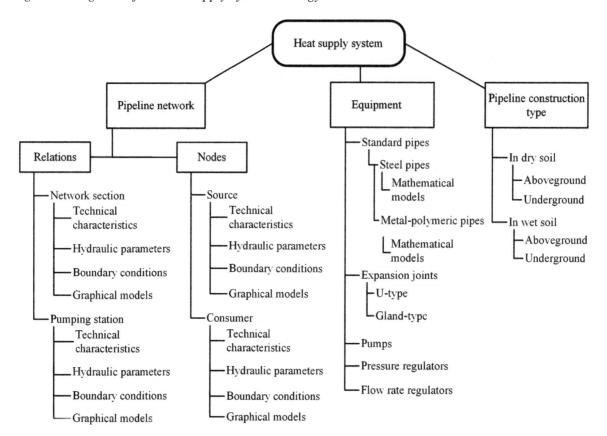

The software ontology is used to provide access to the software components. The control component intended for the organization of computational process traverses the graph of the problem-solving and sequentially calls the components corresponding to the steps of the current problem. The components are called on the basis of data of the software ontology.

The ontology of databases is used to organize interaction between the software and different databases when the configuration of access to DBMS is generated automatically. This approach allows one to adjust the software to the randomly structured databases developed by different programmers and intended for the operation under the control of different DBMSs (MS Access, MS SQL Server, Firebird, etc.).

The authors suggest metaprogramming to create an open software architecture flexibly adjustable to the problem of parameter optimization of a concrete HSS in the course of the problem-solving process (Bartlett, 2005; Stennikov, Barakhtenko & Sokolov, 2011). Metaprogramming technologies make it possible to develop the software that changes or creates another software during its operation. When implementing the software package, we use two aspects of metaprogramming:

1. The polymorphic code is applied involving the corresponding mechanisms of modern high-level programming languages (reflection, code interpretation at the program level, etc.).
2. The software code is generated during the software package operation with the aid of software code generators from the formalized description of object domain.

The first aspect of applying metaprogramming involves the dynamic connection of new computation components to the operating program. This allows one to make up the software system from ready-made components during its operation for a concrete HSS. An approach used to solve this problem is based on the software ontology, reflection mechanisms of the Java programming language and design patterns.

Reflection of Java makes it possible to obtain the data on components, create copies and manipulate them during the program operation. It can be found in the package *java.lang.reflect* of the standard library of the Java platform.

Design patterns are the typical solutions to some characteristic software design problems. Patterns allow one to standardize interfaces of the system components and repeatedly use them in the software development. The SOSNA software uses design patterns Factory, Adapter and Command (Gamma, Helm, Johnson & Vlissides, 1995). Factory is a design pattern that suggests using a software component, which provides an interface for the creation of *other* components (based on the logical name of the requested component, this component downloads its instance in the memory and returns the link to it to the subprogram that has called it). Command is a design pattern that represents a subprogram in the form of a component. Adapter is a design pattern aimed at organizing the use of a non-modifiable component through a special interface. A component that implements the design pattern Factory will be called "a factory", the component that implements the design pattern Command – "a command", and the component Adapter will be called "an adapter".

All the software components of the SOSNA software are commands (this allowed one to standardize their interfaces and make the components interchangeable); they are divided into three groups in accordance with their description above. Computational commands are either computation components or adapters providing a single interface for access to other computation components. This approach makes it possible to easily solve the problem of adapting the legacy software, for example, to install legacy software package and provide a single interface for access to its subprograms.

A scheme for integrating the computation components into the software package on the basis of the software ontology is presented in Figure 9. The control component forms the order of problem solving by requesting a link to the instance of a respective component from the factory in accordance with the logical name of the problem. When downloading the description of the system components from the software ontology, the factory finds a component, which corresponds to the current problem, and prepares data structures required to call it. To create a component instance, the factory uses the reflection

Figure 9. Scheme for integrating the computation component into the software package using the software ontology

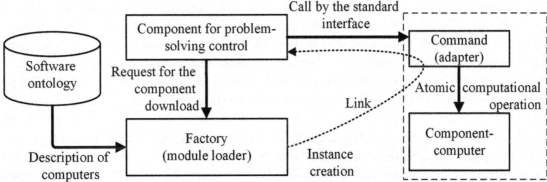

of the Java language. After the component instance is downloaded, the factory transmits the respective link to this instance to the control component, which, in turn, passes the computation parameters and calls the subprogram through a standardized interface. This subprogram serves as a starting point for the component operation.

The second aspect of applying metaprogramming is a compilation of components from the code generated in the course of program operation. The Java programming language has a number of means necessary to generate the Java source code as the operation result of the program itself, code compilation into the software components and their connection to the software system at the stage of the program execution. To this end, we use the components from package *javax.tools* added into Java SE 6 as a standard interface of applied programming (API) for the compilation of the source code in the Java language.

Metaprogramming techniques allow us to adjust the software package to the schemes with different mix of equipment (steel and metal-polymeric pipes of different diameters, pumps and fittings) and types of network construction (underground, aboveground, etc.) that are represented by various mathematical relationships. The models of the network components that are contained in the HSS ontology are used to dynamically generate software components-models for the set of equipment of the calculated HSS. Let us consider a model of a standard steel heat pipeline as an example.

The model has a number of parameters: pipe diameter, its specific weight, parameters and costs of construction, heat losses, hydraulic resistance, etc. Moreover, the model has mathematical relationships among the parameters. For instance, the value of the head loss in a steel pipe is calculated by the formula (Merenkov, 1992):

$$h_i = s_i \mid x_i \mid x_i,$$

where h_i – head loss in a section; s_i – hydraulic resistance of the pipeline; x_i – heat carrier flow rate.

Resistance s_i at the density of the heat carrier equal to $\rho = 0.975$ t/m³ is calculated by the formula (Sennova & Sidler, 1987):

$$s_i = 0.74 \times 10^{-9} \frac{(1+\alpha_i) L_i k_i^{0.25}}{d_i^{5.25}},$$

where α_i – coefficient of local losses, L_i – length, d_i – diameter, k_i – equivalent pipeline roughness.

Application of the metaprogramming techniques makes it possible to automate the process of creating the software components-models for the calculation of a concrete HSS. This approach offers new opportunities to solve the problems of mathematical modeling and optimization of HSS, because it provides an engineer with a tool to conduct studies changing only the model parameters. The approach is universal since the programmer does not have to manually create a set of components anymore.

Each type of HSS components is assigned a unique number, which also belongs to the corresponding component model stored in the HSS ontology. The developed software package uses a bi-level model of data representation (Figure 10): the first level corresponds to the abstract network model, in which each component stores the number of the respective component model, and the second level contains the software components that represent the component models. Mapping from the first level to the second one is performed on the basis of hashing mechanisms (Aho, Hopcroft & Ullman, 1983). This approach

Figure 10. Scheme of a bi-level data representation model

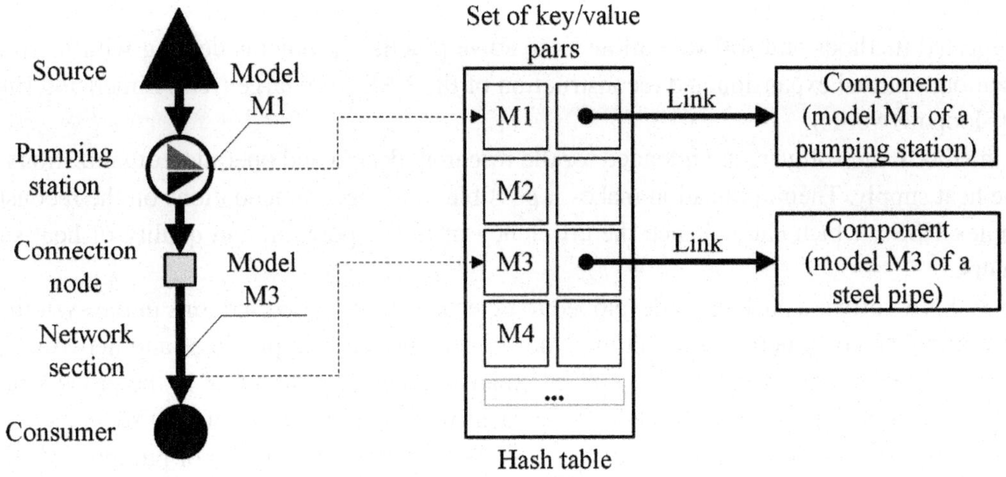

allows us to obtain the link to the component-model using the model number as a key. When forming the initial data, the user chooses the equipment operating in a certain system component (section, pumping station, fittings, etc.), and the software package for the current component environment provides a unique number of the respective model. As a result, once implemented the model can be repeatedly used to describe components of the same types.

During the software package operation the optimizer obtains the numbers of component models used in the scheme and requests software components that correspond to the given models (Figure 11) from the factory. The factory launches the Java code generator, which creates the source code for the compilation, using the component model downloaded from the HSS ontology. Then we launch the Java Compiler, which creates a software component that implements the model of the given component. Using the mechanisms of the Java virtual machine, the factory downloads the created component and returns the link to the component-model ready for making calculations to computer.

Figure 11. Scheme for integration of the software component-computer with the software component representing the network component model

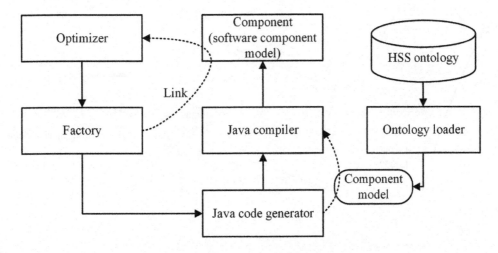

SOLUTIONS AND RECOMMENDATIONS

The suggested methods and software allow us to solve practical problems dealing with the parameter optimization, optimal expansion and reconstruction of the HSSs of a large (real) dimension (including of multi-loop networks).

The methods and software are intended for the research, design and operation organizations dealing with the heat supply. Their application makes it possible to get recommendations on the reconstruction of complex HSSs, which can enhance the efficiency of their operation and quality of heat supply to consumers.

The SOSNA software package is developed to determine overloaded sections in the system, choose the ways of their reconstruction, and optimize newly constructed heat pipelines and network facilities. It can be used for solving the problems of HSS expansion and reconstruction in the process of system operation or design. This software package helps determine the diameters of new sections of the heat network, types of reconstruction of the existing sections, sites for installation of pumping stations, and available heads of heat carrier at the sources and at the network pumping stations, test the system operability, and give an economic estimate of the suggested solution.

The SOSNA is used to conduct studies and make calculations of real HSSs in cities and towns. Figure 12 presents a schematic diagram of heat supply to urban consumers. An expected increase in heat demand makes it necessary to solve the problems of system reconstruction and expansion. Promising directions for the HSS reorganization are formed on the basis of the optimization calculations performed with the developed software package. These calculations help determine:

Figure 12. Enlarged scheme of heat supply system to a town in Irkutsk region

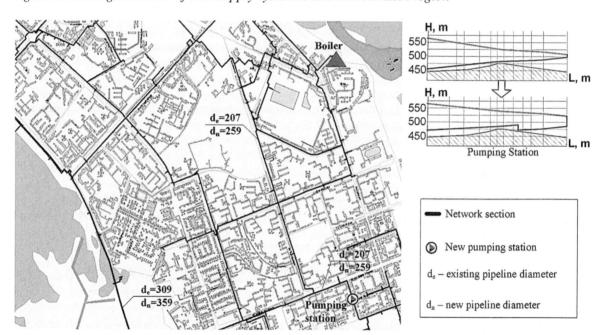

Problems of Modeling and Optimization of Heat Supply Systems

- "Bottlenecks" in the system with insufficient capacity and significant pressure losses of heat carrier;
- Reconstruction types for the overloaded sections (either the laying of heat pipes of a larger diameter or the extension of the existing pipes);
- Optimal pressure of heat carrier in supply and return mains (at sources, at nodes connecting sections and consumers);
- Sites for new pumping stations and their parameters;
- Optimal service zones for the sources in the process of their operation.

The main solutions on the HSS expansion are presented in Figure 12. Heat network sections that need reconstructing due to their insufficient capacity are given in bold. Types of their reconstruction and new pipeline diameters are determined. The selected new pumping station is designated by the corresponding sign. The suggested recommendations for the reconstruction of the HSS fully take into account its existing part and adapt the system to increased heat loads. Moreover, the recommendations will facilitate the optimal operation of the system.

FUTURE RESEARCH DIRECTIONS

Further research into the development of methods and software for the optimization of heat supply network has the following objectives:

- To suggest a mathematical statement of the parameter optimization problem of branched and multi-loop HSSs, considering variable conditions of their operation.
- To improve the existing methods and algorithms for a new problem of the parameter optimization of the branched and multi-loop HSSs, considering their changing operating conditions.
- To implement the developed methods and algorithms in software.

CONCLUSION

1. The problem of optimal HSS reconstruction and expansion is an urgent problem, since it is always necessary to solve its constituent problems in order to satisfy the increasing heat demand and meet the system modernization requirements. Modern mathematical methods and information technologies offer significantly more opportunities, and make the process of solving these problems more efficient and rational.
2. The problem of parameter optimization even of branched HSSs, not to mention the multi-loop ones, is a complex non-convex programming problem. To solve this problem we have further developed the methodological approach and suggested a new algorithm enabling us to make calculations for HSSs of any complexity with any set of nodes, sections and loops. These developments take into account the hierarchical structure of the systems and make it possible to calculate several subsystems of the unified HSS.
3. Expandable architecture of the software package helps create a flexible adaptive model of control of the computational process, while representation of the package in the form of software compo-

nents makes it universal and suitable for repeated use in different software implementations in the problems of HSS design.

4. In this paper we suggest a technique for organizing the computational process of solving the problems of HSS parameter optimization on the basis of concurrent calculations. Application of the technique significantly reduces the time of calculations, when we solve the applied problems of parameter optimization on computers. The paper considers the ways of organizing the computational process on the basis of concurrent calculations.

5. The suggested approaches underlie the software package SOSNA intended for solving the parameter optimization problems of the HSS. This software can be applied in the research, design and operation organizations dealing with the heat supply issues.

6. The developed methodology and software make it possible to generate recommendations on the reconstruction of complex HSSs, which are aimed at enhancing the efficiency of their operation and quality of heat supply to consumers.

REFERENCES

Aho, A. V., Hopcroft, J. E., & Ullman, J. D. (1983). *Data Structures and Algorithms* (1st cd.). Addison Wesley.

Appleyard, J. R. (1978). Optimal design of distribution networks. *The Building Services Engineer, 45*(11), 191-204.

Barakhtenko, E. A., Oshchepkova, T. B., Sokolov, D. V., & Stennikov, V. A. (2013). New Results in Development of Methods for Optimization of Heat Supply System Parameters and Their Software Implementation. *International Journal of Energy Optimization and Engineering, 2*(4), 80–99. doi:10.4018/ijeoe.2013100105

Barakhtenko, E. A., Sokolov, D. V., & Stennikov, V. A. (2014). New Results in Developing of Algorithms Intended for Parameters Optimization of Heat Supply Systems. In *Proceeding of International Conference Mathematical and Informational Technologies, MIT-2013* (pp. 58-66). Beograd, Serbia.

Barreau, A., & Moret-Bailly, J. (1977). Prèsentation de deux methods d'optimization de rèseaux de transport d'eau chaude a grande distans. *Entropie, 13*(75), 21–28.

Bartlett, J. (2005). *The art of metaprogramming, Part 1: Introduction to metaprogramming*. Retrieved from http://www-128.ibm.com/developerworks/linux/library/l-metaprog1.html

Bellman, R. (1957). *Dynamic Programming*. Princeton, NJ: Princeton University Press.

Brambilla, M., Cabot, J., & Wimmer, M. (2012). *Model Driven Software Engineering in Practice. Synthesis Lectures on Software Engineering #1*. Morgan & Claypool.

Cai, T. (2015). Application of Soft Computing Techniques for Renewable Energy Network Design and Optimization. In P. Vasant (Ed.), *Handbook of Research on Artificial Intelligence Techniques and Algorithms* (pp. 204–225). Hershey, PA: Information Science Reference; doi:10.4018/978-1-4666-7258-1.ch007

Caizergues, R., & Franenberg, H. (1977). Application des mèthodes d'optimisation des rèseaux de transport d'eau chaude a quelques cas concretes. *Entropie, 13*(75), 29–38.

Castillo-Villar, K. K., & Smith, N. R. (2014). Supply Chain Design Including Quality Considerations: Modeling and Solution Approaches based on Metaheuristics. In P. Vasant (Ed.), *Handbook of Research on Novel Soft Computing Intelligent Algorithms: Theory and Practical Applications* (pp. 102–140). Hershey, PA: Information Science Reference. doi:10.4018/978-1-4666-4450-2.ch004

Cunha, M., & Sousa, J. (1999). Water distribution network design optimization: Simulated annealing approach. *J. Water Res. Plan. Manage. Div. Soc. Civ. Eng.*, *125*(4), 215–221. doi:10.1061/(ASCE)0733-9496(1999)125:4(215)

Dantzig, G. B. (1963). *Linear Programming and Extensions*. Princeton, NJ: Princeton University press.

Davidson, P. L. (1934). *Methodology for basic calculations on heat networks*. Moscow: Transzheldorizdat. (in Russian)

Eckel, B. (2006). *Thinking in Java* (4th ed.). Prentice-Hall PTR.

Gamma, E., Helm, R., Johnson, R., & Vlissides, J. (1995). *Design Patterns: Elements of Reusable Object-Oriented Software*. Addison-Wesley.

Garbai, L., & Molnar, L. (1974). Optimization of urban public utility networks by discrete dynamic programming. Colloquia mathematical society János Bolyai. 12: Progress in operations research (pp. 373-390). Eger, Hungary.

Gavrilova, T. A., & Khoroshevsky, V. F. (2000). *Knowledge bases of intelligent systems*. Saint Petersburg: Piter. (in Russian)

George, A., & Liu, J. (1981). *Computer Solution of Large Sparse Positive Definite Systems*. Englewood Cliffs, NJ: Prentice Hall.

Guarino, N. (1997). *Understanding, Building, and Using Ontologies*. Retrieved April 29, 2015, from http://www.academia.edu/516503/Understanding_building_and_using_ontologies

Heindrun, C. (1968). Dimensionierung von Wärmeverteilungsnetz nach wirtschaftlichen Gesichtspunkten mit electronischen Rechenanlagen. *Energie (BRD)*, *20*(11), 356–359.

Horstmann, C. S., & Cornell, G. (2004). Core Java™ 2: Vol. II. *Advanced Features* (7th ed.). Prentice Hall PTR.

Kally, E. (1972). Computerized Planning of the Least Cost Water Distribution Network. *Water & Sewage Works*, 121–127.

Khanh, D. V. K., Vasant, P., Elamvazuthi, I., & Dieu, V. N. (2014). Optimization of thermo-electric coolers using hybrid genetic algorithm and simulated annealing. *Archives of Control Sciences*, *24*(2), 155–176. doi:10.2478/acsc-2014-0010

Khasilev, V. Ya. (1957). Issues about technical and economic calculation of heat networks. *Proektirovaniye gorodskikh teplovykh setei*. M.-L.: Gosenergoizdat. (in Russian)

Levitin, A. (2006). *Introduction to the Design & Analysis of Algorithms* (2nd ed.). Addison Wesley.

Mellor, S. J., Scott, K., Uhl, A., & Weise, D. (2004). *MDA Distilled*. Addison-Wesley.

Merenkov, A. P. (Ed.). (1992). Mathematical modeling and optimization of heat, water, oil and gas supply systems. Novosibirsk: VO "Nauka", Sibirskaya izdatelskaya firma. (in Russian)

Merenkov, A. P., & Khasilev, V. Y. (1985). *Theory of hydraulic circuits*. Moscow, Russia: Nauka.

Morgan, D., & Goulter, I. (1985). Optimal urban water distribution design. *Water Resources Research, 21*(5), 642–652. doi:10.1029/WR021i005p00642

Pissanetzky, S. (1984). *Sparse Matrix Technology*. London: Academic Press Inc.

Savic, D., & Walters, G. (1997). Genetic algorithms for least-cost design of water distribution networks. *J. Water Res. Plan. Manage. Div. Soc. Civ. Eng., 123*(2), 67–77. doi:10.1061/(ASCE)0733-9496(1997)123:2(67)

Sennova, E. V., & Sidler, V. G. (1987). *Mathematical modeling and optimization of developing heat supply systems*. Novosibirsk: Nauka. (in Russian)

Shifrinson, B. L. (1940). *Main calculation of heat networks*. Gosenergoizdat. (in Russian)

Sokolov, D. V., Stennikov, V. A., Oshepkova, T. B., & Barakhtenko, E. A. (2012). The New Generation of the Software system Used for the Schematic–Parametric Optimization of Multiple-Circuit Heat Supply Systems. *Thermal Engineering, 59*(4), 337–343. doi:10.1134/S0040601512040106

Stennikov, V. A., Barakhtenko, E. A., & Sokolov, D. V. (2011). Metaprogramming in the software for solving the problems of heat supply system schematic and parametric optimization. *Programmnaya Inzheneriya, 6*, 31-35. (in Russian)

Stennikov, V. A., Barakhtenko, E. A., & Sokolov, D. V. (2011). Methods of integrated development and reconstruction of heat supply systems on the basis of modern information technologies. *Promyshlennaya Energetika, 4*, 17-22. (in Russian)

Sumarokov, S. V. (1976). A method of solving the multiextremal network problem. *Ekonomika i Matematicheskiye Metody, 12*(5), 1016-1018. (in Russian)

Tewarson, R. P. (1973). *Sparse Matrices (Part of the Mathematics in Science & Engineering series)*. New York: Academic Press Inc.

Todini, E., & Pilati, S. (1988). A gradient algorithm for the analysis of pipe networks. In *Computer Applications in Water Supply* (Vol. 1, pp. 1–20). London: John Wiley & Sons.

Vasant, P., Barsoum, N., & Webb, J. (2012). *Innovation in Power, Control, and Optimization: Emerging Energy Technologies*. Hershey, PA: IGI Global; doi:10.4018/978-1-61350-138-2

Vasant, P. M. (2013). *Meta-Heuristics Optimization Algorithms in Engineering, Business, Economics, and Finance*. Hershey, PA: IGI Global; doi:10.4018/978-1-4666-2086-5

Völter, M., Stahl, T., Bettin, J., Haase, A., & Helsen, S. (2006). *Model-Driven Software Development: Technology, Engineering, Management*. Wiley.

Wenzhong, X., & Jiayou, L. (2009). Genetic Algorithm's Optimization for Parameters of Ring-Shaped Heat-Supply Network with Multi-Heat Sources. *Proceedings of the 2009 International Conference on Energy and Environment Technology (ICEET '09)* (Vol. 1, pp. 176-180). doi:10.1109/ICEET.2009.49

Yakimov, L. K. (1931). Maximum operating radius of cogeneration-based district heating. *Teplo i Sila, 9*, 8-10. (in Russian)

Zanfirov, A. M. (1933). Technical economic calculation of water and heat networks. *Teplo i Sila, 11*, 4-10. (in Russian)

Zanfirov, A. M. (1934). On economic calculation of heat networks. *Teplo i Sila, 12*, 38-39. (in Russian)

KEY TERMS AND DEFINITIONS

Dynamic Programming: An optimization approach that transforms a complex problem into a sequence of simpler problems; its essential characteristic is the multistage nature of the optimization procedure.

Heat Supply System: A system for distributing heat generated in a centralized location for residential and commercial heating requirements such as space heating and water heating.

Metaprogramming: A writing of computer programs with the ability to treat programs as their data. It means that a program could be designed to read, generate, analyze and/or transform other programs, and even modify itself while running.

Model-Driven Engineering: A software development methodology which focuses on creating and exploiting domain models, which are conceptual models of all the topics related to a specific problem. Hence it highlights and aims at abstract representations of the knowledge and activities that govern a particular application domain, rather than the computing concepts.

Multi-Loop Optimization Method: A method devised in the framework of the theory of hydraulic circuits by S.V. Sumarokov. The method makes it possible to simultaneously determine the optimal diameters of pipelines in a multi-loop system, identify the sites for the installation and parameters of pumping and throttle stations, find optimal flow distribution in the system, and solve the problem of its existing part reconstruction.

Ontology: In computer science and information science, an ontology is a formal naming and definition of the types, properties, and interrelationships of the entities that really or fundamentally exist for a particular domain of discourse.

Theory of Hydraulic Circuits: Scientific and technical discipline providing the unique language and methodology for solving the problems of modeling, calculation, identification, and optimization of pipeline and hydraulic systems of different type and purposes.

Chapter 5
Problems of Modeling and Optimization of Heat Supply Systems:
Methodological Support for a Comprehensive Analysis of Fuel and Heat Supply Reliability

Valery A. Stennikov
Energy Systems Institute, Siberian Branch of the Russian Academy of Sciences, Russia

Ivan V. Postnikov
Energy Systems Institute, Siberian Branch of the Russian Academy of Sciences, Russia

ABSTRACT

This chapter deals with the problem of comprehensive analysis of heat supply reliability for consumers. It implies a quantitative assessment of the impact of all stages of heat energy production and distribution on heat supply reliability for each consumer of the heat supply system. A short review of existing methods for the analysis of fuel and heat supply reliability is presented that substantiates the key approaches to solving the problem of comprehensive analysis of heat supply reliability. A methodological approach is suggested, in which mathematical models and methods for nodal evaluation of heat supply reliability for consumers are developed and the studies on the impact of different elements of fuel and heat supply systems on its level are described. Mathematical modeling is based on the Markov random processes, models of flow distribution in a heat network, deterministic dependences of thermal processes of heat energy consumption and some other models.

INTRODUCTION

Heat supply is the most important component in support of vital activity of population and development of virtually all economic branches. High socio-economic significance of the heat supply sphere imposes heavy demands to reliability and economic efficiency of heat supply systems (HSSs) that combine heat

DOI: 10.4018/978-1-4666-9755-3.ch005

supply sources (HSs) and heat networks (HNs) in the unified structure. During the long period of formation and development these systems have acquired complex structure and large sizes. Such specific features of current HSSs as a multitude of diverse heat sources, large length of heat networks (hundreds of kilometers), combined into complex multi-loop schemes, numerous heat energy consumers, multi-level structure of heat supply system management all stipulate complexity of the problems to be studied, provision and improvement of their reliability. These problems aim primarily at determining the reliability level of heat supply to consumers and considering the normative requirements to the values of reliability indices at system designing and operation. Increasing scales of HSSs and the corresponding complication of their structure caused by the growing number of consumers and their loads necessitated creation of effective methods for assessment of their reliability and measures on reliable heat supply. The situation is aggravated by disunity in HSS management structure, technical imperfection of the systems, essential wear of their pipelines and equipment, low operation technologies of some system components, thermohydraulic misadjustment of heat networks and consumer installations. Moreover, the methodological base on standardization of heat supply reliability should be developed and improved.

The first studies on HSS reliability by the methods of reliability theory were carried out at the late 1960s – early 1970s (Khasilev & Takaishvili, 1972; Sennova, 2000). Initially improvement of HSS reliability was based only on enhancement of the quality of manufacturing the elements and structures and the corresponding theoretical approaches did not take into account many specific features in operation of these systems. The reliable elemental base, however, has not been created and development of large centralized systems became the main trend in the sphere of heat supply. In these conditions many specialists considered redundancy of heat network schemes as one of the basic methods for reliability control. Later on a great number of studies on reliability of HSS and its elements have been performed, the techniques and models for solving the problems of reliability analysis for heat sources and heat networks have been worked out that characterize their different specific features, the algorithm of constructing HSS with the required reliability level has been developed, the system of standards needed for its solving has been substantiated (Sennova, 2000).

HSS operation starts with fuel provision for sources. Hence, fuel supply reliability is inseparably linked with reliability and quality of heat output to consumers and consideration of fuel component is an integral element in the system approach to solving the problems of analysis and synthesis of heat supply reliability alongside with the reliability of heat production and transportation.

Application of all methodological developments in the area of heat supply reliability depends to a great extent on available information on the reliability indices of heat and fuel supply system elements and classification of information on fuel use conditions depending on different external conditions. The long-term period of HSS operation made it possible to accumulate a deal of statistical data on failures and restorations of HSS elements and as a result of their processing the reliability characteristics of these elements, in particular normative ones, were obtained. However, new systems are designed and existing ones are reconstructed based on introduction of the up-to-date energy saving equipment, whose reliability characteristics can not be objectively estimated because of insufficient experience in its operation. Therefore, the necessity arises to create systems for automatic information collection from the objects of the whole process of heat energy production and distribution.

Recently the works on heat supply reliability have become more topical due to a high extent of HSS equipment tear and wear (for example, capital equipment of many cogeneration plants in Russia was installed more than 50 years ago, wear of heat networks on the average for the country reaches 60%), creation of new management forms for the heat economy of cities with participation of numerous owners in the unified system and also future trends in HSS development on the basis of distributed generation principles, which requires new approaches to the reliability analysis, standardization and provision for systems under reconstruction.

BACKGROUND

Specific Features of Fuel Supply Processes and their Modeling

A group of nodes for fuel production, its storage and transportation to heat sources is a fuel supply system (FSS). The fuel supply systems differ in fuel kind, structure and scale. The aggregate process flow of fuel supply consists of a set of fuel production nodes (mines, wells), a system of fuel transportation to processing enterprises (dressing works, oil and gas processing plants), a system of secondary transportation of finished products (public transport, product pipelines), objects for fuel storage (coal bases and warehouses, petroleum storage depots, underground gas storage tanks) and a system of fuel distribution among consumers (heat sources and other enterprises).

Reliability assessment and maintenance of these systems are complex problems, whose solving is complicated by the distributed structure of objects operating under the influence of numerous internal and external factors (including random) and by insufficient information on reliability of system elements even more than heat sources and heat networks. The problem of reliable fuel supply to heat sources during the heating season is also caused by the irregular heat loads of random character because of changes in meteorological conditions. Since fuel supply to heat sources includes different stages of fuel production, transportation and storage, its reliability should be analyzed integrally, i.e. in terms of interaction of all elements involved in this process.

The main geophysical index that influences the change in fuel demand for heat supply is fluctuation of outdoor temperature; changes in wind loads and amount of solar radiation are less significant factors (Nekrasov & Velikanov, 1986; Nekrasov, Velikanov & Gorunov, 1990). The technologies are conditioned by the character of production processes of enterprises that produce, transport and consume fuel: fuel production technologies, fuel grade and methods of its combustion, interchangeability of fuels among heat sources within one system, possibility for fuel warehousing, capacity of transport systems, etc. Besides there is a class of factors involving organizational measures that influence operation conditions of fuel producing and processing enterprises, changes in the plan targets, violations in the rates of fuel and energy consumption, plans of provision with transport facilities, etc.

All the indicated factors affect the conditions of fuel supply and consumption in a complex way, leading thus to their regular and random fluctuations in the daily, seasonal and long-term periods. Irregular conditions between production and consumption of some kinds of fuel constantly mismatch in time (Mazur, 1986; Zorkaltsev & Ivanova, 1989; Zorkaltsev & Ivanova, 1990).

Reliability of fuel supply to heat sources is characterized by the relationship between fuel demands and supplies for a certain time period in terms of available reserves. The problem of reliable fuel provision is solved first of all by means of fuel stock management at different technological and time levels on the basis of fuel interchangeability among its consumers as well as creation of standby capacities for fuel production and processing.

The existing methods for the reliability analysis of fuel supply to heat sources are, as a rule, intended to determine aggregate indices of fuel provision for the regional energy sector or its individual nodes. The problem of reliable fuel supply to heat sources is solved primarily for reliability assessment, since the shortages determining demands for reserves and stocks are calculated at the considered stage. Because of practical complexity to construct the calculated FSS schemes in terms of reliability characteristics of its elements the method of statistical tests that does not require information about failures and restorations of elements is the most sound method for the analysis of fuel supply reliability. The method of statistical tests (Monte Carlo method) is widely used to model different systems for both their reliability assessment and for solving other problems (Claudio & Rocco, 2003; Naess, Leira & Batsevych, 2009).

The technique for fuel supply reliability analysis that is based on the model of statistical tests is suggested in (Zorkaltsev & Kolobov, 1984). The simulation model serves to study fuel supply reliability for heat-generating units considering the effect of the following factors on reliability of fuel provision: the volumes of seasonal fuel stocks and capacity of warehouses, volumes of fuel supplies planned for a year and validity of these supplies, etc. However, the model does not take into account irregularities of fuel supply within a year. Fuel demand, resources, supplies, demand and shortage are given as the annual indices. The volumes of assigned fuel resources and annual demand are described as random quantities varying about some target level with the specified probability densities of their implementation. The algorithm is constructed on the basis of multi-variant procedures of generating these random quantities that simulate a real process of fuel supply, creation and use of fuel stocks in the long run. The rate of occurrence and the average volume of fuel shortage and surplus, probabilities of shortage occurrence and the expectation of fuel shortage are determined as a result of calculations. The studies in the presented work allow the estimation of the required fuel stocks to maintain the given level of fuel supply reliability for the entire energy system in less detail. In order to coordinate these solutions with the results of joint reliability assessment for heat sources and heat networks, to determine their total impact on heat supply to consumers the suggested methods call for further development.

Methods for the Analysis of Heat Source Reliability

There exist different methods for the analysis of heat source reliability that can be divided into analytical ones based on the Markov (Gnedenko, Belyaev & Soloviev, 1965; Haghifam & Manbachi, 2011; Lisnianski, Elmakias, Laredo & Hanoch, 2012; Polovko, 1964; Sennova, 2000; Ushakov, 1985), semi-Markov models (Korolyuk & Turbin, 1976; Rudenko, Ushakov & Cherkesov, 1984; Woo & Nam, 1991), logical-and-probabilistic methods and methods of statistical modeling (Claudio & Rocco, 2003; Naess, Leira & Batsevych, 2009).

The methods of the first group treat heat sources as a combination of functionally connected elements integrated into a scheme for reliability calculation. Application of the Markov model for the reliability analysis of complex heat source schemes is conditioned by a high dimension of systems of equations of a random process, whose solving is not difficult on the current computers. However, in some cases the decomposition method that is known from the theory of technical system reliability is used to reduce the volume of calculations and construct more compact models (Gnedenko, Belyaev & Soloviev, 1965). In particular, according to the technique for heat source reliability analysis in (Ushakov, 1985) the initial scheme to be calculated is divided into individual and independent subsystems (blocks of elements or branches) and the reliability indices are calculated based on some specified algorithm for each subsystem.

Then the reliability indices of an entire heat source are determined on the basis of simple dependences in terms of series connection of subsystems. Such an approach is applicable to the object, whose elements are connected by the technologically common process and considerably simplifies the calculation, though the assumption on independence of subsystems introduces some error.

Application of the semi-Markov processes allows the description of heat source operation that is more close to real conditions. Their main difference from the Markov models lies in failure formulation. The latter for the models of the semi-Markov processes is understood as an event that involves not only a decreased volume of heat energy output by a heat source below the required one and its stay in such a condition for the time exceeding the time reserve that is needed basically for heat accumulation, hot water reserves and other factors. The authors of the approach (Smirnov, 1990) call it functionally technological. It suggests joint application of the probabilistic models of functioning and deterministic models of thermophysical processes, which essentially complicates calculations, but results in a more accurate reliability characteristic.

The alternative methods for reliability assessment of heat sources are based on the application of simulation mathematical algorithms. Some similar methods are suggested in (Buslenko, 1978; Buslenko, Kalashnikov & Kovalenko, 1983). The method of statistical tests that is extensively used in the studies on energy system reliability results in sufficiently objective evaluations of heat source operation reliability. It requires, however, a sizable volume of preliminary works on initial information acquisition and preparation.

Methods for the Analysis of Heat Network Reliability

The existing heat networks of large centralized heat supply systems are complex multi-loop structures distributed over the city territory that include elements of a linear part (pipelines) and control elements (pumping stations, stop and control valves), consumer facilities. The methods of their reliability analysis are based on the combination of hydraulic modeling and probabilistic methods for description of their state in time.

The general principles of calculating reliability and redundancy of heat networks were formulated in 1972 in (Khasilev & Takaishvili, 1972). The approach that is based on the evaluation of nodal indices was developed later at Siberian Energy Institute (SEI), now it is Energy Systems Institute (ESI) of Siberian Branch of RAS (Merenkov & Khasilev, 1985; Sennova & Sidler, 1985; Sennova, 2000). The general concepts of the suggested approach stem from the network characteristics and particularities of heat network operation. The technique for heat network reliability assessment is based on the combination of deterministic methods for the analysis of normal and emergency conditions for heat networks with probabilistic methods for calculating heat supply reliability indices for each consumer (nodal reliability indices).

The key principles of the method are due to the specific properties of heat networks determined by their purpose, structure and operation conditions. Failure in operation is formulated from the consumer standpoint as decrease in air temperature in buildings below the boundary (minimum admissible) value for the set reliability level. In accordance with some reliability levels there are also several levels of heat supply to consumers: normative and lower (standby). The lower level should be supported in the system during elimination of emergency situations arising in the redundant network part. It is characterized by a lower amount of heat supplied to consumers as against the normative amount that is called a standard of emergency supply. The problem of heat network reliability analysis is to determine nodal indices that evaluate the considered heat supply reliability levels based on the reliability characteristics of network elements.

Along with the nodal approach to heat network reliability assessment there is another concept that is based on application of the integral reliability index (Ionin, 1978; Ionin, 1989). The integral index characterizes the extent of solving the problem of reliable heat supply during the heating season. However, it is not interpreted from the viewpoint of quality of performing the functions by heat networks with respect to consumers. This fact does not allow the use of reliability analysis results for separation of vulnerabilities and the most severe emergencies leading to long-run tripping of consumers. Hence, only the nodal assessment can serve as the base for taking further decisions on heat network reconstruction to improve reliability of heat supply to consumers. The methods and models suggested below also apply the nodal approach to heat supply reliability assessment.

A TECHNIQUE FOR COMPREHENSIVE ANALYSIS OF CONSUMER HEAT SUPPLY RELIABILITY

Technological Structure of the Heat Supply Complex (HSC) and Starting Points in its Reliability Analysis

In the practice of designing and operation the heat supply system (HSS) reliability is assessed separately at the level of a heat source and a heat network. Such separation of problems leads to independent results that do not take into consideration system properties of heat supply systems and their mutual effect on heat supply reliability. They in turn most significantly manifest themselves, when several heat sources operate in one heat network. The estimates of fuel shortages that are obtained in the analysis of heat supply system reliability can not quantitatively evaluate the influence of fuel supply violations on the resulting reliability of heat supply. The presented work suggests approaches that take into account connections between these systems from the viewpoint of their joint impact on reliable provision of consumers with heat energy for the calculated period.

Connectivity in technology, regularity and mutual effect of internal and external processes of fuel production and supply, heat energy production, transportation and consumption make it necessary to comprehensively consider reliability of these processes. This implies joint modeling, calculation and analysis of the whole process flow of heat energy production and distribution. Aggregation of the corresponding processes allows their representation as two relatively independent systems: a fuel supply system (FSS) consisting of subsystems of fuel production (FPS) and transportation (FTS) and a heat supply system (HSS) including heat sources (HSs) and heat networks (HNs). In total they form a heat supply complex (HSC), whose simplified structure is shown in Figure 1.

The main task of the comprehensive analysis of reliable heat supply to consumers (or HSC reliability) is to obtain a quantitative estimation of the integral effect of all HSC elements on reliability of heat supply to consumers and determine an extent of influence of each of them. Its solving is a methodologically complex problem because of the necessity to have sound reliability assessments and support many auxiliary problems of reliability analysis that accumulate and process information about the properties of system elements, estimate the impact of external factors on the performance indices for HSC sub-

Figure 1. Technological structure of HSC

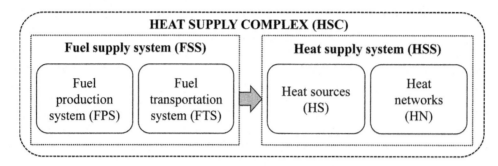

systems, develop normative requirements and others. Many of these particular cases are the subject of independent studies. However, the described below methodological concepts form the basis for their statement and solving.

The network specific character of heat supply systems and the necessity to individually consider each consumer stipulate application of the following basic principles for heat supply reliability calculation (Khasilev & Takaishvili, 1972; Sennova, 2000):

- reliability assessment on the basis of nodal reliability indices;
- consideration of two or more levels in heat supply reliability provision – normal (normative) and emergency (lower);
- combination of probabilistic methods for reliability calculation with deterministic methods for the analysis of operation conditions of HSC elements, in particular emergency thermohydraulic conditions in heat networks (Merenkov & Khasilev, 1985; Sennova & Sidler, 1985).

The schemes of fuel transportation and heat energy distribution in heat supply systems have a network structure, which makes it possible to apply the methods of graph theory and "redundant" design diagrams to solve the problems of HSC reliability (Sennova & Sidler, 1985). They allow representation of the HSC process flow as a common network graph taking account of individual features of all its elements.

An Algorithm for Assessment of HSC Reliability

The general algorithm of comprehensive analysis of HSC reliability is illustrated in an aggregate way in Figure 2. Its computational scheme is presented by three levels – at the level of FSS, HSS and the whole HSC (Voropai, 2014). The set of problems considered can be divided into two subsets – probabilistic modeling of HSC operation and analysis (calculation) of operating conditions of its subsystems.

The topological base for solving the first subset of problems is the functional scheme for HSC calculation that is formed by interconnection of the schemes of all its subsystems that are reduced to the uniform representation of the structure and links of elements. The analysis of technologically possible events in the HSC subsystems allows formation of a set of HSC states as a combination of its subsystem states. This set can also take into consideration a family of external factors (external disturbances). This, however, invites preliminary study and statistical analysis of their effect on the performance indices of the considered objects. The probabilities of HSC states are estimated on the basis of the specified characteristics of element reliability (rates of failures and restorations) on the Markov model.

Figure 2. General algorithm of HSC reliability assessment

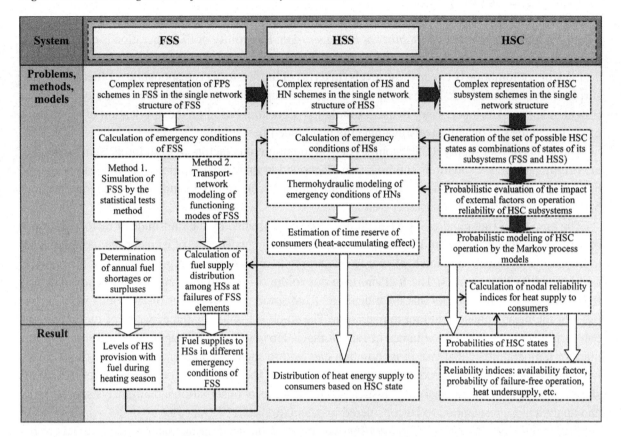

The methods for analysis of emergency conditions of the HSC subsystems involve their technological features and differ in mathematical models. The fuel supply reliability can be assessed by one of the two suggested methods – structural modeling of fuel transportation flows and the method of statistical tests. The latter method does not require an elemental scheme of the fuel supply system to be applied, therefore the Markov model of HSC can be comprehensively represented and constructed combined with heat sources and heat networks. The levels of heat source provision with fuel (shortage or surplus) that were calculated at the stage of fuel supply system modeling determine their performance for the considered period (heating season). These indices are of independent character for the analysis of fuel supply system reliability and can be applied for modeling of the flow distribution in heat supply systems subject to possible failures of equipment at heat sources and in heat networks. And the levels of emergency heat energy supply to consumers at all the states of HSC including violations in operation of all its subsystems are estimated by the results of modeling.

The probabilities of HSC states and the levels of heat energy supply to consumers at these states are used to calculate the nodal reliability indices that characterize reliability of the whole HSC operation (see Figure 1) for each consumer. When determining these indices account is also taken of time redundancy in the system because of heat storage capacities of buildings.

Simulation Modeling of FSS

Specific features of studies on reliability of FSS operation. Reliability of FSS operation is determined by the relationships between the levels of fuel demand and supply for the considered time period (Section 2). Lack of coordination between these levels (in terms of buffer stock) is responsible for fuel shortage or surplus (accumulated stock). The mentioned essential complexity in obtaining the reliability indices for FSS elements makes it necessary to use a simulation algorithm based on the application of the method of statistical tests to quantitatively evaluate the guaranteed fuel supply.

The levels of fuel supply to heat sources depend on numerous factors of both intrasystem and external origin: emergency situations at production places and in fuel transportation lines that can be caused by both equipment failures and the environmental impact, economic risks induced by the interrelations between fuel suppliers and consumers, etc.

The amount of fuel required by heat sources is also determined by the multitude of factors, the main of them being change in the outdoor temperature. The off-design decrease of its mean annual value leads to increase in thermal loads and heat energy consumption and correspondingly to increasing amount of fuel needed for heat sources. The fuel shortage due to the off-design temperature fall is covered from the annual and current stocks that are calculated from some indications for the previous years on the basis of the mean value of outdoor temperature. The stock is limited by the capacity of fuel producing enterprises and the available volumes of fuel storages. However, it can happen so that decrease in the outdoor temperature or increase in its duration can lead to such a situation, at which the fuel demand will exceed the current fuel production volumes and stocks. The probability of occurrence of like events should be estimated on the basis of statistical data of long-term observations, their analysis, processing and analytical representation to be considered in calculations.

Fuel supply modeling for FSS reliability assessment. Construction of mathematical models to describe conditions of fuel supply to and consumption at heat sources and to take into account the effect of the enumerated and other factors on them seems to be a rather complex problem. At the same time the statistical processing of data on changes in fuel supply and demand that depend on the impact of the most significant disturbances allows the distributions of random quantities for the levels of fuel supply and demand for each heat source in the system to be determined. The obtained distributions are used for simulation modeling of FSS operation based on the method of statistical tests. This approach takes into consideration specific features of the system at numerous external disturbances influencing it and compensates for the lack of information about the reliability parameters of its elements. Positive experience of using this method for FSS is confirmed by some studies carried out on this subject. The work (Zorkaltsev & Kolobov, 1984), for example, presents a simulation model of fuel supply that assesses the FSS operation reliability by using the annual indices.

The suggested model for assessment of FSS reliability is intended to determine fuel demand and supply for the specified typical time intervals of the heating season (or distribution of fuel shortages or surpluses by interval of this period). Hence, it becomes possible to examine the heat supply complex states at different fuel supply conditions.

The levels of fuel supply and demand for some time interval n can be represented by quantities a_n and b_n, respectively. Continuous or discrete distributions of these quantities are given as initial data (the sums are used in the case of discrete distribution):

$$\int_{a_n^{\min}}^{a_n^{\max}} f(a_n)da_n = 1, \quad \int_{b_n^{\min}}^{b_n^{\max}} f(b_n)db_n = 1; \qquad (1)$$

$$f(a_n) = 0, \; f(b_n) = 0, \text{ if } a_n \notin [a_n^{\min}, a_n^{\max}], \; b_n \notin [b_n^{\min}, b_n^{\max}]; \qquad (2)$$

where $f(a_n)$ and $f(b_n)$ – densities of distribution of random quantities a_n and b_n, $a_n^{\min}, a_n^{\max}$ and $b_n^{\min}, b_n^{\max}$ – ranges of their change.

The calculated time period τ_0 is taken, as a rule, equal to the heating season. It is divided into N time intervals with duration τ_n, $n = \overline{1, N}$. The total number of heat sources in HSC corresponds to set I. Fuel substitution among the heat sources is possible by compensation for fuel shortages at some heat sources by surpluses at the others both in practice and modeling. The sources operating on substituting fuels are combined into groups I_g, where g – index denoting the group membership.

An algorithm for FSS reliability assessment. An algorithm for evaluation of heat source provision with fuel (the FSS reliability analysis) is presented schematically in Figure 3 and consists of three basic stages.

Stage 1: Preparation of initial information. For each interval n we determine:
- An average fuel demand of the i-th source $\overline{a}_n^i$, $i = \overline{1, I}$ that is calculated based on the average outdoor temperature for the n-th interval in accordance with the source capacity;
- An average volume of fuel supply to the i-th source $\overline{b}_n^i$, $i = \overline{1, I}$ that is assumed from the retrospective data about the heat source operation for a long term;
- Distributions of random quantities of fuel demand and supply a_n^i and b_n^i, and also the ranges of their change $a_n^i \in [a_n^{i\min}, a_n^{i\max}]$, $b_n^i \in [b_n^{i\min}, b_n^{i\max}]$ that satisfy conditions (1) and (2);
- Initial (seasonal) fuel stock (c_1^i) for each source i in interval $n = 1$. (To represent real conditions more precisely account may be taken of the volumes of current stocks ($c_{(\text{cur})n}^i$) as a standardized surplus with respect to the average volume of fuel supply to the heat source).

Stage 2: Determination of fuel demand and supply to the heat source in each interval n of the studied period. The levels of fuel demand and supply a_n^i and b_n^i will be obtained by the special model of statistical tests (the random number generator) for each i-th heat source in interval n.

Stage 3: Distribution of fuel flows in FSS. A sequence of calculation procedures that involves management of fuel reserves in the system in terms of potential fuel substitution among heat sources is applied within the set time interval ($n = \overline{1, N}$) and for each source i.

For each considered time interval n the following sequence of logic procedures and calculations is performed.

The relationship between the volumes of fuel supply and demand is tested. Here two cases are possible: either $a_n \leq b_n$ (a) or $a_n > b_n$ (b).

Figure 3. Algorithm for FSS reliability assessment

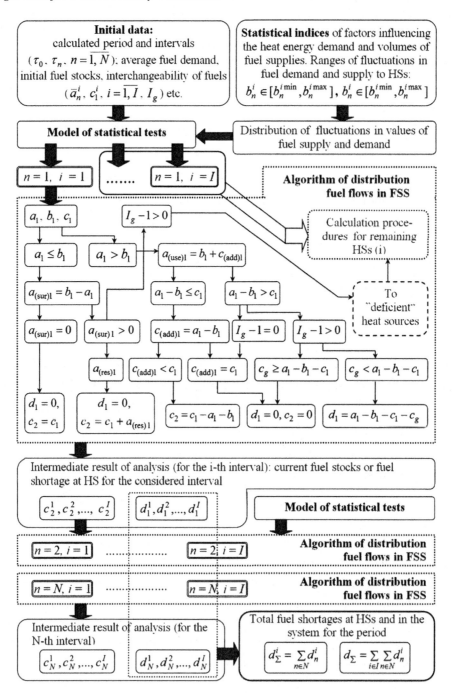

In the first case (a) the quantity of consumed fuel is equal to its demand $a_{(use)n} = a_n$, and its surplus is equal to $a_{(sur)n} = b_n - a_n$ and delivered to fuel-deficient sources $i \in I_g$. When it is used incompletely, the formed residue $a_{(res)n} = a_{(sur)n} - \sum_{i \in I_g}(a_n^i - b_n^i - c_n^i)$ is taken as a fuel stock for the following time interval (c_{n+1}). In the second case (b) the used fuel includes its supplied volume and additional volume from stocks $a_{(\text{use})n} = b_n + c_{(\text{add})n}$. If $a_n - b_n \leq c_n$, then $c_{(\text{add})n} = a_n - b_n$. When the stocks are used fully, $c_{n+1} = 0$. If $a_n - b_n > c_n$, two variants for use of a substituting fuel in the system are possible. This fuel is determined by the expression $c_g = \sum_{i \in I_g}(a_n^i - b_n^i - c_n^i)$. Provided that $c_g \geq a_n - b_n - c_n$, the used reserves of a substituting fuel cover shortage completely ($c_{(\text{add})n} \leq c_g$); if $c_g < a_n - b_n - c_n$, then $c_{(\text{add})n} = c_g$ and the heat source suffers fuel shortage $d_n = a_n - a_{(\text{use})n} = a_n - b_n - c_n - c_g$. When a group of sources I_g includes only the considered heat source, $d_n = a_n - b_n - c_n$.

The presented calculations are made sequentially for each heat source i, resulting in the quantities of fuel shortages and stocks for the subsequent interval $n+1$, for which the above calculation is made by the algorithm of stage 3 on the basis of data obtained by simulation modeling (stage 2). The last step corresponds to interval $n = N$.

As a result N pairs of quantities (shortages and stocks) are formed for each heat source. The average total estimate of heat energy undersupply can be obtained from the values of annual shortage (the expectation of shortage) for each heat source and for the whole system:

$$d_\Sigma^i = \sum_{n \in N} d_n^i, \qquad (3)$$

$$d_\Sigma = \sum_{i \in I} \sum_{n \in N} d_n^i. \qquad (4)$$

In the case of off-design fuel supply to heat sources the obtained values of fuel shortages are used to determine heat undersupply to consumers due to FSS malfunctions (see Figure 3).

Probabilistic Modeling of HSS Operation

From the standpoint of probabilistic description the heat supply system (HSS) operation is characterized by a sequence of events: failures and restorations occurring at a certain rate in all processes of heat energy production and distribution. This sequence is described by using the model of the Markov random process that is a most valid and universal tool to assess reliability of the systems to be restored (Ushakov, 1985).

A set of HSS states is generated from the modeling of possible combinations of states for heat sources and heat networks in the single structure of events (Stennikov & Postnikov, 2008, 2010, 2011, 2014). The stated goals and expected results of calculations determine the application of the stationary or non-stationary model of HSS operation. Reliability of system elements is assumed to remain invariable during the studied period (heating season), since the equipment aging and wear have a crucial importance in the longer run and are estimated at the level of constructing the model of data. The stationary Markov model of HSS operation is represented by the following system of linear algebraic equations (Kolmogorov's equations for the stationary condition):

$$p_s \sum_{z \in E_1} \nu_{sz} = \sum_{z \in E_2} p_z \nu_{zs}, \; s \in E, \tag{5}$$

$$\sum_{s \in E} p_s = 1, \tag{6}$$

where s and z – numbers of HSS state; p_s and p_z – probabilities of HSS states; E – all possible system states; E_1 – subset of system states, into which it can directly pass from state s; E_2 – subset of system states, from which it can directly pass into state s; ν_{sz} – system transition rates from state s into z, where $z \in E_1$; ν_{zs} – transition rates into state s from states z, where $z \in E_2$.

The condition of stationarity of reliability characteristics for elements and probabilities of the HSC states is adequate enough to real properties of the considered subsystems. Major changes in consequences of emergency situations and their probabilities because of variations in climatic conditions are taken into consideration when calculating reliability indices on the basis of heat load curves during the studied period. At the same time there are problems of reliability analysis and synthesis to consider changes in probabilities of the HSC states for a certain time period. They can be solved by the Markov model that describes the HSC operation with variable probabilities of states on the basis of Kolmogorov's equations for the non-stationary condition:

$$p_s(t) \sum_{z \in E_1} \nu_{sz} = p_s'(t) + \sum_{z \in E_2} p_z(t) \nu_{zs}, \; s \in E, \tag{7}$$

$$p_0(0) = p_0, \tag{8}$$

where $p_s(t)$ and $p_z(t)$ – probabilities of states as a function of time, p_0 – probability of state $s = 0$ at time instant $t = 0$ that is specified as an initial condition for solving the system of differential equations (7).

Conformity of the results of stationary and non-stationary modeling is achieved at the time instant, when the system starts to operate with constant reliability characteristics of elements and other factors influencing reliability of subsystems. Therefore, the calculations are made basically on the stationary model.

Modeling of Post-Emergency Conditions of HSC

Modeling of HSC post-emergency conditions is to determine the level of heat supply to consumers at each possible state of HSC. This process proceeds sequentially for FSS and then for HSS with the corresponding agreement of results (see Figure 2).

The levels of heat source provision with fuel that are calculated at the stage of FSS simulation modeling make it possible to determine capacities of heat sources at different system states. The level of heat carrier supply to each consumer j at different states s of HSS is determined by the model for calculation of hydraulic conditions of a heat network (the flow distribution model) based on the methods of hydraulic circuit theory.

The nodal model of flow distribution (hydraulic conditions) in heat networks that is represented in the matrix form is written as (Merenkov & Khasilev, 1985; Sennova & Sidler, 1985):

$$A_s x = g_{sj},\qquad(9)$$

$$\overline{A_s}^{\mathrm{B}} p = h - H,\qquad(10)$$

$$SXx = h,\qquad(11)$$

where A_s – matrix of coupling for linearly independent nodes for state s, x – vector of heat carrier flow rates on the network sections; g_{sj} – vector of heat carrier flow rates at nodes of the calculated scheme for state s; $\overline{A_s}^{\mathrm{B}}$ – transposed matrix $\overline{A_s}$ ($\overline{A_s}$ – complete matrix of coupling of nodes and branches of the network scheme); p – complete vector of pressures at the network nodes; h, H – vectors of pressure head loss and effective heads in the branches; S, X – diagonal matrices of hydraulic resistances of branches that are generated from the values of hydraulic resistances of branches and the absolute values of flow rates in them.

As a result of calculation of the whole system (9)-(11) the heat carrier flow rates g_{sj} are determined at each consumption node j for each HSC state s, and they are used to calculate the heat energy supply q_{sj} by the following relation (Sokolov, 1999):

$$q_{sj} = g_{sj} C(t_{\mathrm{sup}} - t_{\mathrm{ret}}),\qquad(12)$$

where C – thermal capacity of the heat carrier; t_{sup} and t_{ret} – temperatures of the heating water in the supply and return pipelines, respectively.

Determination of Nodal Reliability Indices

The results of probabilistic modeling of the HSS operation (probabilities of system states) and the levels of heat supply to consumers that were calculated at the previous stage are applied for obtaining the final assessment of the HSC reliability on the basis of nodal reliability indices of heat supply to each consumer.

In accordance with the above methodological concepts the reliability of the normative (first) level is determined by the availability factor (K_j). It corresponds to the period in the heating season, during which the normative value of the indoor temperature is maintained for a specified consumer. The reliability of the lower (second) level is evaluated by the probability index of failure-free operation (R_j), i.e. the probability that during the heating season the indoor temperature in buildings will not fall below some boundary value $t_{j\min}$ (Stennikov, Postnikov & Sennova, 2009).

The availability factor and the probability of failure-free operation are determined for each consumer j in terms of their time reserve specified by the heat storage capacity of buildings by the following formulas:

$$K_j = 1 - \sum_{s \in E} p_s \left[\frac{1}{1-\omega} \left(1 - \frac{1}{q_{hnj}} \left(\varphi_j t_{sj} - \varphi_j \left(\frac{C_1 - C_2 \exp B}{C_3(1 - \exp B)} \right) - q_{hwj} \right) \right) \right]^\alpha, \qquad (13)$$

$$R_j = \exp \left[-\sum_{z \in E} \sum_{s \in E_1} \nu_{zs} p_0 \tau_0 \left(\frac{1}{1-\omega} \left(1 - \frac{1}{q_{hnj}} \left(\varphi_j t_{sj} - \varphi_j \left(\frac{C_1 - C_2 \exp B}{C_3(1 - \exp B)} \right) - q_{hwj} \right) \right) \right)^\alpha \right], \qquad (14)$$

$$\omega = q_{hsj} / q_{hnj}, \; \delta = q_{havj} / q_{hnj}, \; \alpha = (1-\delta)/(\delta - \omega), \qquad (15)$$

$$C_1 = t_{0j}(1 - \overline{q}_{sj}), \; C_2 = t_{j\min} - t_{0j}\overline{q}_{sj}, \; C_3 = 1 - \overline{q}_{sj}, \; B = \beta^{-1}(\nu_{szj}^{\max})^{-1}, \qquad (16)$$

where ω – heat load curve irregularity factor; q_{hnj}, q_{hsj}, q_{havj} – heating loads: normative for the heating season start, and average for the heating season; q_{hwj} – normative heat load of hot water supply; $\overline{q}_{sj} = q_{sj}/q_{0j}$ – relative decrease in heat supply at state s (here q_{0j} – total normative heat load of a consumer); φ_j – constant coefficient that depends on the thermophysical properties of consumer building j (Sokolov, 1999); p_0 – probability of fully serviceable system state; τ_0 – heating season duration; t_{0j} – normative indoor temperature; $\nu_{szj}^{\max}$ – maximum rate of transition for the specified consumer to the serviceable state z that is possible for the given state s.

Decomposition Reliability Analysis of HSC Subsystems

Analysis of the HSC reliability performed to take subsequently an effective decision on its redundancy makes it necessary to obtain information about the extent of influence of the HSC subsystems or their individual elements on reliable heat supply to consumers. The reliability indices can be distributed among subsystems or their elements by a methodological approach that is based on the principle of decomposition (elementwise) analysis of the HSC states. Its idea is to separate in the calculated HSC scheme a subsystem to be analyzed with respect to reliability, and the reliability is assessed, provided that no failures of elements of other subsystems decrease the level of heat energy supply. The probabilistic system estimation is performed on the basis of initial reliability parameters of elements. The reliability indices calculated in such a way take into consideration consequences of failures only for the considered subsystem on the basis of the mix of events that integrates elements of all subsystems.

The extent of influence of any element of the HSC scheme on heat supply reliability is estimated, provided that the failure of the specified element is assumed impossible (the element with absolute reliability), and its performance (throughput capacity) remains at the normative level. This condition is

Problems of Modeling and Optimization of Heat Supply Systems

assumed at the stage of probabilistic modeling of the HSC operation by zeroing the rates of transitions because of failure of the considered element. The relationship between the reliability index value determined at reliability assessment for the whole system and its value obtained with respect to the considered element shows, to what degree this element has an effect on reliable heat supply to consumers. The lower is this relationship, the more essential is the decrease in heat supply reliability by the considered element.

SOLUTIONS AND RECOMMENDATIONS

Preparation of Initial Data

The suggested methodological approaches are universal enough and can be applied for reliability calculations of different HSC schemes. Figure 4-a presents an initial HSC scheme that includes a fuel supply system, two heat sources (HS1, HS2), a group of consumers (nodes 1-75) and a loop heat network consisting of 142 sections. The fuel supply system is presented in the scheme for clarity, transport links for fuel supply to HS1 and HS2 are shown in Figure 4-a by the dotted line.

Figure 4. Stages in construction of the calculated HSC scheme: I – equivalenting of the initial scheme; II – construction of the calculated HSS scheme for the analysis of HSC reliability; a) initial HSS scheme; b) equivalented HSS scheme; c) simplified heat source scheme; d) calculated HSS scheme

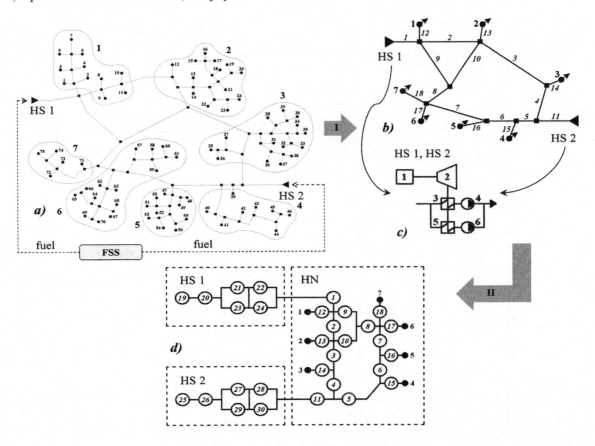

The scheme equivalenting in Figure 4-a allows its simplification to the form in Figure 4-b. The obtained FSS scheme consists only of 18 heat network sections and 7 generalized consumers that integrate groups of consumers, as is shown in Figure 4-a. The schematic elemental diagram of the heat source is presented in Figure 4-c. It includes generalized basic elements – boiler 1, turbine 2, network heater 3 and pump 4. The last two elements are duplicated (5 and 6) and represent a technological reserve of the heat source.

In accordance with the suggested methodological approach to study heat supply reliability (Stennikov & Postnikov, 2008) the transformed calculated HSS scheme in Figure 4-d is represented as a single structure connecting elements of the heat network and heat source schemes. Elements 1-18 are network sections, elements 19-30 correspond to heat sources.

The initial data in Table 1 include engineering parameters of the heat network (diameters and length of pipelines), consumer loads, heat source capacities and reliability characteristics of the scheme, including the rates of failures and restorations of heat network and heat source elements. The restoration rates of elements are calculated according to the dependences of restoration rate of pipelines on their diameter that are presented in (Sennova, 2000), the failure rates of network sections are taken at a level of 0.00002 1/(km·h). Similar characteristics of the heat source elements are assumed based on the data of analysis on operational reliability of power units at a cogeneration plant.

Table 1. Initial data for HSC reliability analysis

HN parameters and reliability characteristics of its elements					Consumer loads and HS capacity			
HN					Consumer	Load, GJ/h	Heat source	Capacity, GJ/h
HN section (Fig. 4-a, b)	Diameter, m	Length, m	Restoration rate, 1/h	Failure rate, 1/h				
1	0.65	2449	0.0236	0.00004898	1	281	HS1	1257
2	0.6	5864	0.0246	0.00011728	2	344	HS2	1467
3	0.55	7811	0.0257	0.00015622	3	268	-	-
4	0.5	3867	0.0269	0.00007734	4	503	-	-
5	0.6	1909	0.0246	0.00003818	5	352	-	-
6	0.55	2799	0.0257	0.00005598	6	402	-	-
7	0.5	6147	0.0269	0.00012295	7	197	-	-
8	0.55	2887	0.0257	0.00005775	Sum	2346	Sum	2724
9	0.55	5540	0.0257	0.00011080	Reliability characteristics of HS1 and HS2 elements			
10	0.5	5580	0.0269	0.00011159				
11	0.7	3128	0.0227	0.00006256	Element of HS1, HS2 (Fig. 4-c)		Restoration rate, 1/h	Failure rate, 1/h
12	0.35	1844	0.0312	0.00003688				
13	0.4	1860	0.0296	0.00003720				
14	0.35	1508	0.0312	0.00003016				
15	0.5	1962	0.0269	0.00003925	19, 25		0.0100	0.0001065
16	0.4	2620	0.0296	0.00005239	20, 26		0.0089	0.0000330
17	0.45	1932	0.0282	0.00003864	21, 23, 27, 29		0.0074	0.0000558
18	0.35	2575	0.0312	0.00005150	22, 24, 28, 30		0.0115	0.0000535

The information needed to calculate reliability of FSS operation is given in Table 2. It contains the average volumes of fuel demands and supplies for each heat source for the calculated interval of the heating season. They are assumed based on the cogeneration plant performance in Irkutsk region. The heating season is divided into 8 calculated intervals equal to one month. In Table 2 they are given in numbers starting from the 10 month (October).

The initial climatic characteristics (outdoor temperature, heating season length, etc.) are assumed for conditions of Irkutsk city, the heat storage capacity of buildings corresponds to the value $\beta = 60$ h. In accordance with the described technique the sequence of calculations to be performed is presented below.

Probabilistic Modeling of HSS Operation

Probabilistic description of HSS operation is based on the following principles. Every element can be in two states – in service and in failure. The flow of events within one subsystem (HN, HS1 and HS2) is the simplest. This condition assumes a simultaneous failure of several elements from different HSS subsystems. Here the states of joint failure of no more than two elements are considered for illustrative purposes. Therefore, a set of states is formed by possible states of HN, HS1, HS2 and their combinations: HN+HS1, HN+HS2 and HS1+HS2. The graph of states representing this structure of system events is demonstrated in Figure 5 in a shortcut form. The number of graph element corresponds to the number of the failed element denoted in Figure 4-d. The states of a simultaneous failure of two elements are denoted by their numbers with the sign «+».

The Markov random process satisfying the given conditions is described by a system of 283 equations of type (5)-(6), and the values of system state probabilities are obtained as a result of solving these equations. Because of a large body of information these results are not presented here.

Calculation of Reliability Indices and their Decomposition

The levels of heat energy supply to consumers at different HSS states are determined on the basis of multi-variant calculations of flow distribution in the heat network on model (9)-(11) and relation (12), the failures of heat source equipment and heat undersupply to consumers due to fuel shortage calculated in item 1 being also taken into account.

The reliability indices are calculated by formulas (13)-(16). The availability factor is calculated for the normative level of heat supply to consumers at $t^{(1)}_{j\min} = 20°C$, and the probability of failure-free operation – for the lower level corresponding to $t^{(2)}_{j\min} = 16°C$.

Table 2. Fuel demands and supplies to HSs, thousand tce

Source		Month							
		10	11	12	01	02	03	04	05
HS1	Demands	31.0	37.7	47.5	45.8	38.8	38.1	27.9	24.4
	Supplies	30.4	37.1	46.6	45.0	38.1	37.4	27.4	24.0
HS2	Demands	32.6	44.3	51.5	47.2	41.8	40.6	31.3	25.6
	Supplies	32.0	43.4	50.5	46.3	40.9	39.8	30.7	25.1

Figure 5. Graph of HSS states

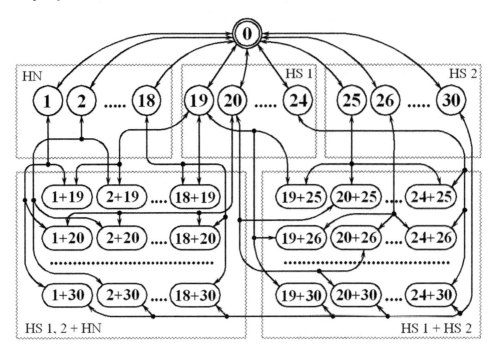

The results of integrated reliability assessment for HSC that involves an impact of its subsystems on reliability of heat supply to consumers are presented in Table 3 under HSC.

The requirements to reliable heat supply to consumers correspond to the following standard values on reliability indices: 0.97 for the probability of failure-free operation of the heat source, 0.9 for the heat network and 0.86 for the whole HSS, about 0.97 for the availability factor of HSS. Comparison of these values and the design values of reliability indices of Table 3 for HSC indicates that the obtained

Table 3. Nodal reliability indices of heat supply to consumers (K – availability factor, R – probability of failure-free operation)

Index		Consumer						
		1	2	3	4	5	6	7
HSC	K	0.8612	0.8629	0.8707	0.8518	**0.8565**	0.8439	0.8660
	R	0.7745	0.7945	0.7852	0.7688	**0.7674**	0.7678	0.7878
FSS	K	0.8890	0.8980	0.8983	0.8782	**0.8890**	0.8791	0.8980
	R	0.8890	0.8980	0.8983	0.8782	**0.8890**	0.8791	0.8980
HSS	K	0.9687	0.9609	0.9693	0.9699	**0.9634**	0.9599	0.9644
	R	0.8712	0.8848	0.8742	0.8754	**0.8632**	0.8734	0.8773
HS	K	0.9897	0.9824	0.9904	0.9906	**0.9841**	0.9811	0.9850
	R	0.9801	0.9795	0.9800	0.9804	**0.9794**	0.9797	0.9804
HN	K	0.9788	0.9782	0.9787	0.9791	**0.9790**	0.9784	0.9791
	R	0.8889	0.9033	0.8920	0.8929	**0.8814**	0.8915	0.8948

availability factor does not exceed 90% (0.8707) of its required value and the probability of failure-free operation – 92% (0.7945). This is the evidence of inadequate reliability of the considered system.

The appropriate comparison of the results of reliability assessment with its normative requirements differentiated for individual subsystems was carried out by the decomposition analysis according to the mentioned above principles. The results obtained by such "distributed" reliability assessment are shown in Table 3 under FSS, HSS, HS and HN. The diagrams in Figures 6-8 give a vivid idea of the relationship between the values of reliability indices that are calculated for different subsystems as compared to their levels for the whole HSC.

The integral assessments of reliability indices are lower than those obtained by separate calculations. If the availability factor for the whole HSC, for example, does not exceed 0.87 (Figure 6), which makes up less than 90% of the standard value for consumers, then with separation of the HSS share the same index for consumers 3, 4 virtually reaches its standard level equal to 0.97. The standard levels of reliability indices are shown in Figures by the dashed lines. A similar relationship between the calculation results is also true for the probability index of failure-free operation, whose maximum value for the whole HSC corresponds to consumer 2 and equals 0.7945 (Figure 7), making up 92% of the standard. At the same time its value for HSS only is higher than the integral assessment and satisfies the standard value for all consumers.

Decomposition of HSS into its subsystems (heat sources and heat networks) also represents a more detailed reliability analysis of each subsystem and provides a more ready base for decision making on HSC development and operation. Figure 8 shows the probability values for failure-free operation for the whole HSS and separately for heat sources and heat networks. When the normative requirements are satisfied for the whole HSS, the reliability index calculated only for the heat network does not correspond to the standard besides consumer 2 and the probability of failure-free operation of heat sources exceeds the standard value for all consumers.

Figure 6. Distribution of the availability factor among HSC subsystems

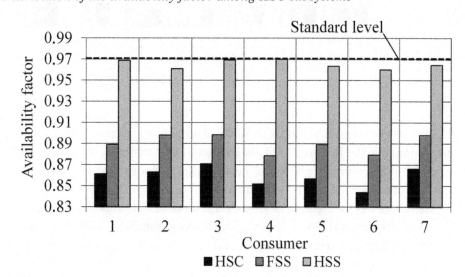

Figure 7. Distribution of the probability of failure-free operation among HSC subsystems

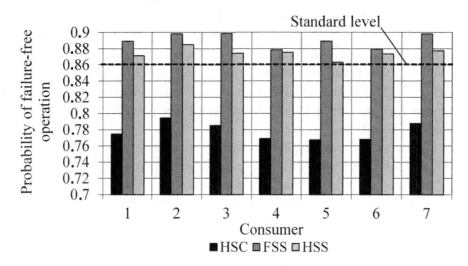

Figure 8. Distribution of the probability of failure-free operation among HSS subsystems

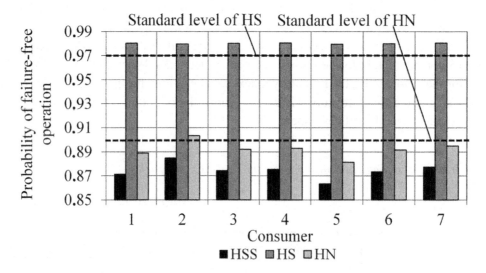

Comparative analysis of reliability indices in Table 3 and Figures 6-8 makes it possible to formulate the priority lines to improve the HSC reliability in less detail. For example, the availability factor takes on the least values for FSS (Figure 6), and for the heat network the normative requirements on the probability of failure-free operation are not met (Figure 8).

Hence, increase in fuel stocks, creation of structural reserve in FSS as additional fuel sources, more reliable system of its transportation all will considerably improve reliability of the calculated heat supply to consumers in the considered system. A complex of measures on the functional and structural redundancy of heat networks being realized, the required level of lower heat supply to consumers in emergency situations will be achieved. A more detailed composition of measures to improve reliability of the HSC subsystems is validly determined when solving henceforth the problem of the optimal synthesis of its reliability. The results of studies that are obtained in the comprehensive analysis on reliability of heat supply to consumers are the initial base for solving the problem.

FUTURE RESEARCH DIRECTIONS

The investigations and their findings provide new scientific and methodological level of solving the problems of heat supply reliability, reveals the advantages of comprehensive properties of the fuel supply systems, heat sources and heat networks, as well as other factors and conditions that affect the reliability of these systems.

Further application of the proposed methods and models in practical calculations increase redundancy subsystems of HSC with the most complete use of the elements interchangeability and structural interconnections.

Further research will be directed at the development of methods submitted, development on the basis of their scientific and methodological support for an optimal reliability synthesis for subsystems of HSC. The problem of optimal reliability synthesis for HSC is to identify the most efficient combination of functional and structural reserve in its subsystems to provide the required level of heat supply reliability. Within the problem of optimal reliability synthesis will also be investigated methods for optimizing fuel inventories on heat sources, as well as the regulation of these reserves during the heating period. The result should be a scientific and methodological base for building efficient and reliable heat supply systems of any topology and scale.

CONCLUSION

The comprehensive approach to the HSC reliability analysis provides a new level of studies, increases capabilities of redundancy of subsystems, with the maximum use of interrelations and interchangeability of their elements. This approach becomes the most topical at joint operation of heat sources in one heat network, though at their individual operation it results in the most rational solutions on reliability. The suggested models and methods for the quantitative assessment of heat supply reliability considerably extend a scope of problems to be solved and have the following features:

- Representation of all the stages of heat energy production and distribution in the common single technological structure allows a system evaluation of reliable heat supply to consumers to be determined;
- Evaluation of HSC states with respect to a concrete consumer is characterized by a simple logic of events (failure or serviceability of the complex), which makes application of the Markov models to the probabilistic description of HSC operation possible;
- Application of the methodological approach based on the principles of elemental decomposition of the calculated scheme to study the effect of each HSC subsystem (groups of elements, individual elements) on the level of heat supply to consumers enables the formulation of rational lines to improve reliability of HSC as a whole and its component subsystems at the stage of reliability analysis;

- The suggested models and methods are universal enough and applicable to HSC of any structure, complexity and scale.

Practical calculations on HSC reliability assessment show that the required reliability of heat supply to consumers can be provided by different subsystems (FSS, HS, HN). Their extent of impact and possible contribution to improving reliability depend in many ways on the specific features and operation conditions of the whole HSC.

REFERENCES

Buslenko, N. P. (1978). *Modeling of complex systems*. Moscow: Nauka. (in Russian)

Buslenko, N. P., Kalashnikov, V. V., & Kovalenko, I. N. (1983). *Lectures on the theory of complex systems*. Moscow: Sovetskoye radio. (in Russian)

Claudio, M., & Rocco, S. (2003). A rule induction approach to improve Monte Carlo system reliability assessment. *Reliability Engineering & System Safety*, *82*(1), 85–92. doi:10.1016/S0951-8320(03)00137-6

Gnedenko, B. V., Belyaev, Yu. K., & Soloviev, A. D. (1965). *Mathematical methods in reliability theory*. Moscow: Nauka. (in Russian)

Haghifam, M., & Manbachi, M. (2011). Reliability and availability modelling of combined heat and power (CHP) systems. *International Journal of Electrical Power & Energy Systems*, *33*(3), 385–393. doi:10.1016/j.ijepes.2010.08.035

Ionin, A. (1978). Criteria for assessment and calculation of heat network reliability. *Vodosnabzhenie i Sanitarnaya Tekhnika*, *12*, 9-10. (in Russian)

Ionin, A. A. (1989). *Reliability of heat network systems*. Moscow: Strojizdat. (in Russian)

Khasilev, V., & Takaishvili, M. (1972). About fundamentals of the technique for calculation and redundancy of heat networks. *Teploenergetika*, *4*, 14–19.

Korolyuk, V. S., & Turbin, A. F. (1976). *Semi-Markov processes and their applications*. Kiev, Ukrainian SSR: Naukova Dumka. (in Russian)

Lisnianski, A., Elmakias, D., & Hanoch, B. H. (2012). A multi-state Markov model for a short-term reliability analysis of a power generating unit. *Reliability Engineering & System Safety*, *98*(1), 1–6. doi:10.1016/j.ress.2011.10.008

Mazur, Yu. Ya. (1986). *Problems of maneuverability in energy development*. Moscow: Nauka. (in Russian)

Merenkov, A. P., & Khasilev, V. Ya. (1985). *Theory of hydraulic circuits*. Moscow: Nauka. (in Russian)

Naess, A., Leira, B., & Batsevych, O. (2009). System reliability analysis by enhanced Monte Carlo simulation. *Structural Safety*, *31*(5), 349–355. doi:10.1016/j.strusafe.2009.02.004

Nekrasov, A. S., & Velikanov, M. A. (1986). Long-term regulation of fuel consumption for heating and ventilation. Progress and prospects. *Energetika*, *46*, 85–98.

Nekrasov, A. S., Velikanov, M. A., & Gorunov, P. V. (1990). *Reliability of fuel supply to power plants: methods and models of studies*. Moscow: Nauka. (in Russian)

Polovko, A. M. (1964). *Fundamentals of reliability theory*. Moscow: Nauka. (in Russian)

Rudenko, Yu. N., Ushakov, I. A., & Cherkesov, G. N. (1984). Theoretical and methodical aspects of reliability analysis of energy systems. *Energy Reviews: Scientific and Engineering Problems of Energy System Reliability, 3*, 82–94.

Sennova, E. (Ed.). (2000). *Reliability of heat supply systems*. Novosibirsk, Russia: Nauka. (in Russian)

Sennova, E. V., & Sidler, V. G. (1985). *Mathematical modeling and optimization of developing heat supply systems*. Novosibirsk: Nauka. (in Russian)

Smirnov, A. V. (1990). *Functional and technological approach to reliability of heat sources in heat supply systems. Heat and energy saving, measurements* (pp. 50–57). Kiev: Institute of Energy Saving Problems of the Ukrainian Academy of Sciences. (in Russian)

Sokolov, V. Ya. (1999). *Cogeneration and heat networks*. Moscow, Russia: Izdatelstvo MEI. (in Russian)

Stennikov, V. A., & Postnikov, I. V. (2008). *Integrated assessment of reliability of heat supply to consumers. Pipeline energy systems. Development of theory and methods of mathematical modeling and optimization* (pp. 139–157). Novosibirsk, Russia: Nauka. (in Russian)

Stennikov, V. A., & Postnikov, I. V. (2010). Development of methods of reliability analysis heating. *Problemy Energetiki, 5-6*, 28-40. (in Russian)

Stennikov, V. A., & Postnikov, I. V. (2011). Comprehensive analysis of the heat supply reliability. Izvestya RAN. *Energrtika, 2*, 107–121.

Stennikov, V. A., & Postnikov, I. V. (2014). Methods for the integrated reliability analysis of heat supply. *Power Technology and Engineering, 6* (47), 446-453.

Stennikov, V. A., Postnikov, I. V., & Sennova, E. V. (2009). Development of methods for analysis of heat supply reliability. In *Proceedings of the VI International Conference "Mathematical Methods in Reliability Theory. Theory. Methods. Applications"* (pp.459-463). Moscow: Russian Peoples' Friendship University.

Ushakov, I. (Ed.). (1985). Reliability of technical systems. Moscow: Radio i svyaz'. (in Russian)

Voropai, N. I. (Ed.). (2014). *Reliability of energy systems: problems, models and methods for solving them*. Novosibirsk, Russia: Nauka. (in Russian)

Woo, S., & Nam, Z. (1991). Semi-Markov reliability analysis of three test/repair policies for standby safety systems in a nuclear power plant. *Reliability Engineering & System Safety, 31*(1), 1–30. doi:10.1016/0951-8320(91)90033-4

Zorkaltsev, V., & Ivanova E. (1990). Intensity and synchronism of fluctuations in fuel demand for heating by economic region of the country. Izvestiya AN SSSR. *Energetika i Transport, 6*, 14-22. (in Russian)

Zorkaltsev, V. I., & Ivanova, E. N. (1989). *Analysis of intensity and synchronism of fluctuations in fuel demand for heating*. Syktyvkar: Komi Nauchny Tsentr UrO AN SSSR. (in Russian)

Zorkaltsev, V. I., & Kolobov, Yu. I. (1984). A simulation model to study reliability of fuel supply to heat generating units. *Trudy Komi filiala Akademii Nauk SSSR* (pp. 33-39). (in Russian)

KEY TERMS AND DEFINITIONS

Conflict of Reserve: The contradiction that arises with increasing low (alarm) level of reliability by adding redundant elements that reduce the reliability of expected level supply due to the increasing number of elements that can fail.

Failure Rate of Element: The number of element failures per unit of time (for example, failures per hour).

Level of Heat Supply Reliability: The criteria determining the value of the minimum allowable inside temperature for this consumer category.

Nodal Reliability Indices: Reliability indices, giving a quantitative measure of a property reliability of the system with respect to a single consumer (or other node of system).

Post-Emergency Conditions (in the Heat Network): Mode, which is installed in the heat network after disconnecting the failed element at the time of his restoration.

Restoration Rate of Element: Indicator, the return value of the restoration time element.

Theory of Hydraulic Circuits: Intersectoral scientific and technical discipline providing the unique language and methodology for solving the problems of modeling, calculation, identification and optimization of pipeline and hydraulic systems of different types and purposes.

Time Reserve of Consumer: Time interval equal to the internal cooling of the indoor air to a minimum level (for this consumer) after total or partial outage heat.

Chapter 6
Fuzzy Random Regression–Based Modeling in Uncertain Environment

Nureize Arbaiy
University Tun Hussein Onn Malaysia, Malaysia

Junzo Watada
Waseda University, Japan

Pei-Chun Lin
Waseda University, Japan

ABSTRACT

The parameter value determination is important to avoid the developed mathematical model is troublesome and may yield inappropriate results. However, estimating the weights of the parameter or objective functions in the mathematical model is sometimes not easy in real situations, especially when the values are unavailable or difficult to decide. Additionally, various uncertainties include in the statistical data makes common mathematical analysis is not competent to deal with. Hence, this paper presents the Fuzzy Random Regression approach to determine the coefficient whereby statistical data used contain uncertainties namely, fuzziness and randomness. The proposed methods are able to provide coefficient information in the model setting and consideration of uncertainties in the evaluation process. The assessment of coefficient value is given by Weight Absolute Percentage Error of Fuzzy Decision. It clarifies the results between fuzzy decision and non-fuzzy decision that shows the distance of different between both approaches. Finally, a real-life application of production planning models is provided to illustrate the applicability of the proposed algorithms to a practical case study.

INTRODUCTION

Real-world decision-making is often referred to rely on human judgment and consideration. Human contributes directly to the decision making process as many evaluations depend on human judgment, which is usually based on experience and intuition. The expression of accurate values in mathematical models has been a complicated problem yet the conventional theory of decision making and problem

DOI: 10.4018/978-1-4666-9755-3.ch006

solving methodologies are developed within the crisp interpretation to obtain the best solution (Zeleny, 1981; Jones and Tamiz, 2010; Jubril, 2012, Vasant, 2012). However, research in decision making methodologies is influential to address the existence of uncertainties that are occurring naturally from the real problem or during the model setting. For example, the information available to a decision maker has been often imprecise because of inaccurate attribute measurements and inconsistency in priorities. Until recently, the decision-making process still utilized subjective judgments when considering human evaluations for certain cases, such as resource planning problems. Hence, it is crucial to embark upon uncertainty to obtain a feasible solution. For this reason, fuzzy sets (Zadeh, 1965) are useful for representing uncertain and imprecise information in mathematical programming, as fuzzy sets reflect these uncertainties and can therefore play a significant role in dealing with such circumstances. Thus, fuzzy programming is valuable for dealing with uncertainties for cases in which the problem model's parameters cannot be estimated precisely from the real situation.

Interpreting uncertainties or incomplete information to the precise numerical values requires an appropriate approach that is difficult and stills a challenge (Zeleny, 1981; Inuiguchi et al., 1990; Bouyssou, 1989; Hasuike and Ishii, 2009). In practical systems, coefficient values should typically be considered to be uncertain values. The uncertainty arises in probability and/or vague situations, such as in cases of predictions of future profits, incomplete historical data, and/or replacement of decision makers, all of which result in uncertain information. Various approaches can be utilized to accommodate the uncertainty. For example, probability distribution, fuzzy numbers, and thresholds types (Bouyssou, 1989). Fuzzy sets (Zadeh, 1965) have made remarkable significant role and are useful for representing uncertain and imprecise information. Probability and possibility theories are also introduced to treat the random and fuzzy information, respectively. Such theories are significant to tackle uncertain information which exists apparently in real-world situation. Additionally, stochastic programming model (i.e., Dantzig, 1955; Sengupta, 1979) and the fuzzy programming model (i.e., Zimmermann, 1976; Zimmermann, 1978; Inuiguchi et al., 1990; Sakawa, 1993; Vasant, 2003; Vasant, 2010) has been introduced to address various real-world problems. The application of fuzzy set and possibility theories to decision-making allows decisions based on imprecise information. In linear programming, possibility theory can be used to include imprecise parameters in the problem formulation (Inuiguchi and Sakawa, 1996; Julien, 1994). Apparently, fuzzy theories and other related theory have made successful to work on the uncertainty problem in modeling and dealing with real world problem.

The elementary aspects of decision making are model setting and goal achievement. Developing a mathematical model from real world application should be carefully undertaken. The developed model is always an abbreviated view of an actual system and typically requires pre-determined and well-defined model parameters. Thus, determining the parameter values is important or the developed mathematical model may yield improper results. However, estimating the coefficient is influenced by the existence of uncertain information. That is, the uncertain information should be treated properly when estimating the coefficient. This is to avoid unawareness of important value presents in the uncertain information if such uncertainties didn't being modeled properly (Inuiguchi and Sakawa, 1996; Möller et al., 2012; Dubois and Prade, 2012; Part et al., 2012). Decisions regarding these coefficients are crucial and influential for the accuracy of the model's results, and the occurrence of errors in the determination of the model's coefficients might ruin the model formulation.

Various fuzzy regression models have been introduced in the past centuries to address fuzzy data. A fuzzy regression model requires information from fuzzy statistics, fuzzy numbers, and/or fuzzy arithmetic (Watada and Tanaka, 1987; Tanaka et al., 1982). Possibilistic programming which describes the type of fuzzy programming also contributes by introducing the concept of possibility theory in the regression analysis. Here in the research, a fuzzy regression model is reinterpreted in the context of possibility (Tanaka and Watada, 1988; Yabuuchi and Watada, 1996; Watada and Toyoura, 2002). Even though various regression analyses may deal with fuzzy data, such analysis is incapable to work with probability and stochastic information simultaneously. In realization of existence of simultaneous fuzzy random information, the fuzzy random variable concept was introduced which extends the classical definition of random variable (Feron, 1976) and other variants were also proposed (Kwarkernaak, 1978; Kwarkernaak, 1979; Puri and Ralescu, 1986; Kruse and Meyer, 1987). A description of probabilistic concept is researched to the case when outcomes of a random experiment are represented by a fuzzy set (Datta, 2015). The variants of fuzzy based regression are significant to support various type of problem solving requirement under uncertain situation.

In real-world applications, statistical data is useful to find hidden information. The data may include both stochastic and fuzzy information at the same time. Such a situation may occur in real-life decision-making simultaneously, and to handle these types of data requires an appropriate approach. Fuzziness is usually originated from imprecise measurement and randomness is captured from the variability of the measured value ((Datta, 2015). To model the uncertainties, fuzziness is represented by possibilistic distribution and randomness is presented using probability distribution (Datta, 2015; Inuiguchi and Sakawa, 1996). Therefore, fuzzy random variables ideal to be utilized as a basic tool for modeling optimization problems containing such uncertainties. Additionally, statistical analysis for fuzzy random variable are widely studied and applied to the social sciences. Many research works have emphasized the error assessment in their proposed method to evaluate the validity of the results produce by the model. The error assessment of the fuzzy demand is explained by Lin et al., (2014) in their procedure of solving location problems. The formulation of error assessment is based on the mean absolute percentage error of fuzzy demand. Another strategy of error assessment is proposed which was mean absolute percentage difference rate of the fuzzy demand. These formulations are used to assess the difference between fuzzy data and non-fuzzy data based and particularly simulated on the game theorem of two-echelon logistic model (Lin et al., 2013; Lin and Nureize, 2014). The information of error measurement information is useful to evaluate the accuracy of the model.

In this study, fuzzy random regression approach is utilized to estimate the coefficient from raw data which contains fuzziness and randomness. This paper presents the fuzzy random regression approach to estimate the coefficient for multi - objective problems. The paper addresses the difficulties to estimate the coefficient whereby fuzziness and randomness uncertainties contained in the data. The assessment to the estimated coefficient is also provided by using Weight Absolute Percentage Error of Fuzzy Decision. A simulation of the proposed algorithm is accompanied using crop industrial data. In the main company level, all data collected from the subordinates should be accumulated and further used to approximate the coefficient to construct the model. The difference between data which comes from different resource is treated as random. Thus, the solution presented in this paper embraces two objectives. First, fuzzy random regression model is used to estimate the multi-objective coefficients. Second, the additive fuzzy goal programming is then used to solve multi-objective linear problem where the goal constraints are developed by means of fuzzy random regression models. The remainder of this paper is divided into five sections. Section II explains the prerequisite mathematics of fuzzy random regression model. Section III describes the formulation for multi-objective problem through goal programming where the goal constraints are developed by fuzzy random regression model. Numerical experiments are portrayed in Section IV and, Section V gives the conclusions.

BACKGROUND

This section is spent explaining the important study in this paper. The Fuzzy Random Regression model is constructed on the basis of fuzzy random variables and its confidence intervals. In this section, the fuzzy random variable, fuzzy random regression model and weight assessment are described.

a) Fuzzy Random Variable: Expected Value and Variance

The integration of probability, possibility, and credibility spaces plays major role in the development of the fuzzy random regression model. Probability measures random event and its probability. Meanwhile possibility theory includes imprecise parameters in the problem formulation by expressing an impression by means of a possibility distribution. There are kinds of fuzzy random variables which are mostly cited in the literature, that is, Kwarkernaak (1978), Puri and Ralescu (1986), and Liu and Liu (2002, 2003). The distinction of these models can be seen through the model setting of a fuzzy random variable but yet can complement each other.

A function $Y : \Gamma \to \Re$ is said to be a fuzzy variable defined on Γ universe which, where the power set $P(\Gamma)$ of a universe Γ which is defined by a possibility measure. Let $\Re$ be the set of real numbers. The possibility of event $\{Y = t\}$ is described as follows: the possibility distribution μ_Y of Y is defined by $\mu_Y(t) = Pos\{Y = t\}$ and t is real number. The possibility and necessity of the event $\{Y \leq r\}$ are given in the following for the fuzzy variable Y with possibility distribution μ_Y.

$$Pos\{Y \leq r\} = \sup_{t \leq r} \mu_Y(t),$$
$$Nec\{Y \leq r\} = 1 - \sup_{t > r} \mu_Y(t). \tag{1}$$

In this study, the expectation value was set based on average of possibility and necessity (Liu and Liu, 2002). The expected value of a fuzzy variable is presented as follows:

Definition 1: Let X be a fuzzy variable and two integrals are finite. The expected value of X is defined as follows

$$E[X] = \int_0^\infty \left(0.5 \left[\sup_{t \geq r} \mu_Y(t) + 1 - \sup_{t < r} \mu_Y(t) \right] \right) dr \\ - \int_{-\infty}^0 \left(0.5 \left[\sup_{t \leq r} \mu_Y(t) + 1 - \sup_{t > r} \mu_Y(t) \right] \right) dr. \tag{2}$$

The event Y is a nonnegative fuzzy variable. The expected value of Y is $E[X] = \dfrac{2c + a^l + a^r}{4}$ when X is a triangular fuzzy number denoted as (c, a^l, a^r). c is central value, and a^l, a^r are the boundaries.

Definition 2: Let $(\Omega, \Sigma, \Pr)$ is a probability space and F_v is a collection of fuzzy variables defined on possibility space $(\Gamma, P(\Gamma), \text{Pos})$. A fuzzy random variable is a map $X : \Omega \to F_v$ such that for any Borel subset B of $\Re$, $\text{Pos}\{X(\omega) \in B\}$ is a measureable function of ω. The Borel, B is the smallest sigma-algebra that contains the set of all open intervals in real numbers. If X be a fuzzy random variable on Ω, for each $\omega \in \Omega$ then $X(\omega)$ is a fuzzy variable.

Let V be a random variable on probability space $(\Omega, \Sigma, \Pr)$. For every $\omega \in \Omega$,

$$X(\omega) = (V(\omega) - 2, V(\omega) + 2, V(\omega) + 6)_\Delta$$

is a triangular fuzzy variable on some possibility space $(\Gamma, P(\Gamma), \text{Pos})$. Therefore, X is a triangular fuzzy random variable.

It makes for any fuzzy random variable X on Ω, the expected value of the fuzzy variable $X(\omega)$ is denoted by $E[X(\omega)]$. Given this, the expected value of the fuzzy random variable X is defined as the mathematical expectation of the random variable $E[X(\omega)]$.

Definition 3: Let X be a fuzzy random variable defined on a probability space $(\Omega, \Sigma, \Pr)$. Assume that,

$$y = \sup_{t \geq r} \mu_{Z(\omega)(t)} + 1 - \sup_{t < r} \mu_{Z(\omega)(t)}, \text{ and } z = \sup_{t \leq r} \mu_{Z(\omega)(t)} + 1 - \sup_{t > r} \mu_{Z(\omega)(t)}.$$

The expected value of X is defined as

$$E[X] = \int_\Omega \left[\int_0^\infty \left(0.5[y]\right) dr - \int_{-\infty}^0 \left(0.5[z]\right) dr \right] \Pr(d\omega) \qquad (3)$$

Definition 4: Let X be a fuzzy random variable defined on a probability space $(\Omega, \Sigma, \Pr)$ with expected value e. The variance of X is defined as

$$Var[X] = E\left[(X - e)^2\right] \qquad (4)$$

The description of expectation value and its variance is useful to build the fuzzy random regression model.

b) Fuzzy Random Data Format

The data format to accommodate fuzzy random variable is explained. Fuzzy random data are formed by output data which is denoted as Y_i and input data denoted as X_{ik} for all $i = 1, \cdots, N$ and $k = 1, \cdots, K$, and are defined as follows

$$Y_i = \bigcup_{t=1}^{M_{Y_i}} \left\{ \left(Y_i^t, Y_i^{t,l}, Y_i^{t,r}\right)_\Delta, pr_i^t \right\} \tag{5}$$

$$X_{ik} = \bigcup_{t=1}^{M_{X_{ik}}} \left\{ \left(X_{ik}^t, X_{ik}^{t,l}, X_{ik}^{t,r}\right)_\Delta, pr_i^t \right\} \tag{6}$$

All input and output values are given fuzzy variables with probabilities. Fuzzy variables $(Y_i^t, Y_i^{t,l}, Y_i^{t,r})_\Delta$ and $(X_{ik}^t, X_{ik}^{t,l}, X_{ik}^{t,r})_\Delta$ are obtained with probability pr_i^t for $i = 1, \cdots, N$, $k = 1, \cdots, K$ and $t = 1, \cdots, M$, respectively.

c) Fuzzy Random Regression Model

The difference between the data and the inferred value obtained from the model is taken to be observational error in standard regression models. In contrast, fuzzy regression models assumed that the gap between the data and the model is an ambiguity in the structure of the system that gives the input and output. Subsequently, fuzzy random regression models consider the coefficient which expresses the system's ambiguity. Also, the models deal with systems for which coefficient gives the possibilities for the coefficient.

A fuzzy linear model with fuzzy coefficients $\overline{A}_i^*, \cdots, \overline{A}_K^*$ as follows:

$$\overline{Y}_i^* = \overline{A}_i^* X_{i1} + \cdots + \overline{A}_K^* X_{iK}, \tag{7}$$

where each $\overline{Y}_i^*$ denotes an estimate of the output and $\overline{A}_k^* = (\frac{[\overline{A}_k^{*l} + \overline{A}_k^{*r}]}{2}, \overline{A}_k^{*l}, \overline{A}_k^{*r})_\Delta$ are symmetric triangular fuzzy coefficients when triangular fuzzy random data X_{ik} are given for $i = 1, \ldots, N$, $k = 1, \ldots, K$.

The input data $X_{ik} = (x_{ik}, x_{ik}^l, x_{ik}^r)_\Delta$ and output data $Y_i = (y_k, y_i^l, y_i^r)_\Delta$ for all $i = 1, \cdots, N$ and $k = 1, \cdots, K$, are fuzzy random variables. The following relation is written using fuzzy random inclusion relation $\supset_{FR}$:

$$\overline{Y}_i^* = \overline{A}_i^* X_{i1} + \ldots + \overline{A}_K^* X_{iK} \supset_{FR} Y_i, \quad i = 1, \ldots, N. \tag{8}$$

The confidence intervals which are induced by the expectation and variance of a fuzzy random variable are then expressed as $(1 \times \sigma)$ interval as follows:

$$I[e_X, \sigma_X] \triangleq \left[E(X) - \sqrt{\mathrm{var}(X)},\ E(X) + \sqrt{\mathrm{var}(X)} \right] \tag{9}$$

Hence, the fuzzy random regression model with σ − confidence intervals (Watada et al., 2009) is described as follows:

$$\min_{\bar{A}} \quad J(\bar{A}) = \sum_{k=1}^{K}(\bar{A}_k^r - \bar{A}_k^l)$$
$$\bar{A}_k^r \geq \bar{A}_k^l,$$
$$\left(\bar{A}_K^r I[e_{X_{iK}} + \sigma_{X_{iK}}]\right)^T \geq \bar{Y}_i^r + (e_{X_{iK}} + \sigma_{X_{iK}}) \qquad (10)$$
$$\left(\bar{A}_K^l I[e_{X_{iK}} - \sigma_{X_{iK}}]\right)^T \leq \bar{Y}_i^l - (e_{X_{iK}} - \sigma_{X_{iK}})$$
$$for \quad i = 1,\ldots,N; \quad k = 1,\ldots,K.$$

Thus, the fuzzy random regression model with confidence interval is given in the following form.

$$\bar{Y}_i^* = \bar{A}_K^* I[e_{X_{iK}} + \sigma_{X_{iK}}], \quad i = 1,\ldots,N \qquad (11)$$

The formulation of fuzzy random regression model (11) is said to retreat the fuzziness and randomness in the statistical data and further used to build the problem model.

d) True Weight Value

The true weight value can be computed by using the Moments and Center of Mass. A planar lamina is a thin, flat plate of material of uniform density.

Definition 5: Moments and Center of Mass of a Planar Lamina

Let f and g be continuous functions such that $f(x) \geq g(x)$ on $[a,b]$, and consider the planar lamina of uniform density ρ bounded by the graphs of $y = f(x), y = g(x)$, and $a \leq x \leq b$ (Larson and Hostetler, 2008).

The moments about the x - axis and y - axis are as follows:

$$M_x = \rho \int_a^b \left[\frac{(f(x) + g(x))}{2}\right][f(x) + g(x)] dx \qquad (12)$$

$$M_y = \rho \int_a^b x[f(x) + g(x)] dx \qquad (13)$$

The center of mass $(\bar{x}, \bar{y})$ is given by $\bar{x} = \dfrac{M_y}{m}$ and $\bar{y} = \dfrac{M_x}{m}$, where $m = \rho \int_a^b [f(x) + g(x)] dx$ is the mass of the lamina.

A planar lamina is a closed surface of mass m and surface density ρ. It can be used to determine moments of inertia of the center of mass.

Property 1: Let $x = [a, b]$ be an interval value, then its membership function is as follows:

$$f(x) = \begin{cases} 1 & a \ll x \ll b \\ 0 & otherwise \end{cases} \quad (14)$$

Moreover, $o = \dfrac{\int_a^b x dx}{\int_a^b dx} = \dfrac{a+b}{2}$ and $l = \dfrac{\int_a^b dx}{2} = \dfrac{b-2}{2}$.

Using this formulation, true value of coefficient can be computed and then is used to evaluate the difference between the estimated coefficients.

MODELING OF MULTI-OBJECTIVE PROBLEM WITH FUZZY RANDOM BASED COEFFICIENT IN UNCERTAINTY

A multi-objective programming problem is a problem involving the simultaneous optimization of various which usually face the conflicting objectives and uncertain information to find compromise solutions (Zeleny, 1981; Jones and Tamiz, 2010; Romero, 1991). Goal programming model (Charnes and Cooper, 1961) is one of the conventional methods to address the mathematical programming problem with multiple objectives which is developed based on a satisfaction philosophy. The variant of goal programming which deals with fuzzy data is known as Fuzzy Goal Programming.

The solution model based on goal programming is presented in this section. The coefficients for the model are derived from fuzzy random regression model. Let us regard the following notations for model definition.

- G_j goal constraints
- $V(\mu)$ fuzzy achievement function
- A system constraint's coefficient
- b right hand side value for system's constraints
- X_i decision variables
- L_i tolerance
- g_i goal target
- ξ_{ij} fuzzy random coefficient
- $[a_j^l, a_j^r]$ interval numbers

The general model of Additive Fuzzy Goal Programming (Tiwari et al., 1987) is as follows:

$$\max \quad V(\mu) = \sum_{i=1}^{m} \mu_i$$
$$\text{subject to} \quad \mu_i = \frac{G_i \cdot X - L_i}{g_i - L_i} \quad (15)$$
$$AX \leq b, \quad \mu_i \leq 1,$$
$$X, \mu_i \geq 0, \quad i = 1, 2, \ldots, m.$$

The additive fuzzy model with fuzzy random goal constraints is then described from the Model (15) as follows:

$$\max \quad V(\mu) = \sum_{i=1}^{m} \mu_i$$
$$\text{subject to} \quad \mu_i = \frac{\xi_i \cdot X_i - L_i}{g_i - L_i} \quad (16)$$
$$AX \leq b, \quad \mu_i \leq 1,$$
$$X, \mu_i \geq 0; \quad i = 1, 2, \ldots, m,$$

where $V(\mu)$ is a fuzzy decision function, and ξ_{ij} is fuzzy random coefficient.

In the Model (16), coefficient ξ_{ij} for decision parameters are inferred by fuzzy random regression. The solution for fuzzy random variables results in an interval closed by the bracketed numbers $\left[a_j^l, a_j^{rl}\right]$. The intervals numbers with a^l and a^r are the lower and upper boundaries is denoted as $\tilde{A} = \left[a_j^l, a_j^{rl}\right] = \{x \mid a^l \leq x \leq a^r, x \in R\}$. The additive fuzzy goal programming model (16) is rewritten into the following model:

$$\max \max V(\mu) = \sum_{i=1}^{m} \mu_i$$
$$\text{subject to}: \mu_i = \frac{\xi_i \cdot X_i - L_i}{g_i - L_i} \quad (17)$$
$$AX < b, \mu_i \leq 1$$
$$X, \mu_i \geq 0, i = 1, \ldots, m.$$

where coefficient $\tilde{\xi}_{ij} = 0.5\left(a^l + a^{lr}\right)$ is the centered value.

The proposed solution of the model can be summarized as follows.

Phase I: Estimating the Coefficient

Step 1: Problem description.
Step 2: Determine the respective goals to construct the goal constraints.

Step 3: For each goal constraints, fuzzy random regression Model (10) is used to approximate the coefficient of decision variables and to model the goal constraints.

Step 4: Construct fuzzy random goal constraints based on estimated coefficients which are approximated from Step 3.

Phase II: Formulating Fuzzy Random Based Goal Programming

Step 1: Substitute a fuzzy random-based goal constraints derived from Phase I Step 4 into the fuzzy goal program model.
Step 2: Solve multi-objective problem through fuzzy goal program method.
Step 3: Perform weight assessment.

Hence, the two phases of the solving process are explained from the solution steps. First, the model's coefficient is estimated by using fuzzy random regression approach. Second, the solution of multi-objective problem is presented by using fuzzy random based goal programming method.

After proposing the fuzzy random based goal programming, the weight assessment (Lin et al., 2013; Lin and Nureize, 2014) for the results from this model is undertaken. This is due to the use of estimated value of the coefficient for decision parameter whereby the difference or error estimate is expected. Hence a weight assessment is provided, which is called the weight absolute percentage error for fuzzy decision (WAPE-FD).

Definition 6: Weight Absolute Percentage Error for Fuzzy Decision

The weight absolute percentage error for fuzzy decision (WAPE-FD) is a measure of accuracy of a method for constructing fitted values in statistics, specifically in trend estimation. It usually expresses accuracy as a percentage, and is defined by the formula:

$$WAPE - FD = 100\% * \left| \frac{WE_k - EWE_k}{WE_k} \right|, \qquad (18)$$

where WE_k is the actual weight expected value and EWE_k is the estimated weight of expected value.

The weight absolute value in this calculation is summed for every fitted or estimated point and multiplying by 100 makes a percentage error.

EXPERIMENTAL RESULTS

A rice crops industry is selected and the milled rice manufacturing data is used as a case study in this experiment. While rice is the essential staple food commodity, continuous rice supply is desirable (Fahmi et al., 2013; Daño and Samonte, 2005). It makes the production of milled rice are constantly in demand. The number of milled rice production is in concerned to support the national needs. Due to this reason, many researcher and institution are contributing a research to assist the competitiveness and profitability enhancement in crops sector.

The problem is modeled as multi-objective problem in which the manager aspires to achieve higher volume of production and better profit returns. The main resources that govern the production target are such as the raw input (paddy), the cultivated land, the mills capacity, and the labor. The assumed $\mp 5\%$ variations in the data from the actual data are denoted as fuzziness in the historical data. Additionally, randomness is presented by the probability assigned to each data, and is observed as the proportional production of different mills distributed in the states. Table 1 shows the sample of production data.

The following step involves the data preparation to estimate the decision coefficients by using a fuzzy random-based regression model. Fuzzy random regression models are used to estimate the coefficient of decision parameters x_1 and x_2. The optimal solution (coefficient) for production and profit are obtained by solving the linear program problem according to Model (10). The interval $\left[A_i^l, A_i^r\right]$ shows the estimated coefficient for decision parameters.

The result depicts that the production of product type A has significant contribution with coefficient of $Z_1 = (x_1, x_2) = [1.637, 1.671]$ compared to product type B with $Z_2 = (x_1, x_2) = [0.810, 0.830]$. For the profit returns, the product A and B have the coefficients $Z_1 = (x_1, x_2) = [0.856, 0.856]$ and $Z_2 = (x_1, x_2) = [1.205, 1.215]$, respectively.

Let us denote Z_1 and Z_2 as the production model and profit model respectively. Making use of the central values of fuzzy intervals, the fuzzy random regression models are written with confidence interval as follows:

$$Z_1 = 0.5\left(A_1^l, A_1^{lr}\right) I\left[e_{x_1}, \sigma_{x_1}\right] = (1.654)_T I\left[e_{x_1}, \sigma_{x_1}\right] + (0.820)_T I\left[e_{x_1}, \sigma_{x_1}\right]$$

$$Z_2 = 0.5\left(A_2^l, A_2^{lr}\right) I\left[e_{x_2}, \sigma_{x_2}\right] = (0.856)_T I\left[e_{x_2}, \sigma_{x_2}\right] + (1.210)_T I\left[e_{x_2}, \sigma_{x_2}\right]$$

Table 1. Data samples for production

Type	Type A			Type B			Total		
Source	M_1	M_2	M_3	M_1	M_2	M_3	M_1	M_2	M_3
Data samples	1.509	1.588	1.667	0.392	0.412	0.433	1.901	2.001	2.101
	1.503	1.582	1.66	0.389	0.409	0.430	1.892	1.992	2.092
	1.269	1.335	1.402	0.344	0.362	0.380	1.613	1.698	1.783
	⋮	⋮	⋮	⋮	⋮	⋮	⋮	⋮	⋮
	⋮	⋮	⋮	⋮	⋮	⋮	⋮	⋮	⋮
	1.421	1.496	1.571	0.376	0.396	0.416	1.798	1.893	1.987
	1.328	1.398	1.467	0.347	0.366	0.384	1.676	1.764	1.852

*M_i denotes the mill

These two models represent the goal constraints in fuzzy goal programming.

Multi-objective problem is now built. Let us assume two functional objectives are investigated under four system constraints. The first objective is to maximize the production volume and the second one is to maximize the profit returns.

The goal constraints are then constructed as follows:

$$Z_1 = [1.637, 1.671] x_1 + [0.810, 0.830] x_2$$

$$Z_2 = [0.856, 0.856] x_1 + [1.205, 1.215] x_2$$

The system constraints and its coefficients are such listed in the Model (19) where the coefficient is the central value of interval $[A_i^l, A_i^r]$ for decision parameters. Therefore the fuzzy goal program model is then re-written as follows:

$$\begin{aligned}
&find x\, to\ satisfy: \\
&\max Z_1: \quad 1.654 x_1 + 0.820 x_2 \leq 6.075 \\
&\max Z_2: \quad 0.856 x_1 + 1.210 x_2 \leq 5.500 \\
&subject\ to\ constraints: \\
&Raw\ input \quad 4.82 x_1 + 1.12 x_2 \leq 28.18 \\
&Cultivated\ land \quad 1.95 x_1 + 1.32 x_2 \leq 14.49 \\
&Mill\ Capacity \quad 15.73 x_1 + 3.13 x_2 \leq 63.64 \\
&Labor \quad 1.05 x_1 + 0.41 x_2 \leq 40.21 \\
&Non-negative \quad \mu_i \leq 1, x_i, \mu_i \geq 0
\end{aligned} \quad (19)$$

The LINGO computer software is employed to run the equivalent ordinary linear programming Model (19).

The second stage meanwhile, concerns with solving the multi-objective problem. The conventional additive fuzzy goal programming is employed by adding the goal constraints consisting of fuzzy random model which was developed in the first stage. The results obtained by the proposed fuzzy goal programming method are $x_1 = 2.18$, $x_2 = 2.99$ with achieved goal values $G_1 = 6.056$, $G_2 = 5.483$ and membership values $\mu_1, \mu_2 = 1.00$. Hence, both goals are successfully achieved.

The WAPE_FD can be utilized so as to compare the non fuzzy and fuzzy weight value in order to check the sensitivity of the results. The triangular fuzzy random coefficient which is deduced from Fuzzy Random Regression is presented in a form of interval $[A_1^l, A_1^r]$. Consequently, the interval value of fuzzy random coefficient is denoted as $\langle \tilde{c}_1, s_1 \rangle$ with center $\tilde{c}_1$ and spread s_1 for the purpose to calculate to true value.

Based on *Definition 5*, the true weight for Problem (19) is written as following:

$$Z_1 = 1.654, 0.017 x_1 + 0.820, 0.010 x_2$$

$Z_2 = 0.856, 0.000x_1 + 1.210, 0.005x_2$

The central true points are computed as follows:

$$z_1 = \frac{(1.654x_1 + 0.820x_2) + (0.017x_1 + 0.010x_2)}{2} = \frac{1.671x_1 + 0.830x_2}{2}$$

$$z_2 = \frac{(0.856x_1 + 0.000x_2) + (1.210x_1 + 0.005x_2)}{2} = \frac{2.066x_1 + 0.005x_2}{2}$$

Crisp multi-objective linear programming problem is formulated as follows:

$$\left. \begin{array}{l} find\ x\ to\ satisfy \\ \max \max Z_1 : \quad 0.835x_1 + 0.415x_2 \\ \max \max Z_2 : \quad 1.033x_1 + 0.002x_2 \\ subject\ to: \\ \quad 4.82x_1 + 1.12x_2 \leq 28.18 \\ \quad 1.95x_1 + 1.32x_2 \leq 14.49 \\ \quad 15.73x_1 + 3.13x_2 \leq 63.64 \\ \quad 1.05x_1 + 0.41x_2 \leq 40.21 \\ \quad \mu_i \leq 1, x_i, \mu_i \geq 0 \end{array} \right\} \quad (20)$$

Problem (20) is solved with optimal solution $(x_1, x_2) \approx (2.33, 3.30)$. From the analysis, the error measurement for coefficient analysis are is 0.06% and 0.09% for x_1 and x_2, respectively. Using these values, it indicates that the estimated weight obtain nearly same value and accurate as the true weight value. The WAPE_FD indicates the risk which decision maker should consider in this problem. It also could clarify the results between fuzzy decision and non-fuzzy decision that shows the distance of different between both approaches. Therefore, the results could be described that the fuzzy results were differentiate from the crisp number's results in this problem.

FUTURE RESEARCH DIRECTIONS

Although studies in decision making and evaluations have circulated steadily in the literature, there are numerous opportunities for further research beyond the solution presented in this paper. From a theoretical perspective, it may be valuable to include risk analysis in the proposed methodology to ensure that the decision maker is aware of the consequences of the selected solution. In this case, more information can be presented to the decision maker to assist in selecting the best choice and while minimizing the risk of particular decision.

The multi-objective evaluation problems discussed in this dissertation are solved with a satisficing-based technique, a term based on the words 'satisfy' and 'suffice'. That is, this approach describes decision makers' behavior, in which reaching their goals suffices them in that decision situation and hence, they are satisfied. In future research, heuristic or evolutionary optimization can be studied to investigate the optimum solution which decision maker can possibly have, based on the optimization point of view.

CONCLUSION

In this paper, the estimation of multi-objective model's coefficient is presented by using fuzzy random regression model. This research emphasizes the fuzzy random regression model to alleviate the difficulties to determine the coefficient in mathematical model setting. Moreover the existence of the randomness and fuzziness in the historical data influence the estimation whereby existing method is incapable to treat such simultaneous uncertainties. The property of fuzzy random regression model is used to take into consideration the co-existence of fuzziness and randomness in the data. The analytical results demonstrated that the proposed method using the central values of fuzzy intervals can achieve the fuzzy additive goal programming. This shows that the fuzzy random regression used in the first stage of solution enables us to reduce the difficulty of determining the coefficient value for the multi-objective model as produced by the conventional ones.

This study additionally presents the error assessment to the estimated coefficient value. The assessment is necessary to evaluate the accuracy of the solution model. The WAPE_FD is uses to capture and measure the effects of the estimated weight on the total solution. In summary, it is important in this paper to note that the enhanced solution approach has been proposed under the consideration of fuzzy random based model's coefficient. The enhancement of the existing model is provided by including the determination of model coefficient value to develop and solve multi-objective mathematical model which include treatment to the fuzzy random data. The preservation of the uncertainties in the mathematical programming model is important to ensure that the developed model expresses the decision maker's intention appropriately.

ACKNOWLEDGMENT

The author expresses her appreciation to the University Tun Hussein Onn Malaysia (UTHM), Research Acculturation Grant Scheme (RAGS) Vot R044 and GATES IT Solution Sdn Bhd.

REFERENCES

Bouyssou, D. (1989). Modeling inaccurate determination, uncertainty, imprecision using multiple criteria. In A. G. Lockett & G. Islei (Eds.), *Improving Decision Making in Organizations, Lecture Notes in Economics and Mathematical Systems*. Springer-Verlag.

Charnes, A., & Cooper, W. (1962). Programming with linear fractional functions. *Naval Research Logistics Quarterly*, *9*(3-4), 181–186. doi:10.1002/nav.3800090303

Daño, E. C., & Samonte, E. D. (2005). Public sector intervention in the rice industry in Malaysia. *State intervention in the rice sector in selected countries: Implications for the Philippines.*

Dantzig, G. B. (1955). Linear programming under uncertainty. *Management Science, 1*(3-4), 197–206. doi:10.1287/mnsc.1.3-4.197

Datta, D. (2015). Implementation of Fuzzy Random Variable in Imprecise Modeling of Contaminant Migration through a Soil Layer. *International Journal of Scientific & Engineering Research, 6*(3), 271–281.

Dubois, D., & Prade, H. (2012). Gradualness, uncertainty and bipolarity: Making sense of fuzzy sets. *Fuzzy Sets and Systems, 192,* 3–24. doi:10.1016/j.fss.2010.11.007

Fahmi, Z., Samah, B. A., & Abdullah, H. (2013). Paddy Industry and Paddy Farmers Well-being: A Success Recipe for Agriculture Industry in Malaysia. *Asian Social Science, 9*(3), 177. doi:10.5539/ass.v9n3p177

Feron, R. (1978). Ensembles aleatoires flous. *C.R. Acad. Sci. Paris Ser. A, 282,* 903–906.

Hasuike, T., & Ishii, H. (2009). Robust expectation optimization model using the possibility measure for the fuzzy random programming problem, Applications of Soft Computing: From Theory to Praxis. Advances in Intelligent and Soft Computing, 58, 285 – 294.

Inuiguchi, M., Ichihashi, H., & Tanaka, H. (1990). Fuzzy programming: A survey of recent developments. In R. Slowinski & J. Teghem (Eds.), *Stochastic versus Fuzzy Approaches to multiobjective mathematical programming under uncertainty* (pp. 45–68). Dordrecht: Kluwer Academics. doi:10.1007/978-94-009-2111-5_4

Inuiguchi, M., & Sakawa, M. (1996). Possible and necessary efficiency in possibilistic multiobjective linear programming problems and possible efficiency test. *Fuzzy Sets and Systems, 78*(2), 231–241. doi:10.1016/0165-0114(95)00169-7

Jones, D., & Tamiz, M. (2010). *Practical goal programming.* Heidelberg, Germany: Springer New York. doi:10.1007/978-1-4419-5771-9

Jubril, A. M. (2012). A nonlinear weights selection in weighted sum for convex multiobjective optimization. *Facta universitatis-series. Mathematics and Informatics, 27*(3), 357–372.

Julien, A. (1994). An extension to possibilistic linear programming. *Fuzzy Sets and Systems, 64*(2), 195–206. doi:10.1016/0165-0114(94)90333-6

Kruse, R., & Meyer, K. D. (1987). *Statistics with vague data.* D. Reidel Publishing Company. doi:10.1007/978-94-009-3943-1

Kwakernaak, H. (1978). Fuzzy random variables – I. definitions and theorems. *Information Sciences, 15*(1), 1–29. doi:10.1016/0020-0255(78)90019-1

Kwakernaak, H. (1979). Fuzzy random variables – II. Algorithms and examples for the discreet case. *Information Sciences, 17*(3), 253–278. doi:10.1016/0020-0255(79)90020-3

Larson, R., Hostetler, R., & Edwards, B. H. (2008). *Essential Calculus: Early Transcendental Functions.* Boston: Houghton Mifflin Company.

Lin, P.-C., & Nureize, A. (2014). Two-Echelon logistic model based on game theory with fuzzy variable. *Recent Advances on Soft Computing and Data Mining. Advances in Intelligent Systems and Computing, 287*, 325–334. doi:10.1007/978-3-319-07692-8_31

Lin, P.-C., Watada, J., & Wu, B. (2013). Identifying the distribution difference between two populations of fuzzy data based on a nonparametric statistical method. *IEEJ Transactions on Electronics. Information Systems, 8*(6), 591–598.

Lin, P.-C., Watada, J., & Wu, B. (2014). A parametric assessment approach to solving facility location problems with fuzzy demands. *IEEJ Transactions on Electronics. Information Systems, 9*(5), 484–493.

Liu, B., & Liu, Y.-K. (2002). Expected value of fuzzy variable and fuzzy expected value models. *IEEE Transactions on Fuzzy Systems, 10*(4), 445–450. doi:10.1109/TFUZZ.2002.800692

Liu, B., & Liu, Y.-K. (2003). Fuzzy random variable: A scalar expected value operator. *Fuzzy Optimization and Decision Making, 2*(2), 143–160. doi:10.1023/A:1023447217758

Marler, R. T., & Arora, J. S. (2010). Weighted sum method for multi-objective optimization: New insights. *Structural and Multidisciplinary Optimization, 41*(6), 853–862. doi:10.1007/s00158-009-0460-7

Moller, B., Graf, W., & Beer, M. (2003). Safety assessment of structures in view of fuzzy randomness. *Computers & Structures, 81*(15), 1567–1582. doi:10.1016/S0045-7949(03)00147-0

Möller, B., Graf, W., Beer, M., & Sickert, J. U. (2002). Fuzzy randomness-towards a new modeling of uncertainty. *Fifth World Congress on Computational Mechanics*, Vienna, Austria.

Nureize, A., & Watada, J. (2010). Constructing fuzzy random goal constraints for stochastic goal programming.Integrated Uncertainty Management and Applications, 68, 293–304.

Nureize, A., & Watada, J. (2014). Linear fractional programming for fuzzy random based possibilistic programming problem. *International Journal of Simulation Systems. Science & Technology, 14*(1), 24–30.

Park, H. J., Um, J. G., Woo, I., & Kim, J. W. (2012, January). Application of fuzzy set theory to evaluate the probability of failure in rock slopes. *Engineering Geology, 125*, 92–101. doi:10.1016/j.enggeo.2011.11.008

Puri, M. L., & Ralescu, D. A. (1986). Fuzzy random variables. *Journal of Mathematical Analysis and Applications, 114*(2), 409–422. doi:10.1016/0022-247X(86)90093-4

Romero, C. (1991). *Handbook of critical issues in goal programming*. Oxford, UK: Pergamon Press.

Saaty, L. (1980). *The analytic hierarchy process: planning, priority setting, resource allocation*. New York: McGraw-Hill.

Sakawa, M. (1993). *Fuzzy sets and interactive multi-objective optimization. Applied Information Technology*. New York: Plenum Press. doi:10.1007/978-1-4899-1633-4

Sengupta, J. K. (1979). Stochastic goal programming with estimated parameters. *Journal of Economics, 39*(3-4), 225–243. doi:10.1007/BF01283628

Tanaka, H., Uejima, S., & Asai, K. (1982). Linear regression analysis with fuzzy model. *IEEE Transactions on Systems. Man and Cybernetics SMC., 12*(6), 903–907. doi:10.1109/TSMC.1982.4308925

Tanaka, H., & Watada, J. (1988). Possibilistic linear systems and their application to the linear regression model. *Fuzzy Sets and Systems, 27*(3), 275–289. doi:10.1016/0165-0114(88)90054-1

Tanaka, H., Watada, J., & Hayashi, I. (1986). Three formulations of fuzzy linear regression analysis. *Transactions of the Society of Instrument and Control Engineers, 22*(10), 1051–1057. doi:10.9746/sicetr1965.22.1051

Tiwari, R. N., Dharmar, S., & Rao, J. R. (1987). Fuzzy goal programming – an additive model. *Fuzzy Sets and Systems, 24*(1), 27–34. doi:10.1016/0165-0114(87)90111-4

Vasant, P. (2012). Novel Meta-Heuristic Optimization Techniques for Solving Fuzzy Programming Problems. In M. Khan & A. Ansari (Eds.), *Handbook of Research on Industrial Informatics and Manufacturing Intelligence: Innovations and Solutions* (pp. 104–131). Hershey, PA: Information Science Reference; doi:10.4018/978-1-4666-0294-6.ch005

Vasant, P. (2013). Hybrid Optimization Techniques for Industrial Production Planning. In Z. Li & A. Al-Ahmari (Eds.), *Formal Methods in Manufacturing Systems: Recent Advances* (pp. 84–111). Hershey, PA: Engineering Science Reference; doi:10.4018/978-1-4666-4034-4.ch005

Vasant, P., Elamvazuthi, I., & Webb, J. F. (2010). Fuzzy technique for optimization of objective function with uncertain resource variables and technological coefficients. *International Journal of Modeling, Simulation, and Scientific Computing, 1*(3), 349–367. doi:10.1142/S1793962310000225

Vasant, P. M. (2003). Application of fuzzy linear programming in production planning. *Fuzzy Optimization and Decision Making, 2*(3), 229–241. doi:10.1023/A:1025094504415

Watada, J., & Tanaka, H. (1987). Fuzzy quantification methods. *Proceedings, the 2nd IFSA Congress, at Tokyo*, (pp. 66-69).

Watada, J., & Toyoura, Y. (2002). Formulation of fuzzy switching auto-regression model. *International Journal of Chaos Theory and Applications, 7*(1&2), 67–76.

Watada, J., Wang, S., & Pedrycz, W. (2009). Building confidence interval-based fuzzy random regression model. *IEEE Transactions on Fuzzy Systems, 11*(6), 1273–1283. doi:10.1109/TFUZZ.2009.2028331

Yabuuchi, Y., & Watada, J. (1996). Fuzzy robust regression analysis based on a hyper elliptic function. *Journal of the Operations Research Society of Japan, 39*(4), 512–524.

Zadeh, L. A. (1965). Fuzzy sets. *Information and Control, 8*(3), 338–353. doi:10.1016/S0019-9958(65)90241-X

Zeleny, M. (1981). The pros and cons of goal programming. *Computers & Operations Research, 8*(4), 357–359. doi:10.1016/0305-0548(81)90022-8

Zimmermann, H.-J. (1976). Description and optimization of fuzzy systems. *International Journal of General Systems, 2*(4), 209–215. doi:10.1080/03081077608547470

Zimmermann, H.-J. (1978). Fuzzy programming and linear programming with several objective functions. *Fuzzy Sets and Systems*, *1*(1), 45–55. doi:10.1016/0165-0114(78)90031-3

ADDITIONAL READING

Arenas, M., Bilbao, A., Perez, B., and Rodriguez, M. V. (2005). Solving a multiobjective possibilistic problem through compromise programming. *European Journal of Operational Research, Recent Advances in Scheduling in Computer and manufacturing Systems*, *164(3)*, 748-759.

Arenas, M., Bilbao, A., & Rodriguez, M. V. (2001). A fuzzy goal programming approach to portfolio selection. *European Journal of Operational Research*, *133*(2), 287–297. doi:10.1016/S0377-2217(00)00298-8

Baky, I. A. (2010). Solving multi-level multi-objective linear programming problems through fuzzy goal programming approach. *Applied Mathematical Modelling*, *34*(9), 2377–2387. doi:10.1016/j.apm.2009.11.004

Bhattacharya, A., Vasant, P., & Susanto, S. (2008). Simulating theory-of constraint problem with novel fuzzy compromise linear programming model. In A. El Sheikh, A. Al Ajeeli, & E. Abu-Taieh (Eds.), *Simulation and Modeling: Current Technologies and Applications* (pp. 307–336). Hershey, PA: IGI Publishing; doi:10.4018/978-1-59904-198-8.ch011

Hasuike, T., & Ishii, H. (2009). Robust expectation optimization model using the possibility measure for the fuzzy random programming problem. Applications of Soft Computing: From Theory to Praxis (Editors: J. Mehnen et al.), Advances in Intelligent and Soft Computing, 58, 285 – 294. doi:10.1007/978-3-540-89619-7_28

Hasuike, T., & Ishii, H. (2009). Product mix problems considering several probabilistic conditions and flexibility of constraints. *Computers & Industrial Engineering*, *56*(3), 918–936. doi:10.1016/j.cie.2008.09.006

Islam, S., Keung, J., Lee, K., & Liu, A. (2012). Empirical prediction models for adaptive resource provisioning in the cloud. *Future Generation Computer Systems*, *28*(1), 155–162. doi:10.1016/j.future.2011.05.027

Jahan, A., Mustapha, F., Sapuan, S. M., Ismail, M. Y., & Bahraminasab, M. (2012). A framework for weighting of criteria in ranking stage of material selection process. *International Journal of Advanced Manufacturing Technology*, *58*(1-4), 411–420. doi:10.1007/s00170-011-3366-7

Jiménez, M., Arenas, M., Bilbao, A., & Rodríguez, M. V. (2000). Solving a possibilistic linear program through compromise programming. *Mathware and Soft Computing.*, *7*(2–3), 175–184.

Katagiri, H., & Sakawa, M. (2003). A study on fuzzy random linear programming problems based on possibility and necessity measure. In T. Bilgic, B. D. Baets, & O. Kaynak (Eds.), *Fuzzy Sets and Systems – IFSA 2003* (pp. 725–732). Lecture Notes in Computer Science Springer.

Li, D.-F., & Sun, T. (2007). Fuzzy linear programming approach to multi-attribute decision-making with linguistic variables and incomplete information. *Advances in Complex Systems*, *10*(4), 505–525. doi:10.1142/S0219525907001306

Nureize, A., & Watada, J. (2011). Fuzzy goal programming for multi-level multi-objective problem: an additive model J.M. Zain et al. (Eds.): ICSECS 2011, Part II, CCIS 180, pp. 81–95, Springer.

Nureize, A., & Watada, J. (2011). Building multi-attribute decision model based on kansei information in environment with hybrid uncertainty, Watada et al. (Eds.): Smart Innovation, Systems and Technologies, 10(1), 103-112.

Nureize, A., & Watada, J. (2012). Multi-objective top-down decision making through additive fuzzy goal programming [SICE]. *Transactions of the Society of Instrument and Control Engineers, 5*(2), 63–69.

Nureize, A., and Watada, J. (2012). Building fuzzy goal programming with fuzzy random linear programming for multi-level multi-objective problem, *International Journal of New Computer Architectures and their Applications, 1(4),* 911-925.

Nureize, A., & Watada, J. (2013). A fractile optimisation approach for possibilistic programming problem in fuzzy random environment, *Int. J. Artificial Intelligence and Soft Computing, 3*(4), 330–343. doi:10.1504/IJAISC.2013.056829

Nureize, A., & Watada, J. (2015). Multi-granular evaluation model through fuzzy random regression to improve information granularity, I*nformation Granularity. Big Data, and Computational Intelligence Studies in Big Data, 8,* 231–245.

Vasant, P. (2006). Fuzzy decision making of profit function in production planning using S-curve membership function. *Computers & Industrial Engineering, 51*(4), 715–725. doi:10.1016/j.cie.2006.08.017

Vasant, P. (2015). *Handbook of research on artificial intelligence techniques and algorithms* (pp. 1–873). Hershey, PA: IGI Global; http://www.hindawi.com/28921463/ http://www.igi-global.com/journal/international-journal-energy-optimization-engineering/47187, doi:10.4018/978-1-4666-7258-1

Vasant, P., Barsoum, N., & Webb, J. (2012). *Innovation in power, control, and optimization: Emerging Energy Technologies* (pp. 1–396). Hershey, PA: IGI Global; doi:10.4018/978-1-61350-138-2

Vasant, P., Elamvazuthi, I., & Webb, J. F. (2010). Fuzzy technique for optimization of objective function with uncertain resource variables and technological coefficients. *International Journal of Modeling, Simulation, and Scientific Computing, 1*(3), 349–367. doi:10.1142/S1793962310000225

Vasant, P., Ganesan, T., Elamvazuthi, I., & Webb, J. F. (2011). Fuzzy linear programming for the production planning: The case of textile firm. *International Review on Modelling and Simulations, 4*(2), 961–970.

Vasant, P. M. (2014). *Handbook of research on novel soft computing intelligent algorithms: Theory and Practical Applications* (pp. 1–1018). Hershey, PA: IGI Global; doi:10.4018/978-1-4666-4450-2

KEY TERMS AND DEFINITIONS

Coefficient: A multiplicative factor in some term of a polynomial, a series, or any expression.

Fuzzy Decision: Fuzzy decision represents a value which is derived within fuzzy definition (i.e., fuzzy coefficient).

Fuzzy Random Regression: A fuzzy regression models is defined based on fuzzy random data.

Fuzzy Random Variable: A fuzzy random variable is characterized as a vague perception of a crisp but unobservable random variable and also as a random fuzzy set.

Goal Programming: Goal programming is a branch of multiobjective optimization, which is an extension or generalisation of linear programming to handle multiple, and conflicting objective measures.

Multi-Objective Evaluation: Is an area of multiple criteria decision making, that is concerned with mathematical optimization problems involving more than one objective function to be optimized simultaneously.

Uncertainties: The state that is uncertain or that causes uncertain.

Chapter 7
A Novel Optimization Algorithm for Transient Stability Constrained Optimal Power Flow

Sourav Paul
Dr. B. C. Roy Engineering College, India

Provas Kumar Roy
Jalpaiguri Government Engineering College, India

ABSTRACT

Optimal power flow with transient stability constraints (TSCOPF) becomes an effective tool of many problems in power systems since it simultaneously considers economy and dynamic stability of power system. TSC-OPF is a non-linear optimization problem which is not easy to deal directly because of its huge dimension. This paper presents a novel and efficient optimisation approach named the teaching learning based optimisation (TLBO) for solving the TSCOPF problem. The quality and usefulness of the proposed algorithm is demonstrated through its application to four standard test systems namely, IEEE 30-bus system, IEEE 118-bus system, WSCC 3-generator 9-bus system and New England 10-generator 39-bus system. To demonstrate the applicability and validity of the proposed method, the results obtained from the proposed algorithm are compared with those obtained from other algorithms available in the literature. The experimental results show that the proposed TLBO approach is comparatively capable of obtaining higher quality solution and faster computational time.

1 INTRODUCTION

Optimal power flow (OPF) (Dommel & Tinney, 1968; Chen, Tada, Okamoto, Tanabe, & Ono, 2001; El-Hawari, 1996; Laufenberg & Pai, 1998; Yan, Xiang, & Zhang, 1996). has become an important issue to the researchers over past two decades and has established its position as one of the main tools for optimal operation and planning of modern power systems. The prime objective of the OPF problem is to

DOI: 10.4018/978-1-4666-9755-3.ch007

optimize generation cost, by optimal adjusting the power system control variables and satisfying various system operating such as power flow equations and inequality constraints, simultaneously

OPF with transient stability constraints is an extension of the traditional OPF problems. In addition to the common constraints of OPF, the TSCOPF problems consider the dynamic stability constraints of power system. When any of a specified set of disturbances occurs, a feasible operation point should withstand the fault and ensure that the power system moves to a new stable equilibrium after the clearance of the fault without violating equality and inequality constraints even during transient period. These conditions for all of the specified credible contingencies are called as transient stability constraints.

Because of larger power transfers over longer distances, complex coordination, difficult interaction among various system controllers and less power reserves, complexity is gradually increasing for planning and operation purpose of modern power system network. Secure and reliable operation of the power system has always been a top priority for system operators. Power system is said to be secure when it is able to withstand sudden disturbances with minimum loss of the quality of service i.e. whenever disturbance occurs, the system survives the ensuing transient and moves into a suitable stable condition where all operating constraints are within limits.

Evolutionary Algorithms (EA) such as such as genetic algorithms (GA) (Lai, & Ma, 1997), evolutionary programming (Yuryevich, & Wong, 1999; Wang, Zheng, & Tang, 2002), simulated annealing (Roa-Sepulveda & Pavez-Lazo, 2003), tabu search (Abido, 2002a; Lin, Cheng, & Tsay, 2002) and particle swarm optimization (Abido, 2002b), a hybrid tabu search and simulated annealing (TS/TA) (Ongsakul & Bhasaputra, 2002) have become the method of choice for optimization problems that are too complex to be solved using deterministic techniques such as linear programming (LP), non-linear programming (NLP), quadratic programming (QP), Newton method and interior point methods (IPM) (Momoh, El-Hawary, & Adapa, 1999; Habiabollahzadeh and Luo, 1989; Burchet, Happ, & Vierath, 1984; Mota-Palomin, & Quintana, 1986; Sun, Ashley, Brewer, Hughes, & Tinney, 1984; Yan & Quintana, 1999).

Among various OPF problems, the OPF with transient stability constraints research is an interesting problem, since it not only considers some optimal strategies but also includes all security and stability constraints. Transient stability constraints describe about the limits that system dynamics must satisfy during transient conditions to be stable. The cost of losing synchronous through a transient instability is extremely high in modern power systems.

Transient stability constrained optimal power flow (TSC-OPF) is an effective measure to coordinate the security and economic of power system. The main aim behind the transient stability-constrained OPF (TSC-OPF) is to integrate economic objectives and both steady-state and stability constraints in one unique formulation. TSCOPF is a large-scale nonlinear optimization problem with both algebraic and differential equations developed by Sauer and Pai (1998), Kundur (1994), and Stott (1979). One method is to simplify TSCOPF problem by converting the differential equations into equivalent algebraic equations as inequality constraints and then integrated the TSCOPF as an algebraic optimization problem which can be solved by Gan, Thomas, and Zimmerman (2000) through conventional optimization techniques. Based upon this, multicontingency concept has come from Yuan, Kubokawa, and Sasaki (2003) which gave a modified formulation for integrating transient stability model into conventional OPF that reduces the calculation load considerably. But it faces some obligations that due to a large number of variables not only convergence problem but also computation inaccuracy occur for approximation. In Xia and Wei (2012) authors adopted numerous method to convert the differential equations to a series of algebraic equations in order to solve TSCOPF problem and the robustness and effectiveness of the proposed method were tested on the four test systems. For solving this infinitesimal dimension of TSCOPF problem to finite dimensional problem, a new technique, functional transformation technique is proposed by Chen, Tada, Okamoto, Tanabe, and Ono (2001). Shiwei, Chan, and Guo (2014) proposed interior point method (IPM) to solve TSCOPF problem. The novelty of the proposed approach was incorporated into OPF model as a single stability constraint derived from the minimum kinetic energy for normal unstable case or the minimum accelerating power distance for extreme unstable case using the time domain simulation based single machine equivalent (SIME) method.

Single machine equivalent (SIME) and trajectory sensitivity methods were proposed by Pizano-Martinez et al. (2014) to formulate TSCOPF in the Euclidian space, where only one single stability constraint is necessary in the optimization problem to represent all dynamic and transient stability constraints of the multi-machine system. The proposed method resulted in a tractable approach to the preventive control of transient stability in realistic power systems. The validity and effectiveness of the proposed method were numerically demonstrated in a 10-machine 39-bus New England system and Mexican 46-machine 190-bus system. Tu et al. introduced a global TSCOPF based dynamic reduction method (Tu, Dessaint, & Kamwa, 2013) which decomposed the power system into several coherent areas and represents the original system by a reduced equivalent system. In Ahmadi, Ghasemi, Haddadi, and Lesani (2013) the author proposed two approaches namely, maximum relative rotor angle deviation (MRRAD) method and the transient stability margin of the system based on generators output power (GOP) to include transient stability constrained in the OPF problem considering detailed dynamic models for generator. The proposed methods were examined in WSCC 9-bus, the New England 39-bus and the IEEE 300-bus test systems.

But unfortunately, these methods are infeasible in practical systems because of high sensitivity problem of initial conditions. Owing to the computational difficulties, the degree of freedom in the objective functions and the types of constraints in OPF problems such as transient stability limit are restricted. Thus, it becomes necessary to develop optimization techniques that are capable of overcoming these drawbacks and handling such difficulties. In recent years, computational intelligence-based techniques such as such as genetic algorithms (GA) (Lai, & Ma, 1997), evolutionary programming (Yuryevich, & Wong, 1999), simulated annealing (SA) (Roa-Sepulveda, & Pavez-Lazo, 2003), tabu search (Abido, 2002a), particle swarm optimization (PSO) (Abido, 2002a) hybrid evolutionary programming (HEP) (Swain, & Morris, 2000), chaotic ant swarm optimization (CASO) (Cai, Ma, Li, Yang, Peng, & Wang, 2007), Bacteria foraging optimization (BFO) (Tripathy, & Mishra, 2007), differential evolution (DE) (Ela, Abido, & Spea, 2010; Sivasubramani, & Swarup, 2012), biogeography based optimization (BBO) (Roy, Ghosal, & Thakur, 2010), harmony search algorithm (HSA) (Khazali, & Kalantar, 2011), artificial bee colony optimization (ABC) (Ayan, Kılıc, 2012), TLBO (Mandal, & Roy, 2014) and gravitational search algorithm (GSA) (Roy, Mandal, & Bhattacharya, 2012) have been proposed as an alternative for solving OPF problem. Among the above-mentioned methods, the TLBO algorithm developed by Roa et al. (Rao, Savsani, & Vakharia, 2011) proofs its superiority by successful application in various fields of power system such as economic load dispatch (ELD) (Roy, & Bhui, 2013), short term hydro-thermal scheduling (HTS) (Roy, Sur, & Pradhan, 2013), combined heat and power dispatch (CHPD) (Roy, Paul, & Sultana, 2014), OPF (Mandal, & Roy, 2014), optimal reactive power dispatch (ORPD) (Ghasemi, Ghanbarian, Ghavidel, Rahmani, & Moghaddam, 2014; Ghasemi, Taghizadeh, Ghavidel, Aghaei, Abbasian, 2015), optimal capacitor placement in radial distribution system (Sultana, & Roy, 2014). TLBO algorithm is a teaching–learning process inspired algorithm based on the effect of influence of a teacher on the learners in a classroom. It requires only common controlling parameters like population size and number of generations for its working. TLBO algorithm has excellent capability for global searching at the beginning of the run and a local search near the end of the run. Therefore, for solving problems with more local optima, TLBO algorithms can explore local optima at the end of each run.

Some of the population-based optimization methods that have been successfully applied to TSCOPF problems are improved genetic algorithm by Chan, Ling, Chan, Iu, and Pong (2007), particle swarm optimization by Mo, Zou, Chan, and Pong (2007), differential evolution by Cai, Chung, and Wong (2008) and the artificial bee colony (ABC) by Karaboga (2005). A group search optimization (IGSO) algorithm based on backward searching strategy, Cauchy mutation and inheritance operator to solve TSCOPF was introduced in Xia, Zhou, Chan, and Guo (2015). The effectiveness of the proposed method was tested for four comprehensive case studies considering the nonlinearities such as valve-point effects and prohibited operation zones (POZs) of generators to prove the superiority of the proposed method compared to seven representative artificial intelligence algorithms. A chaotic artificial bee colony (CABC) algorithm based on chaos theory was implemented in Ayan, Kilic, and Barakli (2015) for solving TSCOPF problem of IEEE 30-bus test system and New England 39-bus test systems. The obtained results were compared to those obtained from previous studies in literature and the comparative results were given to show validity and effectiveness of proposed method.

2 MOTIVATION

Despite the fact that most of the above mentioned techniques have good convergence characteristics for solving non-linear optimization problem, but the performances of most of these techniques are sometimes largely depends on the initial population, i.e. they might converge to local solutions instead of global ones. Moreover, most of the above mentioned heuristics techniques require longer simulation times and larger iterations to reach to the optimal solution.

Very recently, a new optimization technique, known as teaching learning based optimization (TLBO) has been developed by Rao et al. (2011) It is one of the newest evolutionary algorithms and is based on the natural phenomenon of teaching and learning process. Teaching learning based optimization (TLBO) is relatively a new, much simpler and more robust optimization algorithm compared to the many other well popular optimization methods proposed by past scholars. Though this technique has the ability to search near global optimal.

Most of the evolutionary and population based techniques in the literature are probabilistic techniques and necessitate some common controlling parameters. GA involves determination of some algorithm-specific parameters such as crossover rate and mutation rate. PSO has its own parameters like inertia weight, social and cognitive parameters. In case of HS, determination of harmony memory consideration rate, pitch adjusting rate, and number of improvisations are necessary. The global solution of any function is only achieved with the proper tuning of these algorithm-specific parameters. Improper tuning may lead to local optimal solution or increase in convergence time. However, unlike the other population based techniques, TLBO does not require any input controlling parameters for its functionality. It makes the proposed TLBO method superior to other algorithms. This motivates the authors to implement TLBO to solve TSCOPF problem. Dealing of TLBO with TSCOPF problem is a totally new concept and has not been used before. It improves the quality of the solution and the corresponding computational results show the superiority of the proposed algorithm.

In this paper TLBO algorithm is successfully applied to the TSCOPF problem for IEEE 30-bus, IEEE 118-bus, WSCC 3 generator, 9 bus system and New England 10 generator, 39 bus system. The detailed mathematical formulation of the TSCOPF problem is given in Section 2. The details of the TLBO method and algorithm are explained in Section 4 and section 5. The test result of the TLBO algorithm is given in section 6. Conclusions and discussion related to simulation results are given in Section 7.

3 MATHEMATICAL PROBLEM FORMULATION

3.1 Objective Function

The objective of OPF is to minimize the objective function while satisfying all the equality and inequality constraints. The general optimal power flow problem can be expressed as a constrained optimization problem as follows:

$$\text{minimize } F_C(u, v) \tag{1}$$

$$\text{subject to } \begin{cases} g(u, v) = 0 \\ h(u, v) \leq 0 \end{cases} \tag{2}$$

Here, F_C is the objective function to be minimized and u is the vector of dependent variables consisting of generator active power output at slack bus P_{G1}, load voltages $\left(V_{L1}, ..., V_{LN_{PQ}}\right)$, generators' reactive powers $\left(Q_{G1}, ..., Q_{GN_{PV}}\right)$, transmission line loadings $\left(S_{L1}, ..., S_{LN_L}\right)$. N_{PQ} is the number of load buses, N_{PV} is the number of generator buses and N_L is the number of transmission lines respectively.

Therefore u can be expressed as follows:

$$u^T = \left[P_{G1}, V_{L1}, ..., V_{LN_{PQ}}, Q_{G1}, ..., Q_{GN_{PV}}, S_{L1}, ..., S_{LN_L}\right] \tag{3}$$

v is the vector of independent variables consisting of generators' active powers except slack bus $\left(P_{G2}, ..., P_{GN_{PV}}\right)$, generators' voltages $\left(V_{G1}, ..., V_{GN_{PV}}\right)$, transformers' tap settings $\left(T_1, ..., T_{N_T}\right)$, reactive power injection $\left(Q_{C1}, ..., Q_{CN_C}\right)$.

Therefore v can be expressed as:

$$v^T = \left[P_{G2}, ..., P_{GN_{PV}}, V_{G1}, ..., V_{GN_{PV}}, T_1, ..., T_{N_T}, Q_{C1}, ..., Q_{CN_C}\right] \tag{4}$$

Here, N_T is the number of tap setting transformer branches and N_C is the number of capacitor banks respectively.

In this simulation study, the total generation cost is considered as the objective function. For quadratic cost function, the total fuel cost of generating units is given by:

$$F_C(u,v) = \sum_{p=1}^{N_{PV}} \left(a_p P_{G_p}^2 + b_p P_{G_p} + c_p \right) \tag{5}$$

Here, a_p, b_p, c_p are fuel cost coefficients, and P_{G_p} is the generation of the pth generator.

However, in practical generator multiple valve steam turbines are present to achieve more accurate and flexible operation and therefore, the valve point effect should also be considered to represent accurate cost function. The total cost of generating units with valve point loading is given by:

$$F_C(u,v) = \sum_{p=1}^{N_{PV}} a_p P_{G_p}^2 + b_p P_{G_p} + c_p + d_p \sin\left[e_p \left(P_{G_p}^{\min} - P_{G_p} \right) \right] \tag{6}$$

Here, d_p, e_p are the fuel cost coefficients with valve point loading effect and $P_{G_p}^{\min}$ is the minimum generation of the pth generator.

3.2 Constraints

The OPF problem has two categories of constraints as mentioned below:

3.2.1 Equality Constraints

The non-linear power flow equations are the equality constraints which may mathematically be expressed as follows:

$$\Delta P_p = P_{G_p} - P_{D_p} - \sum_{q=1}^{N} V_p V_q Y_{pq} \cos\left(\delta_p - \delta_q - \theta_{pq} \right) \tag{7}$$

$$\Delta Q_p = Q_{G_p} - Q_{D_p} - \sum_{q=1}^{N} V_p V_q Y_{pq} \sin\left(\delta_p - \delta_q - \theta_{pq} \right) \tag{8}$$

Here, V_p, V_q are the voltages at bus p and q, Y_{pq} is the pq-th element of admittance matrix amplitude, δ_p, δ_q are the voltage angles at bus p and q, P_{G_p} and Q_{G_p} is the active and reactive power generation at bus p, P_{D_p}, Q_{D_p} is the active and reactive power demand at bus p, N is the total number of bus bars.

3.2.2 Inequality Constraints

(a) The Generator Real and Reactive Power Outputs

$$P_{G_p}^{\min} \leq P_{G_p} \leq P_{G_p}^{\max}, p = 1, ..., N_{PV} \tag{9}$$

$$Q_{G_p}^{\min} \leq Q_{G_p} \leq Q_{G_p}^{\max}, p = 1, ..., N_{PV} \tag{10}$$

Here, $P_{G_p}^{\min}$ and $P_{G_p}^{\max}$ are the minimum and maximum active power, $Q_{G_p}^{\min}$ and $Q_{G_p}^{\max}$ are the minimum and maximum reactive power of the *p-th* generating unit.

(b) Voltage Magnitudes at Each Bus in the Network

$$V_p^{\min} \leq V_p \leq V_p^{\max}, p = 1, ..., N \tag{11}$$

Here, $V_p^{\min}$ and $V_p^{\max}$ are the minimum and maximum voltage of the *p-th* bus.

(c) The Discrete Transformer Tap Settings

$$T_p^{\min} \leq T_p \leq T_p^{\max}, p = 1, ..., N_T \tag{12}$$

Here, $T_p^{\min}$ and $T_p^{\max}$ minimum and maximum tap setting limits of the *p-th* transformer.

(d) The Discrete Reactive Power Injections Due to Capacitor Banks

$$Q_{Cp}^{\min} \leq Q_{Cp} \leq Q_{Cp}^{\max}, p = 1, ..., N_C \tag{13}$$

Here, $Q_{Cp}^{\min}$ and $Q_{Cp}^{\max}$ are minimum and maximum Var injection limits of the *p-th* shunt capacitor.

(e) The Transmission Lines Loading

$$S_{Lp} \leq S_{Lp}^{\max}, p = 1, ..., N_L \tag{14}$$

Here, S_{Lp} and $S_{Lp}^{\max}$ are the apparent power flow of the *p-th* branch and maximum apparent power flow limit of the *p-th* branch.

4 TRANSIENT STABILITY ASSESSMENT

This paper proposes the combination of time-domain simulation and TLBO to evaluate the system transient stability assessment. For describing the transient behavior of a system incorporating transient stability, the generator rotor angles are expressed here with respect to the inertial center of all the generators. So, the position of center of inertia (COI) is proposed by Athay, Podmere, and Virmani (1979),

$$\delta_{COI} = \frac{\sum_{k=1}^{N_{PV}} M_k \delta_k}{\sum_{k=1}^{N_{PV}} M_k} \tag{15}$$

Here, M_k is the inertia constant of kth generator, δ_k is the rotor angle of the k-th generator. In view of inequality constraints, the rotor angle constraint may be expressed as follows:

$$\delta' \leq \delta_k - \delta_{COI} \leq \delta'' \tag{16}$$

$$k \subseteq S_G$$

Here, S_G is the set of generators and δ', δ'' are the lower and upper limits of generator rotor angles respectively.

5 TEACHING LEARNING BASED OPTIMIZATION (TLBO)

The TLBO algorithm is a newly introduced evolutionary algorithm which is a population based technique, using a population of solutions to reach the global solution. Teaching-learning is an important process where every individual tries to learn something from other individuals to improve themselves. TLBO (Rao, Savsani, & Vakharia, 2011) uses the population of students to achieve the best possible solution of the search space. The backbone of TLBO method is based on the simulation of a classical learning process consisting of two stages: (i) learning through teacher (also known as teacher phase) and (ii) learning through interacting with the other learners (known as learner phase). Basically, this optimization method is based on the effect of the influence of a teacher on the output of learners in a class. In this optimization algorithm a group of learners is considered as population and different subjects offered to the learners are considered as different design variables of the optimization problem and a learner's result is analogous to the 'fitness' value of the optimization problem. The teacher is considered here the most learned person in the society; the best solution so far is analogous to Teacher in TLBO. In a real classroom, the learners try to increase their knowledge using the knowledge of their teacher. Therefore in order to improve their quality (knowledge), the learners move towards their teacher.

5.1 Teacher Phase

This is the initial part of the algorithm where a teacher tries to increase the mean result of the class in the subject taught by him or her depending on his or her capability. However, it also depends on the learning capability of the students present in the class room. The formulated form of the teacher phase can be written as follows:

Suppose the teacher tries to enhance the average result of the class from initial value of Ψ to his own level. If the new mean result of the mth subject at the nth iteration is Ψ_{new}^n, then the difference between the previous and new mean Ψ_{diff}^n may be defined as follows:

$$\Psi_{diff}^n = \text{rand} \times \left(\Psi_{new}^n - T_f . \Psi_m^n \right) \tag{17}$$

Here, $rand$ represents a uniformly distributed random variable within the range $[0,1]$ and T_f is a teaching factor which will be used for deciding the value of mean to be changed and its value can be 1 or 2.

So the updated result of the *m-th* subject of the *i-th* student can be modified as follows:

$$x_{i,m}^{n+1} = x_{i,m}^n + \Psi_{diff}^n \tag{18}$$

Here, $x_{i,m}^n, x_{i,m}^{n+1}$ are the result of the mth subject of the ith student at the nth and $(n+1)^{th}$ iterations, Ψ_{diff}^n is the difference between the mean result of the mth subject at the nth and $(n+1)^{th}$ iterations.

5.2 Learner Phase

This is the second part of the algorithm where the interaction of learners with one another takes place. In this phase, enhancement of the knowledge of an individual is done by peer learning from an optional students and self-organized learning between the individual and other learners, The mutual interaction process tends to enhance the knowledge of the learner if the randomly selected student has better knowledge than him. So, mathematically we can write,

$$x_{i,m}^{n+1} = x_{i,m}^n + \text{rand} \times \left(x_{i,m}^n - x_{i,k}^n \right) \text{ if } f(X_i) < f(X_k) \tag{19}$$

$$x_{i,m}^{n+1} = x_{i,m}^n + \text{rand} \times \left(x_{i,k}^n - x_{i,m}^n \right) \text{ if } f(X_i) > f(X_k) \tag{20}$$

Here, $x_{i,m}^n, x_{i,m}^{n+1}$ are the grade point of the mth subject of the ith student at n^{th} and $(n+1)^{th}$ iterations, $x_{i,k}^n$ is the grade point of the kth subject of the ith student (randomly selected) at n^{th} iteration, $f(X_k)$ is the overall grade point of the kth student.

The flow chart of the proposed TLBO algorithm is shown in Figure 1.

Figure 1. Flow chart of the TLBO algorithm

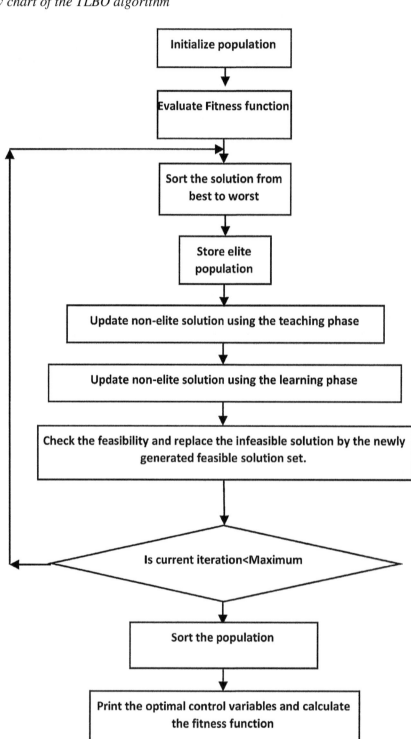

6 TLBO ALGORITHM APPLIED TO TSCOPF PROBLEM

The main idea behind TLBO is the simulation of a classical school learning process. The step size procedure for the implementation of TLBO algorithm is given as:

Step 1: Initialize the population of the entire individuals randomly within their operating limits. For TSCOPF problem, the active power of all the generator buses except the slack bus, voltages of generator buses, tap settings of regulating transformers, reactive power injections are generated randomly satisfying the inequality constraints (9), (11), (12) and (13). Afterward, NR-based load flow is run to determine the dependent variables such as active power of slack bus, reactive power of generator buses, load voltages, line loadings of the transmission lines check whether they satisfy the inequality constraints (10), (11) and (14). Moreover, the transient inequality constraint defined in (16) is verified. If any of these variable does not satisfy the inequality constraints; that particular population set is discarded and it is reinitialized. A feasible solution set represents the grade point of different subjects of a student. Several numbers of students depending upon the population size are generated. Each set of the student matrix represents a potential solution of the OPF.

Step 2: Evaluate the fuel cost using (6) of all the individuals of the current population. The fuel cost of individual solution represents the overall grade of individual student.

Step 3: Based on the overall grade point, sort them from best to worst. Identify the student having best grade point and assume it as a teacher and upgrade the design variables based on best solution using (17) and (18).

Step 4: Modified the grade point of all students in each subject using learning phase concept using (19) and (20).

Step 5: Run the load flow. Obtain the active power of slack bus, reactive power of generator buses, load voltages, line loadings of the transmission lines and rotor angle deviation and check whether they satisfy the inequality constraints (10), (11), (14) and (16) or not. The infeasible solutions are replaced by the by randomly generated new solutions set.

Step 6: Stop if the maximum no of iteration is achieved. Otherwise repeat from Step 2.

6.1 Mathematical Model of TLBO Algorithm Applied to TSCOF Problem

Shown in Figure 2.

7 SIMULATION RESULTS AND DISCUSSION

In this section, the TLBO algorithm is tested on IEEE-30 bus and IEEE-118 bus systems OPF for base load condition and it is also proposed for the solution of the TSCOPF problem on a Western Systems Coordinating Council (WSCC) 3-generator, 9-bus system and New England 10-generator, 39-bus system. In both systems, a classical generator model is used as synchronous generators and a constant impedance model is used for the loads. All simulations are carried out using MATLAB 2008a and implemented on a personal computer with Intel Pentium Dual-Core 1.73 GHz processor with 2GB-RAM. Integration time step is taken as 0.01s and total simulation period is 3.0s. The stopping criterion is chosen as 100 iterations.

Figure 2. Mathematical model of TLBO algorithm applied to TSCOPF problem

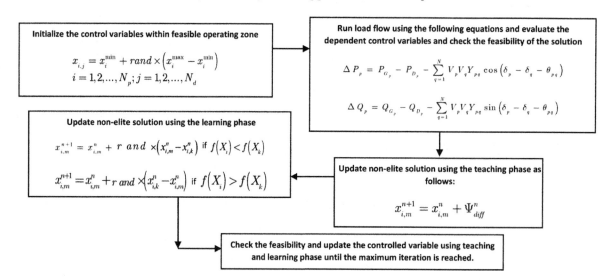

7.1 Description of Test Systems

7.1.1 Description of IEEE-30 Bus System

Firstly, the standard IEEE 30-bus system consisting of 6 generator buses at buses 1, 2, 5, 8, 11, 13 and 24 load bus shown in Figure 3 is used to evaluate the correctness and the relative performance of the proposed TLBO method. In this case study, the OPF problem is solved without considering the transient stability constraint of the power system. The power system is interconnected by 41 transmission lines of which four branches (6–9), (6–10), (4–12) and (28–27) are with the tap changing transformer and the total system demand for the 24 load buses is 2.834 p.u. In this study, the voltage magnitude limits of all buses are set to 0.95 p.u for lower bound and to 1.1 p.u for upper bound. The lower and upper limits of the transformers tap settings are considered as 0.9 p.u. and 1.1 p.u., respectively. To assess the potential of the proposed approach, a comparison between the results of fuel cost obtained by the proposed TLBO approach and those reported in the literature is carried out.

7.1.2 Description of IEEE-118 Bus System

In order to show the effectiveness of the proposed TLBO approach in larger scale power systems, a standard IEEE 118-bus test system is considered in this case study. This test system consists of 54 generator buses, 186 transmission lines, 12 shunt capacitors and 9 branches under load tap setting transformer branches. The shunt reactive power sources are considered at buses 34, 44, 45, 46, 48, 74, 79, 82, 83, 105, 107, and 110 and the tap changing transformers are connected between the transmission lines 5-8, 25-26, 17-30, 37-38, 59-63, 61-64, 65-66, 68-69, and 80-81. The total load demand of system is 4242 MW and 1438 MVAR. The bus data, the line data and the cost coefficients and minimum and maximum limits of real power generations are taken from (IEEE 118-bus test system). The maximum and minimum values of voltages of all generators are set at 1.1–0.9 in p.u. The maximum and minimum values for voltages of all load buses are defined as 1.06 and 0.94 in p.u, respectively and the maximum and minimum values of

Figure 3. The IEEE-30 bus test system

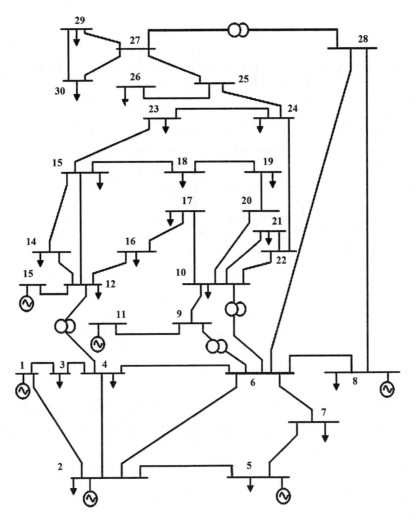

shunt reactive power sources are 0.0 and 0.4 in p.u., respectively. The tap settings are considered within the interval 0.9–1.05. To validate the performance and superiority of the proposed TLBO algorithm, the test results are compared with those obtained by other algorithms available in the literature.

7.1.3 Description of the WSCC 3-Machine, 9-Bus Test System

In order to further demonstrate the effectiveness of the proposed TLBO method, the OPF problem considering transient stability constraint is performed in the Western Systems Coordinating Council (WSCC) 9-Bus system. The single line diagram of the WSCC system is shown in the Figure 4. The fuel cost parameters and the ratings of generators are taken from Nguyen and Pai (2003) and the system data are available in Sauer and Pai (1998).

Figure 4. The WSCC three-generator, nine-bus test system

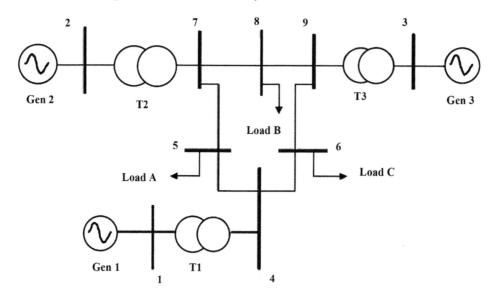

7.1.4 Description of the New England 10-Machine, 39-Bus Test System

The single line diagram of the system is shown in the Figure 5. The system data are available from (Pai, 1989), the cost co-efficient data and power generation limits of different generators of New England 39-bus system are adopted from (Xu, Dong, Meng, Zhao, & Wong, 2012) and also given in Table 6. The system comprises of 10 generator buses and 19 load buses. Bus 39 is taken here as slack bus. A 3-phase to ground fault occurs at bus 29 and between lines 28-29 in the system. The above fault is cleared by opening the contacts of the circuit breakers by 100ms.

7.2 Discussion

7.2.1 Discussion on IEEE 30-Bus Bus System

Table 1 lists the optimal settings of control variables, minimum fuel cost obtained by using different methods namely TLBO, MDE (Sayah, & Zehar, 2008), BBO (Bhattacharya, & Chattopadhyay, 2011) and GSA (Duman, Guvenc, Sonmez, & Yorukeren, 2012). From the results it is quite clear that the best fuel cost obtained from proposed TLBO technique is less compared to other discussed evolutionary algorithms. Figure 6 shows the fuel cost convergence characteristics obtained by the proposed TLBO method. It is clear from the figure that the solution obtained by TLBO converges quickly to optimal solution.

7.2.2 Discussion on IEEE 118-Bus Bus System

The simulation results in Table 2 show that the TLBO applied in this paper performs better than the BBO and QOBBO algorithms reported in the literature. Figure 7 shows the convergence test result obtained using TLBO approach for this test system. The graph illustrated in Figure 7 shows that the rate

A Novel Optimization Algorithm

Figure 5. The New England 10-machine, 39-bus test system

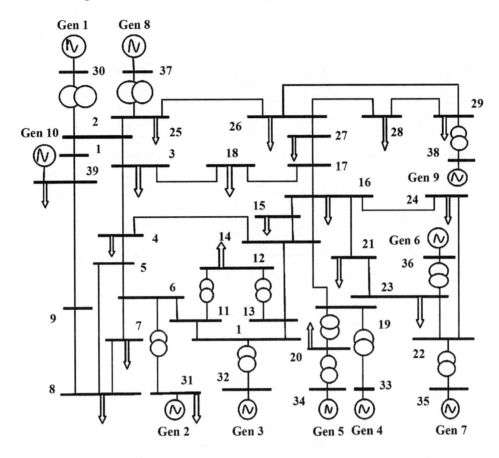

Figure 6. Cost convergence characteristics using TLBO (IEEE 30-bus system)

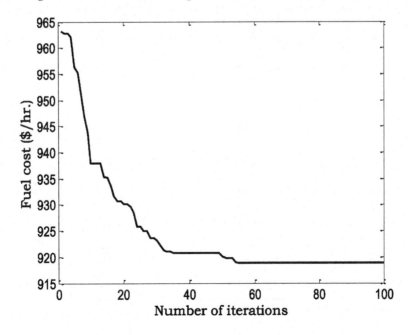

Table 1. Simulation results obtained by different techniques for Case 1 (Base Load Condition)

Algorithms	MDE	BBO	GSA	TLBO
P_{g1} (MW)	197.426	199.99	199.599	199.60
P_{g2} (MW)	52.037	37.812	51.946	20.12
P_{g5} (MW)	15.000	20.251	15.000	21.00
P_{g8} (MW)	10.000	14.375	10.000	26.82
P_{g11} (MW)	10.001	10.035	10.000	11.55
P_{g13} (MW)	12.000	12.001	12.000	13.87
V_{g1} (p.u.)	1.0371	1.0500	1.0990	1.0960
V_{g2} (p.u.)	1.0130	1.0358	1.0180	1.0777
V_{g5} (p.u.)	0.9648	1.0086	1.0522	1.0468
V_{g8} (p.u.)	1.0320	1.0137	0.9500	1.0612
V_{g11} (p.u.)	1.0982	1.0958	0.9634	1.0768
V_{g13} (p.u.)	1.0890	1.0814	0.9507	1.0872
T_{11}	1.0969	1.0308	0.9090	1.0730
T_{12}	1.0909	0.91443	0.9182	1.0144
T_{15}	1.0991	0.99142	0.9256	1.0995
T_{36}	1.0021	0.93864	0.9459	0.9774
FC ($/hr)	930.793	919.7647	929.7240	918.8692

Table 2.

Control Variables	TLBO	QOBBO	BBO	Control Variables	TLBO	QOBBO	BBO
P_{G1}	28.49	0.00	28.61	V_{G27}	1.0236	1.0635	1.0102
P_{G4}	0.09	1.93	0.00	V_{G31}	1.001	1.0518	0.9975
P_{G6}	1.77	9.96	1.80	V_{G32}	1.0143	1.0584	1.0048
P_{G8}	0.29	2.00	0.00	V_{G34}	1.0351	1.0071	1.0253
P_{G10}	391.64	394.75	399.97	V_{G36}	1.0382	0.9981	1.0258
P_{G12}	86.81	86.37	85.39	V_{G40}	1.0261	0.9954	1.0138
P_{G15}	16.01	16.95	15.43	V_{G42}	1.0286	1.0053	1.0118
P_{G18}	10.2	18.39	7.66	V_{G46}	1.0369	1.0381	1.0175
P_{G19}	19.24	10.24	18.07	V_{G49}	1.0467	1.0525	1.0264
P_{G24}	0.63	1.98	0.00	V_{G54}	1.0221	1.0233	0.9745
P_{G25}	195.98	195.82	193.74	V_{G55}	1.0228	1.0221	0.9721
P_{G26}	284.15	287.14	279.26	V_{G56}	1.0207	1.0237	0.9719
P_{G27}	8	8.91	9.21	V_{G59}	1.0307	1.0521	0.9983
P_{G31}	7.73	8.15	7.23	V_{G61}	1.0425	1.0703	1.0151
P_{G32}	19.37	22.60	16.46	V_{G62}	1.0379	1.0670	1.0136
P_{G34}	5.17	5.66	4.93	V_{G65}	1.0962	1.0971	1.0618

continued on following page

A Novel Optimization Algorithm

Table 2. Continued

Control Variables	TLBO	QOBBO	BBO	Control Variables	TLBO	QOBBO	BBO
P_{G36}	9.5	8.67	11.01	V_{G66}	1.0611	1.0752	1.0398
P_{G40}	45.11	46.64	50.81	V_{G69}	1.0874	1.0793	1.0819
P_{G42}	35.56	44.21	43.01	V_{G70}	1.0647	1.0486	1.0588
P_{G46}	18.6	18.49	19.15	V_{G72}	1.0484	1.0507	1.0429
P_{G49}	200.05	185.54	193.79	V_{G73}	1.065	1.0501	1.0563
P_{G54}	47.78	50.43	49.62	V_{G74}	1.055	1.0368	1.0549
P_{G55}	39.9	35.29	30.40	V_{G76}	1.0484	1.0368	1.0483
P_{G56}	33.84	29.57	29.65	V_{G77}	1.0617	1.0461	1.0625
P_{G59}	147.79	152.71	148.91	V_{G80}	1.0722	1.0595	1.0726
P_{G61}	149.57	150.35	147.76	V_{G85}	1.0633	1.0512	1.0689
P_{G62}	0.04	0.04	0.00	V_{G87}	1.0645	1.0546	1.0684
P_{G65}	356.87	342.32	353.72	V_{G89}	1.0694	1.0647	1.0835
P_{G66}	346.34	349.63	353.93	V_{G90}	1.0472	1.0447	1.0567
P_{G69}	457.8573	460.78	455.82	V_{G91}	1.0455	1.0393	1.0552
P_{G70}	0.51	2.26	2.49	V_{G92}	1.0535	1.0501	1.0641
P_{G72}	0.68	2.17	2.37	V_{G99}	1.0466	1.0507	1.0486
P_{G73}	1.86	1.13	2.29	V_{G100}	1.0436	1.0516	1.0466
P_{G74}	14.35	15.80	19.98	V_{G103}	1.0283	1.0457	1.0278
P_{G76}	21.82	23.22	25.44	V_{G104}	1.0101	1.0324	1.0129
P_{G77}	0.22	0.35	0.00	V_{G105}	1.0004	1.0311	1.0069
P_{G80}	424.35	429.12	433.87	V_{G107}	0.9992	1.0283	0.9982
P_{G85}	0.79	2.68	0.00	V_{G110}	1.003	1.0461	1.0035
P_{G87}	3.61	3.67	3.57	V_{G111}	1.0096	1.0537	1.0103
P_{G89}	501.86	489.60	497.28	V_{G112}	0.9957	1.0395	0.9935
P_{G90}	0.01	0.13	0.00	V_{G113}	1.0096	1.0570	1.0048
P_{G91}	0.57	1.22	0.00	V_{G116}	1.0851	1.0947	1.0479
P_{G92}	0.09	1.68	0.00	Q_{C34}	0.2708	0.54	22.25
P_{G99}	0.33	0.92	0.00	Q_{C44}	0.0255	4.89	4.11
P_{G100}	230.49	232.74	232.10	Q_{C45}	0.2362	20.81	21.33
P_{G103}	36.99	36.38	38.27	Q_{C46}	0.1284	1.27	6.89
P_{G104}	2.03	6.04	0.00	Q_{C48}	0.0797	10.18	6.70
P_{G105}	8.31	7.78	1.8	Q_{C74}	0.0917	2.63	7.01
P_{G107}	27.93	31.30	27.62	Q_{C79}	0.2813	28.10	29.86
P_{G110}	9.26	13.57	6.5	Q_{C82}	0.295	29.38	27.58
P_{G111}	35.71	34.31	34.97	Q_{C83}	0.0625	8.61	9.22
P_{G112}	31.73	34.09	36.25	Q_{C105}	0.1978	9.54	21.59
P_{G113}	0.38	0.96	0.00	Q_{C107}	0.0869	24.07	26.74

continued on following page

Table 2. Continued

Control Variables	TLBO	QOBBO	BBO	Control Variables	TLBO	QOBBO	BBO
P_{G116}	0.24	0.68	0.00	Q_{C110}	0.0817	29.43	28.35
V_{G1}	1.0106	1.0349	0.9761	T_{8-5}	1.0464	1.0171	1.0826
V_{G4}	1.0282	1.0494	0.9927	T_{26-25}	1.0965	1.0972	1.1000
V_{G6}	1.0187	1.0442	0.9820	T_{30-17}	1.0665	10229	1.0678
V_{G8}	1.0996	1.0914	1.0958	T_{38-37}	1.0171	1.0554	1.0261
V_{G10}	1.0976	1.0998	1.1000	T_{63-59}	1.038	1.0259	1.0432
V_{G12}	1.0133	1.0381	0.9791	T_{64-61}	1.0459	1.0272	1.0450
V_{G15}	1.0032	1.0422	0.9949	T_{65-66}	0.9	0.9039	0.9000
V_{G18}	1.0063	1.0482	0.9976	T_{68-69}	0.9699	0.9700	0.9331
V_{G19}	1.0004	1.0426	0.9953	T_{81-80}	1.0007	1.0206	0.9679
V_{G24}	1.037	1.0630	1.0350	Fuel Cost ($/hr.)	129626.6503	129633.0745	129686.3547
V_{G25}	1.0572	1.0988	1.0572				
V_{G26}	1.0947	1.0995	1.1000	Loss (KW)	76.4973	75.3150	78.1412

Simulation results for individual minimization of cost and emission (IEEE 118-bus system)

of convergence of TLBO is quick and it locates the best fuel cost so far obtained. Therefore from the simulation results and convergence graph it may be concluded that TLBO is well balanced between fast convergence and solution accuracy.

7.2.3 Discussion on WSCC 3-Machine, 9-Bus Test System

For this test system, the OPF and TSCOPF problems are solved for 3 fault cases.
The 3 fault cases for the test system are described in the following.

Figure 7. Cost convergence characteristic using TLBO (IEEE 118-bus system)

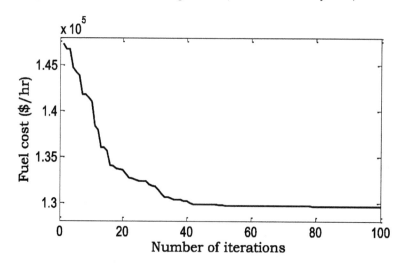

Case 1: Base Load Condition (OPF Without Transient Stability Constraint) Equality Constraints

The optimal power flow solution without transient stability limits has been obtained using TLBO algorithm. The main objective is to minimize the total fuel cost of the entire system. The minimum fuel cost values and power generations are compared with the other optimization methods, Base Case Solution (BPD) by Ahmadi, Ghasemi, Haddadi, and Lesani (2013), Artificial Bee Colony Optimization (ABC) by Kursat and Ulas (2013), Differential Evolution (DE) by Cai, Chung, and Wong (2008), Trajectory Sensitivities (TS) by Nguyen and Pai (2003), Time Domain Simulation (TDS) by Zarate-Minano, Cutsem, Milano, and Conejo (2010). The comparison tables of the simulation results by different optimization techniques for Case 1 are given in Table 3. From this table we can say that the value of the fuel cost obtained by the proposed algorithm is closer to the results obtained from others in the above problem.

Case 2: Opf With Transient Stability Constraints (Three Phase to Ground Fault at Bus 7)

In this case, an asymmetrical fault i.e. 3-phase to ground fault at bus 7 and between lines 7-5 in the system. The above fault was cleared by opening the contacts of the circuit breakers by 0.35 sec. The solution, obtained from this case satisfies transient stability limit. The minimum fuel cost values and power generations are compared with other optimization methods, TSCOPF Detailed Model (TSCOPF-DM) and TSCOPF Detailed Model Well Tuned (TSCOPF-DMWT) by Ahmadi, Ghasemi, Haddadi, and Lesani (2013), Artificial Bee Colony Optimization (ABC) by Kursat and Ulas (2013), Differential Evolution (DE) by Cai, Chung, and Wong (2008), Trajectory Sensitivities (TS) by Nguyen and Pai (2003), Time Domain Simulation (TDS) by Zarate-Minano, Cutsem, Milano, and Conejo (2010) are given in Table 4. The corresponding rotor angles of the generators (generator 2 and generator 3) are shown in Figure 8 and the cost convergence characteristics in Figure 9. From these figures, it is observed that, both these curves are within the safe limits and the system is stable and entirely secure.

Case 3 Opf With Transient Stability Constraints (Three Phase to Ground Fault at Bus 9)

In this case, an asymmetrical fault i.e. 3-phase to ground fault occurs at bus 9 and in lines 9-6 in the system. The above fault was cleared at 0.3sec by opening the contacts of the nearby circuit breakers. Here also, the minimum fuel cost values and power generations are compared with other optimization methods, Artificial Bee Colony Optimization (ABC) by Kursat and Ulas (2013), Differential Evolution

Table 3. Simulation results obtained by different techniques for Case 1 (WSCC 9-bus system)

Algorithms	BPD	ABC	DE	TS	TDS	TLBO
P_{g1} (MW)	105.94	107.15	105.94	106.19	105.94	105.3200
P_{g2} (MW)	113.04	114.18	113.04	112.96	113.04	113.1800
P_{g3} (MW)	99.23	96.74	99.29	99.20	99.24	99.3700
V_{g1} (p.u.)	1.05	1.033	1.050	1.000	1.05	1.0998
V_{g2} (p.u.)	1.05	1.024	1.050	1.000	1.05	1.1000
V_{g3} (p.u.)	1.04	1.028	1.040	1.000	1.04	1.0999
FC ($/hr)	1132.17	1131.87	1132.30	1132.59	1132.18	1131.0340

Table 4.

Algorithms	TSCOPF-DM	TSCOPF_DMWT	ABC	DE	TS	TDS	TLBO
P_{g1} (MW)	138.47	116.25	117.69	130.94	170.20	117.85	105.3100
P_{g2} (MW)	94.20	106.26	105.89	94.46	48.94	103.50	113.3300
P_{g3} (MW)	85.01	95.48	94.23	93.09	98.74	96.66	99.2400
V_{g1} (p.u.)	1.05	1.05	1.025	0.9590	1.000	1.05	1.0988
V_{g2} (p.u.)	1.05	1.05	1.070	1.0139	1.000	1.05	1.0987
V_{g3} (p.u.)	1.04	1.04	1.070	1.0467	1.000	1.04	1.0999
FC ($/hr)	1143.42	1133.34	1133.18	1140.06	1179.95	1134.01	1131.1908

Simulation results obtained by different techniques for Case 2 (WSCC 9-bus system)

Figure 8. Relative rotor angles of generator2 and generator 3 for fault at bus 7 (Case 2 of WSCC 9-bus system)

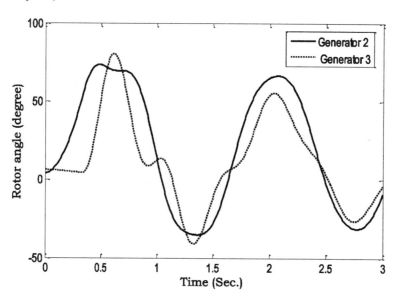

(DE) by Cai, Chung, and Wong (2008), Trajectory Sensitivities (TS) by Nguyen and Pai (2003), Time Domain Simulation (TDS) by Zarate-Minano, Cutsem, Milano, and Conejo (2010) given in Table 5 and corresponding rotor angles of the generators are shown in Figure 10 and cost convergence characteristics in Figure 11. It is observed that both curves are within limits and the system is entirely stable and secure.

7.2.4 Discussion on New England 10-Machine, 39-Bus Test System

The simulation results that we are getting from TLBO are compared with the classical model (CM), dynamic model (DM) and dynamic simulation algorithm (DSA) model. The active power generation and voltage of different generating units, fuel cost and transmission loss obtained from Classical model (CM) and Dynamic model (DM) (Ahmadi, Ghasemi, Haddadi, & Lesani, 2013), Dynamic simulation algorithm (DSA) model (Tu, Dessaint, & Duc, 2013) are also given in Table 7. From that we can say,

Figure 9. Cost convergence characteristics using TLBO (Case 2 of WSCC 9-bus system)

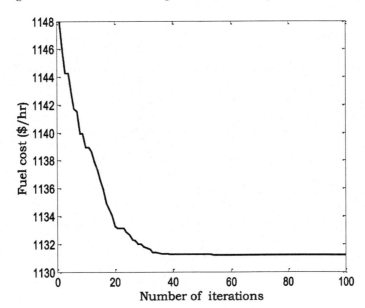

Table 5. Simulation results obtained by different techniques for Case 3 (WSCC 9-bus system)

Algorithms	ABC	DE	TS	TDS	TLBO
P_{g1} (MW)	121.23	130.01	164.38	120.01	105.21
P_{g2} (MW)	120.63	127.17	112.44	121.13	121.45
P_{g3} (MW)	75.94	60.72	41.00	76.84	99.32
V_{g1} (p.u.)	1.019	1.0495	1.000	1.05	1.0990
V_{g2} (p.u.)	1.041	1.0481	1.000	1.05	1.0988
V_{g3} (p.u.)	1.044	1.0327	1.000	1.04	1.0998
FC ($/hr)	1137.78	1148.58	1179.95	1137.82	1132.0151

the fuel cost obtained by the proposed TLBO method is significantly less compared to that obtained by other techniques. The relative position of the rotor angles of the generators with respect to generator 1 are shown in Figure 12. Here, it is observed that, all the rotor angle curves are within the safe limits and simulation results obtained with TLBO satisfy transient stability limit. The cost convergence characteristic of the proposed TLBO method is illustrated in Figure 13. Therefore, it may finally be concluded that the proposed TLBO algorithm produces minimum fuel cost without violating transient stability constraints.

8 CONCLUSION

A robust and efficient method for solving TSCOPF problems based on TLBO algorithm has been developed in this paper to meet the pressing need of the modern power systems. In this paper, the TLBO algorithm is successfully applied for the solution to the TSCOPF problem. Four test systems, IEEE 30-

Figure 10. Relative rotor angles of generator 2 and generator 3 for fault at bus 9
(Case 3 of WSCC 9-bus system)

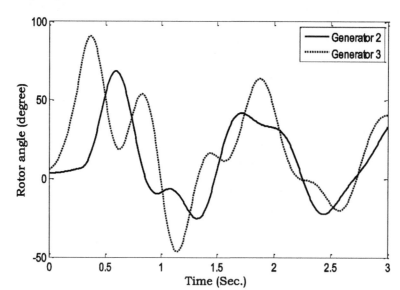

Figure 11. Cost convergence characteristics using TLBO for (Case 3 of WSCC 9-bus system)

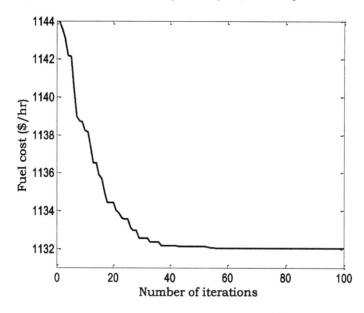

bus, IEEE 118-bus, WSCC 3-generator 9-bus system and New England 10 generator, 39 bus system are considered for practical generator operation in the proposed method. The computation results of TLBO algorithm is compared to other methods reported in the literature. It is clear from the simulation results that TLBO is able to find the optimal solutions in all cases without violating any operating constraint. Moreover, to verify the superiority, TLBO is compared with other state-of-the art algorithms and found

Table 6. Cost Co-efficient data and power generation limits of New England 39-bus system

Generator	α_i	β_i	λ_i	P_{min} (MW)	P_{max} (MW)
G_1	0.0193	6.9	0	0	350
G_2	0.0111	3.7	0	0	650
G_3	0.0104	2.8	0	0	800
G_4	0.0088	4.7	0	0	750
G_5	0.0128	2.8	0	0	650
G_6	0.0094	3.7	0	0	750
G_7	0.0099	4.8	0	0	750
G_8	0.0113	3.6	0	0	700
G_9	0.0071	3.7	0	0	900
G_{10}	0.0064	3.9	0	0	1200

Table 7. Simulation results obtained by TLBO, CM, DM and DSA for fault at bus number 29 of New England 39-bus system

Active Power Generation (in MW)	TLBO	CM	DM	DSA	Generated Voltages (in p.u.)	TLBO	CM	DM	DSA
PG_{30}	2.4422	248.7300	249.4470	247.8300	V_{30}	1.0996	1.0150	1.0150	0.9840
PG_{31}	5.6007	577.8400	578.3590	577.2300	V_{31}	1.0924	1.0870	1.0870	1.0740
PG_{32}	6.4363	654.4700	654.3560	653.4100	V_{32}	1.0863	1.0290	1.0290	1.0080
PG_{33}	6.3298	645.0000	641.7600	643.2800	V_{33}	1.0985	1.0160	1.0160	1.0140
PG_{34}	5.1098	518.8200	517.4100	517.7800	V_{34}	1.0997	1.0220	1.0220	1.0190
PG_{35}	6.5306	664.3200	660.7310	662.4600	V_{35}	1.0999	1.0620	1.0620	1.0670
PG_{36}	5.6103	571.3700	568.1810	569.5900	V_{36}	1.0999	1.0900	1.0900	1.0870
PG_{37}	5.3286	547.8100	547.6950	543.8800	V_{37}	1.0994	1.0470	1.0470	1.0120
PG_{38}	8.4015	752.0200	754.6180	774.5400	V_{38}	1.0995	1.0380	1.0380	1.0510
PG_{39}	9.5845	995.6000	1003.1810	1000.3500	V_{39}	1.0998	1.0530	1.0530	1.0190
Cost ($/hr)	60873.5574	61600.7600	61597.7600	61799.6800	Loss (MW)	43.15359	NA	NA	NA

that it performs significantly better than all other algorithms discussed in this paper in terms of the solution quality, speed and stability of the final solution. The proposed algorithm also possesses good convergence property. It may finally be concluded that novelty basis, the proposed algorithm is a very promising evolutionary optimization algorithm for the global optimization of any TSCOPF problems.

Figure 12. Relative rotor angles with respect to generator 1 for the fault at bus 29

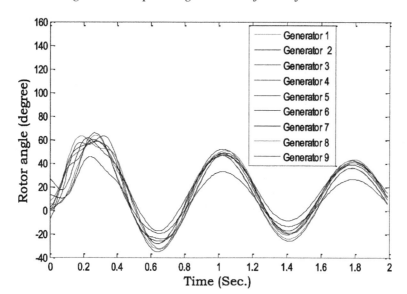

Figure 13. Cost convergence characteristics using TLBO

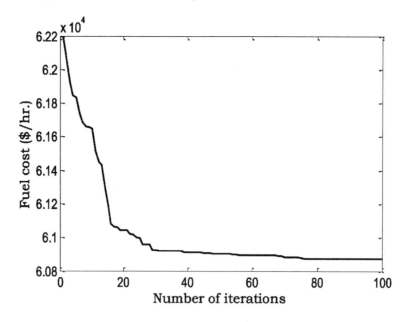

9 FUTURE RESEARCH DIRECTIONS

Though in this article, the proposed TLBO method is used to solve transient stability constrained OPF problem without incorporating any FACTS devices, in future, the work a can be extended by applying various FACTS devices such as UPFC, TCSC, STATCOM, etc in the proposed area. As the proposed method is proved as an efficient optimization technique for the proposed areas, it may be recommended as a promising algorithm for solving other complex engineering optimization problems. Though TLBO technique has the ability to search near global optimal solution, nevertheless the algorithm requires further improvement to produce probable global optimal solutions in reasonable time. To improve the convergence speed and solution quality of the TLBO algorithm, some modification may be made in the basic TLBO algorithm in future study.

REFERENCES

Abido, M. A. (2002a). Optimal power flow using tabu search algorithm. *Electric Power Components Systems*, *30*(5), 469–483. doi:10.1080/15325000252888425

Abido, M. A. (2002b). Optimal power flow using particle swarm optimization. *International Journal of Electrical Power & Energy Systems*, *24*(7), 563–571. doi:10.1016/S0142-0615(01)00067-9

Ahmadi, H., Ghasemi, H., Haddadi, A. M., & Lesani, H. (2013). Two approaches to transient stability-constrained optimal power flow. *Electrical Power and Energy Systems*, *47*, 181–192. doi:10.1016/j.ijepes.2012.11.004

Anitha, M., Subramanian, S., & Gnanadass, R. (2007). FDR PSO-Based transient stability constrained optimal power flow solution for deregulated power industry. *Electrical Power Components and Systems*, *35*(11), 1619–1632. doi:10.1080/15325000701351641

Athay, T., Podmere, R., & Virmani, S. (1979). A practical method for the direct analysis of Transient Stability. *IEEE Transactions on Power Apparatus and Systems*, *98*(2), 573–584. doi:10.1109/TPAS.1979.319407

Ayan, K., Kilic, U., & Barakli, B. (2015). Chaotic artificial bee colony algorithm based of security and transient stability constrained optimal power flow. *International Journal of Electrical Power & Energy Systems*, *64*, 136–147. doi:10.1016/j.ijepes.2014.07.018

Bhattacharya, A., & Chattopadhyay, P. K. (2011). Application of biogeography based optimization to solve different optimal power flow problems, IET generation. *Transmission and Distribution*, *5*, 72–80.

Burchet, R. C., Happ, H. H., & Vierath, D. R. (1984). Quadratically convergent optimal power flow. *IEEE Transactions on Power Apparent Systems*, *PAS-103*(11), 3267–3276. doi:10.1109/TPAS.1984.318568

Cai, H. R., Chung, C. Y., & Wong, K. P. (2008). Application of differential evolution algorithm for transient stability constrained optimal power flow. *IEEE Transactions on Power Systems*, *23*(2), 719–728. doi:10.1109/TPWRS.2008.919241

Cai, J., Ma, M., Li, L., Yang, Y., Peng, H., & Wang, X. (2007). Chaotic ant swarm optimization to economic dispatch. *Electric Power Systems Research*, *77*(10), 1373–1380. doi:10.1016/j.epsr.2006.10.006

Chan, K. Y., Ling, S. H., Chan, K. W., Iu, H. H. C., & Pong, G. T. Y. (2007), Solving multi-contingency transient stability constrained optimal power flow problems with an improved GA. *IEEE Congress on Evolutionary Computation* (pp. 2901–2908). doi:10.1109/CEC.2007.4424840

Chen, L., Tada, Y., Okamoto, H., Tanabe, R., & Ono, A. (2001). Optimal operation solutions of power systems with transient stability constraints, IEEE Transactions On Circuits and Systems—I: Fundamental Theory. *Appl.*, *48*, 327–339.

Dommel, H., & Tinny, W. (1968). Optimal power flow solution. *IEEE Transactions on Power Apparatus and Systems*, *PAS-87*(10), 1866–1876. doi:10.1109/TPAS.1968.292150

Duman, G., & Yorukeren, S. (2012). Optimal power flow using gravitational search algorithm. *Energy Conversion and Management*, *59*, 86–95. doi:10.1016/j.enconman.2012.02.024

El-Hawary, M. E. (Ed.). (1996). *Optimal Power Flow: Solution Techniques, Requirement and Challenges, IEEE Tutorial Course*. Piscataway, NJ: IEEE Service Center.

Ela, A. A. A. E., Abido, M. A., & Spea, S. R. (2010). Optimal power flow using differential evolution algorithm. *Electric Power Systems Research*, *80*(7), 878–885. doi:10.1016/j.epsr.2009.12.018

Fang, D. Z., Xiaodong, Y., Jingqiang, S., Shiqiang, Y., & Yao, Z. (2007). An optimal generation re-scheduling approach for transient stability enhancement. *IEEE Transactions on Power Systems*, *22*(1), 386–394. doi:10.1109/TPWRS.2006.887959

Fouad, A. A., & Vittal, V. (1992). *Power System Transient Stability Analysis Using the Transient Energy Function Method*. Englewood Cliffs, NJ: Prentice-Hall.

Gan, D., Thomas, R. J., & Zimmerman, R. D. (2000). Stability constrained optimal power flow. *IEEE Transactions on Power Systems*, *15*(2), 535–540. doi:10.1109/59.867137

Geng, G., & Jiang, Q. (2012). A Two-Level Parallel Decomposition Approach for Transient Stability Constrained Optimal Power Flow. *IEEE Transactions on Power Systems*, *27*(4), 2063–2073. doi:10.1109/TPWRS.2012.2190111

Ghasemi, M., Ghanbarian, M. M., Ghavidel, S., Rahmani, S., & Moghaddam, E. M. (2014). Modified teaching learning algorithm and double differential evolution algorithm for optimal reactive power dispatch problem: A comparative study. *Information Sciences*, *278*, 231–249. doi:10.1016/j.ins.2014.03.050

Ghasemi, M., Taghizadeh, M., Ghavidel, S., Aghaei, J., & Abbasian, A. (2015). Solving optimal reactive power dispatch problem using a novel teaching–learning-based optimization algorithm. *Engineering Applications of Artificial Intelligence*, *39*, 100–108. doi:10.1016/j.engappai.2014.12.001

Habiabollahzadeh, H., & Luo, S. A. (1989). Hydrothermal optimal power flow based on combined linear and nonlinear programming methodology. *IEEE Transactions on Power Apparent Systems*, *PWRS-4*(2), 530–537. doi:10.1109/59.193826

IEEE. (n.d.). *IEEE 118-bus test system*. Available at: http://www.ee.washington.edu/research/pstca/pf118/pg_tca118bus.htm

Jian, Q., & Huang, Z. (2008). An enhanced numerical discretization method for transient stability constrained optimal power flow. *IEEE Transactions on Power Systems*, *25*(4), 1790–1797. doi:10.1109/TPWRS.2010.2043451

Karaboga, D. (2005). *An idea based on honey bee swarm for numerical optimization, Technical Report-TR06*. Erciyes University, Engineering Faculty, Computer Engineering Department.

Khazali, A. H., & Kalantar, M. (2011). Optimal reactive power dispatch based on harmony search algorithm. *International Journal of Electrical Power Systems*, *33*(3), 684–692. doi:10.1016/j.ijepes.2010.11.018

Kundur, P. (1994). Power System Stability and Control, McGraw-Hill. Kursat, Y., Ulas, K., Solution of transient stability-constrained optimal power flow using artificial bee colony algorithm. *Turkish Journal of Electrical Engineering & Computer Sciences*, *21*, 360–372.

Lai, L. L., Ma, J. T., Yokoyama, R., & Zhao, M. (1997). Improved genetic algorithms for optimal power flow under both normal and contingent operation states. *International Journal of Electrical Power Systems*, *19*(5), 287–292. doi:10.1016/S0142-0615(96)00051-8

Laufenberg, M. J., & Pai, M. A. (1998). A new approach to dynamic security assessment using trajectory sensitivities. *IEEE Transactions on Power Systems*, *13*(3), 953–958. doi:10.1109/59.709082

Layden, D., & Jeyasurya, B. (2004). Integrating security constraints in optimal power flow studies. *IEEE Power Engineering Society General Meeting, 1*, 125-129.

Lin, W. M., Cheng, F. S., & Tsay, M. T. (2002). An improved tabu search for economic dispatch with multiple minima. *IEEE Transactions on Power Systems*, *17*(1), 851.

Mandal, B., & Roy, P. K. (2014). Multi-objective optimal power flow using quasi-oppositional teaching learning based optimization. *Applied Soft Computing*, *21*, 590–606. doi:10.1016/j.asoc.2014.04.010

Mo, N., Zou, Z. Y., Chan, K. W., & Pong, T. Y. G. (2007). Transient stability constrained optimal power flow using particle swarm optimization, IET Generation. *Transmission & Distribution*, *1*(3), 476–483. doi:10.1049/iet-gtd:20060273

Momoh, J., El-Hawary, M., & Adapa, R. (1999). A review of selected optimal power flow literature to 1993, Parts I and II. *IEEE Transactions on Power Systems*, *14*(1), 96–111. doi:10.1109/59.744492

Mota-Palomino, R., & Quintana, V. H. (1986). Sparse reactive power scheduling by a penalty function linear programming technique. *IEEE Transactions on Power Systems*, *1*(3), 31–39. doi:10.1109/TPWRS.1986.4334951

Nguyen, T. B., & Pai, M. A. (2003). Dynamic security-constrained rescheduling of power systems using trajectory sensitivities. *IEEE Transactions on Power Systems*, *18*(2), 848–854. doi:10.1109/TPWRS.2003.811002

Ongsakul, W., & Tantimaporn, T. (2006). Optimal power flow by improved evolutionary programming. *Electric Power Components and Systems*, *34*(1), 79–95. doi:10.1080/15325000691001458

Pablo, Y., Juan, M. R., & Carlos, R. C. (2008). Optimal power flow subject to security constraints solved with a particle swarm optimizer. *IEEE Transactions on Power Systems*, *23*(1), 33–40. doi:10.1109/TPWRS.2007.913196

Pai, M. A. (1989). *Energy Function Analysis for Power System Stability*. Norwell, MA: Kluwer. doi:10.1007/978-1-4613-1635-0

Pizano-Martinez, A., Fuerte-Esquivel, C. R., & Ruiz-Vega, D. (2011). A new practical approach to global transient stability-constrained optimal power flow. *IEEE Transactions on Power Systems, 26*(3), 1686–1696. doi:10.1109/TPWRS.2010.2095045

Pizano-Martinez, A., Fuerte-Esquivel, C. R., Zamlra-Cardenas, E., & Ruiz-Vega, D. (2014). Selective transient stability-constrained optimal power flow using a SMIE and trajectory sensitivity unified analysis. *Electric Power Systems Research, 109*, 32–44. doi:10.1016/j.epsr.2013.12.003

Rao, R. V., Savsani, V. J., & Vakharia, D. P. (2011). Teaching–learning-based optimization: A novel method for constrained mechanical design optimization problems. *Computer Aided Design, 43*(3), 303–315. doi:10.1016/j.cad.2010.12.015

Roa-Sepulveda, C. A., & Pavez-Lazo, B. J. (2003). A solution to the optimal power flow using simulated annealing. *International Journal of Electrical Power Systems, 25*(1), 47–57. doi:10.1016/S0142-0615(02)00020-0

Roy, P. K., & Bhui, S. (2013). Multi-objective quasi-oppositional teaching learning based optimization for economic emission load dispatch problem. *International Journal of Electrical Power & Energy Systems, 53*, 937–948. doi:10.1016/j.ijepes.2013.06.015

Roy, P. K., Ghosal, S. P., & Thakur, S. S. (2010). Biogeography based optimization for multi-constraint optimal power flow with emission and non-smooth cost function. *Expert Systems with Applications, 37*(12), 8221–8228. doi:10.1016/j.eswa.2010.05.064

Roy, P. K., Mandal, B., & Bhattacharya, K. (2012). Gravitational search algorithm based optimal reactive power dispatch for voltage stability enhancement, Electric. Power Components and. *Systems, 40*(9), 956–976.

Roy, P. K., Paul, C., & Sultana, S. (2014). Oppositional teaching learning based optimization approach for combined heat and power dispatch. *International Journal of Electrical Power & Energy Systems, 57*, 392–403. doi:10.1016/j.ijepes.2013.12.006

Roy, P. K., Sur, A., & Pradhan, D. (2013). Optimal short-term hydro-thermal scheduling using quasi-oppositional teaching learning based optimization. *Engineering Applications of Artificial Intelligence*.

Ruiz-Vega, D., & Pavella, M. (2003). A comprehensive approach to transient stability control. I. Near optimal preventive control. *IEEE Transactions on Power Systems, 18*(4), 1446–1453. doi:10.1109/TPWRS.2003.818708

Sauer, P. W., & Pai, M. A. (1998). *Power System Dynamics and Stability*. Upper Saddle River, NJ: Prentice-Hall.

Sayah, S., & Zehar, K. (2008). Modified differential evolution algorithm for optimal power flow with non-smooth cost functions. *Energy Conversion and Management, 49*(11), 3036–3042. doi:10.1016/j.enconman.2008.06.014

Shiwei, X., Chan, K. W., & Guo, Z. (2014). A novel margin sensitivity based method for transient stability constrained optimal power flow. *Electric Power Systems Research, 108*, 93–102. doi:10.1016/j.epsr.2013.11.002

Sivasubramani, S., & Swarup, K. S. (2012). Multiagent based differential evolution approach to optimal power flow. *Applied Soft Computing*, *12*(2), 735–740. doi:10.1016/j.asoc.2011.09.016

Stott, B. (1979). Power system dynamic response calculations. *Proceedings of the IEEE*, *67*(2), 219–241. doi:10.1109/PROC.1979.11233

Sultana, S., & Roy, P. K. (2014). Optimal capacitor placement in radial distribution systems using teaching learning based optimization. *International Journal of Electrical Power Systems*, *54*, 387–398. doi:10.1016/j.ijepes.2013.07.011

Sun, D. I., Ashley, B., Brewer, B., Hughes, A., & Tinney, W. F. (1984). Optimal powerflowby Newton approach. *IEEE Transactions on Power and Apparent Systems*, *PAS-103*(10), 2864–2875. doi:10.1109/TPAS.1984.318284

Swain, A. K., & Morris, A. S. (2000). A novel hybrid evolutionary programming method for function optimization. In *Proceedings of 2000 congress on evolutionary computation*, (pp. 699–705). doi:10.1109/CEC.2000.870366

Tripathy, M., & Mishra, S. (2007). Bacteria foraging-based solution to optimize both real power loss and voltage stability limit. *IEEE Transactions on Power Systems*, *22*(1), 240–248. doi:10.1109/TPWRS.2006.887968

Tu, X., Dessaint, A. L., & Nguyen-Duc, H. (2013). Transient stability constrained optimal power flow using independent dynamic simulation, IET Generation. *Transmission & Distribution*, *7*(3), 244–253. doi:10.1049/iet-gtd.2012.0539

Tu, X., Dessaint, L. A., & Kamwa, I. (2013). Fast approach for transient stability constrained optimal power flow based on dynamic reduction method. *IET Gen. Trans. Distr*, *8*(7), 1293–1305. doi:10.1049/iet-gtd.2013.0404

Wang, L., Zheng, D. Z., & Tang, F. (2002). An improved evolutionary programming for optimization. In *Proceedings of the Fourth World Congress on Intelligent Control and Automation*. doi:10.1109/WCICA.2002.1021386

Xia, S. W., Zhou, B., Chan, K. W., & Guo, Z. Z. (2015). An improved GSO method for discontinuous non-convex transient stability constained optimal power flow with complex system model. *International Journal of Electrical Power & Energy Systems*, *64*, 483–492. doi:10.1016/j.ijepes.2014.07.051

Xia, X., & Wei, H. (2012). Transient stability constrained optimal power flow based on second-order differential equations. *Procedia Engineering.*, *29*, 874–878. doi:10.1016/j.proeng.2012.01.057

Xu, Y., Dong, Z. Y., Meng, K., Zhao, J. H., Wong, K. P. (2012). A hybrid method for transient stability-constrained optimal power flow computation. *IEEE Trans. Power Syst.*, *27*(4), 1769-1777.

Yan, X., & Quintana, V. H. (1999). Improving an interior point based OPF by dynamic adjustments of step sizes and tolerances. *IEEE Transactions on Power Systems*, *14*(2), 709–717. doi:10.1109/59.761902

Yan, Z., Xiang, N. D., Zhang, B. M., Wang, S. Y., & Chung, T. S. (1996). A hybrid decoupled approach to optimal power flow. *IEEE Transactions on Power Systems*, *11*(2), 947–954. doi:10.1109/59.496179

Yuan, Y., Kubokawa, J., & Sasaki, H. (2003). A solution of optimal power flow with multicontingency transient stability constraints. *IEEE Transactions on Power Systems, 18*(3), 1094–1102. doi:10.1109/TPWRS.2003.814856

Yuryevich, J., & Wong, K. P. (1999). Evolutionary programming based optimal power flow algorithm. *IEEE Transactions on Power Systems, 14*(4), 1245–1250. doi:10.1109/59.801880

Zarate-Minano, R., Cutsem, T. V., Milano, F., & Conejo, A. J. (2010). Securing transient stability using time-domain simulations within an optimal power flow. *IEEE Transactions on Power Systems, 25*(1), 243–253. doi:10.1109/TPWRS.2009.2030369

Zhang, X., Dunn, R. W., & Li, F. (2003). Stability constrained optimal power flow for the balancing market using genetic algorithms. *IEEE Power Engineering Society General Meeting, 2*, 932-937.

Chapter 8
Improved Pseudo-Gradient Search Particle Swarm Optimization for Optimal Power Flow Problem

Jirawadee Polprasert
Asian Institute of Technology, Thailand

Weerakorn Ongsakul
Asian Institute of Technology, Thailand

Vo Ngoc Dieu
Ho Chi Minh City University of Technology, Vietnam

ABSTRACT

This paper proposes an improved pseudo-gradient search particle swarm optimization (IPG-PSO) for solving optimal power flow (OPF) with non-convex generator fuel cost functions. The objective of OPF problem is to minimize generator fuel cost considering valve point loading, voltage deviation and voltage stability index subject to power balance constraints and generator operating constraints, transformer tap setting constraints, shunt VAR compensator constraints, load bus voltage and line flow constraints. The proposed IPG-PSO method is an improved PSO by chaotic weight factor and guided by pseudo-gradient search for particle's movement in an appropriate direction. Test results on the IEEE 30-bus and 118-bus systems indicate that IPG-PSO method is superior to other methods in terms of lower generator fuel cost, smaller voltage deviation, and lower voltage stability index.

INTRODUCTION

Optimal power flow (OPF) is to determine the optimal settings of control variables including real power generation outputs, generator bus voltages, tap setting of transformer and shunt VAR compensators outputs to minimize the generator fuel cost function subject to power balance, and generator operating and network constraints. The objective functions are generation fuel cost with valve-point loading effects and

DOI: 10.4018/978-1-4666-9755-3.ch008

voltage deviation for voltage profile improvement while satisfying the power flow equations, generator bus voltage, real power generation output and reactive power generation output, transformer tap setting, shunt VAR compensators, load bus voltages, and transmission line loadings constraints. OPF has long been developed for on-line operation and control of power system.

So far, various conventional programming techniques such as Newton's method, gradient-based methods, linear programming (LP), nonlinear programming (NLP), quadratic programming (QP), and interior point methods (IPMs) (Lai & Ma,1997; Abido, 2002; Roa-Sepulveda & Pavez-Lazo, 2003; Kennedy & Eberhart, 1995; Abou El Elaa & Abido, 1992; Thanushkodi et.al., 2008) have been applied to solve OPF problems. Even though these methods can quickly find a solution, they are highly sensitive to the starting points and may converge prematurely. Moreover, these methods cannot handle non-convex objective function. Therefore, these techniques may not be practical because of nonlinear characteristics of objective function and constraints. To overcome these difficulties, the artificial intelligence and evolutionary based methods including improved genetic algorithm (IGA) (Lai & Ma, 1997), improved evolutionary programming (IEP), tabu search (TS) (Abido, 2002), simulated annealing (SA) (Roa-Sepulveda & Pavez-Lazo, 2003), gravitational search algorithm (GSA) (Duman et.al., 2012), biogeography-based optimization (BBO) (Bhattacharya & Chattopadhyay, 2011) were proposed to solve OPF problems. However, the solutions were still far from the optimal solutions.

Recently, PSO has been proposed for solving economic dispatch (ED) (Abou El Elaa & Abido, 1992; Thanushkodi et.al., 2008; Park et.al., 2006; Chen & Yeh, 2006), reactive power dispatch (RPD) (Chaturvedi et.al., 2008; Yoshida et.al., 2001; Krost, Venayagamoorthy, & Grant, 2008), optimal power flow (OPF), and optimal location of FACTs devices (Swarup, 2006; El-Ela, & El-Sehiemy, 2007; Sutha & Kamaraj, 2008; Abido, 2002). PSO with time-varying inertia weighting factor (PSO-TVIW), PSO with time-varying acceleration coefficients (PSO-TVAC), self-organizing hierarchical PSO with TVAC (SPSO-TVAC) (Ratnaweera & Halgamuge, 2004), PG-PSO (Le & Dieu, 2012), and stochastic weight trade-off PSO (SWT-PSO) (Chalermchaiarbha & Ongsakul, 2013) have been proposed to obtain better and faster solutions of OPF problem. In addition, the pseudo-gradient based PSO (PG-PSO) method was also applied for solving optimal reactive power dispatch (ORPD) (Le & Dieu, 2012) and ED problem (Dieu et.al., 2013). The PG-PSO is based on SPSO-TVAC with guiding position by using pseudo-gradient (PG) for searching the suitable direction towards the optimal solution. The PG-PSO method uses constant weighting factor which may not be effective in solving optimal power flow analysis.

In this paper, an improved pseudo-gradient search particle swarm optimization (IPG-PSO) algorithm is proposed with new linearly chaotic weighting factor and pseudo gradient search algorithm. The proposed IPG-PSO method is tested on the IEEE 30-bus and IEEE 118-bus systems with three different objective functions including quadratic fuel cost function with valve-point loading effects, quadratic fuel cost function with voltage deviation, and quadratic fuel cost function with voltage stability index. The obtained results are compared with PSO-TVIW, PSO-TVAC, SPSO-TVAC, PG-PSO, SWT-PSO, ACO (Allaoua & Laoufi, 2009), IEP (Ongsakul & Tantimaporn, 2006), EP (Yuryevich & Wong, 1999), gravitational search algorithm (GSA) (Duman et.al., 2012), differential evolution (DE) (Abou El Elaa & Abido, 2010), modified differential evolution (MDE) (Sayah & Zehar, 2008), evolving ant direction PSO based approach (EADPSO) (Vaisakh et.al., 2013), and biogeography-based optimization (BBO) (Bhattacharya & Chattopadhyay, 2011).

The organization of this paper is as follows: Section II describes the OPF problem formation. Improved Pseudo Gradient Search Particle Swarm Optimization is proposed in Section III. Numerical results of the non-convex OPF problem using the proposed IPG-PSO method on the IEEE 30-bus and 118-bus systems are demonstrated in Section IV. Finally, Section V concludes the paper.

OPF PROBLEM FORMULATION

Mathematically, the OPF problem can be formulated as follows:

Case 1: Minimization of Generator fuel cost with valve point loading. The objective function is as *followings (Polprasert et.al, 2013; Dieu & Ongsakul, 2012):*

$$F_1 = \sum_{i=1}^{NG} \left(a_i + b_i P_{Gi} + c_i P_{Gi}^2 + \left| d_i \times \sin\left\{ e_i \times \left(P_{Gi}^{\min} - P_{Gi} \right) \right\} \right| \right) \tag{1}$$

Case 2: Minimization of Generator fuel cost with voltage deviation

$$F_2 = \sum_{i=1}^{NG} \left(a_i + b_i P_{Gi} + c_i P_{Gi}^2 \right) + w_f \sum_{i=1}^{NLB} \left| V_i - 1.0 \right| \tag{2}$$

In this objective function, weighting factor (w_f) is selected as 100.

Case 3: Minimization of Generator fuel cost with voltage stability index

$$F_3 = \sum_{i=1}^{NG} \left(a_i + b_i P_{Gi} + c_i P_{Gi}^2 \right) + w_f \times L_{\max} \tag{3}$$

$$L_{\max} = \max\left(L_j \right) = \max\left(\left| 1 - \sum_{i=1}^{NG} F_{ji} \frac{V_i}{V_j} \right| \right) \tag{4}$$

In this objective function, weighting factor (w_f) is selected as 100.
Subject to:

Equality Constraints

- Power balance constraints

$$P_i(V,\theta) = P_{Gi} - P_{Di} = \sum_{j=1}^{NB} V_i V_j \left(G_{ij} \cos\theta_{ij} + B_{ij} \sin\theta_{ij} \right) \tag{7}$$

$$Q_i(V,\theta) = Q_{Gi} - Q_{Di} = \sum_{j=1}^{NB} V_i V_j \left(G_{ij} \sin\theta_{ij} + B_{ij} \cos\theta_{ij} \right) \quad ; \quad i = 1,...,NB \tag{8}$$

Inequality Constraints

The system operating and security constraints which can be specified within their maximum and minimum limits as follows:

- Generator constraints

$$V_{Gi}^{\min} \leq V_{Gi} \leq V_{Gi}^{\max}, \quad i = 1,\ldots,NG \tag{9}$$

$$P_{Gi}^{\min} \leq P_{Gi} \leq P_{Gi}^{\max}, \quad i = 1,\ldots,NG \tag{10}$$

$$Q_{Gi}^{\min} \leq Q_{Gi} \leq Q_{Gi}^{\max}, \quad i = 1,\ldots,NG \tag{11}$$

- Transformer tap setting constraints

$$t_i^{\min} \leq t_i \leq t_i^{\max}, \quad i = 1,\ldots,NT \tag{12}$$

- Shunt VAR compensator constraints

$$Q_{ci}^{\min} \leq Q_{ci} \leq Q_{ci}^{\max}, \quad i = 1,\ldots,NC \tag{13}$$

- Voltage limit and line flow constraints

$$V_{Li}^{\min} \leq V_{Li} \leq V_{Li}^{\max}, \quad i = 1,\ldots,NLB \tag{14}$$

$$S_{l_i} \leq S_{l_i}^{\max}, \quad i = 1,\ldots,NTL \tag{15}$$

The vectors of dependent/state variables (x) can be represented as:

$$x = \left[P_{G1}, V_{L1}\ldots V_{LNLB}, Q_{G1}\ldots Q_{GNG}, S_{l1}\ldots S_{lNTL}\right] \tag{16}$$

Similarly, the vector of control variables (u) can be expressed as:

$$u = \left[P_{G2}\ldots P_{GNG}, V_{G1}\ldots V_{GNG}, Q_{C1}\ldots Q_{CNC}, t_1\ldots t_{NT}\right] \tag{17}$$

To handle the inequality constraints, the dependent or state variables including power generation output at slack bus (P_{G1}), voltages at load buses (V_{Gi}), reactive power generation output (Q_{Gi}), and transmission line loadings (S_{li}) are incorporated in the objective function as a quadratic penalty function method. Thus, the fitness function (FF_i) with the extended objective function can be represented as:

$$FF_i = F_i + K_S \left(P_{GS} - P_{GS}^{\lim}\right)^2 + K_V \sum_{i=1}^{NLB} \left(V_{Li} - V_{Li}^{\lim}\right)^2 + K_Q \sum_{i=1}^{NG} \left(Q_{Gi} - Q_{Gi}^{\lim}\right)^2 + K_T \sum_{i=1}^{NTL} \left(S_{li} - S_{li}^{\lim}\right)^2 \qquad (18)$$

The limits of dependent or state or control or output variables ($x^{\lim}$) are given as:

$$x^{\lim} = \begin{cases} x^{\max}; & x > x^{\max} \\ x^{\min}; & x < x^{\min} \end{cases} \qquad (19)$$

IMPROVED PSEUDO-GRADIENT SEARCH PARTICLE SWARM OPTIMIZATION FOR OPF PROBLEM

In the conventional PSO method, a population of particles moves in the search space. During the process, each particle is randomly adjusted according to its own experiences and neighbor's experience of particles. The velocity of particles is changed over time and their positions will be updated accordingly. Mathematically, in d-dimension optimization problem, the velocity and position of each particle can be updated as follows:

$$v_{id}^{k+1} = \omega^k v_{id}^k + c_1 \times rand_1 \times \left(pbest_{id}^k - x_{id}^k\right) + c_2 \times rand_2 \times \left(gbest_d^k - x_{id}^k\right) \qquad (20)$$

$$x_{id}^{k+1} = x_{id}^k + v_{id}^{k+1} \qquad (21)$$

$$i = 1,....,np; \quad d = 1,....,nd$$

To enhance the convergence characteristics, suitable selection of the inertia weighting in (22) provides a balance between global and local explorations for finding the optimal solutions. Generally, the inertia weight is designed to decrease linearly from about 0.9 to 0.4 during run as follows:

$$\omega^k = \omega_{\max} - \left(\omega_{\max} - \omega_{\min}\right) \times \frac{k}{k_{\max}} \qquad (22)$$

a. Improved PSO with Chaotic Weighting Factor

Chaos is the appearance of unstable behavior or stochastic that occurs in a deterministic nonlinear system under deterministic conditions (Köse & Arslan, 2014; Bousson & Velosa, 2015). For determining the inertia weight factor, the inertia weight is modified by using *logistic map* equation as following (Caponetto et.al, 2003):

$$\eta^k = 4 \times \eta^{k-1} \times \left(1 - \eta^{k-1}\right) \qquad (23)$$

To improve capability of global searching solution and escape the local minimum, the new weighting factor is applied to chaotic sequence which can be modified as follows:

$$c\omega^k = \omega^k \times \eta^k \qquad (24)$$

In Eq.(23), the initial η^0 should be in range [0,1] and $\eta^0 \notin \{0, 0.25, 0.5, 0.75, 1.0\}$.

b. Concept of Pseudo-Gradient Search

Pseudo-gradient is a search algorithm determining the search direction of each particle in population based methods (Le & Dieu, 2012). The merit of pseudo gradient is to provide the good direction in the multidimensional search space and also the objective function is not required to be differentiable. Thus, the proposed pseudo-gradient method is suitable for solving non-convex problems.

In d-dimension optimization problem with objective function, $f(x)$ to be non-differentiable where $x = [x_1, x_2, \ldots, x_d]$, a pseudo-gradient $p_g(x)$ for objective function can be classified into two cases as following:

Suppose that the movement of particle in each d-dimension, $x_a = [x_{a1}, x_{a2}, \ldots, x_{and}]$ is a point in the search space and moves to another point x_b.

Case 1: If $f(x_b) < f(x_a)$, the direction from x_a to x_b is positive direction. The pseudo-gradient at point x_b can be determined by:

$$p_g(x_b) = \left[\phi(x_{b1}), \phi(x_b), \ldots, \phi(x_{bnd})\right]^T \qquad (25)$$

$$\phi(x_{bd}) = \begin{cases} 1 & if \quad x_{bd} > x_{ad} \\ 0 & if \quad x_{bd} = x_{ad} \\ -1 & if \quad x_{bd} < x_{ad} \end{cases} \qquad (26)$$

Case 2: If $f(x_a) \geq f(x_b)$, the direction from x_a to x_b is negative direction. The pseudo-gradient at point x_b can be determined by:

$$p_g(x_b) = 0 \qquad (27)$$

If the pseudo gradient at point b, $p_g(x_b)$ is not equal to zero, it implies that the pseudo gradient could find a better solution in the next step which is based on direction indicator. Otherwise, the search direction at that point should not be changed.

c. Pseudo-Gradient PSO method

The main objective of the proposed IPG-PSO method is to employ chaotic weighting factor to find the proper velocity updating for accelerating the search for global optimal solution. Here, the combined pseudo-gradient and PSO with chaotic weight factor is proposed. The updating position of particles between two points, x_a and x_b corresponding to x^k and x^{k+1}, respectively can be rewritten as follows:

$$x_{id}^{k+1} = \begin{cases} x_{id}^k + \alpha \times \phi\left(x_{id}^{k+1}\right) \times \left|v_{id}^{k+1}\right| & if \quad p_g\left(x_{id}^{k+1}\right) \neq 0 \\ x_{id}^k + v_{id}^{k+1} & otherwise \end{cases} \quad (28)$$

From (28), if the pseudo-gradient is non-zero, the involved particle's position is speeded up to the global optimal solution by its enhanced velocity; otherwise, the position is generally updated by (21). The value of α can be adjusted so that a particle can move faster or slower depending on the characteristic of individual problem. In fact, if the value of α is too large, it may not to lead to optimal solution because particles are stuck at their limit positions. On the other hand, too small value of α may lead particles being trapped in local minima in the search space. In this paper, the proper value of α should be tuned in the range [1,10].

d. Procedure of IPG-PSO for Nonconvex OPF Problem

The proposed IPG-PSO procedure can be described as follows:

Step 1: Initialization of swarm

In the implementation of IPG-PSO, each generator output is taken as the encoded parameter. For the system consists of *NG* generators, positions of each particle representing for control variables can be defined as follows:

$$x_i = \left[P_{G2}....P_{GNGnp}, V_{G1}....V_{GNGnp}, Q_{C1}....Q_{CNCnp}, T_1....T_{NTnp}\right] i = 1,...,np \quad (29)$$

The upper and lower limits for velocity of each particle are given based on their limit of position as following:

$$v_i^{max} = R \times \left(x_i^{max} - x_i^{min}\right) \quad (30)$$

$$v_i^{min} = -v_i^{max} \quad (31)$$

Both of particle positions and velocities are initialized with their limits represented by:

$$x_i^{(0)} = x_i^{\min} + rand_3 \times \left(x_i^{\max} - x_i^{\min}\right) \tag{30}$$

$$v_i^{(0)} = v_i^{\min} + rand_4 \times \left(v_i^{\max} - v_i^{\min}\right) \tag{31}$$

Step 2: Run power flow and evaluate fitness function

Run MATPOWER power flow analysis. The fitness function to be minimized is based on different objective functions and state or output variables including power generation at slack bus, reactive power generation, load bus voltages and transmission line flow which are restricted by adding as quadratic penalty terms. The fitness function is calculated from (18). The limits of state variables are determined based on their calculated values as (19).

Step 3: Determining *pbest* and *gbest*

From the step 1, all particles are defined as *pbest* and the best value among fitness function is referred to *gbest*.

Step 4: Updating velocity

The updated velocity is computed using (20) and (24). The velocity limits will be adjusted as follows:

$$v_{id} = \begin{cases} v_{id}^{\max}, v_{id} > v_{id}^{\max} \\ v_{id}^{\min}, v_{id} < v_{id}^{\min} \end{cases} \tag{32}$$

Step 5: Update position

The particle position is updated by using (28). If any particle position is beyond the operating limit, it will be adjusted as follows:

$$x_{id} = \begin{cases} x_{id}^{\max}, x_{id} > x_{id}^{\max} \\ x_{id}^{\min}, x_{id} < x_{id}^{\min} \end{cases} \tag{33}$$

Step 6: Update pbest and gbest

The *pbest* of each particle i at iteration $k+1$, ($pbest_i^{k+1}$) is updated. If the update position value of each particle is better than the previous *pbest*, ($pbest_i^k$), the current value is set to be *pbest*. Also, *gbest* at iteration $k+1$ is set as the best value among all $pbest_i^{k+1}$.

Step 7: Stopping criteria

The proposed IPG-PSO method will be stopped if the maximum iterations are reached. Otherwise, go back to step 4.

The flowchart of the proposed IPG-PSO method is shown in Figure 1.

NUMERICAL RESULTS

The proposed IPG-PSO method has been tested on the IEEE 30-bus and 118-bus systems for solving non-convex OPF problem considering quadratic generator fuel cost function with valve point loading effects, voltage deviation, and voltage stability index. The proposed method is run 50 independent trials and the obtained results are compared to those from different PSO and other methods. The coding have been written and implemented in MATLAB R2013a computing environment and executed on a 2.0 GHz Intel(R) Core™ i5-3210M CPU with 8.00 GB RAM. The data for IEEE 30-bus system can be found in (Alsac & Stot, 1974).

1. IEEE 30-Bus System

The characteristics and data for the base case of the test system are given in Table 1. The load bus voltages are limits in the range [0.95-1.10] p.u.

Figure 1. Flow Chart of Proposed IPG-PSO Algorithm

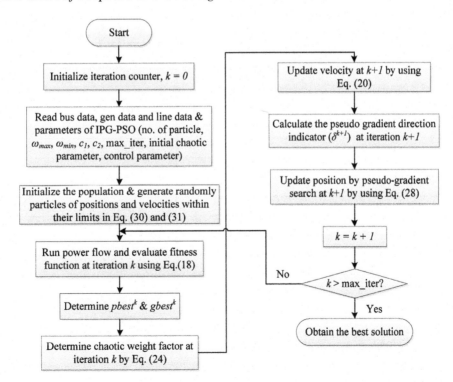

Table 1. Characteristics of IEEE 30-bus system

No. of generators	No. of transformers	No. of branches	No. of capacitor banks	No. of control variables	IEEE 30-bus system
6	4	41	9	19	
$\sum P_{Di}$ (MW)	$\sum Q_{Di}$ (MVar)	$\sum P_{Gi}$ (MW)	$\sum Q_{Gi}$ (MVar)	P_{loss} (MW)	Q_{loss} (MVar)
283.4	126.2	287.93	66.32	4.533	23

In this paper, the power flow solutions are obtained from MATPOWER toolbox (MATPOWER). For comparison, five PSO methods including PSO-TVIW, PSO-TVAC, SPSO-TVAC, PG-PSO, and SWT-PSO methods are applied to solve non-convex OPF problems.

Base Case: Minimizing generator fuel cost

From the Appendix Table 19, the generator fuel cost characteristics for all generating units are represented as the quadratic cost function.

Table 2 shows the proposed IPG-PSO dispatch of the generators. Note that all the outputs of generator are within its permissible limits. The obtained results of the proposed IPG-PSO method render lower cost than PG-PSO, SWT-PSO, SPSO-TVAC, PSO-TVAC, and PSO-TVIW in Table 3. In Table 4, it is observed that the proposed IPG-PSO method renders lower total cost than the other PSO methods and IGA (Lai & Ma, 1997), ACO (Allaoua & Laoufi, 2009), IEP (Ongsakul & Tantimaporn, 2006), GSA (Duman et.al., 2012), DE (Abou El Elaa & Abido, 2010), MDE (Sayah & Zehar, 2008), EADPSO (Vaisakh et.al., 2013), BBO (Bhattacharya & Chattopadhyay, 2011), GA (Bouktir et.al., 2004), and EP (Sinha et.al., 2003) in a faster computing time.

Table 2. The Proposed IPG-PSO Solution on the IEEE 30-bus system for base case

Control Variable	Min	Max	Best Solution
P_{G1} (MW)	50	200	177.0961
P_{G2} (MW)	20	80	48.6864
P_{G5} (MW)	15	50	21.3043
P_{G8} (MW)	10	35	21.0156
P_{G11} (MW)	10	30	11.8815
P_{G13} (MW)	12	40	12.0000
V_{G1} (p.u.)	0.90	1.10	1.1000
V_{G2} (p.u.)	0.90	1.10	1.0877
V_{G5} (p.u.)	0.90	1.10	1.0614
V_{G8} (p.u.)	0.90	1.10	1.0691
V_{G11} (p.u.)	0.90	1.10	1.1000
V_{G13} (p.u.)	0.90	1.10	1.1000
T_{11} (p.u.)	0.90	1.10	1.0410
T_{12} (p.u.)	0.90	1.10	0.9000

continued on following page

Table 2. Continued

Control Variable	Min	Max	Best Solution
T_{15} (p.u.)	0.90	1.10	0.9792
T_{36} (p.u.)	0.90	1.10	0.9622
Q_{c10} (MVar)	0.90	1.10	5.0000
Q_{c12} (MVar)	0.90	1.10	5.0000
Q_{c15} (MVar)	0.90	1.10	5.0000
Q_{c17} (MVar)	0.90	1.10	5.0000
Q_{c20} (MVar)	0.90	1.10	4.2597
Q_{c21} (MVar)	0.90	1.10	5.0000
Q_{c23} (MVar)	0.90	1.10	2.6878
Q_{c24} (MVar)	0.90	1.10	5.0000
Q_{c29} (MVar)	0.90	1.10	2.3300
Total Cost ($/h)			798.6153
P_{loss} (MW)			8.5839
CPU Time (s)			9.636

Table 3. Result comparison with other PSO methods on the IEEE 30-bus system for base case

Control Variable	PSO-TVIW	PSO-TVAC	SPSO-TVAC	PG-PSO	SWT-PSO	IPG-PSO
P_{G1}(MW)	191.2484	166.3323	155.3859	177.2272	177.1513	177.0961
P_{G2}(MW)	38.2831	53.705	60.9115	48.6351	48.0627	48.6864
P_{G5}(MW)	19.5684	21.6504	27.8668	21.34	21.3361	21.3043
P_{G8}(MW)	10.0874	25.6504	17.5885	21.1757	20.8543	21.0156
P_{G11}(MW)	21.1311	13.9794	11.5827	11.6486	12.5556	11.8815
P_{G13}(MW)	13.6331	12.1386	18.5272	12.0000	12.0671	12.0000
Min. Cost ($/h)	809.536	806.146	811.248	799.0462	799.1635	798.6153
Avg. Cost ($/h)	850.2606	848.7793	844.9720	802.1020	800.9710	799.04223
Max. Cost ($/h)	893.1734	904.9051	941.8581	809.2248	803.1791	800.6904
Std. of cost	25.3050	24.0530	25.9717	1.8222	0.8702	0.2678
P_{loss} (MW)	10.5514	9.6626	8.4626	8.6265	8.6271	8.5839
CPU Time(s)	13.369	14.321	15.304	14.508	12.355	9.636

Case 1: Minimizing quadratic cost function with valve point loading effects

In this case, only generating units at bus 1 and bus 2 include the valve point loading effects. The fuel cost coefficients of generating units are given in Appendix Table 20.

Table 5 shows the generation outputs of the IPG-PSO solution. Test results indicates that the IPGPSO algorithm gives better results than other PSO and IEP (Ongsakul & Tantimaporn, 2006), DE (Abou

Table 4. Result comparison with other methods on the IEEE 30-bus system for base case

Methods	Total cost ($/h)	P_{loss} (MW)	CPU time (s)
IGA [1]	800.805	NA	NA
ACO [16]	803.123	9.462	20
IEP [17]	802.465	9.511	99.013 min.
GSA [19]	798.675	8.386	10.758
DE [20]	799.289	NA	NA
MDE [21]	802.376	NA	NA
EADPSO [22]	800.228	8.870	3.74
BBO [23]	799.112	8.630	11.02
GA [26]	803.699	9.518	315
EP [27]	802.620	9.370	51.4
IPG-PSO	798.615	8.584	9.636

Table 5. IPG-PSO Solution on the IEEE 30-bus system for Case 1

Control Variable	Min	Max	Best Solution
P_{G1} (MW)	50	200	198.1117
P_{G2} (MW)	20	80	51.5809
P_{G5} (MW)	15	50	15.0000
P_{G8} (MW)	10	35	10.0000
P_{G11} (MW)	10	30	10.0000
P_{G13} (MW)	12	40	12.0000
V_{G1} (p.u.)	0.90	1.10	1.0629
V_{G2} (p.u.)	0.90	1.10	1.0415
V_{G5} (p.u.)	0.90	1.10	0.9500
V_{G8} (p.u.)	0.90	1.10	0.9613
V_{G11} (p.u.)	0.90	1.10	1.0223
V_{G13} (p.u.)	0.90	1.10	1.0654
T_{11} (p.u.)	0.90	1.10	0.9194
T_{12} (p.u.)	0.90	1.10	0.9343
T_{15} (p.u.)	0.90	1.10	0.9259
T_{36} (p.u.)	0.90	1.10	0.9570
Q_{c10} (MVar)	0	19	1.4884
Q_{c24} (MVar)	0	4.3	4.3000
Total Cost ($/h)			924.0752
P_{loss} (MW)			9.4111
CPU Time (s)			11.669

El Elaa & Abido, 2010), MDE (Sayah & Zehar, 2008), and GSA (Bouktir et.al, 2004) and methods as shown in Table 6 and 7. Therefore, the proposed IPGPSO is effective for solving the OPF problem with valve point loading effects in a shorter computing time.

Case 2: Minimizing quadratic cost function with voltage deviation

In this case, the objective function is to minimize the total fuel cost with the voltage deviation for improving voltage profile.

Table 8 shows the OPF solution of IPG-PSO. Note that all the outputs of generator are within its permissible limits. In Table 9 and Table 10, it is observed that the proposed IPG-PSO method yields lower total generation cost and voltage deviation than DE (Abou El Elaa & Abido, 2010), BBO (Bhattacharya & Chattopadhyay, 2011), and the other PSO methods in a faster computing manner.

Case 3: Minimizing quadratic cost function with voltage stability index

In this case, the objective function is to minimize the total fuel cost with the voltage stability index for enhancing voltage stability.

In Table 11 and Table 12, it is observed that the proposed IPG-PSO method yields lower minimum and average L-index than PSO (Abido, 2002), BBO (Bhattacharya & Chattopadhyay, 2011), TLBO (Bouchekara et.al, 2014), and other PSO methods in a slightly faster computing time.

2. IEEE 118-Bus System

To evaluate the effectiveness on a lager system, the proposed IPG-PSO method for solving OPF problem has been applied to IEEE 118-bus system. The characteristics of the IEEE 118-bus system are shown in Table 13.

Table 6. Result comparison with other PSO methods on the IEEE 30-bus system for Case 1

Control Variable	PSO-TVIW	PSO-TVAC	SPSO-TVAC	PG-PSO	SWT-PSO	IPG-PSO
P_{G1}(MW)	192.6015	193.9116	192.6173	149.7381	191.8841	198.1117
P_{G2}(MW)	20.0000	20.0000	20.0000	52.0718	20.0878	51.5809
P_{G5}(MW)	27.4342	26.1228	29.4405	24.9827	28.0371	15.0000
P_{G8}(MW)	22.3731	26.1552	21.3138	28.4109	22.3463	10.0000
P_{G11}(MW)	14.0312	11.0893	10.0000	17.8773	11.2294	10.0000
P_{G13}(MW)	16.3711	16.0022	19.9591	17.3562	19.0381	12.0000
Min. Cost ($/h)	927.7455	927.219	932.6041	951.9248	928.9073	924.0752
Avg. Cost ($/h)	958.1128	959.5723	959.8046	959.0725	953.1293	957.5613
Max. Cost ($/h)	1035.6125	1027.922	1038.106	1037.280	960.2557	1020.218
Std. of cost	17.9339	20.7018	13.3693	15.2189	8.8473	13.5068
P_{loss} (MW)	13.2926	9.8811	9.9308	7.0370	9.2227	9.4111
CPU Time(s)	11.856	12.652	11.872	12.886	12.683	11.669

Table 7. Result comparison with other methods on the IEEE 30-bus system for Case 1

Methods	Total cost ($/h)	P_{loss} (MW)	CPU time (s)
IEP [17]	953.573	7.6458	93.583 min.
DE [20]	931.085	12.605	44.96
MDE [21]	930.793	13.064	41.85
GSA [26]	929.724	15.1458	9.8374
IPG-PSO	924.075	9.4111	11.669

Table 8. IPG-PSO Solution on the IEEE 30-bus system for Case 2

Control Variable	Min	Max	Best Solution
P_{G1} (MW)	50	200	174.4208
P_{G2} (MW)	20	80	48.6046
P_{G5} (MW)	15	50	21.5684
P_{G8} (MW)	10	35	22.5019
P_{G11} (MW)	10	30	10.0069
P_{G13} (MW)	12	40	12.0007
V_{G1} (p.u.)	0.90	1.10	1.0548
V_{G2} (p.u.)	0.90	1.10	1.0308
V_{G5} (p.u.)	0.90	1.10	1.0021
V_{G8} (p.u.)	0.90	1.10	1.0045
V_{G11} (p.u.)	0.90	1.10	0.9987
V_{G13} (p.u.)	0.90	1.10	1.0309
T_{11} (p.u.)	0.90	1.10	1.0095
T_{12} (p.u.)	0.90	1.10	0.9001
T_{15} (p.u.)	0.90	1.10	1.0150
T_{36} (p.u.)	0.90	1.10	0.9686
Q_{c10} (MVar)	0.90	1.10	4.5069
Q_{c12} (MVar)	0.90	1.10	2.4262
Q_{c15} (MVar)	0.90	1.10	3.1824
Q_{c17} (MVar)	0.90	1.10	2.0484
Q_{c20} (MVar)	0.90	1.10	1.8676
Q_{c21} (MVar)	0.90	1.10	3.7289
Q_{c23} (MVar)	0.90	1.10	5.0000
Q_{c24} (MVar)	0.90	1.10	5.0000
Q_{c29} (MVar)	0.90	1.10	2.5259
Total Cost ($/h)	-	-	802.8874
VD (p.u.)	-	-	0.1065
P_{loss} (MW)	-	-	9.7033
CPU Time (s)	-	-	12.012

Table 9. Result comparison with other PSO methods on the IEEE 30-bus system for Case 2

Control Variable	PSO-TVIW	PSO-TVAC	SPSO-TVAC	PG-PSO	SWT-PSO	IPG-PSO
P_{G1} (MW)	180.0830	176.4723	177.5702	175.5152	175.6289	174.4998
P_{G2} (MW)	49.2563	50.2107	48.6056	49.6265	48.9087	48.2879
P_{G5} (MW)	22.1780	20.7805	21.3464	21.7218	21.5602	21.5618
P_{G8} (MW)	18.6345	21.6057	20.7642	21.2494	21.7980	22.4085
P_{G11} (MW)	10.0000	12.0748	12.6466	11.9020	13.1859	11.4227
P_{G13} (MW)	13.1486	12.0000	12.2075	12.9492	12.0000	12.0007
Min. $Cost+VD$ ($/h)	803.3443	803.2655	803.1800	802.9855	803.4104	802.4653
Avg. $Cost+VD$ ($/h)	805.0690	804.7403	803.7530	804.8278	803.9711	803.9976
Max. $Cost+VD$ ($/h)	822.0907	808.4271	804.6783	816.1438	805.8837	804.6731
SD	2.5948	1.3124	0.2891	0.5176	2.3717	0.4717
Min VD (p.u.)	0.1648	0.1975	0.1630	0.1629	0.1443	0.1014
P_{loss} (MW)	9.9005	9.7440	9.7405	9.5640	9.6817	9.7007
CPU Time(s)	12.121	12.418	12.215	12.184	13.400	11.983

Table 10. Result comparison on the IEEE 30-bus system for Case 2

Methods	Total cost ($/h)	VD (p.u.)	P_{loss} (MW)	CPU time (s)
DE [20]	805.2619	0.1357	10.4412	NA
BBO [23]	804.9982	0.1020	9.95	NA
IPG-PSO	802.4653	0.1014	9.7007	11.983

Table 11. Result comparison with other PSO methods on the IEEE 30-bus system for Case 3

Control Variable	PSO-TVIW	PSO-TVAC	SPSO-TVAC	PG-PSO	SWT-PSO	IPG-PSO
Min. $Cost+L_{max}$ ($/h)	809.7497	809.8941	809.3397	809.2690	801.7026	798.5600
Avg. $Cost+L_{max}$ ($/h)	809.8829	810.0331	809.4664	809.3841	801.8355	799.6737
Max. $Cost+L_{max}$ ($/h)	810.0436	811.1892	809.5946	809.5080	802.9721	801.7935
SD	0.0056	0.0067	0.0036	0.0074	0.0064	0.0034
Min L_{max}	0.1222	0.1230	0.1214	0.1120	0.1279	0.1083
P_{loss} (MW)	16.1567	16.1933	16.0528	16.260	16.6513	16.0349
CPU Time(s)	10.53	10.67	10.34	10.42	10.34	10.06

Table 12. Result comparison with other methods on the IEEE 30-bus system for Case 3

Methods	Total cost ($/hr)	L_{max} (p.u.)	CPU time (sec)
PSO [15]	801.16	0.1246	-
TLBO [31]	799.978	0.1131	-
BBO [23]	-	0.1104	16.29
DE [20]	-	0.1219	-
IPG-PSO	798.56	0.1083	10.06

Table 13. Characteristics of IEEE 118-bus system

No. of generators	No. of transformers	No. of branches	No. of capacitor banks	No. of control variables	IEEE 30-bus system
54	9	186	14	77	
$\sum P_{Di}$ (MW)	$\sum Q_{Di}$ (MVar)	$\sum P_{Gi}$ (MW)	$\sum Q_{Gi}$ (MVar)	P_{loss} (MW)	Q_{loss} (MVar)
4,242	1,438	4,374.86	795.68	132.863	783.79

Base Case: *Minimization of generation fuel cost*

The objective function is the same as case 1 which is minimizing the total generation fuel cost for all generating units. Using the proposed IPG-PSO method, the minimum, average and maximum total costs with 50 independent trail runs are shown in Table 14. It can be seen that the proposed method obtain the minimum total generation cost with 129,154.1193 $/hr which is less than the other PSO methods and other methods (Bouchekara et.al, 2014; Dieu & Schegner, 2012; Dinh et.al, 2013) in Table 15. In addition, the obtained solution of optimal control variables by using the proposed IPG-PSO method is shown in Appendix Table 22.

Table 14. Result comparison with other PSO methods on the IEEE 118-bus system for base case

Method	PSO-TVIW	PSO-TVAC	SPSO-TVAC	PG-PSO	SWT-PSO	IPG-PSO
Min. Cost	132,852.0317	134,361.9242	132,130.2206	130,791.8432	138,117.6785	129,154.1193
Avg. Cost	134,682.8339	137,699.3456	133,234.0938	132,641.7614	142,667.3284	130,032.3273
Max. Cost	138,859.9008	145,486.9981	134,741.6894	147,276.8675	150,683.1968	133,993.0514
SD	1,158.7284	1,977.7646	588.9416	2,659.5152	2,746.4379	875.0017
P_{loss} (MW)	91.4606	96.1244	72.1032	82.4166	76.4388	82.8730
VD	1.3803	1.4604	1.9369	1.2584	1.0334	1.506
L_{max}	0.0685	0.0677	0.0629	0.0713	0.0687	0.0674
CPU time (sec)	23.509	20.842	23.541	22.339	20.889	20.746
Avg. CPU time (sec)	21.311	21.60	22.285	22.360	21.296	21.143

Table 15. Comparison of IPG-PSO with other methods on the IEEE 118-bus system for base case

Methods	Total cost ($/h)	Ploss (MW)	CPU time (s)
GA [31]	132,746.3517	-	-
DSA [31]	129,756.2275	-	-
TLBO [31]	129,682.8440	-	-
IPSO [32]	145,520.0109	-	163.300
ABC [33]	138,886.5482	93.7142	69.75
IPG-PSO	129,154.1193	82.8730	21.143

Case 1: Minimizing quadratic cost function with valve point loading effects

The fuel cost coefficient of generator and power generation limits are shown in Appendix Table 21. From Appendix Table 23, the obtained solutions of control variables on the IEEE 118-bus system for case 1 are shown. Note that all the outputs of generator are within its permissible limits. In Table 16, it can be seen that the proposed IPG-PSO method provides lower cost than other PSO methods. Hence, the proposed IPG-PSO is very effective for solving the OPF problem with valve point loading effects.

Case 2: Minimizing quadratic cost function with voltage deviation

Table 17 shows the result comparison of IPG-PSO method with other PSO methods for solving OPF solution for Case 2. It is observed that the proposed IPG-PSO method yields lower total generation cost and voltage deviation than the other PSO methods.

Case 3: Minimizing quadratic cost function with voltage stability index

Table 18 shows the result comparison between proposed IPG-PSO method and other PSO methods for Case 3. The obtained result indicates that the proposed IPG-PSO method gives lower total generation cost and minimum voltage stability index than the other PSO methods in a faster computational time.

Table 16. Comparison of IPG-PSO with other PSO methods on the IEEE 118-bus system for Case 1

Control Variable	PSO-TVIW	PSO-TVAC	SPSO-TVAC	PG-PSO	SWT-PSO	IPG-PSO
Min. Cost ($/h)	77,002.6582	86,755.8253	85,578.4587	76,812.9695	104,341.6176	76,390.7488
Avg. Cost ($/h)	83,519.6981	96,984.2948	90,392.5013	84,087.8549	113,612.8140	80,340.1059
Max. Cost ($/h)	100,569.1972	109,098.4635	96,179.2609	95,412.5351	131,518.4449	91,312.8696
SD	4,414.1066	4,462.9421	2,205.3592	3,877.2868	6,101.7031	3,045.1170
P_{loss} (MW)	557.0289	432.0129	440.0383	570.5509	206.3971	560.6368
CPU Time(s)	24.289	24.726	23.79	23.868	23.962	23.525
Avg.CPU Time (s)	24.499	26.018	23.70	25.050	22.304	24.119

Table 17. Result comparison with other PSO methods on the IEEE 118-bus system for Case 2

Method	PSO-TVIW	PSO-TVAC	SPSO-TVAC	PG-PSO	SWT-PSO	IPG-PSO
Min. Cost+VD	171,448.5026	151,343.0646	163,761.1163	150,928.5276	154,454.3591	145,684.0767
Avg. Cost+VD	172,475.5326	152,400.7646	164,797.1863	151,959.0076	157,542.4191	147,707.3367
Max. Cost+VD	174,529.5426	154,484.1746	165,846.4763	153,032.0676	159,674.3091	148,738.3867
SD	0.0609	0.0992	0.0479	0.0923	0.1845	0.0288
P_{loss} (MW)	95.8808	85.5346	99.4237	92.5649	72.8757	70.3895
Min VD	0.1999	0.4154	0.2617	0.1986	0.5905	0.1846
CPU time (sec)	24.430	23.322	25.179	22.932	24.570	22.917
Avg. CPU time (sec)	24.575	23.354	24.531	23.754	25.272	23.027

Table 18. Result comparison with other PSO methods on the IEEE 118-bus system for Case 3

Method	PSO-TVIW	PSO-TVAC	SPSO-TVAC	PG-PSO	SWT-PSO	IPG-PSO
Min. Cost+ L_{max}	169,044.0278	163,247.8809	161,508.0212	166,239.3487	155,397.3742	148,736.7518
Avg. Cost+ L_{max}	170,105.2278	164,254.0007	162,414.1612	167,245.5287	156,403.7142	149,242.8518
Max. Cost+ L_{max}	171,111.8678	165,460.7007	163,720.5512	169,625.7387	157,410.4642	149,448.9618
SD	0.0009	0.0009	0.0007	0.0002	0.0013	0.0001
P_{loss} (MW)	129.7572	121.6188	135.3612	151.1544	66.8234	149.4374
Min L_{max}	0.0611	0.0610	0.0611	0.0614	0.0612	0.0609
CPU time (sec)	27.22	29.66	27.39	26.99	29.36	25.86
Avg. CPU time (sec)	29.65	28.63	29.71	27.43	28.29	26.66

FUTURE RESEARCH DIRECTIONS

The future research directions are as follows.

- The proposed IPG-PSO method can be applicable for other optimization problems in power systems such as economic dispatch (ED), and optimal reactive power dispatch (ORPD). In addition, it can be applied ORPD problem for minimizing transmission line loss, voltage stability index or voltage deviation for enhancing voltage stability and improving voltage profile.
- The proposed IPG-PSO method is applicable to multi-objective OPF considering wind and solar power uncertainty.
- The proposed IPG-PSO based methods may be combined with other methods for solving stochastic UC problem with wind and solar forecast uncertainty.

CONCLUSION

In this chapter, an improved pseudo-gradient particle swarm optimization (IPG-PSO) method has been efficiently solving OPF problems considering objective functions including generator fuel cost function, generator fuel cost function with valve point loading, generator fuel cost function with voltage deviation, and generator fuel cost function with voltage stability index. The features of IPG-PSO include linearly chaotic weighting factor to avoid trapping in the local minimum and pseudo-gradient to better guide particles in the search space. Test results on the IEEE 30-bus and 118-bus systems indicate that the proposed IPG-PSO method can obtain a higher solution quality than other methods, leading to generator fuel cost savings, voltage profile and voltage stability enhancements.

ACKNOWLEDGMENT

The authors are grateful to Greater Mekong Subregion Academic and Research Network (GMSARN) for partial financial support.

REFERENCES

Abido, M. A. (2002). Optimal power flow using tabu search algorithm. *Electric Power Components and Systems, 30*(5), 469–483. doi:10.1080/15325000252888425

Abido, M. A. (2002). Optimal power flow using particle swarm optimization. *Electrical Power and Energy Systems, 24*(7), 563–571. doi:10.1016/S0142-0615(01)00067-9

Abou El Elaa, A. A., & Abido, M. A. (1992). Optimal Operation Strategy for Reactive Power Control Modeling. Simul. Control. *Part A, 41*(3), 19–40.

Abou El Elaa, A. A., & Abido, M. A. (2010). Optimal power flow using differential evolution algorithm. *Electric Power Systems Research, 80*(7), 878–885. doi:10.1016/j.epsr.2009.12.018

Allaoua, B., & Laoufi, A. (2009). Optimal power flow solution using ant manners for electrical network. *Advances in Electrical and Computer Engineering, 9*(1), 34–40. doi:10.4316/aece.2009.01006

Alsac, O., & Stot, B. (1974). Optimal power flow with steady-state security. *IEEE Transactions on Power Apparatus and Systems, 93*(3), 745–754. doi:10.1109/TPAS.1974.293972

Bhattacharya, A., & Chattopadhyay, P. K. (2011). Application of biogeography-based optimization to solve different optimal power flows. *IET Gener. Transm. Distrib., 5*(1), 70–80. doi:10.1049/iet-gtd.2010.0237

Bouchekara, H. R. E. H., Abido, M. A., & Bouchekara, M. (2014). Optimal Power Flow using Teaching-Learning-Based Optimization Technique. *Electric Power Systems Research, 114*, 49–59. doi:10.1016/j.epsr.2014.03.032

Bouktir, T., Slimani, L., & Belkacemi, M. (2004). A genetic algorithm for solving the optimal power flow problem. *Leonardo Journal of Sciences, 4*, 44–58.

Bousson, K., & Velosa, C. (2015). Robust control and synchronization of chaotic systems with actuator constraints. In P. Vasant (Ed.), Handbook of Research on Artificial Intelligent and Algorithms (Vols. 1–2, pp. 1–43). Hershey, PA: IGI Global. doi:10.4018/978-1-4666-7258-1.ch001

Caponetto, R., Fortuna, L., Fazzino, S., & Xibilia, M. G. (2003). Chaotic sequences to improve the performance of evolutionary algorithms. *IEEE Transactions on Evolutionary Computation, 7*(3), 289–304. doi:10.1109/TEVC.2003.810069

Chalermchaiarbha, S., & Ongsakul, W. (2013). Stochastic Weight Trade-Off Particle Swarm Optimization for Nonconvex Economic Dispatch. *Energy Conversion and Management, 70*, 66–75. doi:10.1016/j.enconman.2013.02.009

Chaturvedi, K. T., Pandit, M., & Srivastava, L. (2008). Self-organizing hierarchical particle swarm optimization for nonconvex economic dispatch. *IEEE Transactions on Power Systems, 23*(3), 1079–1087. doi:10.1109/TPWRS.2008.926455

Chen, C. H., & Yeh, S. N. (2006). Particle swarm optimization for economic power dispatch with valve-point effects. In *Proceedings of IEEE PES Transmission and Distribution Conference and Exposition Latin America*. IEEE. doi:10.1109/TDCLA.2006.311397

Dieu, V. N., & Ongsakul, W. (2012). Hopfield lagrange network for economic load dispatch. In P. Vasant, B. Barsoum, & J. Webb (Eds.), *Innovation in Power, Control, and Optimization: Emerging Energy Technologies* (pp. 57–94). Hershey, PA: IGI Global. doi:10.4018/978-1-61350-138-2.ch002

Dieu, V. N., & Schegner, P. (2012). An Improved particle swarm optimization for optimal power flow. In P. Vasant (Ed.), *Meta-Heuristics Optimization Algorithms in Engineering, Business, Economics and Finance* (pp. 1–40). Hershey, PA: IGI Global.

Dieu, V. N., Schegner, P., & Ongsakul, W. (2013). Optimization Method for Nonconvex Economic Dispatch. In Lecture Notes. *Electrical Engineering*, *239*, 1–27.

Dinh, L. L., Dieu, V. N., & Vasant, P. (2013). Artificial Bee Colony Algorithm for Solving Optimal Power Flow Problem. Research Article. *TheScientificWorldJournal*, *2013*, 1–9. doi:10.1155/2013/159040

Duman, S., Guvenc, U., Sonmez, Y., & Yörükeren, N. (2012). Optimal power flow using gravitational search algorithm. *Energy Conversion and Management*, *59*, 86–95. doi:10.1016/j.enconman.2012.02.024

El-Ela, A. A. A., & El-Sehiemy, R. A-A. (2007). Optimized generation costs using modified particle swarm optimization version. *WSEAS Trans. Power Systems*, 225-232.

Kennedy, J., & Eberhart, R. C. (1995). Particle swarm optimization. In *Proceedings of IEEE International Conference on Neural Networks*. Perth, Australia: IEEE. doi:10.1109/ICNN.1995.488968

Köse, U., & Arslan, A. (2014). Chaotic systems and their recent implementations on improving intelligent systems. In P. Vasant (Ed.), Handbook of Research on Novel Soft Computing Intelligent Algorithms: Theory and Practical Applications (Vols. 1–2, pp. 69–101). Hershey, PA: IGI Global. doi:10.4018/978-1-4666-4450-2.ch003

Krost, G., Venayagamoorthy, G. K., & Grant, L. (2008). *Swarm intelligence and evolutionary approaches for reactive power and voltage control*. IEEE Swarm Intelligence Symposium.

Lai, L. L., Ma, J. T., Yokoyama, R., & Zhao, M. (1997). Improved genetic algorithms for optimal power flow under both normal and contingent operation states. *International Journal of Electrical Power & Energy Systems*, *19*(5), 287–292. doi:10.1016/S0142-0615(96)00051-8

Le, D. A., & Dieu, V. N. (2012). Optimal Reactive Power Dispatch by Pseudo-Gradient Guided Particle Swarm Optimization. In *Proceedings of IPEC, 2012 Conference on Power & Energy*, (pp. 7-12). doi:10.1109/ASSCC.2012.6523230

MATPOWER. (n.d.). Retrieved from http://www.ee.washington.edu/reserach/pstca

Ongsakul, W., & Tantimaporn, T. (2006). Optimal Power Flow by Improved Evolutionary Programming. *Electric Power Components and Systems*, *34*(1), 79–95. doi:10.1080/15325000691001458

Park, J.-B., Jeong, Y.-W., & Lee, W.-N. (2006). *An improved particle swarm optimization for economic dispatch problems with non-smooth cost functions*. IEEE Power Engineering Society General Meeting.

Polprasert, J., Ongsakul, W., & Dieu, V. N. (2013). A new improved particle swarm optimization for solving nonconvex economic dispatch problems. *International Journal of Energy Optimization and Engineering*, *2*(1), 66–77. doi:10.4018/ijeoe.2013010105

Ratnaweera, A., Halgamuge, S. K., & Watson, H. C. (2004). Self-Organizing Hierarchical Particle Swarm Optimizer with Time-Varying Acceleration Coefficients. *IEEE Transactions on Evolutionary Computation*, *8*(3), 240–255. doi:10.1109/TEVC.2004.826071

Roa-Sepulveda, C. A., & Pavez-Lazo, B. J. (2003). A solution to the optimal power flow using simulated annealing. *International Journal of Electrical Power & Energy Systems*, *25*(1), 47–57. doi:10.1016/S0142-0615(02)00020-0

Sayah, S., & Zehar, K. (2008). Modified differential evolution algorithm for optimal power flow with non-smooth cost functions. *Int. J. Energy Convers. Manage.*, *49*(11), 3036–3042. doi:10.1016/j.enconman.2008.06.014

Sinha, N., Chakrabarti, R., & Chattopadhyay, P. K. (2003). Evolutionary programming techniques for economic load dispatch. *IEEE Transactions on Evolutionary Computation*, *7*(1), 83–94. doi:10.1109/TEVC.2002.806788

Sutha, S., & Kamaraj, N. (2008). Optimal location of multi type FACTS devices for multiple contingencies using particle swarm optimization. *International Journal of Electrical Systems Science and Engineering*, *1*(1), 16–22.

Swarup, K. S. (2006). Swarm intelligence approach to the solution of optimal power flow. Indian Institute of Science.

Thanushkodi, K., Pandian, S. M. V., & Apragash, R. S. D. (2008). An efficient particle swarm optimization for economic dispatch problems with non-smooth cost functions. *WSEAS Trans. Power Systems*, *4*(3), 257–266.

Vaisakh, K., Srinivas, L. R., & Meah, K. (2013). Genetic evolving ant direction particle swarm optimization algorithm for optimal power flow with non-smooth cost functions and statistical analysis. *Applied Soft Computing*, *13*(12), 4579–4593. doi:10.1016/j.asoc.2013.07.002

Yoshida, H., Kawata, K., Fukuyama, Y., Takayama, S., & Nakanishi, Y. (2001). A particle swarm optimization for reactive power and voltage control considering voltage security assessment. *IEEE Transactions on Power Systems*, *15*(4), 1232–1239. doi:10.1109/59.898095

Yuryevich, J., & Wong, K. P. (1999). Evolutionary programming based optimal power flow algorithm. *IEEE Transactions on Power Systems*, *14*(4), 1245–1250. doi:10.1109/59.801880

APPENDIX

Nomenclature

a_i, b_i, c_i Fuel cost coefficients of unit i

c_1, c_2 Cognitive and social acceleration coefficients, respectively

$c\omega^k$ Chaotic weight factor at iteration k

d_i, e_i Fuel cost coefficients of unit i that represent the valve-point loading effects

F_i i^{th} Objective function

FF_i i^{th} Fitness finction

gbest Global best position among all the particles in the group

$G_{ij}, B_{ij}, \theta_{ij}$ Transfer conductance, susceptance and different phase angle of voltage between bus i and bus j, respectively

k Current iteration

k_{max} Maximum iteration

K_S, K_V, K_Q, K_T Penalty factors for active power generation at slack bus, load bus voltages, reactive power generation outputs and transmission line loadings, respectively

L_j No load and voltage collapse conditions of bus j in the range [0,1]

L_{max} Voltage stability index

nd Number of dimension in a particle

np Number of particles in the group

NC Number of shunt VAR compensators

NG Number of committed generating unit

NLB Number of load buses

NT Number of transformers

NTL Number of transmission lines

pbest$_i$ Best position of ith particle

P_{G1} Real power generation output at slack bus

P_i, Q_i Net real and reactive power output at bus i, respectively

P_{Gi}, Q_{Gi} Active and reactive power generation outputs at bus i, respectively

$P_{Gi}^{min}, P_{Gi}^{max}$ Minimum and maximum values of active power generation outputs at bus i, respectively

P_{Di}, Q_{Di} Active and reactive power demands at bus i, respectively

P_{loss} Active power transmission line loss

Q_{ci} Reactive power of shunt *VAR* compensator at unit i

$Q_{ci}^{min}, Q_{ci}^{max}$ Minimum and maximum values of shunt VAR compensator, respectively

Q_{loss} Reactive power transmission line loss

Q_{Gi} Reactive power generation output at unit i

$Q_{Gi}^{min}, Q_{Gi}^{max}$ Minimum and maximum values of reactive power generation output at bus i, respectively

$rand_i$ Uniformly distributed random numbers in range [0,1]

R Limit factor of particle velocity

$p_g(x)$ Pseudo-gradient at point x

S_{li} Transmission line flow at branch i
S_{li}^{max} Maximum of transmission line flow at branch i
t_i^{min}, t_i^{max}, Minimum and maximum values of transformer tap setting at bus i, respectively
v_{id}^k, x_{id}^k Velocity and position of particle i in the dth dimension at iteration k, respectively
V_{Gi}^{min}, V_{Gi}^{max} Minimum and maximum values of generator bus voltage at bus i, respectively
V_{Li} Load bus voltage at bus i
V_i, V_j Voltage magnitude at bus i and bus j
V_{Li}^{min}, V_{Li}^{max} Minimum and maximum values of load bus voltage at bus i, respectively
w_f Weighting factor
x^{lim} Limits of dependent or state or control or output variables
x^{max}, x^{min} Upper and lower limits of the variables (x), respectively
η^0 Initial chaotic parameter in the range [0,1]
η^k Chaotic parameter at iteration k
ω^k Inertia weighting factor at iteration k
ω_{min}, ω_{max} Minimum and maximum values of inertia weighting factor
$\phi(x_{bd})$ Direction indicator of element x_d moving from point a to point b
α Acceleration factor for updating position of particle

Tables

Table 19. Generator cost coefficients on the IEEE 30-bus system for base case

Bus no.	P_{Gi}^{min} (MW)	P_{Gi}^{max} (MW)	Q_{Gi}^{min} (MVar)	S_{Gi}^{max} (MVA)	Cost coefficients		
					a	b	c
1	50	200	-20	250	0	2.00	0.00375
2	20	80	-20	100	0	1.75	0.01750
5	15	50	-15	80	0	1.00	0.06250
8	10	35	-15	60	0	3.25	0.00834
11	10	30	-10	50	0	3.00	0.02500
13	12	40	-15	60	0	3.00	0.02500

Table 20. Generator cost coefficients for Case 1

Bus no.	From MW	To MW	Cost coefficients				
			a	b	c	d	e
1	50	200	150.00	2.00	0.0016	50.00	0.0630
2	20	80	25.00	2.50	0.0100	40.00	0.0980
5	15	50	0	1.00	0.06250	0	0
8	10	35	0	3.25	0.00834	0	0
11	10	30	0	3.00	0.02500	0	0
13	12	40	0	3.00	0.02500	0	0

Table 21. Generator cost coefficients on the IEEE 118-bus system for Case 1

Bus no.	From MW	To MW	Cost coefficients				
			a	b	c	d	e
1	0	680	550	8.10	0.00028	300	0.035
4	0	360	309	8.10	0.00056	200	0.042
6	0	360	307	8.10	0.00056	150	0.042
8	60	180	240	7.74	0.00324	150	0.063
10	60	180	240	7.74	0.00324	150	0.063
12	60	180	240	7.74	0.00324	150	0.063
15	60	180	240	7.74	0.00324	150	0.063
18	60	180	240	7.74	0.00324	150	0.063
19	60	180	240	7.74	0.00324	150	0.063
24	40	120	126	8.60	0.00284	100	0.084
25	40	120	126	8.60	0.00284	100	0.084
26	55	120	126	8.60	0.00284	100	0.084
27	55	120	126	8.60	0.00284	100	0.084

Table 22. The Proposed IPG-PSO Solution on the IEEE 118-bus system for base case

Control Variable	Min	Max	Best Solution Base case	Best Solution Case 1
P_{G1} (MW)	0	100	37.1674	680.0000
P_{G4} (MW)	0	100	0	360.0000
P_{G6} (MW)	0	100	0	360.0000
P_{G8} (MW)	0	100	4.3914	153.3078
P_{G10} (MW)	0	550	356.5333	180.0000
P_{G12} (MW)	0	185	87.1771	180.0000
P_{G15} (MW)	0	100	8.7178	180.0000
P_{G18} (MW)	0	100	48.3397	180.0000
P_{G19} (MW)	0	100	0	180.0000
P_{G24} (MW)	0	100	0	120.0000
P_{G25} (MW)	0	320	183.1095	120.0000
P_{G26} (MW)	0	414	286.2731	120.0000
P_{G27} (MW)	0	100	52.169	120.0000
P_{G31} (MW)	0	107	0	0.0000
P_{G32} (MW)	0	100	14.4314	0.0000
P_{G34} (MW)	0	100	28.6668	0.4207
P_{G36} (MW)	0	100	0	0.0000
P_{G40} (MW)	0	100	64.1268	0.0000
P_{G42} (MW)	0	100	0.0103	0.2309

continued on following page

Table 22. Continued

Control Variable	Min	Max	Best Solution Base case	Best Solution Case 1
P_{G46} (MW)	0	119	17.6077	0.0000
P_{G49} (MW)	0	304	189.8905	136.6461
P_{G54} (MW)	0	148	47.0016	0.0000
P_{G55} (MW)	0	100	11.5398	0.0000
P_{G56} (MW)	0	100	15.8572	0.0000
P_{G59} (MW)	0	255	163.0423	0.0000
P_{G61} (MW)	0	260	149.1176	0.0000
P_{G62} (MW)	0	100	0	0.0000
P_{G65} (MW)	0	491	353.3733	222.7684
P_{G66} (MW)	0	492	340.2501	235.9711
P_{G69} (MW)	0	805.2	453.7219	308.2664
P_{G70} (MW)	0	100	28.1188	0.0000
P_{G72} (MW)	0	100	6.0562	0.0000
P_{G73} (MW)	0	100	40.9646	0.0000
P_{G74} (MW)	0	100	19.0034	0.0000
P_{G76} (MW)	0	100	13.9438	0.1624
P_{G77} (MW)	0	100	15.8002	0.0000
P_{G80} (MW)	0	577	418.5705	296.9816
P_{G85} (MW)	0	100	0	0.0000
P_{G87} (MW)	0	104	1.1939	0.2751
P_{G89} (MW)	0	707	488.7021	423.5223
P_{G90} (MW)	0	100	0	0.0000
P_{G91} (MW)	0	100	0	0.0000
P_{G92} (MW)	0	100	0.0109	0.0021
P_{G99} (MW)	0	100	0	0.0000
P_{G100} (MW)	0	352	226.5758	168.8745
P_{G103} (MW)	0	140	40.4084	0.0000
P_{G104} (MW)	0	100	18.7861	0.0000
P_{G105} (MW)	0	100	0	0.0302
P_{G107} (MW)	0	100	0	19.9632
P_{G110} (MW)	0	100	29.2755	0.0000
P_{G111} (MW)	0	136	34.6613	40.4323
P_{G112} (MW)	0	100	33.3938	0.0000
P_{G113} (MW)	0	100	2.0801	0.0000
P_{G116} (MW)	0	100	0	0.0000
V_{G1} (p.u.)	0.95	1.05	1.0128	1.0116
V_{G4} (p.u.)	0.95	1.10	1.0433	1.0342
V_{G6} (p.u.)	0.95	1.10	0.9907	1.0093
V_{G8} (p.u.)	0.95	1.10	0.9967	1.0453

continued on following page

Table 22. Continued

Control Variable	Min	Max	Best Solution Base case	Best Solution Case 1
V_{G10} (p.u.)	0.95	1.10	0.989	1.0036
V_{G12} (p.u.)	0.95	1.10	1.0076	1.0208
V_{G15} (p.u.)	0.95	1.10	1.0141	1.0625
V_{G18} (p.u.)	0.95	1.10	1.0285	1.1
V_{G19} (p.u.)	0.95	1.10	1.0248	1.088
V_{G24} (p.u.)	0.95	1.10	1.0467	1.0849
V_{G25} (p.u.)	0.95	1.10	1.0437	1.0994
V_{G26} (p.u.)	0.95	1.10	0.9905	1.0505
V_{G27} (p.u.)	0.95	1.10	1.0053	0.9809
V_{G31} (p.u.)	0.95	1.10	0.9976	0.9835
V_{G32} (p.u.)	0.95	1.10	1.0117	0.977
V_{G34} (p.u.)	0.95	1.10	1.0375	1.0917
V_{G36} (p.u.)	0.95	1.10	1.0413	1.1
V_{G40} (p.u.)	0.95	1.10	1.0423	1.0424
V_{G42} (p.u.)	0.95	1.10	1.0357	0.9997
V_{G46} (p.u.)	0.95	1.10	0.9749	1.003
V_{G49} (p.u.)	0.95	1.10	1.0188	0.993
V_{G54} (p.u.)	0.95	1.10	1.0007	1.0117
V_{G55} (p.u.)	0.95	1.10	1.003	1.0204
V_{G56} (p.u.)	0.95	1.10	0.9973	1.0112
V_{G59} (p.u.)	0.95	1.10	1.0382	1.046
V_{G61} (p.u.)	0.95	1.10	1.0509	1.0042
V_{G62} (p.u.)	0.95	1.10	1.0367	0.9878
V_{G65} (p.u.)	0.95	1.10	1.0379	1.0071
V_{G66} (p.u.)	0.95	1.10	1.0351	0.95
V_{G69} (p.u.)	0.95	1.10	0.99	1.0056
V_{G70} (p.u.)	0.95	1.10	0.9902	1.0389
V_{G72} (p.u.)	0.95	1.10	1.0418	1.0394
V_{G73} (p.u.)	0.95	1.10	0.9967	1.017
V_{G74} (p.u.)	0.95	1.10	0.9718	0.95
V_{G76} (p.u.)	0.95	1.10	0.9715	1.0755
V_{G77} (p.u.)	0.95	1.10	0.999	1.0356
V_{G80} (p.u.)	0.95	1.10	1.0089	1.0533
V_{G85} (p.u.)	0.95	1.10	1.0221	0.9999
V_{G87} (p.u.)	0.95	1.10	0.9965	1.1
V_{G89} (p.u.)	0.95	1.10	1.0632	0.95
V_{G90} (p.u.)	0.95	1.10	1.0308	0.95
V_{G91} (p.u.)	0.95	1.10	1.0598	0.9841
V_{G92} (p.u.)	0.95	1.10	1.0434	1.0059

continued on following page

Table 22. Continued

Control Variable	Min	Max	Best Solution Base case	Best Solution Case 1
V_{G99} (p.u.)	0.95	1.10	0.998	1.1
V_{G100} (p.u.)	0.95	1.10	1.0497	1.0641
V_{G103} (p.u.)	0.95	1.10	1.0424	1.1
V_{G104} (p.u.)	0.95	1.10	1.0381	1.0041
V_{G105} (p.u.)	0.95	1.10	1.0238	1.0039
V_{G107} (p.u.)	0.95	1.10	1.0133	0.9767
V_{G110} (p.u.)	0.95	1.10	1.0414	1.0518
V_{G111} (p.u.)	0.95	1.10	1.0588	1.0382
V_{G112} (p.u.)	0.95	1.10	1.0222	1.0576
V_{G113} (p.u.)	0.95	1.10	1.0097	0.9956
V_{G116} (p.u.)	0.95	1.10	1.0101	1.0119
Q_{c5} (p.u.)	-40	0	-40	-40
Q_{c34} (p.u.)	0	14	14	14
Q_{c37} (p.u.)	-25	0	-25	-25
Q_{c44} (p.u.)	0	10	10	10
Q_{c45} (p.u.)	0	10	10	10
Q_{c46} (p.u.)	0	10	10	10
Q_{c48} (p.u.)	0	15	15	15
Q_{c74} (p.u.)	0	12	12	12
Q_{c79} (p.u.)	0	20	20	20
Q_{c82} (p.u.)	0	20	20	20
Q_{c83} (p.u.)	0	10	10	10
Q_{c105} (p.u.)	0	20	20	20
Q_{c107} (p.u.)	0	6	6	6
Q_{c110} (p.u.)	0	6	6	6
T_8 (p.u.)	0.90	1.10	0.991	1.0398
T_{32} (p.u.)	0.90	1.10	0.9396	0.9986
T_{36} (p.u.)	0.90	1.10	1.019	1.0332
T_{51} (p.u.)	0.90	1.10	0.9547	1.0538
T_{93} (p.u.)	0.90	1.10	0.9771	0.9193
T_{95} (p.u.)	0.90	1.10	0.9779	0.9
T_{102} (p.u.)	0.90	1.10	0.979	1.0045
T_{107} (p.u.)	0.90	1.10	1.0237	1.0364
T_{127} (p.u.)	0.90	1.10	1.0204	1.1
Total Cost ($/h)			129,154.12	76,390.75
P_{loss} (MW)			82.8730	560.6368
CPU Time (s)			20.746	23.525

Table 23. The Proposed IPG-PSO Solution on the IEEE 118-bus system for case 1

Control Variable	Min	Max	Best Solution Case 1
P_{G1} (MW)	0	680	680.0000
P_{G4} (MW)	0	360	360.0000
P_{G6} (MW)	0	360	360.0000
P_{G8} (MW)	60	180	153.3078
P_{G10} (MW)	60	180	180.0000
P_{G12} (MW)	60	180	180.0000
P_{G15} (MW)	60	180	180.0000
P_{G18} (MW)	60	180	180.0000
P_{G19} (MW)	60	180	180.0000
P_{G24} (MW)	40	120	120.0000
P_{G25} (MW)	40	120	120.0000
P_{G26} (MW)	55	120	120.0000
P_{G27} (MW)	55	120	120.0000
P_{G31} (MW)	0	107	0.0000
P_{G32} (MW)	0	100	0.0000
P_{G34} (MW)	0	100	0.4207
P_{G36} (MW)	0	100	0.0000
P_{G40} (MW)	0	100	0.0000
P_{G42} (MW)	0	100	0.2309
P_{G46} (MW)	0	119	0.0000
P_{G49} (MW)	0	304	136.6461
P_{G54} (MW)	0	148	0.0000
P_{G55} (MW)	0	100	0.0000
P_{G56} (MW)	0	100	0.0000
P_{G59} (MW)	0	255	0.0000
P_{G61} (MW)	0	260	0.0000
P_{G62} (MW)	0	100	0.0000
P_{G65} (MW)	0	491	222.7684
P_{G66} (MW)	0	492	235.9711
P_{G69} (MW)	0	805.2	308.2664
P_{G70} (MW)	0	100	0.0000
P_{G72} (MW)	0	100	0.0000
P_{G73} (MW)	0	100	0.0000
P_{G74} (MW)	0	100	0.0000
P_{G76} (MW)	0	100	0.1624
P_{G77} (MW)	0	100	0.0000
P_{G80} (MW)	0	577	296.9816
P_{G85} (MW)	0	100	0.0000

continued on following page

Table 23. Continued

Control Variable	Min	Max	Best Solution Case 1
P_{G87} (MW)	0	104	0.2751
P_{G89} (MW)	0	707	423.5223
P_{G90} (MW)	0	100	0.0000
P_{G91} (MW)	0	100	0.0000
P_{G92} (MW)	0	100	0.0021
P_{G99} (MW)	0	100	0.0000
P_{G100} (MW)	0	352	168.8745
P_{G103} (MW)	0	140	0.0000
P_{G104} (MW)	0	100	0.0000
P_{G105} (MW)	0	100	0.0302
P_{G107} (MW)	0	100	19.9632
P_{G110} (MW)	0	100	0.0000
P_{G111} (MW)	0	136	40.4323
P_{G112} (MW)	0	100	0.0000
P_{G113} (MW)	0	100	0.0000
P_{G116} (MW)	0	100	0.0000
V_{G1} (p.u.)	0.95	1.05	1.0116
V_{G4} (p.u.)	0.95	1.10	1.0342
V_{G6} (p.u.)	0.95	1.10	1.0093
V_{G8} (p.u.)	0.95	1.10	1.0453
V_{G10} (p.u.)	0.95	1.10	1.0036
V_{G12} (p.u.)	0.95	1.10	1.0208
V_{G15} (p.u.)	0.95	1.10	1.0625
V_{G18} (p.u.)	0.95	1.10	1.1
V_{G19} (p.u.)	0.95	1.10	1.088
V_{G24} (p.u.)	0.95	1.10	1.0849
V_{G25} (p.u.)	0.95	1.10	1.0994
V_{G26} (p.u.)	0.95	1.10	1.0505
V_{G27} (p.u.)	0.95	1.10	0.9809
V_{G31} (p.u.)	0.95	1.10	0.9835
V_{G32} (p.u.)	0.95	1.10	0.977
V_{G34} (p.u.)	0.95	1.10	1.0917
V_{G36} (p.u.)	0.95	1.10	1.1
V_{G40} (p.u.)	0.95	1.10	1.0424
V_{G42} (p.u.)	0.95	1.10	0.9997
V_{G46} (p.u.)	0.95	1.10	1.003
V_{G49} (p.u.)	0.95	1.10	0.993
V_{G54} (p.u.)	0.95	1.10	1.0117
V_{G55} (p.u.)	0.95	1.10	1.0204

continued on following page

Table 23. Continued

Control Variable	Min	Max	Best Solution Case 1
V_{G56} (p.u.)	0.95	1.10	1.0112
V_{G59} (p.u.)	0.95	1.10	1.046
V_{G61} (p.u.)	0.95	1.10	1.0042
V_{G62} (p.u.)	0.95	1.10	0.9878
V_{G65} (p.u.)	0.95	1.10	1.0071
V_{G66} (p.u.)	0.95	1.10	0.95
V_{G69} (p.u.)	0.95	1.10	1.0056
V_{G70} (p.u.)	0.95	1.10	1.0389
V_{G72} (p.u.)	0.95	1.10	1.0394
V_{G73} (p.u.)	0.95	1.10	1.017
V_{G74} (p.u.)	0.95	1.10	0.95
V_{G76} (p.u.)	0.95	1.10	1.0755
V_{G77} (p.u.)	0.95	1.10	1.0356
V_{G80} (p.u.)	0.95	1.10	1.0533
V_{G85} (p.u.)	0.95	1.10	0.9999
V_{G87} (p.u.)	0.95	1.10	1.1
V_{G89} (p.u.)	0.95	1.10	0.95
V_{G90} (p.u.)	0.95	1.10	0.95
V_{G91} (p.u.)	0.95	1.10	0.9841
V_{G92} (p.u.)	0.95	1.10	1.0059
V_{G99} (p.u.)	0.95	1.10	1.1
V_{G100} (p.u.)	0.95	1.10	1.0641
V_{G103} (p.u.)	0.95	1.10	1.1
V_{G104} (p.u.)	0.95	1.10	1.0041
V_{G105} (p.u.)	0.95	1.10	1.0039
V_{G107} (p.u.)	0.95	1.10	0.9767
V_{G110} (p.u.)	0.95	1.10	1.0518
V_{G111} (p.u.)	0.95	1.10	1.0382
V_{G112} (p.u.)	0.95	1.10	1.0576
V_{G113} (p.u.)	0.95	1.10	0.9956
V_{G116} (p.u.)	0.95	1.10	1.0119
Q_{c5} (p.u.)	-40	0	-40
Q_{c34} (p.u.)	0	14	14
Q_{c37} (p.u.)	-25	0	-25
Q_{c44} (p.u.)	0	10	10
Q_{c45} (p.u.)	0	10	10
Q_{c46} (p.u.)	0	10	10
Q_{c48} (p.u.)	0	15	15

continued on following page

Table 23. Continued

Control Variable	Min	Max	Best Solution Case 1
Q_{c74} (p.u.)	0	12	12
Q_{c79} (p.u.)	0	20	20
Q_{c82} (p.u.)	0	20	20
Q_{c83} (p.u.)	0	10	10
Q_{c105} (p.u.)	0	20	20
Q_{c107} (p.u.)	0	6	6
Q_{c110} (p.u.)	0	6	6
T_8 (p.u.)	0.90	1.10	1.0398
T_{32} (p.u.)	0.90	1.10	0.9986
T_{36} (p.u.)	0.90	1.10	1.0332
T_{51} (p.u.)	0.90	1.10	1.0538
T_{93} (p.u.)	0.90	1.10	0.9193
T_{95} (p.u.)	0.90	1.10	0.9
T_{102} (p.u.)	0.90	1.10	1.0045
T_{107} (p.u.)	0.90	1.10	1.0364
T_{127} (p.u.)	0.90	1.10	1.1
Total Cost ($/h)			76,390.75
P_{loss} (MW)			560.6368
CPU Time (s)			23.525

Chapter 9
Engineering QoS and Energy Saving in the Delivery of ICT Services

Alessandra Pieroni
Guglielmo Marconi University of Study, Italy

Giuseppe Iazeolla
Guglielmo Marconi University of Study, Italy

ABSTRACT

ICT service-providers are to daily face the problem of delivering ICT services (data processing (Dp) and/ or telecommunication (Tlc) services) assuring the best compromise between Quality of Service (QoS) and Energy Optimization. Indeed, any operation of saving energy involves waste in the QoS. This holds both for Dp and for Tlc services. This paper introduces models the providers may use to support their decisions in the delivery of ICT services. Dp systems totalize millions of servers all over the world that need to be electrically powered. Dp systems are also used in the government of Tlc systems, which also require Tlc-specific power, both for mobile networks and for wired networks. Research is thus expected to investigate into methods to reduce ICT power consumption. This paper investigates ICT power management strategies that look at compromises between energy saving and QoS. Various optimizing ICT power management policies are studied that optimize the ICT power consumption (minimum absorbed Watts), the ICT performance (minimum response-time), and the ICT performance-per-Watt.

INTRODUCTION

The growth in ICT energy consumption is driven by the growth of demand for greater data processing (Dp) and larger access to telecommunications (Tlc), within almost every organization.

This growth has a number of important implications, including (U.S. Environmental Protection Agency, 2007):

- Increased energy costs for business and government.
- Increased emissions, including greenhouse gases, from electricity generation.

DOI: 10.4018/978-1-4666-9755-3.ch009

- Increased strain on the existing power grid to meet the increased electricity demand.
- Increased capital costs for expansion of data center capacity and construction of new data centers.
- Increased capital costs for expansion of wired and wireless access to communications.

Made 100 the total electrical power consumption for the ICT, around 14% is taken by the mobile *Tlc*, 74% by the wired *Tlc* and the remaining 12% by the *Dp* technology. *Dp*, however, is only apparently the less powered sector, since *Tlc* is itself a *Dp* consumer, and so any effort to reduce the *Dp* power consumption may produce cascade effects that also reduce the *Tlc* one. Studying ways to save *Dp* power in thus central to any study for ICT power control and optimization.

In the US, power absorbed by data centers is estimated in more than 100 Billion kW, for an expenditure of $ 8 Billions a year, that corresponds to the expenditure in electricity of about 17 Million homes (Harchol-Balter, 2010). This same US-local data-center problem becomes a global one when seeing at power consumption by web companies, say Google, Yahoo etc. The number of Google servers will reach an estimated 2,376,640 units by the end of 2013 (Dean, 2009).

Assuming a busy server absorbs around 240 W of power, Google will need about 600 MW of electrical power by the end of year 2013.

In *Tlc* systems, about 90% of *Tlc*-specific power consumption is concentrated in the routers. The links only absorb 10%. Current routers consume between 0.01 and 0.1 W/Mbps (Vereecken, VanHeddeghem, et al. 2010).

IT research is thus expected to investigate into methods to reduce power absorbed by *Dp* and *Tlc* systems. To do that, one may decide to adopt policies to periodically switch-off *Dp* servers or *Tlc* routers when they are in an idle-state. Such policies, however, are to be sufficiently intelligent not to degrade the system Quality of Service (QoS). Indeed, returning an off-server or an off router to its on state requires spending a non-negligible amount of setup time that makes the server or router slower to respond to customer requests. This may turn into low-quality services such as low response to web queries, unsatisfactory VoIP communications and streaming of data. Any research in power management should thus look at compromises between energy saving and QoS.

In this paper, studies on *Tlc* power management policies and then on *Dp* power management policies will be discussed.

ENERGY MANAGEMENT IN TLC SYSTEMS

Tlc systems may consist of wired or wireless access networks or of a combination thereof. In addition to the basic *Dp* infrastructure, *Tlc* systems also include *Tlc*-specific subsystems: cell-phone, towers with associated base stations, subscriber stations, switching nodes, etc., for the wireless part, and communication processors, routers, gateways, switches, etc., for the wired part. Power management in *Tlc* systems, thus, includes not only power optimization of their *Dp* infrastructure, but also power optimization of *Tlc*-specific subsystems.

In this Section we will only deal with *Tlc*-specific subsystems, since the power optimization of *Dp* infrastructure is dealt with in Sect. "Power management in *Dp* systems".

Figure 1 describes a typical *Tlc* architecture, which combines wired and wireless communication networks. In the wired part, three main types of connections are found: 1) the twisted pair copper cable connection based on the DSL (Digital Subscriber Line) technology; 2) the coax cable connection based

Figure 1. Typical Tlc network architecture (Vereecken, VanHeddeghem, et al., 2010).

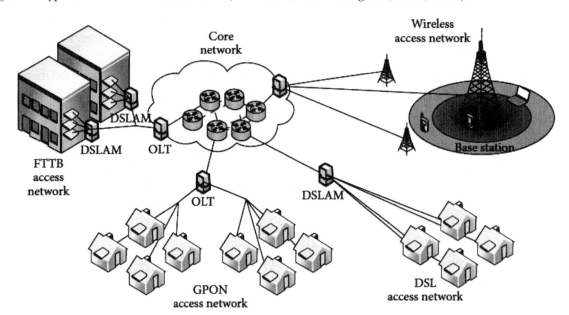

on the DOCS (Data Over Cable Service) technology and 3) the optical fiber connection based on the GPON (Gigabit Passive Optical Network) technology, used when higher bit rates are required. The illustration also shows the DSLAM (DSL Access Multiplexer) nodes, the OLT (Optical Line Termination) nodes and the FTTB (Fiber to the Building) nodes. In order to interconnect different user areas, a core network is used, that consists of a number of core nodes that are interconnected through wavelength-division multiplexed (WDM) optical fibre links, usually in a mesh or ring topology.

In the wireless part of the network we find the Base Transceiver Stations (BTS) to which the user's devices are connected by means of radio signals. Each BS is further connected to the core network through a so-called backhaul network. Different technologies can be found, from WiMAX (Worldwide Interoperability for Microwave Access) (Iazeolla, Pieroni, D'Ambrogio & Gianni, 2010), to HSPA (High Speed Packet Access), and to the most recent LTE (Long Term Evolution). In such a system, about 90% of *Tlc-specific* power consumption is concentrated in the *routers* (with 75% the line cards, 10% the power supply and fans, and 10% the switch fabric) (Iazeolla, Pieroni, 2014). Current routers consume between 0.01 and 0.1 W/Mbps. One can calculate that at ADSL access rates (8 Mbps) the power absorbed per subscriber is of about 0.24W/subs, while at 100Mbps becomes of about 3W/subs (Vereecken, VanHeddeghem, et al. 2010). Currently, *Tlc* networks are designed to handle the *peak* loads. Designing *adaptable networks*, where one can switch *off* line cards when the demand is lower, can lead to lower power consuming networks. In core networks this can be achieved by use of *dynamic topology optimization* algorithms: from all possible topologies that satisfy the required traffic demand, the topologies with lower overall power consumption are chosen. By such algorithms, reductions of power consumption for more than 50% during off-peak hours can be achieved (Puype, Vereecken, Colle, Pickavet and Demeester, 2007).

BTS with differentiated cell sizes are the key in wireless networks optimization if the so-called *hybrid hierarchical BTS deployment* is used. A low layer access network is first created, providing a low bit rate (but large cell sizes) to the users. In the higher layers, BS with higher bit rates (but smaller cell sizes) is utilized to provide high bandwidth connections when required. The advantage is that the higher layers can be switched to the *idle*, and only switched *on* with high traffic demand.

Tlc power optimization, also tries to minimize the power consumption of the home *gateways*. These are individual devices that only need to be *on* when the user is active. At other times, they could be switched *off*. In reality this is rarely operated, but legislations concerning standby power consumption standards of 0.5 W are emerging (Vereecken, VanHeddeghem et al. 2010).

This paper concentrates on the policies for the energy saving in the *BTS* stations, as seen in next section.

Energy Management in BTS Stations

Research on the management of the BTS stations investigates methods to reduce the BTS absorbed power. To do that, one may decide to adopt policies to periodically switch-*off* BTS channels when they are in an *idle* state. Such policies, however, are to be sufficiently intelligent not to degrade the BTS Quality of Service (QoS). Indeed, returning an *off*-antenna to its *on* state requires spending a non-negligible amount of time (called *setup time*) that makes the system slower to respond to customer requests. This may turn into BTS low QoS, such as low response time to incoming calls, unsatisfactory VoIP communications and limited streaming of data. Any research in wireless systems power management should thus look at compromises between energy saving and QoS.

System Assumptions

Let us consider the wireless system illustrated in Figure 2, with a series of cells S each served by a central BTS (e.g.: the Base Station in WiMAX, the eNodeB in LTE, etc.). Let us assume that the confining cells are not overlapping, in other words, that the User Equipment (UEs) residing in a given cell can perform call establishment with only the cell's BTS. Also assume that each BTS serves the 360° degrees of its area S by three sets of antennas, called sectorial-antennas. Each set of sectorial-antennas serves 120° degrees of Area S, called sector, see Figure 3. Let c be the number of antennas per-sector.

Each cell of the system serves number three sectors and thus three User Communities (UC1, UC2, UC3 as illustrated in Figure 3) are served in the cell. Each UCi is the source of calls for the c_i corresponding sectorial antennas.

Figure 2. The considered wide-band wireless system

Figure 3. The considered three sectorial-antennas structure

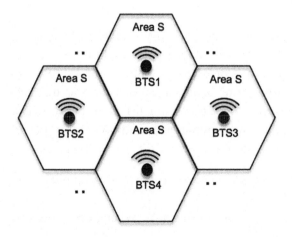
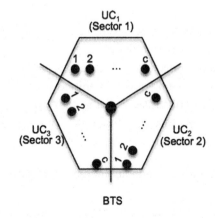

We shall assume the antennas are not always available and can be switched between three possible states (*on, off* and *idle*). The sectorial-antenna *response time E[T]* is defined as the total time in which the call remains in the system. The *power E[P]*, instead, is defined as the power consumption of the c antennas during an observation period.

Let us denote by the pair (π, δ) the sectorial antenna power management strategy, where:

- π is the power management policy, or the criterion according to which the antennas are switched between the *on, off*, or *idle* state.
- δ the queueing discipline, or the criterion according to which antennas peak calls from the call queue when they become idle.

Any power management strategy (π, δ) is a strategy that optimizes the following indices:

- Sectorial antenna average power consumption *E[P]* (in Watts to be minimized),
- Sectorial antenna average response time *E[T]* (in time-units to be minimized),
- Sectorial antenna response-per-Watt *ERP* (in time-units per Watt to be minimized).

Most of the power absorbed by the antennas is wasted. Indeed, due to overprovisioning, antennas are found in the *on* state (i.e. making transmission work) only 20% to 30% of the time (Tombaz, Vastberg and Zander, 2011). Energy saving requires the adoption of power management policies π to avoid powering the antennas when they are not transmitting. In other words, policies to decide in which state (*idle* or *off*) to keep the antennas when not in the *on* state (or busy). So, in order to save energy, one would suggest to keep antennas in the *idle* or in the *off* state when not busy. However, switching an antenna from the *off* to the *on* state implies a time-overhead (the so-called *setup time*). Thus, any power-saving policy may result in a time-wasting problem. As a consequence, the sectorial antenna may lose performance (increased *response time* to the incoming calls, low throughput of VoIP and streaming packets, etc.) and lose QoS. For the sake of conciseness, in next sections we only deal with the investigation of *E[P]*, since the *E[T]* and *ERP* studies, based on queueing simulation models, give results of the same type illustrated in the section that deals with the Power management in Dp systems (see comments on Table 2 at the end of the section).

Evaluation of the BTS Power Consumption

We shall assume, according to (Conte, 2012), that a BTS absorbs about 1500W, 500W of which remain in the rack (air conditioning, power rectifiers, etc.), and 1000W go to the power amplifier to feed $3xc$ antennas (3 sectors x c antennas per sector). According to (Conte, 2012), only 4% of 1500W (i.e.: $1500x0.04 = 60W$) is transmitted in the air. For a BTS powering $3c$ antennas, each antenna transmits (*on* state) $60W / 3c0W$ in the air. Assuming a 12% efficiency of the power amplifier, this means that an antenna absorbs $60 / 3c / 0.12 = (60 / 3c)x8.3W$ from the power amplifier. In addition to this, the antenna will absorb $500 / 3cW$ from the rack. For the sake of an example, for e.g. c = 4 antennas per sector, one antenna in the *on* state will absorbs $\dfrac{60}{3x4} x8.33 \cong 42W$ from the power amplifier and $\dfrac{500W}{3x4} \cong 42W$ from the rack. In total, the antenna, in the *on* state, absorbs 84W. When in stand-by state (*idle* state) we can assume the antenna absorbs only from the rack and thus the power absorption in the idle state is $\frac{500W}{3c}$ (i.e.: 42W in the example). Finally, we shall also assume the antenna can be in the *off* state, in other words that the antenna is totally switched off. In this case, one can assume that it absorbs only part of the rack power, in particular the power absorbed by the air conditioning that amounts to $200W$, according to (Conte, 2012). In other words, in the *off* state, the antenna power absorption amount to $\frac{200W}{3c}$ (i.e.: $200/12 \cong 17W$). In summary, for the $c = 4$ antennas per sector example, one antenna absorbs: $84W$ in the *on* state, $42W$ in the *idle* and $17W$ in the *off* state.

Two different policies are studied to manage the c-antennas of each sector: the o*n/Idle* policy, and the o*n/off* policy. Under the *On/Idle* policy, the antennas are never turned *off*. All antennas are either *on* or *idle*, and remain in the *idle* state when there are no calls to serve. Assume the farm consists of c antennas, if an arrival finds an antenna *idle* it starts serving on the *idle* antenna. Arrival that find all n antennas *on* (busy), join a central queue from which the antennas pick calls when they become *idle*.

The *On/Off* policy consists, instead, of immediately turning *off* the antennas when not in use. As said above, however, there is a setup cost (in terms of time-delay and of additional power penalty) for turning *on* an *off* antenna. Figure 4 compares the *On/Off* and *On/Idle* policies for an example case.

Figure 4 shows the average power consumption for a case study with $c = 4$ antennas and utilization $= 30\%$. This means that per time-unit period, under the *On/Idle* policy, the four sectorial-antennas absorb $(84Wx0.3 + 42Wx0.7)x4 = 218W$. In other words, 84W in the *on* period (30% of time) and 42W in the *idle* period (70% of time). Under the *On/Off* policy, instead, the four sectorial-antennas absorb $(84Wx0.3 + 17Wx0.7)x4 = 148W$. In other words, 84W in the *on* period (30% of time) and only 17W in the *off* period (70% of time). The *On/Idle* policy involves a larger amount of power waste with respect to the *On/Off* policy, since of the amount of power an *idle* antenna absorbs.

POWER MANAGEMENT IN DP SYSTEMS

Data centers have become common and essential to the functioning of business, communications, academic, and governmental systems. During the past years, increasing demand for *Dp* services has led to significant growth in the number of data centers, along with an estimated doubling in the energy used by servers and the power and cooling infrastructure that supports them.

Figure 5 illustrates the way energy is spent in data centers. Heating and services for Ventilation and Air Conditioning/Backup (HVAC/UPS) absorb around 40% of electrical energy and *Dp* services the

Figure 4. Average power consumption E[P] with c = 4 antennas and utilization = 30%

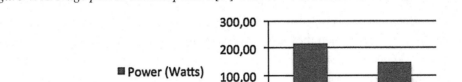

Figure 5. Data Center energy consumption sources (Renzi, 2007).

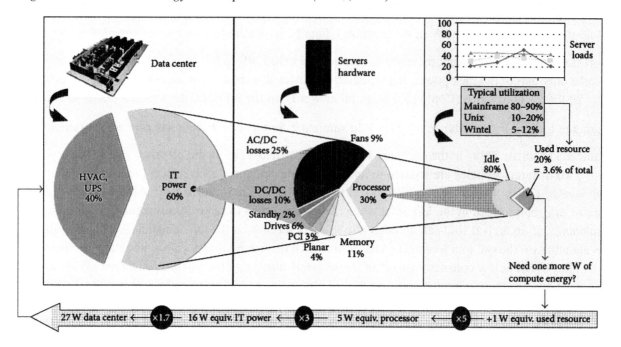

remaining 60%. The latter is in turn divided between AC/DC losses (25%), DC/DC losses (10%), Fans, Drives, PCI, etc., and Memory consumptions (for a total 35%) and the remaining 30% is consumption in server processors.

In other words, the processors consumption totalizes $0.30x0.60$, i.e., 20% of total data center consumption. Such an amount, even though apparently negligible with respect to the total, is the main cause of the remaining 80%. Thus, any effort to reduce the processors 20% may produce cascade effects that also reduce the remaining 80%.

Figure 6 shows that a 1W savings at servers component level (processor, memory, hard disk, etc.) creates a reduction in data center energy consumption of approximately 2.84W.

For this reason, any research in ICT energy saving should concentrate on policy to reduce Dp consumption at server components level. Data centers can be seen as composed of a number of servers that can be organized into *single-farms* or a *multi-farms*. The following sections study the power saving policies in the *single-farm* case, and in the *multi-farm* one.

Energy Saving in Single-Farm Data Centres

Most of power absorbed by the servers of a farm is wasted, since servers are *busy* (i.e. making processing work) only 20% to 30% of the time, on average. So, energy saving requires the adoption of management policies to avoid powering the servers when they are not processing. In other words, policies to decide in which state (*idle* or *off*) to keep the servers when not busy. Two types of server management policies will be considered: *static* and *dynamic* policies.

Figure 6. Cascade effect of energy saving in data centres (Emerson Network Power, 2012).

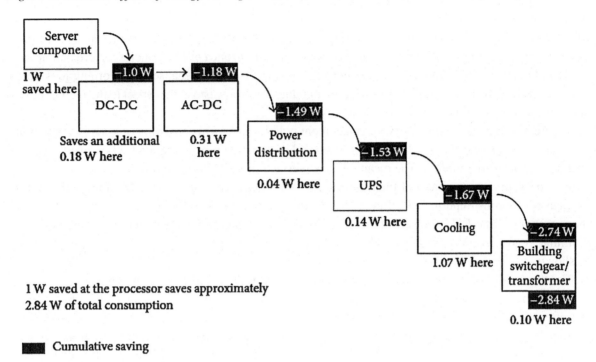

Energy Saving with Static Policies

One may assume a *busy*-server in the *on* state absorbs around 240W (P_{ON}), an *idle*-server about 160W (P_{IDLE}) and an *off*-server 0W (P_{OFF}). So why not to keep in the *idle* state or in the *off* state the servers when not busy? Just since switching a server from *off* to *on* consumes a time-overhead. Thus, a power-saving policy may result in a time-wasting problem. As a consequence, the servers may lose performance (e.g., increased *response time* to the incoming jobs, lower throughput of communication packets, etc.) and its service may become unacceptable to customers.

To turn *on* an *off* server, we must first be put the server in *setup* mode. During the setup period, the server cannot process jobs. The time spent in setup is called *setup time*. In (Adan, Gandhi, Harchol-Balter, 2013) the authors consider server farms with a setup cost. Setup costs always take the form of a time delay, and sometimes there is also a power penalty, since during that entire period the server consumes the same power as being in the *on* state.

In (Adan, Gandhi, Harchol-Balter, 2013) three different policies are studied to manage server farms: *On/Idle* policy, *On/Off* policy and *On/Off/Stag* policy.

Under the *On/Idle* policy, servers are never turned *off*. All servers are either *on* or *idle*, and remain in the *idle* mode when there are no jobs to serve. Assume the farm consists of *n* servers, if an arrival finds a server *idle* it starts serving on the *idle* server. Arrival that find all *n* servers *on* (busy), join a central queue from which the servers pick jobs when they become *idle*.

The *On/Off* policy consists instead of immediately turning *off* the servers when not in use. As said above, however, there is a setup cost (in terms of time-delay and of additional power penalty) for turning *on* an *off* server.

Finally, the *On/Off/Stag* policy is the same as the *On/Off* one, except that at most 1 server can be in setup at any point of time. This policy is known as the "staggered boot up" policy in data centers, or "staggered spin up" in disk farms (Adan, Gandhi, Harchol-Balter, 2013; Corporation, 2004; Storer, Greenan, Miller, Voruganti, 2008). Figure 7 compares the *On/Off* and *On/Idle* policies for an example case.

The *On/Idle* policy proves to be better in terms of *response time*, because the incoming jobs do not suffer by *setup time* delays, but involves a larger amount of power waste with respect to the *On/Off* policy, since of the amount of power an *idle* server absorbs.

Figure 8 compares the three server management policies in a farm consisting of k=10 servers, when the average *setup time* changes from 1 to 100sec and the average processing load λ (i.e. average job arrival rate) from 1 to 7 job/sec. The mean job size (service time) is assumed of 1 sec.

Comparison is on the basis of the resulting mean response time $E[T]$ to the incoming jobs and the average power consumption $E[P]$. In the *On/Idle* case, when λ is low, there is no waiting and thus the mean response time $E[T]$ is of about the mean job service time (1sec) and increases for increasing λ. A similar trend can be observed for the *On/Off/Stag* policy, since: $E[T]_{ON/OFF/STAG} = E[T]_{ON/IDLE} + E[Tsetup]$, as shown in (Adan, Gandhi, Harchol-Balter, 2013).

For the *On/Off* policy, instead, the response time curve follows *bathtub* behaviour.

When the load λ is low, the mean response time is high since almost every arrival finds servers in the *off* state, and thus every job incurs the *setup time*. For medium loads, servers are less frequently switched to the *off* condition and thus jobs are more likely served by available servers in the *on* state, and do not incur in setup times. For high loads, finally, the mean response time increases due to large queueing in the system.

For the power consumption, one can show (Adan, Gandhi, Harchol-Balter, 2013) that $E[P]_{ON/OFF/STAG} < E[P]_{ON/OFF}$, since at most one server can be in setup for the *On/Off/Stag* policy. There also results $E[P]_{ON/OFF} < E[P]_{ON/IDLE}$, since servers are turned *off* in the *On/Off* case. However, for loads λ above the me-

Figure 7. Average power consumption E[P] and average response time E[T] with 4 servers, utilization = 30% average setup time = 200sec, average job size (service time) =7sec (Adan, Gandhi, Harchol-Balter, 2013).

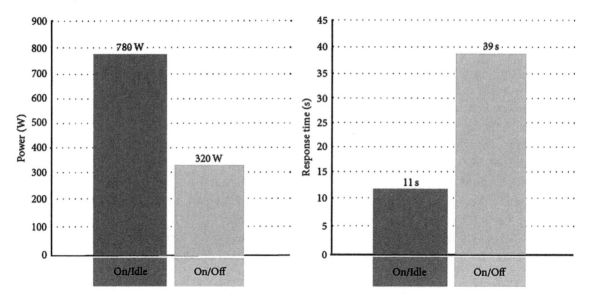

Figure 8. Effects of server management policies and setup time on response time and power consumption (Adan, Gandhi, Harchol-Balter, 2013).

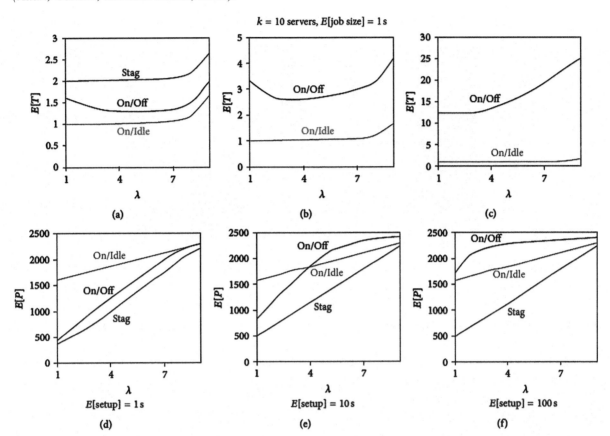

dium, there results $E[P]_{ON/OFF} > E[P]_{ON/IDLE}$ for medium *setup time* i.e. E[setup time] = 10sec, while for large *setup time* i.e. E[setup time] = 100sec there always results $E[P]_{ON/OFF} > E[P]_{ON/IDLE}$, since of the large amount of power wasted in turning servers *on* in the *On/Off* policy. Table 1 gives a synthetic comparison of the three considered policies in terms of *response time* and *power consumption*.

In conclusion, any reduction in power consumption is paid by an increase in response times. So, why not to adopt queueing disciplines that minimize average *response times*? The SPTF (Shortest Processing Time First) (Conway, Maxwell, Miller, 1967) or SJF (Shortest Job First) (Harchol-Balter, 2010) queueing discipline is known to perform better that the common FIFO (First In First Out). Its use can then reduce the amount to pay in terms of *response time* to obtain a given power saving.

Figure 7 (that illustrates the FIFO queueing case) shows that to reduce the power consumption from 780 to 320W we have to pay an increase from 11 to 39 sec in average response time.

Figure 9 illustrates that if the SPTF discipline is instead used, the debt to pay in response time is much smaller (from 39 to 27 sec) as proved in our simulations studies (Iazeolla and Pieroni, 2014).

Under the assumption of Poisson arrivals, exponential service times and deterministic *setup times*, authors in (Gandhi, Gupta, Harchol-Balter, Kozuch, 2010) prove that the optimal, or nearly optimal, combination of (π, δ), with π being one policy from the set {*On/Off, On/Idle*} and δ one queueing discipline from the set {*FIFO, LIFO, RAND*}, means minimizing a new metric called *ERP (Energy-Response time Product)* and defined as: $ERP(\pi, \delta) = E[P](\pi, \delta) \times E[T](\pi, \delta)$.

Figure 9. Experimental results with same parameters of Figure 3a except SPTF or SJF queueing discipline.
Seeking the optimal (π, δ) strategy for the single-farm data centers

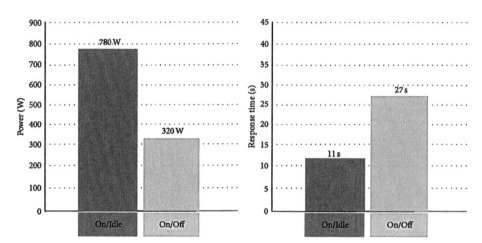

Minimizing $ERP(\pi, \delta)$ can be seen as maximizing the performance per Watt, with performance been defined as the inverse of the mean response time (Gandhi, Gupta, Harchol-Balter, Kozuch, 2010). In other words, according to their results, there is no need to consider other policies than the *On/Off* and the *On/Idle* policy.

They, however, only study the effects of moving from one policy π to another, without paying attention to the effects of also moving from a δ = *FIFO* discipline to another *time-independent* discipline. Under the *FIFO* assumption, however, they find that the *On/Idle* policy is typically superior to the remaining two in terms of $ERP(\pi, \delta)$.

Our aim is to extend such results by studying the effects of the queueing discipline δ, both on the $ERP(\pi, \delta)$ index and on the $E[P](\pi, \delta)$ and $E[T](\pi, \delta)$ indices separately.

More precisely, the following four (π, δ) strategies are investigated in the paper:

1. (*On/Idle, FIFO*)
2. (*On/Idle, SPTF*)
3. (*On/Off, FIFO*)
4. (*On/Off, SPTF*)

And for each of such strategy the $ERP(\pi, \delta)$ product is studied, besides the $E[P](\pi, \delta)$ and $E[T](\pi, \delta)$ indices. Two largely different *setup times* ($E[Tsetup] = 1s$ and $E[Tsetup] = 100s$) will be used to stress the effect of the *setup time* on the *On/Idle* and *On/Off* polices. Similarly, two largely different farm data center loads, low ρ ($\rho \leq 0.5$) and high ρ ($\rho \to 1$) will be used to stress the effect of the queueing disciplines. The following farm data center characteristics are assumed: server mean setup-time $E[Tsetup] = 1sec$ (or 100sec), server $P_{ON} = 240W$, server $P_{SETUP} = 240W$, server $P_{IDLE} = 150W$, server $P_{OFF} = 0W$, mean job E[service time] = 1sec, n = 30 servers.

Table 2 shows simulation results (Iazeolla and Pieroni, 2014) that compare the *Dp* power and QoS indices in the low setup case ($E[T_{setup}] = 1sec$):

- Seeing at the power consumption E[P], we note that there is no effect by the queueing discipline δ on the power consumption E[P], while there is an effect by the policy π for low ρ. Indeed, a drastic reduction can be seen (from 6000W to 4200W, for low ρ) when moving from *On/Idle* to *On/Off*, since when ρ is low, the waiting queue is almost empty and thus a large number of servers is in the *off* state. For high ρ instead, the power consumption E[P] remains unchanged (E[P] = 7100W) with the discipline δ since the queue is always full and thus the servers remains always in the on state.
- Seeing at the *response time* E[T], we note that there is an effect both by the queueing discipline δ and by the policy π. The effects hold both for low ρ and for high ρ. In the *On/Idle* case, when ρ is low, there is no waiting in the *Dp* queue and thus the mean response time T is of about the mean job service time (S = 1sec) while it increases (E[T] = 1.8sec for δ = *FIFO* and E[T] = 1.3sec for δ = *SPTF*) for high ρ. In the *On/Off* case, when ρ is low, the mean *response time* is higher (E[T] = 1.2sec with no effect by the discipline, the queue is empty), since almost every arrival finds servers in the *off* state, and thus every job incurs in the *setup time*. For high ρ, instead, the mean *response time* increases (*E[T]* = 2sec for δ = *FIFO* and *E[T]* = 1.35sec for δ = *SPTF*) due to large queueing. As predicted above, we can see that the benefit in *response time* one may obtain moving from *FIFO* to *SPTF* is larger than the one obtainable moving from *On/Off* to *On/Idle*. Indeed (see high ρ) moving from the (*On/Off*, *FIFO*) strategy to the (*On/Idle*, *FIFO*) the *response time E[T]* changes from 2 to 1.8 (a 10% reduction). Moving, instead, from the (*On/Idle*, *FIFO*) strategy to the (*On/Idle*, *SPTF*) the *response time E[T]* changes from 1.8 to 1.3 (an almost 30% reduction).
- Seeing at the ERP (π, δ) index, its values are a consequence of the *E[P]* and the *E[T]* ones. Table 1 shows that the optimal ERP(π, δ) is obtained for the (*On/Off*, *FIFO*) strategy and for the (*On/Off*, *SPTF*) strategy when ρ is low, while it is obtained for the (*On/Idle*, *SPTF*) strategy only when ρ is high.

Table 1. Synthetic view of the On/Off, On/Idle and On/Off/Stag power optimization policies

	Response Time	**Power Consumption**
On/Idle	*Small response times*	*High waste of power*
On/Off	*Medium response times*	*Medium waste for low setup times, high for increasing setup*
On/Off/Stag	*Large response times*	*Low waste of power*

Table 2. Server farm results for low setup-time (E[Tsetup] = 1sec)

π, δ	E[P](π, δ) W		E[T](π, δ) (sec)		ERP (π, δ)	
	ρ <= 0.5	ρ −>1	ρ = 0.5	ρ −>1	ρ <= 0.5	ρ −>1
On/Idle FIFO	6000	7100	1	1.8	6000	12780
On/Idle SPTF	6000	7100	1	1.3	6000	9230
On/Off FIFO	4200	7100	1.2	2	5040	14200
On/Idle SPTF	4200	7100	1.2	1.35	5040	9585

For the high *setup time* case ($E[Tsetup] = 100$sec), the reader is sent to (Iazeolla and Pieroni, 2014). In summary, making predictions of the *Dp* power management policies, that optimizes:

- The *Dp* power consumption (minimum absorbed Watts), or
- The *Dp* performance (minimum response_time), or
- The *Dp* performance-per-Watt (minimum response_time-per-Watt)

is a non trivial task. The most significant policies π are first to be drawn from the universe of all possible policies. Then, for each such a policy, the effects of time-dependent and time-independent queueing disciplines are to be studied. On the other hand, once the modeling work has been done, the work the server-farm manager has to perform to direct his *Dp* is greatly simplified, since the universe of all possible (π, δ) strategies he needs to choose from is drastically reduced to very a small set of most significant strategies.

Note that the Table 2 comparison of the four (π, δ) strategies, relating to a farm consisting of a given number n of *Dp* servers, can be extended to the case of a BTS sector with n-antennas by assuming for the antenna the same example *E[Tsetup]* time and *E[service time]* of the server.

Energy Saving with Dynamic Policies

In any practical situation, the load λ changes over the time, according to a given pattern $\lambda(t)$. One should then find policies that adapt themselves to changing load patterns. This is not the case of the policies introduced in Sect. "Energy saving with static policies", which are somehow static in nature and remain efficient only for given values of λ, while becoming inefficient for other values. Looking, for example, at the Figure 8 case with $E[T_{setup}] = 10$sec one can see that for changing values of λ there are situations in which the *On/Off* policy consumes less power than *On/Idle* and vice versa.

For this reason, two adaptive versions of the *On/Off* and *On/Idle* policies are known in literature, respectively called *DelayedOff* and *LookAhead*, which *dynamically* adapt themselves to changing loads (Gandhi, Gupta, Harchol-Balter, Kozuch, 2010).

The *DelayedOff* policy is an improvement of the *On/Off*. According to *DelayedOff*, when a server goes idle, rather than turning *off* immediately, it sets a timer of duration t_{wait} and sits in the idle state for t_{wait} seconds. If a request arrives at the server during these t_{wait} seconds, the server goes back to the *busy* state (with zero setup cost); otherwise, the server is turned *off*.

The *LookAhead* policy is an improvement of the *On/Idle*. Under such a policy, the system fixes an optimally chosen number n^* of servers maintained in the *on* or *idle* states. According to the standard *On/Idle*, if an arrival finds a server *idle* it starts serving on the *idle* server. Arrivals that find all n^* server *on* (busy), join a central queue from which servers pick jobs when they become *idle*. The optimal t_{wait} and the optimal n^* of the two policies respectively are chosen to minimize the ERP index. As said above, minimizing *ERP* can be seen as maximizing the performance per Watt, with performance been defined as the inverse of the mean response time (Gandhi, Gupta, Harchol-Balter, Kozuch, 2010).

In the *LookAhead* policy, n^* changes as a function of time. Indeed, the policy calculates $n^*(t)$ for each time t basing on the forecast of the load $\lambda(t)$ at time t.

Figure 10 illustrates the auto-scaling capabilities of the *LookAhead* and *DelayedOff* policies (Gandhi, Gupta, Harchol-Balter, Kozuch, 2010), with respect to the conventional *On/Off*, for poissonian arrivals with $\lambda(t)$ changing sinusoidally with time (period = 6 hrs).

Figure 10. Effects of dynamic policies with respect to the static On/Off (Gandhi, Gupta, Harchol-Balter, Kozuch, 2010).

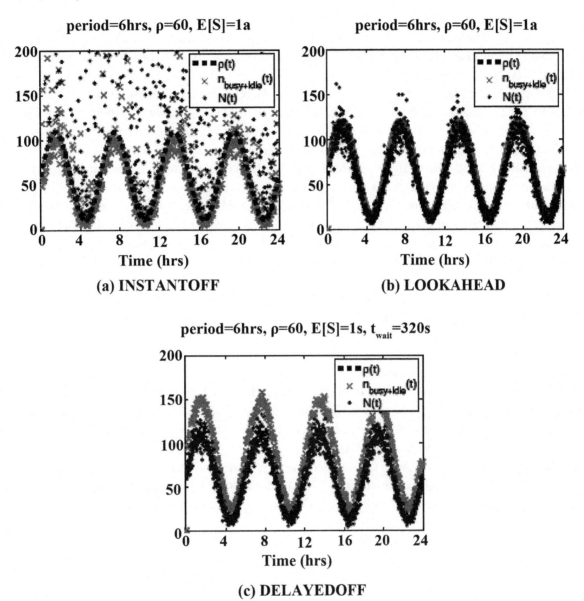

The left graph refers to the *On/Off* policy (called *InstantOff*), the middle graph to the *LookAhead* and the right one to the *DelayedOff*. The dashed line denotes the varying load at time t, $\lambda(t)$. The crosses denote the number $n_{busy+idle}(t)$ of servers that are busy or idle at time t, and the dots denote the number N(t) of jobs in the system at time t.

The illustration shows how, with the two dynamic policies, number $n_{busy+idle}(t)$ and number N(t) almost completely follow the behaviour of the demand pattern $\lambda(t)$, while in the *On/Off* case such numbers are somewhat dispersed, in other words, some servers remain in the *idle* state whereas they should be *busy* and vice versa, with the consequence of waste of power and worsened *response time*. The two dynamic policies above simply try to optimize the *E[P]* by *E[T]* product. In many practical situations, instead, the objective is to meet a given average response time, according to requirements dictated by specific Service Level Agreements (SLAs).

In this case, specific dynamic policies have been introduced, which try to respect the *E[T]* requirement while minimizing the average power consumed by the servers ($\mathbf{P}_{avg}$) and the average number of used servers ($\mathbf{N}_{avg}$).

Such policies are known as the *AutoScale* policy (Harchol-Balter, Gandhi, Gupta, Raghunathan, Kozuch, 2012), the *AlwaysOn* policy (Chen, He, Li et al., 2008), the *Reactive* policy (Urgaonkar and Chandra, 2005) and the *Predictive MWA* policy (Bodìk, Griffith et al., 2009; Grunwald, Morrey III et al., 2000; Pering, Burd and Brodersen, 1998; Verma, Dasgupta et al., 2009). The latter will not be dealt with here, and we will only treat the *AutoScale* policy, which is an evolution of the remaining three.

The *AutoScale* policy generalizes the use of the $\mathbf{t}_{wait}$ time already seen for the *DelayedOff* policy. Differently from this latter, however, in the *AutoScale* case each server decides autonomously when to turn off, setting a timer of duration $\mathbf{t}_{wait}$ and sitting in the *idle* state for $\mathbf{t}_{wait}$ sec. As with the *DelayedOff*, however, if a request arrives at the server during these $\mathbf{t}_{wait}$ sec, then the server goes back to the *busy* state (with zero setup cost). Otherwise, the server is turned *off*.

The *AutoScale* and the three remaining policies have been evaluated in (Harchol-Balter, Gandhi, Gupta, Raghunathan, Kozuch, 2012) according to a specific load pattern $\lambda(t)$ varying over time between 0 and 800req/s (see Figure 11). Such a pattern, known as *Dual Phase* pattern is used to represent the diurnal nature of typical data center traffic, where the request rate is low at the night-time and high at day time.

Figure 12 illustrates the performance of the *AutoScale* policy, when the *time requirement* to meet is a 95- percentile *response time* T goal of 400 to 500ms (denoted T_{95}). In the illustration, the red lines denote the number $k_{busy+idle+setup}(t)$ of *busy+idle+setup* servers and the blue lines the number $k_{busy+idle}(t)$ of *busy+idle* servers at time t. The k_{ideal} line represents the number of servers that should be *on* at any given time to fully satisfy the demand $\lambda(t)$.

The illustration shows how, with the *AutoScale* policy, there is no dispersion in the available servers and the number $k_{busy+idle+setup}(t)$ and number $k_{busy+idle}(t)$ almost totally follow the demand pattern $\lambda(t)$, and a $T_{95} = 491$ms goal is achieved, with $P_{avg} = 1,297$W and $N_{avg} = 7.2$ servers.

Figure 11. Dual Phase pattern (Harchol-Balter, Gandhi, Gupta, Raghunathan, Kozuch, 2012).

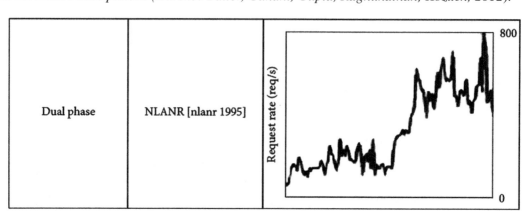

Figure 12. Effects of dynamic AutoScale policy (Harchol-Balter, Gandhi, Gupta, Raghunathan, Kozuch, 2012).

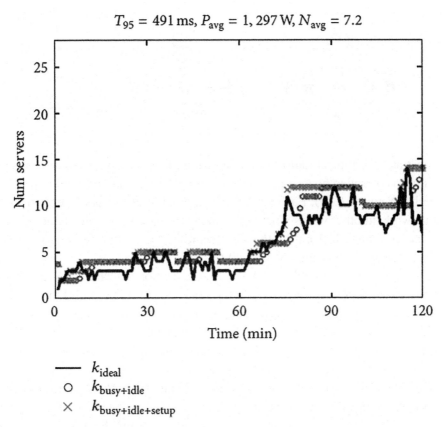

In the mentioned similar policies (*AlwaysOn*, *Reactive* and *Predictive MWA*), instead, the T_{95} requirement can be seen met only at the expense of server dispersion and/or at the expense of P_{avg} and N_{avg} (Harchol-Balter, Gandhi, Gupta, Raghunathan, Kozuch, 2012).

Indeed, in the *AlwaysOn* case the T_{95} requirement is met (T_{95}=291ms) but at the expense of large dispersion of servers and large power consumption (P_{avg}=2,322W and N_{avg}=14).

In the *Reactive* case, instead, a low dispersion of servers is achieved, with low power and low number of servers (P_{avg}=1,281W, N_{avg}=6.2) but the time requirement is absolutely out of range (T_{95} = 11,003ms).

A better time performance (T_{95} = 7,740ms) is found in the *Predictive MWA* with similarly low dispersion of servers, and similarly low power and number of servers (P_{avg}= 1,276W, N_{avg}= 6.3).

Energy Saving in Multi-Farm Data Centers

Energy saving in multi farms is based on so-called *self-organization* and *self-differentiation* algorithms, whose goal is to transfer the load from a server to a less loaded one, to maximize the power efficiency of the whole data center.

These algorithms are widely adopted in the Autonomic Computing field. The term Autonomic indicates systems able to self-manage, self-configure, self-protect, and self-repair, thus systems that have no need of external action to be managed (Barbagallo, Di Nitto, Dubois, and Mirandola, 2019).

Figure 13. Example of a multi-farm data center.

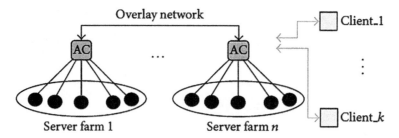

Figure 13 illustrates the typical multi-farm architecture, that consists of a series of server farms (1 through n) each farm controlled by a so-called Autonomic Component (AC), with the ACs interacting through an Overlay Network (ON). Each farm serves a number of Clients (1 trough k). The ON is a self-organized network, in other words a network which is created, maintained and optimized through *self-organization algorithms* which cluster the ACs according their properties or type (Sahuquillo Martínez, Pareta, Otero, Ferrari, Manzalini, Moiso, 2009).

The ACs, in turn, execute a particular kind of self-organization algorithm called *self-differentiation algorithm*, which take decentralized decisions on the state and configuration of the ACs. The AC aims at putting in *idle* state the servers and transferring the load on the other servers to limit performance degradation. Three types of self-differentiation algorithms are known: *Stand-by*, *Load Distribution* and *Wake-up* algorithms whose details can be found in (Sahuquillo Martínez, Pareta, Otero, Ferrari, Manzalini, Moiso, 2009). The algorithms were evaluated by means of simulations of an use-case in which server farms are in charge to serve requests issued by a set of clients. Each client performs several requests, before terminating the connection. The percentage of energy that can be saved in a day goes from about 7% to about 12%, with a debt to pay in terms of *response time* from about 9 units of time (when the power saving is 7%) to about 11 units of time (when the power saving is 12%).

CONCLUSION

The problem of delivering ICT services (data processing (*Dp*) and/or telecommunication (*Tlc*) services) assuring the best compromise between Quality of Service (QoS) and Energy Optimization is the problem the ICT service providers are to daily face. Indeed, any operation of saving energy involves waste in the QoS (high response times and high response time per Watt). This holds both for *Dp* and *Tlc* services.

This paper has introduced models the providers may use to support their decisions in the optimal delivery of ICT services.

By use of the models, the provider may choose between: 1) the optimal power consumption, or minimum absorbed Watts, 2) the optimal performance, or minimum response-time, and 3) the optimal performance-per-Watt, or minimum response-time-per-Watt, or a combination thereof. Indeed, any operation of energy saving yields wastes in the QoS, both in data processing and in telecommunication case. Various ICT power management policies have been investigated and the effect of power management on the QoS (response time and response time per Watt) has been studied. In administering ICT resources, the service-provider can thus apply the best strategy to minimize the power consumption, while maintaining its QoS to its best.

ACKNOWLEDGMENT

This work was partially supported by the *PhD research program* of the University of Roma TorVergata funded by Telecom Italia, and by the Guglielmo Marconi University of Studies in Roma.

REFERENCES

Adan, I., Gandhi, A., & Harchol-Balter, M. (2013). Server farms with setup costs. Performance Evaluation ScienceDirect. Elsevier.

Barbagallo, D., Di Nitto, E., Dubois, D. J., & Mirandola, R. (2009). A Bio-Inspired Algorithm for Energy Optimization in a Self-organizing Data Center. In *Proceedings of the First international conference on Self-organizing architectures*, (pp. 127-151). Academic Press.

Bodìk, P., & Griffith, R. (2009). Statistical machine learning makes automatic control practical for internet data centers. In *Proceedings of the 2009 Conference on Hot Topics in Cloud Computing* (HotCloud'09). Academic Press.

Chen, G., He, W., & Liu, J. (2008). Energy-aware server provisioning and load dispatching for connection intensive internet services. In *Proceedings of the 5th USENIX Symposium on Networked Systems Design and Implementation* (NSDI'08), (pp. 337–350). USENIX.

Conte, A. (2012). *Power consumption of base stations*. Alcatel-Lucent Bell Labs France.

Conway, R. W., Maxwell, W. L., & Miller, L. W. (1967). *Theory of scheduling*. Addison-Wesley.

Corporation, I. (2004). *Serial ATA Staggered Spin-Up*. White Paper.

Dean, J. (2009). *Designs, Lessons and Advice from Building Large Distributed Systems*. Google Fellow.

Emerson Network Power. (2012). *Energy Logic: Reducing Data Center Energy Consumption by Creating Savings that Cascade Across Systems*. A White Paper from the Experts in Business-Critical Continuity.

Gandhi, A., Gupta, V., Harchol-Balter, M., & Kozuch, M. A. (2010). Optimality Analysis of Energy-Performance Trade-off for Server Farm Management. Performance Evaluation, ScienceDirect. Elsevier.

Grunwald, D., & Morrey, C. B. III. (2000). Policies for dynamic clock scheduling. In *Proceedings of the 4th Conference on Symposium of Operating System Design and Implementation (OSDI'00)*. OSDI.

Harchol-Balter M. (2010). *Open Problems in Power Management of Data Centers*. IFIP WG7.3 Workshop, University of Namur, Belgium.

Harchol-Balter, M. (2013). *Performance Modeling and Design of Computer Systems*. Cambridge University Press.

Harchol-Balter, M., Gandhi, A., Gupta, V., Raghunathan, R., & Kozuch, M. A. (2012). AutoScale: Dynamic, Robust Capacity Management for Multi-Tier Data Centers. *ACM Transactions on Computer Systems, 30*(4), 14.

Iazeolla, G., & Pieroni, A. (2014). Energy saving in data processing and communication system. *TheScientificWorldJournal, 2014*, 1–11. doi:10.1155/2014/452863 PMID:25302326

Iazeolla, G., & Pieroni, A. (2014). *Power management of server farms.* doi:10.4028/www.scientific.net/AMM.492.453

Iazeolla, G., Pieroni, A., D'Ambrogio, A., & Gianni, D. (2010). *A distributed approach to wireless system simulation.* Paper presented at the 6th Advanced International Conference on Telecommunications, AICT 2010. doi:10.1109/AICT.2010.66

Iazeolla, G., Pieroni, A., D'Ambrogio, A., & Gianni, D. (2010). *A distributed approach to the simulation of inherently distributed systems.* Paper presented at the Spring Simulation Multiconference 2010, SpringSim'10. doi:10.1145/1878537.1878675

Iazeolla, G., Pieroni, A., & Scorzini, G. (2013). *Simulation study of server farms power optimization.* RI.01.13 T.R. Software Engineering Lab., University of Roma TorVergata.

Marsan, M. A. (2011). Energy-Efficient Networking: the views of leaders of international research projects. CCW 2011, Hyannis.

Meo, M., & Mellia, M. (2010). *Green Networking. IEEE INFOCOM Workshop on Communications and Control for Sustainable Energy Systems*, Orlando, FL.

Pering, T., Burd, T., & Brodersen, R. (1998). The simulation and evaluation of dynamic voltage scaling algorithms. In *Proceedings of the International Symposium on Low Power Electronics and Design (ISLPED'98)* (pp. 76–81). doi:10.1145/280756.280790

Puype, B., Vereecken, W., Colle, D., Pickavet, M., & Demeester, P. (2007). *Power reduction techniques in multilayer traffic engineering.* In Transparent Optical Networks, ICTON '09. 11th International Conference On.

Renzi, F. (2007). *Business E^3: Energy, Efficiency, Economy.* IBM Corporation.

Sahuquillo Martínez, S., Pareta, J. S., Otero, B., Ferrari, L., Manzalini, A., & Moiso, C. (2009). Self-Organized Server Farms for Energy Savings. *ICAC-INDST, 9*(June), 16.

Storer, M. W., Greenan, K. M., Miller, E. L., & Voruganti, K. (2008). Pergamum: replacing tape with energy efficient, reliable, disk-based archival storage. FAST'08.

Tombaz, S., Vastberg, A., & Zander, J. (2011). Energy – and cost-efficient ultra-high-capacity wireless access. IEEE Wireless Communications, 18(5), 18-24.

Urgaonkar, B., & Chandra, A. (2005). Dynamic provisioning of multi-tier Internet applications. *Proceedings of the 2nd International Conference on Automatic Computing (ICAC'05)*, (pp. 217–228). ICAC.

U.S. Environmental Protection Agency ENERGY STAR Program. (2007). *Report to Congress on Server and Data Center Energy Efficiency Public Law 109-431.* Author.

Vereecken, W., & VanHeddeghem, W. (2010). Power Consumption in Telecommunication Networks: Overview and Reduction Strategies. *IEEE Communications Magazine, 4*(6).

Verma, A., & Dasgupta, G. (2009). Server workload analysis for power minimization using consolidation. *Proceedings of the 2009 Conference on USENIX Annual Technical Conference (USENIX'09).* USENIX.

Chapter 10
Mathematical Modelling of the Thermal Process in the Aquatic Environment with Considering the Hydrometeorological Condition at the Reservoir–Cooler by Using Parallel Technologies

Alibek Issakhov
al-Farabi Kazakh National University, Kazakhstan

ABSTRACT

This paper presents the mathematical model of the thermal power plant in reservoir under different hydrometeorological conditions, which is solved by three dimensional Navier - Stokes and temperature equations for an incompressible fluid in a stratified medium. A numerical method based on the projection method, which divides the problem into four stages. At the first stage it is assumed that the transfer of momentum occurs only by convection and diffusion. Intermediate velocity field is solved by fractional steps method. At the second stage, three-dimensional Poisson equation is solved by the Fourier method in combination with tridiagonal matrix method (Thomas algorithm). At the third stage it is expected that the transfer is only due to the pressure gradient. Finally stage equation for temperature solved like momentum equation with fractional step method. To increase the order of approximation compact scheme was used. Then qualitatively and quantitatively approximate the basic laws of the hydrothermal processes depending on different hydrometeorological conditions are determined.

DOI: 10.4018/978-1-4666-9755-3.ch010

INTRODUCTION

Environment - the basis of human life, as mineral resources and energy are produced from them. Moreover they are the basis of modern civilization. However, the current generation of energy cause appreciable harm to the environment, worsening living conditions. The basis of the same energy - are the various types of power plants. But power generation in thermal power plants (TPP), hydro power plant (HPP) and nuclear power plants (NPP) is associated with adverse effects on the environment. The problem of the interaction of energy and the environment has taken on new features, extending the influence of the vast territory, most of the rivers and lakes, the huge volumes of the atmosphere and hydrosphere.

With the increasing use of coastal waters for the economic and social needs the vulnerability of coastal waters due to the harmful effects of excess concentration of natural substances is elevating too. Pollution associated with the release of industrial products, etc. require special attention and management of coastal waters. Chemical plants and power plants use the coastal water in large quantities for cooling and throw away the water in the environment with higher temperature.

The circulation system of cold water from thermal power plant Kilakap is modeled in the paper (Suryaman et al., 2012). It was built and is operating since 2006, which uses the water of the Indian Ocean as a cooling unit. The water passing through the condenser, is poured back into the ocean through the drainage system. The discharged water temperature is higher than the ocean's water temperature. It is important that poured warm water from drainage circulating did not get back into the water absorbent cooling system, otherwise the temperature of water absorbent will be growing, in turn causing a temperature increase in the water absorbent and drainage. Any rise in temperature water absorbent reduces cooling efficiency (capacitors), which eventually leads to a decrease in overall plant efficiency. The heated water interacts with the natural conditions of the sea or rivers, affecting aquatic life. When the heat acts as a contaminant this is called thermal pollution. The temperature rise affects various physical and chemical properties of the sea water such as density, viscosity, dissolved oxygen, salinity, turbidity, aquatic flora. Physical and chemical properties' change directly affects to the lives of marine organisms. Some sea creatures can counteract to a rise in temperature, but not all of them.

The effect of the discharged water depends on the volume and temperature of discharged water, seawater temperatures (near the drain) and on the circulation of the flow around the drain. Heat transfer in coastal waters is part of the physical properties that depends on the tides and ebbs, water depth, sea waves, river flows, salinity, heat source, coastal structures.

Interconnected numerical model is built in the paper (Ali, Fieldhouse & Talbot, 2011). A three-dimensional temperature distribution of hot water thrown to standing water is modeled in the work above. This model is the relationship with the natural experiment, which was carried out on the one side of the British Channel. The thermal camera is used, as a source of shallow and stagnant water. The result of artificial experiment was 10 times smaller, which was held in the laboratory. This model is used for calculating the temperature distribution in the channel when using water as a coolant channel for environmental purposes instead of the use of conventional capacitors. After the use of water as the coolant it is returned again into the channel as a hot thermal dome discharged from the tube end.

The model shows the heat-proportioned profile of the jet, which resets the thermal plume discharging horizontally into shallow and still receiving water.

A mathematical model and numerical simulation to verify operation of the circuit water and thermal circulation is used in the paper (Beckers & Van Ormelingen, 1995). It was imposed by the developer of energy-station seaport Zeebrugge in Belgium, on the North Sea coast. Particular attention was paid to the water cooling, as the classical method of cooling the released warm water in the main river was impossible because the internal port (harbor) Zeebrugge, where the station was actually planned is not supplied with water from the river. This inner reservoir filled with sea water is cut off from the outer harbor and the sea with two gateways (barriers).

The cooling system was the following: the power-station was planned in the inner harbor of Zeebrugge in Belgium with indirect emission into the North Sea. If there is a stratification, the lower part of the water will be cooler while surface water interacting with the environment may lose some of the heat. It is shown that the stratification is created in inland waters. In the main harbor evacuation of heat is mainly due to evaporation and only 20% due to the release to the open sea.

Several works under different atmospheric conditions and operating conditions of the plant have been carried out to determine the increase in temperature in the harbor and at the entrance to energy plant. The expected increase in the inlet temperature of the plant was 1.6 °C and 6.6 °C, but substantially increase in temperature was between 3-4 °C. It was found that the main influence on the circulation of the internal reservoir have atmospheric conditions. But in the case of atmospheric conditions on the coast of Belgium, the temperature rise is not intensive and not serious.

There are many mathematical and numerical models have been developed to simulate distribution temperature in reservoir - cooler after launching TPP (Zhang, 2004; Oey, 2006; Cheng, 2004)

We are considering for numerical simulation Ekibastuz SDPP-1, located in the Pavlodar region, 17 km. to the north-east of the town of Ekibastuz is taken as an example of such an impact of TPP on the cooling reservoir. Technical water supply SDPP-I was carried on the back of the circuit with cooling circulating water. The surface of the reservoir is at 158.5 meters, the area is 19.6 square meters. km, the maximum size of 4×6 km, the average depth of 4.6 m, maximum depth of 8.5 m at the intake, the volume of the reservoir is 80 million cubic meters In the body of water used selective intake and spillway combined type. Waste water enters the pre-channel mixer, then through a filtration dam uniformly enters the cooling pond. Water intake is at a distance of 40 meters from the dam at a depth of 5 m. Design flow 120 cubic meters of water per second, and the actual flow rate varies depending on the mode of TPP within 80-120 cubic meters / sec.

Nomenclature/Abbreviation

g_i – The gravity acceleration.
β – The coefficient of volume expansion.
u_i – Velocity components.
χ – Thermal diffusivity coefficient.
T_0 – The equilibrium temperature.
T – Deviation of temperature from the balance.
ρ – Density.
p – Pressure.
$G(r)$ – Filters.
h_i – Step in the direction of the calculated grid corresponding axes of a Cartesian coordinate system.
Δ – Characteristic length of the filter.

MATHEMATICAL MODELS

Hydro-physical problems associated with discharge into ponds and streams of heated water in the operation of thermal and nuclear power plants, acquires now important. To emerging resetting heated water hydrologic engineering and environmental problems need to be able to predict and control the temperature of the water and the content of impurities in the reservoirs and rivers. Successful study of the processes occurring in the reservoir involves a comprehensive study of the problem: in-situ measurements of hydrothermal parameters, mathematical modeling of the processes in the water, followed by a comparison with the results of the calculation of physical modeling in laboratory and natural conditions. To construct mathematical models to consider the main characteristic of the flow in a reservoir - turbulent fluid motion. This in turn merges with one of the major problems of hydrodynamics - theory of turbulence.

Thermal and nuclear power plants both require reservoirs. Electricity production is increasing worldwide, and in the postwar period, doubling every 7-10 years. To operate these plants require large amounts of water for cooling units, an average of 35-40 cubic meters / sec. 1 million kW of installed capacity. Hence it is evident that for the thermal power plants of 2.4 million kW requires 70-160 cubic meters of water every second. Therefore, when choosing building coal and nuclear power becomes important their water supply. Naturally, the large thermal power stations should be located on the banks of large rivers, ponds and lakes or artificial reservoirs. The creation of artificial reservoirs requires large investment, so power stations tend to have on existing reservoirs and lakes. Often, industrial facilities, located on the shores of lakes and reservoirs disposed of with warm water waste products in the form of impurities. If we consider that in most developed countries for 2010-2020 for cooling thermal power plants and industrial facilities will be used for more than 10% of water resources, the problems of optimal and efficient use of water reservoir for cooling stations are of great importance. In solving these problems need to be able to predict and control the temperature of the water and the spread of pollutants and passive reservoir. The distribution of temperature and passive scalar affect not only the processes of heat and mass, but also the density stratification. Stratification appears in connection with the difference between the density of water discharged from the density of the surrounding water in the pond, or the presence of impurities in the discharged water. For example, the heated water is easier, so it is in the form of a jet or standing stretches near the free surface. Sustainable density stratification of water reduces turbulent exchange between the vertical layers of fluids, especially in the area of the density jump. In general, hydrothermal regime of the reservoir is formed under the influence of uncontrollable natural factors (solar and atmospheric radiation, wind, convective heat transfer, evaporation) and the factors which may be adjusted (the amount and temperature of the discharged water, the presence of impurities, selective sampling, etc.).

In recent years, put tough restrictions associated with the protection of the environment. According to the designer usually has to follow the rules, which limit the size "zone transfer" of fault hot water so that it does not exceed half the width of the river and occupying no more than half of the total cross-sectional area and flow. Not observing these rules may lead to short-term or long-term stopping power, and therefore the accuracy requirements constructive analysis are very strict. In fact, emerging with the hydrodynamic problem can be described as a fully three-dimensional, with irregular borders, with the presence of the mass forces of buoyancy and the velocity of the main flow, which can vary by an order, sometimes so fast that the important role played by the effects of non-stationary. In addition, if certain combinations of conditions when the fault-heated water is almost drawn into the upstream region of cooling water, there are large areas of recycling. The result could be a significant loss of total operating efficiency of the system.

From the above it follows that the construction of a theoretical model relevant to real processes occurring in the water - cooler is quite a challenge.

The impact of thermal and nuclear power plants in the hydrological and biological conditions of the reservoir varied. What matters most is thermal pollution often reaches water heating to 30-35 degrees. This increase in water temperature adversely affects the hydrobiological condition, self-purification of water quality of the reservoir. In the cooling - reservoirs spatial temperature change is small. Therefore, stratified flow in the reservoirs - cooler can be described by equations in the Boussinesq approximation. For the mathematical modeling of the system of equations are considered the equations of motion, the continuity equation and the equation for the temperature. Considers the development of spatial turbulent stratified reservoir - cooler (Fletcher, 1988; Roache, 1972; Peyret, 1983; Tannehill, Anderson, 1997). Three dimensionally model is used for distribution of temperature modelling in a reservoir (Issakhov, 2012c; Lowe, Schuepfer, & Dunning, 2009; Issakhov, 2013; Issakhov, 2014)

$$\frac{\partial u_i}{\partial t} + \frac{\partial u_j u_i}{\partial x_j} = -\frac{1}{\rho}\frac{\partial p}{\partial x_i} + \nu \frac{\partial}{\partial x_j}\left(\frac{\partial u_i}{\partial x_j}\right) + \delta_{i3}\beta g(T - T_0) - \frac{\partial \tau_{ij}}{\partial x_j} \qquad (1)$$

$$\frac{\partial u_j}{\partial x_j} = 0 \qquad (2)$$

$$\frac{\partial T}{\partial t} + \frac{\partial u_j T}{\partial x_j} = \frac{\partial}{\partial x_j}\left(\chi \frac{\partial T}{\partial x_j}\right) \qquad (3)$$

where

$$\tau_{ij} = \overline{u_i u_j} - \overline{u_i}\,\overline{u_j} \qquad (4)$$

g_i – the gravity acceleration, β – the coefficient of volume expansion, u_i - velocity components, χ – thermal diffusivity coefficient, T_0 – the equilibrium temperature, T – deviation of temperature from the balance, ρ – density, p – pressure.

The basic idea of LES method is a mathematical division of the large, and at the same time fragile, from small universal vortices. This procedure can be performed through spatial averaging, i.e. define the field of large-scale quantities by filter.

$$\overline{u}(x,t) = \int_V G(\mathbf{r},\mathbf{x})u(x-r,t)dr \qquad (5)$$

where $\mathbf{u} = (u_1, u_2, u_3)$ – vector of velocity components, sign "dash" denotes averaging, $\mathbf{x} = (x_1, x_2, x_3)$ – vector of coordinate system, $\mathbf{r} = (r_1, r_2, r_3)$ – vector of coordinate system for which doing the integration, V – volume of integration, $G(\mathbf{r}, \mathbf{x})$ – filter function with characteristic length scale such that:

$$\int_V G(\mathbf{r},\mathbf{x})d\mathbf{r} = 1 \qquad (6)$$

A small-scale fluctuations are as follows:

$$u'(x,t) = u(x,t) - \overline{u}(x,t) \qquad (7)$$

in many cases, depending on the filter function

$$\overline{u'}(\mathbf{x},t) \neq 0 \qquad (8)$$

There are different approaches to characterize the filter:

- The filter of the "box"

$$G(\mathbf{r}) = \frac{1}{\Delta^3}\begin{cases} 1, |r| \leq \Delta/2 \\ 0, |r_i| > \Delta/2 \end{cases}, \quad \forall i = 1,2,3; \qquad (9)$$

- Gaussian filter

$$G(\mathbf{r}) = \prod_{i=1}^{3}\left(\frac{6}{\pi\Delta^2}\right)^{\frac{1}{2}} \exp\left(-\frac{6r_i^2}{\Delta^2}\right); \qquad (10)$$

- Cut filter

$$G(\mathbf{r}) = \prod_{i=1}^{3} \frac{\sin(\pi r_i / \Delta)}{\pi r_i}. \qquad (11)$$

Δ – characteristic length of the filter, which is of the order of the mesh size. It is usually taken for (Lesieur, Metais, & Comte, 2005):

$$\Delta = (h_1 h_2 h_3)^{1/3},$$

$$\Delta = (h_1^2 + h_2^2 + h_3^2)^{1/2}, \qquad (12)$$

$$\Delta = \min(h_1, h_2, h_3),$$

where h_i – step in the direction of the calculated grid corresponding axes of a Cartesian coordinate system.

We start with regular LES corresponding to a "bar-filter" of Δx width, an operator associating the function $\overline{f}(\overline{x},t)$. Then we define a second "test filter" tilde of large $2\Delta x$ width associating $\overline{f}(\overline{x},t)$. Let us first apply this filter product to the Navier-Stokes equation. The subgrid-scale tensor of the field $\tilde{u}_i$ is obtained from equation (4) with the replacement of the filter bar by the double filter and tilde filter:

$$T_{ij} = \overline{\widetilde{u_i u_j}} - \overline{\widetilde{u}_i \widetilde{u}_j} \tag{13}$$

$$l_{ij} = \overline{\widetilde{\overline{u}_i \overline{u}_j}} - \overline{\widetilde{\overline{u}}_i \overline{\widetilde{u}}_j} \tag{14}$$

Now we apply the tilde filter to equation (4), which leads to

$$\tilde{\tau}_{ij} = \widetilde{\overline{u_i u_j}} - \widetilde{\overline{u}_i \overline{u}_j} \tag{15}$$

Adding Equations (14) and (15) and using Equation (13), we obtain

$$l_{ij} = T_{ij} = \tilde{\tau}_{ij} \tag{16}$$

We use Smagorinsky model expression for the subgrid stresses related to the bar filter and tilde-filter it to get

$$\tilde{\tau}_{ij} - \frac{1}{3}\delta_{ij}\tilde{\tau}_{kk} = -2C\tilde{A}_{ij} \text{ where } A_{ij} = (\Delta x)^2 \overline{|S|S_{ij}} \tag{17}$$

Further on we have to determine τ_{ij}, the stress resulting from the filter product. This is again obtained using the Smagorinsky model, which yields to

$$T_{ij} - \frac{1}{3}\delta_{ij}T_{kk} = -2CB_{ij} \text{ where } B_{ij} = (2\Delta x)^2 \left|\tilde{S}\right|\tilde{S}_{ij} \tag{18}$$

Subtraction of (17) from (18) with the aid of Germano's identity yields to

$$l_{ij} - \frac{1}{3}\delta_{ij}l_{kk} = 2CB_{ij} - 2C\tilde{A}_{ij} \tag{19}$$

$$l_{ij} - \frac{1}{3}\delta_{ij}l_{kk} = 2CM_{ij} \tag{20}$$

where

$$M_{ij} = B_{ij} - \tilde{A}_{ij} \tag{21}$$

All the terms of equation (21) may now be determined with the aid of $\bar{u}$. Unfortunately, there are five independent equations for only one variable C, and thus the problem is overdetermined. A first solution proposed by Germano is to multiply (21) tensorially by $\overline{S_{ij}}$ to get

$$C = \frac{1}{2} \frac{l_{ij} \overline{S_{ij}}}{M_{ij} \overline{\overline{S_{ij}}}}$$

This provides finally dynamical evaluation of C, which can be used in the LES of the bar field $\bar{u}$ (Lesieur, 2005; Tennekes, 1972).

Initial and boundary conditions are defined for the non-stationary 3D equations of motion, continuity and temperature, satisfying the equations.

NUMERICAL ALGORITHM

Numerical solution of (1) - (4) is carried out on the staggered grid using the scheme against a stream of the second type and compact approximation for convective terms (Tolstykh, 1990; Issakhov, 2011; Zhumagulov, 2012; Issakhov, 2012b). Scheme of splitting on physical parameters is used to solve the problem in view of the above with the proposed model of turbulence (Yanenko, 1979). It is anticipated that at the first stage the transfer of momentum occurs only through convection and diffusion. Intermediate field of speed is handled by using method of fractional steps through the tridiagonal method (Thomas algorithm). The second phase is for pressure which is found by the help of intermediate field of speed. Poisson equation for pressure is solved by Fourier method in combination with the tridiagonal method (Thomas algorithm) that is applied to determine the Fourier coefficients (Issakhov, 2012a). At the third stage, it is supposed that the transfer is carried out only by the pressure gradient. The algorithm was parallelized on the high-performance system (Issakhov, 2011).

$$
\begin{aligned}
&1. \quad \frac{\vec{u}^* - \vec{u}^n}{\tau} = -\left(\nabla \vec{u}^n \, \vec{u}^* - \nu \Delta \vec{u}^*\right) \\
&2. \quad \Delta p = \frac{\nabla \vec{u}^*}{\tau} \\
&3. \quad \frac{\vec{u}^{n+1} - \vec{u}^*}{\tau} = -\nabla p.
\end{aligned}
\tag{22}
$$

The first step for intermediate field of speed is handled by using method of fractional steps through the tridiagonal method (Thomas algorithm).

$$\frac{f^{n+1/3} - f^n}{\tau} = \frac{1}{2}\Lambda_1 f^{n+1/3} + \frac{1}{2}\Lambda_1 f^n + \Lambda_2 f^n + \Lambda_3 f^n,$$

$$\frac{f^{n+2/3} - f^{n+1/3}}{\tau} = \frac{1}{2}\Lambda_2 f^{n+2/3} - \frac{1}{2}\Lambda_2 f^n, \qquad (23)$$

$$\frac{f^* - f^{n+2/3}}{\tau} = \frac{1}{2}\Lambda_3 f^* - \frac{1}{2}\Lambda_3 f^n,$$

where the operators are as follows:

$$\Lambda_1 f = -\frac{H}{L_1}\frac{\partial(\bar{u}_1^n f)}{\partial x_1} + \frac{1}{\mathrm{Re}}\frac{H^2}{L_1^2}\frac{\partial^2 f}{\partial x_1^2} - \frac{H}{L_1}\frac{\partial \tau_{i1}}{\partial x_1},$$

$$\Lambda_2 f = -\frac{H}{L_2}\frac{\partial(\bar{u}_2^n f)}{\partial x_2} + \frac{1}{\mathrm{Re}}\frac{H^2}{L_2^2}\frac{\partial^2 f}{\partial x_2^2} - \frac{H}{L_2}\frac{\partial \tau_{i2}}{\partial x_2}, \qquad (24)$$

$$\Lambda_3 f = -\frac{\partial(\bar{u}_3^n f)}{\partial x_3} + \frac{1}{\mathrm{Re}}\frac{\partial^2 f}{\partial x_3^2} - \frac{\partial \tau_{i3}}{\partial x_3},$$

$$f = \bar{u}_i, \quad i = 1, 2, 3.$$

The second stage is for pressure which is found by the help of intermediate field of speed. Three dimensional Poisson equations for pressure is solved by Fourier method for one coordinate in combination with the tridiagonal method (Thomas algorithm) that is applied to determine the Fourier coefficients (Issakhov, 2012a; Issakhov, 2012b). The numerical algorithm for Poisson equation was parallelized on the high-performance system.

$$p_{i,j,k} = \frac{2}{N_3}\sum_{l=0}^{N_3}\rho_l a_{i,j,k}\cos\frac{\pi kl}{N_3} \qquad (25)$$

where

$$a_{i,j,l} = \sum_{k=0}^{N_3}\rho_k\, p_{i,j,k}\cos\frac{\pi kl}{N_3};$$

$$b_{i,j,l} = \sum_{k=0}^{N_3}\rho_k\, F_{i,j,k}\cos\frac{\pi kl}{N_3}; \qquad (26)$$

Substituting equation above into three dimensional Poisson equations for pressure we obtain the following expression:

$$\frac{H^2}{L_1^2}\frac{2}{N_3}\sum_{l=0}^{N_3}\frac{\rho_l\left(a_{i+1,j,l}-2a_{i,j,l}+a_{i-1,j,l}\right)}{\Delta x_1^2}\cos\frac{\pi kl}{N_3}+\frac{H^2}{L_2^2}\frac{2}{N_3}\sum_{l=0}^{N_3}\frac{\rho_l\left(a_{i,j+1,l}-2a_{i,j,l}+a_{i,j-1,l}\right)}{\Delta x_2^2}x\cos\frac{\pi kl}{N_3}$$
$$+\frac{2}{N_3}\sum_{l=0}^{N_3}\frac{\rho_l a_{i,j,l}}{\Delta x_3^2}\left(\cos\frac{\pi(k+1)l}{N_3}-2\cos\frac{\pi kl}{N_3}+\cos\frac{\pi(k-1)l}{N_3}\right)=\frac{2}{N_3}\sum_{l=0}^{N_3}\rho_l b_{i,j,l}\cos\frac{\pi kl}{N_3}.$$
(27)

Using expression below

$$\cos\frac{\pi(k+1)l}{N_3}+\cos\frac{\pi(k-1)l}{N_3}=2\cos\frac{\pi kl}{N_3}\cos\frac{\pi l}{N_3},\tag{28}$$

Then can write (25) to the following form:

$$\frac{H^2}{L_1^2}\frac{2}{N_3}\sum_{l=0}^{N_3}\frac{\rho_l\left(a_{i+1,j,l}-2a_{i,j,l}+a_{i-1,j,l}\right)}{\Delta x_1^2}\cos\frac{\pi kl}{N_3}+\frac{H^2}{L_2^2}\frac{2}{N_3}\sum_{l=0}^{N_3}\frac{\rho_l\left(a_{i,j+1,l}-2a_{i,j,l}+a_{i,j-1,l}\right)}{\Delta x_2^2}\cos\frac{\pi kl}{N_3}$$
$$+\frac{2}{N_3}\sum_{l=0}^{N_3}\frac{\rho_l a_{i,j,l}}{\Delta x_3^2}\left(2\cos\frac{\pi l}{N_3}-2\right)2\cos\frac{\pi kl}{N_3}=\frac{2}{N_3}\sum_{l=0}^{N_3}\rho_l b_{i,j,l}\cos\frac{\pi kl}{N_3}.$$
(29)

The last expression can be written at a fixed value l and divide by the amount of $\frac{2}{N_3}\rho_l\cos\frac{\pi kl}{N_3}$ and then take:

$$\frac{H^2}{L_1^2}\frac{a_{i+1,j}-2a_{i,j}+a_{i-1,j}}{\Delta x_1^2}+\frac{H^2}{L_2^2}\frac{a_{i,j+1}-2a_{i,j}+a_{i,j-1}}{\Delta x_2^2}+\frac{a_{i,j}}{\Delta x_3^2}\left(2\cos\frac{\pi l}{N_3}-2\right)=b_{i,j}.\tag{30}$$

This equation is transformed to the following form:

$$-\frac{H^2}{L_2^2}\frac{a_{i,j-1}}{\Delta x_2^2}$$
$$+\left[\left(\frac{H^2}{L_1^2}\frac{2}{\Delta x_1^2}+\frac{H^2}{L_2^2}\frac{2}{\Delta x_2^2}-\frac{1}{\Delta x_3^2}\left(2\cos\frac{\pi l}{N_3}-2\right)\right)a_{i,j}-\frac{H^2}{L_1^2}\frac{a_{i+1,j}+a_{i-1,j}}{\Delta x_1^2}\right]-\frac{H^2}{L_2^2}\frac{a_{i,j+1}}{\Delta x_2^2}=-b_{i,j}.\tag{31}$$

In vector form, this equation can be written as follows:

$$-A_j \vec{a}_{j-1} + B_j \vec{a}_j - C_j \vec{a}_{j+1} = \vec{F}_j, \qquad (32)$$

where matrices A_j, B_j, C_j and vectors $\vec{F}_j$, $\vec{a}_j$ take the form:

$$\vec{a}_j = \begin{vmatrix} a_{0,j} \\ \vdots \\ a_{N_1,j} \end{vmatrix};$$

$$A_j = \begin{vmatrix} \dfrac{H^2}{L_2^2}\dfrac{1}{\Delta x_2^2} & & 0 \\ & \ddots & \\ 0 & & \dfrac{H^2}{L_2^2}\dfrac{1}{\Delta x_2^2} \end{vmatrix};$$

$$B_j = \begin{vmatrix} \dfrac{H^2}{L_1^2}\dfrac{2}{\Delta x_1^2} + \dfrac{H^2}{L_2^2}\dfrac{2}{\Delta x_2^2} - \dfrac{1}{\Delta x_2^2}\left(2\cos\dfrac{\pi l}{N_3} - 2\right) & -\dfrac{H^2}{L_1^2}\dfrac{2}{\Delta x_1^2} & & 0 \\ -\dfrac{H^2}{L_1^2}\dfrac{1}{\Delta x_1^2} & \ddots & & -\dfrac{H^2}{L_1^2}\dfrac{1}{\Delta x_1^2} \\ 0 & & -\dfrac{H^2}{L_1^2}\dfrac{2}{\Delta x_1^2} & \dfrac{H^2}{L_1^2}\dfrac{2}{\Delta x_1^2} + \dfrac{H^2}{L_2^2}\dfrac{2}{\Delta x_2^2} - \dfrac{1}{\Delta x_2^2}\left(2\cos\dfrac{\pi l}{N_3} - 2\right) \end{vmatrix};$$

$$C_j = \begin{vmatrix} \dfrac{H^2}{L_2^2}\dfrac{1}{\Delta x_2^2} & & 0 \\ & \ddots & \\ 0 & & \dfrac{H^2}{L_2^2}\dfrac{1}{\Delta x_2^2} \end{vmatrix};$$

$$\vec{F}_j = -\begin{vmatrix} b_{0,j} \\ \vdots \\ b_{N_1,j} \end{vmatrix}.$$

Algorithm of tri diagonal matrix method for (32) is

$$\alpha_{j+1} = (B_j - A_j \alpha_j)^{-1} C_j, \quad \alpha_1 = B_0^{-1} C_0$$

$$\beta_{j+1} = (C_j - A_j\alpha_j)^{-1}(F_j + A_j\beta_j), \quad \beta_1 = B_0^{-1}F_0$$

$$a_j = \alpha_{j+1}a_{j+1} + \beta_{j+1}, \quad a_{N_2} = \beta_{N_2+1}$$

The third is correction stage, it is supposed that the transfer is carried out only by the pressure gradient. After calculating $a_{i,j,k}$, pressure field values are found from (25). To calculate the sum (30) it is necessary to apply the fast Fourier transform, which allows to calculate $O(N\ln N)$ the amount of data for action to reduce the computing time.

RESULTS OF NUMERICAL MODELLING

Initial and boundary conditions were posed to meet the challenges. In the calculation we used the mesh of 100x100x100 size.

Figures 1 and 2 show the solved three dimensional spatial outline and contour of the temperature distribution at different times for after the launch of the SDPP-1, on the surface, from different angles. Figures 3 and 4 show the solved spatial contour, contour of temperature and velocity vectors at different times in the north-west wind after the launch of the SDPP-1, on the surface, from different angles. In figures we can see that temperature vary from $25°C$ to $33°C$. And we can see from numerical simulation that temperature near the Ekibastuz SDPP-1 is higher than far distance from SDPP-1. It means that mathematical model qualitatively descried physical process. All the figures show that the temperature distribution with distance from the flow approaches the isothermal distribution. The results show that the temperature distribution is distributed over a larger area of the reservoir - cooler.

Figure 1. Outline and contours of temperature at 15 and 24 h after launch of the SDPP-1, on the surface, the side view.

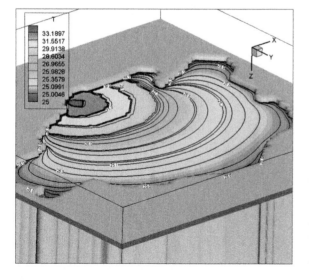
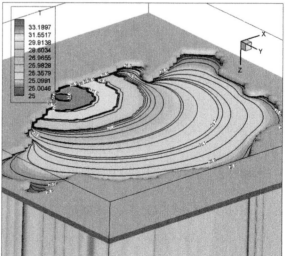

Figure 2. Outline and contours of temperature at 15 and 24 h after launch of the SDPP-1, on the surface, top view.

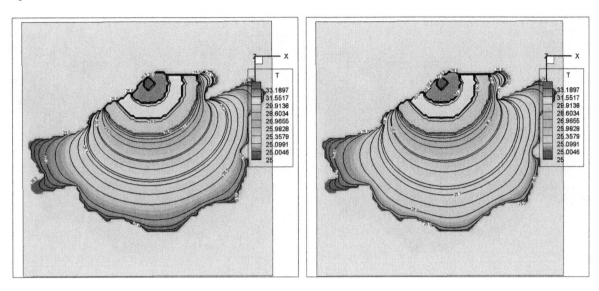

Figure 3. Outline and contours of temperature at 15 and 24 h at the north-west wind after the launch of the SDPP-1, on the surface, the side view.

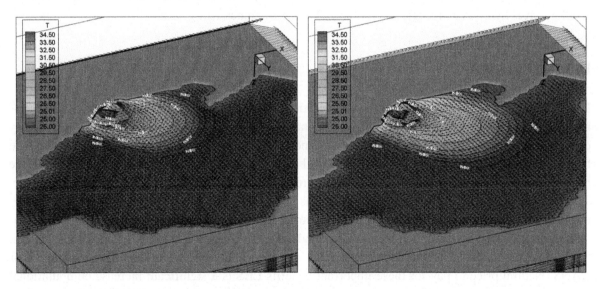

FUTURE RESEARCH DIRECTIONS

Moreover, future increases in computational resources will be accompanied by continuing demands from application scientists for increased resolution and the inclusion of additional physics. Consequently, new approaches are needed in order to decrease simulation computational costs. Such techniques will be required to harness both the underlying model hierarchy and the stochastic hierarchy.

Figure 4. Outline and contours of temperature at 15 and 24 h at the north-west wind after the launch of the SDPP-1 on the surface of the water, the top view.

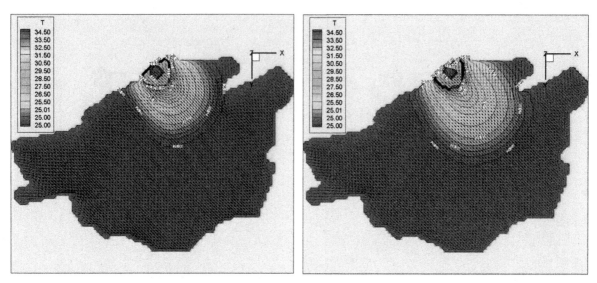

Future research should continue to examine knowledge in qualitatively and quantitatively approximate the basic laws of the hydrothermal processes depending on different hydrometeorological conditions, which will tie to real hydrometeorological conditions. These researches will give more real numerical results.

DISCUSSIONS

LES is a more universal approach to the construction of closure for large-eddy simulation models. A necessary condition for the performance of turbulent closures is the ability to "subgrid" model correctly describes the dissipation of the kinetic energy of smoothed velocity fluctuations - the ability to simulate the circuit direct energy cascade from large to small eddies. This stage is the primary mechanism for the redistribution of energy in the inertial range of three dimensional homogeneous isotropic turbulence. The principal advantage of the LES from RANS is that, due to the relative homogeneity and isotropy of the small-scale turbulence, plotting a subgrid model is much simpler than the construction of turbulence models for RANS, when it is necessary to model the full range of turbulence. For the same reason, the hope for a "universal" subgrid model for LES are much more reasonable than a similar model for RANS. Natural price to pay for these important benefits LES is a significant increase computational cost associated with the need (as in the case of DNS) of three-dimensional time-dependent calculations on sufficiently fine grids, even in cases where direct interest to the practice of the average flow is two-dimensional and stationary. On the other hand, for obvious reasons, the computational resources required to implement the LES, is much smaller than for the DNS The degree of influence of different processes governing the formation of stratified flows and hydrothermal conditions, the entire body of water can be divided into two zones: the first (near), directly adjacent to the water outlet structures, and the second is for the major part of the reservoir. In the near zone forming stratified flow is influenced by the processes of mixing water discharged from the water reservoir and its possible regulation by creating a specific hydraulic regime in the outfalls. In the second zone of hydrothermal regime is formed primarily by the processes of heat transfer. The propagation of heat in this part of the reservoir is more dependent on the wind (direction and speed). When you reset the heated water in a cold environment appears density jump between the upper layer of warm water and cold bottom, where there is a compensation for the direction toward the spillway. This allows the use of a combined intake and outfalls instead of building costly diversion canals to the spillway. This raises the problem of optimal choice of the geometrical and operational parameters of the cooling pond for efficient power plant.

CONCLUSION

Thus, using a mathematical model of three-dimensional stratified turbulent flow can be determined qualitatively and quantitatively approximate the basic laws of the hydrothermal processes occurring in the reservoirs. Performed computational studies of hydrothermal regime of the cooling pond Ekibastuz SDPP-I by measuring the velocity and temperature fields have revealed the basic laws of hydrothermal and thermal fields in the water when stationary and various meteorological conditions. To simulate more accurately thermal process in the reservoir-cooler needs more detailed mathematical model and data analysis are required for future research direction.

LIMITATION

Various limitations may exist in this study. A first limitation is mesh size, to approach good numerical results, we need very large mesh size to simulate, which is now not achievable, because we limited by computer resource. A second limitation is the difficulty of doing and analyzing experimental data from thermal power plants, which include environmental factors, neglected differences between subjects, and unexpected changes during the experiment.

ACKNOWLEDGMENT

This work is supported by grant from the Ministry of education and science of the Republic of Kazakhstan.

REFERENCES

Ali J., Fieldhouse J. & Talbot Ch. (2011). Numerical modeling of three-dimensional thermal surface discharges. *Engineering application of computational fluid mechanics, 5* (2), 201-209.

Beckers, J.-M., & Van Ormelingen, J.-J. (1995). Thermohydrodynamical modelling of a power plant implementation in the Zeebrugge harbor. *Journal of Hydraulic Research, 33*(2), 163–180. doi:10.1080/00221689509498668

Cheng, R. T., & Casulli, V. (2004). Modeling a three-dimensional river plume over continental shelf using a 3D unstructured grid model. *The Proceedings of the 8th Inter. Conf. on Estuarine and Coastal Modeling*.

Fletcher, C. A. (1988). Computational Techniques for Fluid Dynamics: Vol. 2. *Special Techniques for Differential Flow Categories*. Berlin: Springer-Verlag.

Issakhov, A. (2011). Large eddy simulation of turbulent mixing by using 3D decomposition method. *Journal of Physics: Conference Series*, *318*(4), 042051. doi:10.1088/1742-6596/318/4/042051

Issakhov, A. (2012a). Parallel algorithm for numerical solution of three-dimensional Poisson equation. *Proceedings of world academy of science, engineering and technolog*, *64*, 692-694.

Issakhov, A. (2012b). Development of parallel algorithm for numerical solution of three-dimensional Poisson equation. *Journal of Communication and Computer*, *9*(9), 977–980.

Issakhov, A. (2012c). Mathematical modelling of the influence of thermal power plant to the aquatic environment by using parallel technologies. *AIP Conference Proceedings*, *1499*, 15–18. doi:10.1063/1.4768963

Issakhov, A. (2013). Mathematical Modelling of the Influence of Thermal Power Plant on the Aquatic Environment with Different Meteorological Condition by Using Parallel Technologies. *Power, Control and Optimization. Lecture Notes in Electrical Engineering*, *239*, 165–179. doi:10.1007/978-3-319-00206-4_11

Issakhov A. (2014). Mathematical Modelling of Thermal Process to Aquatic Environment with Different Hydrometeorological Conditions. *The Scientific World Journal*. doi:10.1155/2014/678095

Lesieur, M., Metais, O., & Comte, P. (2005). *Large eddy simulation of turbulence*. New York: Cambridge University Press. doi:10.1017/CBO9780511755507

Lowe, S. A., Schuepfer, F., & Dunning, D. J. (2009). Case study: Three-dimensional hydrodynamic model of a power plant thermal discharge. *Journal of Hydraulic Engineering*, *135*(4), 247–256. doi:10.1061/(ASCE)0733-9429(2009)135:4(247)

Oey, L. Y. (2006). An OGCM with movable land–sea boundaries. *Ocean Modelling*, *13*(2), 176–195. doi:10.1016/j.ocemod.2006.01.001

Peyret, R., & Taylor, D. Th. (1983). *Computational Methods for Fluid Flow*. New York, Berlin: Springer-Verlag. doi:10.1007/978-3-642-85952-6

Roache, P. J. (1972). *Computational Fluid Dynamics*. Albuquerque, NM: Hermosa Publications.

Suryaman, D., Asvaliantina, V., Nugroho, S., & Kongko, W. (2012). Cooling Water Recirculation Modeling of Cilacap Power Plant. In *Proceeding of the Second International Conference on Port, Coastal, and Offshore Engineering* (2nd ICPCO).

Tannehill, J. C., Anderson, D. A., & Pletcher, R. H. (1997). *Computational Fluid Mechanics and Heat Transfer* (2nd ed.). New York: McGraw-Hill.

Tennekes, H., & Lumley, J. L. (1972). *A first course in turbulence*. The MIT Press.

Tolstykh, A. I. (1990). *Compact difference scheme and their applications to fluid dynamics problems. M*. Nauka.

Yanenko, N. N. (1979). *The Method of Fractional Steps. New York: Springer-Verlag* (J. B. Bunch & D. J. Rose, Eds.). Space Matrix Computations, New York: Academics Press.

Zhang, Y. L., Baptista, A. M., & Myers, E. P. (2004). A cross-scale model for 3D baroclinic circulation in estuary-plume-shelf systems. I: Formulation and skill assessment. *Continental Shelf Research*, *24*(18), 2187–2214. doi:10.1016/j.csr.2004.07.021

Zhumagulov, B., & Issakhov, A. (2012). Parallel implementation of numerical methods for solving turbulent flows. *Vestnik NEA RK*, *1*(43), 12–24.

KEY TERMS AND DEFINITIONS

Aquatic Environment: Is an ecosystem in a body of water.

Hydrometeorological Conditions: Is a branch of meteorology and hydrology conditions that studies the transfer of water and energy between the land surface and the lower atmosphere.

Mathematical Model: Is a description of a system of equation using mathematical concepts and laws.

Navier–Stokes Equations: Describe the motion of viscous fluid substances. Various limitations may exist in this study. Sample composition, which is one of the most frequently cited threats to external validity, is not considered a limitation in this study. A first limitation is mesh size, approach to good numerical results, we need very large mesh size, which is now not achievable, because we are limited by computer resource. A second limitation is the difficulty of doing and analyzing experimental data from thermal power plants.

Parallel Computing: Is a form of computation in which many calculations are carried out simultaneously on many processors.

Reservoir – Cooler: Is a passive heat exchanger that cools a building or plants by dissipating heat into the surrounding water.

Thermal Power Plant: Is a power plant in which the prime mover is steam driven.

Chapter 11
A Novel Evolutionary Optimization Technique for Solving Optimal Reactive Power Dispatch Problems

Provas Kumar Roy
Jalpaiguri Government Engineering College, India

ABSTRACT

Biogeography based optimization (BBO) is an efficient and powerful stochastic search technique for solving optimization problems over continuous space. Due to excellent exploration and exploitation property, BBO has become a popular optimization technique to solve the complex multi-modal optimization problem. However, in some cases, the basic BBO algorithm shows slow convergence rate and may stick to local optimal solution. To overcome this, quasi-oppositional biogeography based-optimization (QOBBO) for optimal reactive power dispatch (ORPD) is presented in this study. In the proposed QOBBO algorithm, oppositional based learning (OBL) concept is integrated with BBO algorithm to improve the search space of the algorithm. For validation purpose, the results obtained by the proposed QOBBO approach are compared with those obtained by BBO and other algorithms available in the literature. The simulation results show that the proposed QOBBO approach outperforms the other listed algorithms.

1 INTRODUCTION

The optimal reactive power dispatch (ORPD) is a very important aspect in power system planning and operation. It is used to determine a secure operating state of power systems.

It is an effective method to minimize the transmission losses and maintain the power system running under normal conditions. In ORPD, the optimal adjustments of voltage control devices or VAR sources such as voltage of generators, tap ratio of transformers, Var injection of shunt compensators are made for optimizing real power losses, while satisfying all the system operational constraints. It is an effective method to improve voltage level, decrease power losses and maintain the power system running under normal conditions.

DOI: 10.4018/978-1-4666-9755-3.ch011

A Novel Evolutionary Optimization Technique

In earlier stages, a variety of classical optimization algorithms such as Gradient method (GM) (Lee, Park, & Ortiz,1984), Newton's based approach (Bjelogrlic, Calovic, Babic, & Ristanovic, 1990), Newton Raphson approach (Rang-Mow, & Nanming, 1995), linear programming (LP) (Aoki, Fan & Nishikor, 1988), decomposition techniques (Deb, & Shahidehpour, 1990; Granada, Marcos, Rider, Mantovani, & Shahidehpour, 2012), interior point methods (Chebbo, & Irving, 1995; Granville, 1994; Yan, Yu, Yu, & Bhattarai,2006), quadratic Programming (Momeh, Guo, Oghuobiri, & Adapa, 1994) and dynamic programming (DP) (Lu, & Hsu, 1995) were attempted to solve ORPD problems.

Though most of these conventional methods have excellent convergence characteristics but they face difficulties in handling the nonlinear constraints and solution process slows with complex objective functions. Moreover, these classical techniques are local optimizers in nature, i.e., they might converge to local solutions instead of global ones if the initial guess happens to be in the neighborhood of a local solution. DP method may cause the dimensions of the problem to become extremely large, thus requiring enormous computational efforts.

In last two decades, various heuristic evolutionary optimization algorithms such as evolutionary programming (EP) (Ma & Lai, 1996; Wu & Ma, 1995), genetic algorithm (GA) (Iba, 1994; Lee & Park, 1995; Swarup,Yoshimi & Izui, 1994), particle swarm optimization (PSO) (Kawata, Fukuyama,Takayama & Nakanish, 2000; Li, Cao, Liu, Liu & Jiang, 2009; Zhao, Guo and Cao, 2005), Tabu search (TS) (Yiqin, 2010), differential evolution (DE) (Liang, Chung, Wong & Dual, 2007; Ramesh, Kannan, & Baskar, 2012; Varadarajan & Swarup, 2008; Zhang, Wenyin, & Zhihua, 2013; Chen, Dai & Cai 2010), seeker optimization approach (SOA) (Dai, Chen, Zhu & Zhang 2009a; Dai, Chen, Zhu & Zhang 2009b), EP-SED (Titare, Singh, Arya, & Choube, 2014), ant colony optimization (ACO) (Huang, & Huang, 2012), hybrid evolutionary algorithm (Ghasemi, Ghavidel, Ghanbarian, & Habibi, 2014).BBO (Bhattacharya & Chattopadhyay, 2010), quasi-oppositional teaching learning based optimization (QOTLBO) (Mandal, & Roy, 2013), shuffled frog leaping algorithm (SFLA) (Khorsandi, Alimardani, Vahidi, & Hosseinian, 2010), harmony search algorithm (HSA) (Khazali, & Kalantar, 2011), opposition based HSA (OHSA) (Chatterjee, Ghosal, & Mukherjee, 2012) have been proposed because they found to be robust and flexible in solving optimization problem. These methods (Vasant, Barsoum, & Webb, 2012; Vasant, 2013; Vasant, 2014; Vasant, 2015) are widely applied in many scientific and engineering fields, These methods do not depend on convexity assumptions and are capable of handling non-linear optimization problems. Each of these methods has its own characteristics, strengths and weaknesses; but long computational time is a common drawback for most of them, especially when the solution space is hard to explore. Many efforts have been made to accelerate convergence of these methods.

In 2008, a new optimization technique, known as Biogeography-based Optimization (BBO) was developed by Simon (Simon, 2008). It is based on the concept of biogeography of species. BBO has already proven itself a worthy optimization technique. In BBO poor solutions accept a lot of new features from good ones using migration operation which may improve the quality of those solutions. The unique features of BBO over other evolutionary methods are (i) in most of the evolutionary algorithms solution changes rapidly as the optimization process progresses. Previous solutions of most of the evolutionary based algorithms "die" at the end of each generation. Therefore, many solutions whose fitness are initially good, sometimes lose their quality in later stage of the process. However, in BBO; solutions get fine tuned gradually as the process goes on through migration operation. (ii) In most of the heuristic optimization technique, solutions are more likely to clump together in similar groups, while in BBO, solutions do not have any tendency to cluster due to its unique mutation operation. This is an added advantage of BBO in comparison to other algorithms. (iii) BBO irequires few computational steps per iteration as compared to other approaches which results in faster convergence. These versatile properties of BBO algorithm encouraged the present authors to apply this newly developed algorithm for solving OPF problems. However, sometimes, the basic BBO algorithm shows slow convergence rate and may stick to local optimal solution. In order to avoid this short coming of the standard BBO, a quasi-opposition based learning (QOBL) concept is proposed to improve search space in initialization stage and as well as in each generation cycles. In order to speed up the convergence property and to find better results, the authors proposed, quasi-oppositional BBO (QOBBO) combining exploration ability of QOBL with exploitation ability of BBO. The concept of OBL is earlier applied to accelerate PSO (Zhang, Ni, Wu & Gu, 2009), DE (Rahnamayan, Tizhoosh & Salama, 2007) and ant colony optimization (ACO) (Haiping, Xieyon & Baogen, 2010). The main idea behind the OBL is considering the estimate and opposite estimate (guess and opposite guess) at the same time in order to achieve a better approximation for current candidate solution. Purely random selection of solutions from a given population has the chance of visiting or even revisiting unproductive regions of the search space. The chance of this occurring is lower for opposite numbers than it is for purely random ones. A mathematical proof has been proposed (Haiping, Xieyong & Baogen, 2010) to show that, in general, opposite numbers are more likely to be closer to the optimal solution than purely random ones.

The effectiveness of the proposed QOBBO based ORPD algorithm is tested on IEEE 30-bus and IEEE 57-bus systems. The results of QOBBO are compared to those of PSO, comprehensive learning PSO (CLPSO), CGA, real standard version of PSO (SPSO-07), PSO with constriction factor (PSO-cf), PSO with adaptive inertia weight (PSO-w), local search based self-adaptive differential evolution (L-SADE), SOA, and BBO to make it clear that the proposed method is powerful and reliable.

The rest of the presented study is categorized in 8 sections, described as follows: Section 2 explains the typical formulation of an ORPD problem while Section 3 is dedicated to discuss the basic theory of the BBO algorithm, Section 4 of the chapter covers the optimization procedure of the BBO algorithm. Opposition-based learning and quasi-oppositional based BBO applied to the ORPD are addressed in Section 5 and Section 6, respectively. Section 7 illustrates the optimal input parameters of the proposed algorithm. The results and conducts comparison and performance analysis of the proposed QOTLBO approach used to solve the case studies of optimization problem on standard IEEE systems are presented in section 8 and last section of this article presents the conclusion of the present study.

Nomenclature/Abbreviation

ORPD: Optimal reactive power dispatch
BBO: Biogeography based optimization
QOBBO: Quasi-oppositional biogeography based-optimization
OBL: Oppositional based learning
LP: Linear programming
EP: Evolutionary programming
GA: Genetic algorithm
PSO: Particle swarm optimization
TS: Tabu search
DE: Differential evolution
SOA: Seeker optimization approach
ACO: Ant colony optimization

SFLA: Shuffled frog leaping algorithm
QOTLBO: Quasi-oppositional teaching learning based optimization
HAS: Harmony search algorithm
OHSA Opposition based HSA
QOBL: Quasi-opposition based learning
CLPSO: Comprehensive learning PSO
SPSO-07: Real standard version of PSO
PSO-cf: PSO with constriction factor
PSO-w: PSO with adaptive inertia weight
L-SADE: Local search based self-adaptive differential evolution
ELD: Economic load dispatch
OPF: Optimal power flow
SIV: Suitability index variables
$f(u,v)$: objective function
$g(u,v) = 0$: Equality constraints
$h(u,v) = 0$: Inequality constraints
V_{L_i} : Voltage of the i-th load bus
Q_{G_i} : Reactive power generation of the i-th generator bus
S_{L_i} : Apparent power flow of the i-th branch
NL : Number of load buses
NG : Number of generator buses
NTL : Number of transmission lines
NB : Number of buses
NT : Number of regulating transformers
NC : Number of shunt compensators
V_{G_i} : Voltage of the i-th generator bus
T_i : Tap setting of the i-th regulating transformer
Q_{C_i} : Reactive power generation of the i-th Var source
P_{loss} : Total power losses
G_k : Conductance of k-th line connected between i-th and j-th buses
V_i, V_j : Voltage of i-th and j-th buses respectively
∂_{ij} : Phase angle between i-th and j-th bus voltages
P_{G_i}, Q_{G_i} : Active and reactive power of the i-th generator
P_{L_i}, Q_{L_i} : Active and reactive power of the i-th load bus
G_{ij}, B_{ij} : Conductance and susceptance of transmission line connected between the i-th and the j-th bus
$V_{G_i}^{\min}, V_{G_i}^{\max}$: Minimum and maximum voltage of the i-th generating unit
$Q_{G_i}^{\min}, Q_{G_i}^{\max}$: Minimum and maximum reactive power of the i-th generating unit $V_{L_i}^{\min}, V_{L_i}^{\max}$: Minimum and maximum load voltage of the i-th unit

S_{L_i} : Apparent power flow of the i-th branch

$S_{L_i}^{\max}$: Maximum apparent power flow limit of the i-th

$T_i^{\min}, T_i^{\max}$: Minimum and maximum tap setting limits of the i-th transformer

$Q_{C_i}^{\min}, Q_{C_i}^{\max}$: Minimum and maximum Var injection limits of the i-th shunt capacitor

HSI habitat suitability index

λ_j, μ_j : Immigration and emigration rates of the j-th individual

I : Maximum possible immigration rate

E : Maximum possible emigration rate

j : Number of species of the j-th individual

n : Maximum number of species

m_j : Mutation rate for the j-th habitat having j number of species

$m_{\max}$: Maximum mutation rate

$P_{\max}$: Maximum species count probability

P_j : Species count probability for the j-th habitat

λ_{j+1}, μ_{j+1} : Immigration and emigration rate for the j-th habitat contains $j+1$ species λ_{j-1}, μ_{j-1} : Immigration and emigration rate for the j-th habitat contains $j-1$ species

x^o : Opposite number

x^{qo} : Quasi-opposite number

c Center of the interval [a, b]

N_P : Population size

N_D : Number of independent variables.

J_r : Jumping rate

2 PROBLEM DESCRIPTION

The general ORPD problem under normal operating condition may be formulated as follows:

$$Minimize \quad f(u,v) \tag{1}$$

$$Subject\ to \quad \begin{cases} g(u,v) = 0 \\ h(u,v) \leq 0 \end{cases} \tag{2}$$

where $f(u,v)$ is the objective function; $g(u,v) = 0$ are the equality constraints; $h(u,v) = 0$ are the inequality constraints;

u is the vector of dependent variables and may be expressed as:

$$u = [V_{L_1},...,V_{L_i},...V_{L_{NL}}, Q_{G_1},...,Q_{G_i},...,Q_{G_{NG}}, S_{L_1},...,S_{L_i},...,S_{L_{NTL}}] \tag{3}$$

where V_{L_i} is the voltage of the i-th load bus; Q_{G_i} is the reactive power generation of the i-th generator bus; S_{L_i} is the apparent power flow of the i-th branch; NL is the number of load buses; NG is the number of generator buses and NTL is the number of transmission lines.

v is the set of the independent variables, which may be expressed as:

$$v = [V_{G_1}, ..., V_{G_i}, ..., V_{G_{NG}}, T_1, ..., T_i, ..., T_{NT}, Q_{C_1}, ..., Q_{C_i}, ..., Q_{C_{NC}}] \tag{4}$$

where V_{G_i} is the voltage of the i-th generator bus; T_i is the tap setting of the i-th regulating transformer; Q_{C_i} is the reactive power generation of the i-th Var source; NT is the number of regulating transformers and NC is the number of shunt compensators.

2.1 Objective Function

The objective of ORPD problem is to minimize the transmission loss while satisfying all equality and inequality constraints. The transmission loss may be expressed as

$$P_{loss} = \sum_{k=1}^{NTL} G_k \left[V_i^2 + V_j^2 - 2|V_i||V_j|\cos\delta_{ij} \right] \tag{5}$$

where, P_{loss} is the total power losses; G_k is the conductance of k-th line connected between i-th and j-th buses; V_i, V_j are the voltage of i-th and j-th buses respectively; ∂_{ij} is the phase angle between i-th and j-th bus voltages.

2.2 Constraints

2.2.1 Equality Constraints

The equality constraints represent the load flow equations, which are given below for i-th bus:

$$\begin{cases} P_{G_i} - P_{L_i} = \sum_{j=1}^{NB} |V_i||V_j| \left(G_{ij}\cos\delta_{ij} + B_{ij}\sin\delta_{ij} \right) \\ Q_{G_i} - Q_{L_i} = \sum_{j=1}^{NB} |V_i||V_j| \left(G_{ij}\sin\delta_{ij} - B_{ij}\cos\delta_{ij} \right) \end{cases} \tag{6}$$

where, P_{G_i}, Q_{G_i} are the active and reactive power of the i-th generator; P_{L_i}, Q_{L_i} are the active and reactive power of the i-th load bus; G_{ij}, B_{ij} are the conductance and susceptance of transmission line connected between the i-th and the j-th bus; NB is the number of buses.

2.2.2 Inequality Constraints

(I) Generator Constraints

Voltage and reactive power of the i-th generator bus lies between its operating limits as given below:

$$\begin{cases} V_{G_i}^{\min} \leq V_{G_i} \leq V_{G_i}^{\max} & i = 1, 2, \cdots, NG \\ Q_{G_i}^{\min} \leq Q_{G_i} \leq Q_{G_i}^{\max} & i = 1, 2, \cdots, NG \end{cases} \tag{7}$$

where, $V_{G_i}^{\min}, V_{G_i}^{\max}$ are the minimum and maximum voltage of the i-th generating unit; $Q_{G_i}^{\min}, Q_{G_i}^{\max}$ are the minimum and maximum reactive power of the i-th generating unit.

(ii) Load Bus Constraints

$$V_{L_i}^{\min} \leq V_{L_i} \leq V_{L_i}^{\max} \quad i = 1, 2, \cdots, NL \tag{8}$$

where, $V_{L_i}^{\min}, V_{L_i}^{\max}$ are the minimum and maximum load voltage of the i-th unit.

(iii) Transmission Line Constraints

$$S_{Li} \leq S_{L_i}^{\max} \quad i = 1, 2, \cdots, NTL \tag{9}$$

where, S_{L_i} is the apparent power flow of the i-th branch; $S_{L_i}^{\max}$ is the maximum apparent power flow limit of the i-th branch.

(iv) Transformer Tap Constraints

Transformer tap settings are bounded between upper and lower limits as given below:

$$T_i^{\min} \leq T_i \leq T_i^{\max} \quad i = 1, 2, \cdots, NT \tag{10}$$

where, $T_i^{\min}, T_i^{\max}$ are the minimum and maximum tap setting limits of the i-th transformer.

(v) Shunt Compensator Constraints

Shunt compensation are restricted by their limits as follows:

$$Q_{C_i}^{\min} \leq Q_{C_i} \leq Q_{C_i}^{\max} \quad i = 1, 2, \cdots, NC \tag{11}$$

where, $Q_{C_i}^{\min}, Q_{C_i}^{\max}$ are the minimum and maximum Var injection limits of the i-th shunt capacitor.

3 BRIEF INTRODUCTION TO BIOGEOGRAPHY-BASED OPTIMIZATION

Biogeography-based optimization (BBO) developed by Simon (Simon, 2008) is a new biogeography inspired algorithm and is an example of how a natural process can be modeled to solve optimization problems. BBO has already been applied successfully to solve economic load dispatch (ELD) (Roy, Ghoshal & Thakur, 2010a; Bhattacharya, & Chattopadhyay, 2010), optimal power flow (OPF) (Roy, Ghoshal & Thakur, 2010b; Roy, Ghoshal & Thakur, 2010c) and ORPD (Roy, Ghoshal & Thakur, 2011; Roy, Ghoshal & Thakur, 2012) problems. In BBO, each possible solution is an island and their features that characterize habitability are called suitability index variables (*SIV*). The goodness of each solution is called its habitat suitability index (*HSI*). Over evolutionary periods of time, some islands may tend to accumulate more species than others because they posses certain environmental features that are more suitable to sustaining those species than islands with fewer species. Habitats with high *HSI* have large population, high emigration rate, simply by virtue of large number of species that migrate to other habitats. The immigration rate is low for these habitats as these are already saturated with species. On the other hand, habitats with low *HSI* have high immigration and low emigration rate, because of sparse population.

In BBO, each solution is associated with the fitness which is analogous to *HSI* of a habitat. A good solution is analogous to a habitat having high *HSI* and a poor solution represents a habitat having a low *HSI*. Good solutions share their features with poor solutions by means of migration (immigration and emigration). Good solutions have more resistance to change than poor solutions. On the other hand, poor solutions are more dynamic and accept a lot of new features from good solutions.

The immigration rate and emigration rate of the j-th island may be formulated as follows (Simon, 2008):

$$\lambda_j = I \cdot \left(1 - \frac{j}{n}\right) \quad (12)$$

$$\mu_j = \frac{E \cdot j}{n} \quad (13)$$

where λ_j, μ_j are the immigration rate and emigration rate of the j-th individual; I is the maximum possible immigration rate; E is the maximum possible emigration rate; j is the number of species of the j-th individual; and n is the maximum number of species.

In BBO, the mutation is used to increase the diversity of the population to get the good solutions. Mutation operator modifies a habitat's *SIV* randomly based on mutation rate. The mutation rate m_j is expressed as (14) (Simon, 2008).

$$m_j = m_{\max}\left(\frac{1-P_j}{P_{\max}}\right) \quad (14)$$

where m_j is the mutation rate for the j-th habitat having j number of species; $m_{\max}$ is the maximum mutation rate; $P_{\max}$ is the maximum species count probability; P_j is the species count probability for the j-th habitat and is given by (Simon, 2008):

$$P_j = \begin{cases} -(\lambda_j + \mu_j)P_j + \mu_{j+1}P_{j+1} & j = 0 \\ -(\lambda_j + \mu_j)P_j + \lambda_{j-1}P_{j-1} + \mu_{j+1}P_{j+1} & 1 \leq j \leq n-1 \\ -(\lambda_j + \mu_j)P_j + \lambda_{j-1}P_{j-1} & j = n \end{cases} \quad (15)$$

where λ_{j+1}, μ_{j+1} are the immigration and emigration rate for the j-th habitat contains $j+1$ species; λ_{j-1}, μ_{j-1} are the immigration and emigration rate for the j-th habitat contains $j-1$ species.

4 BBO ALGORITHM

Step 1: Initialize the input parameters of the BBO algorithm and generate initial random set of habitats according to the constraints of the problem
Step 2: Evaluate the fitness (*HSI*) of each habitat
Step 3: Based on the fitness value, perform sorting operation of the habitats from best to worst
Step 4: Based on the *HSI* values, identify the elite habitats of the population
Step 5: Mapping the habitat to the number of species based on HSI values
Step 6: Modify each habitat with probabilistic migration operation using immigration rate and emigration rate of each habitat
Step 7: Modify non elite habitats performing probabilistic mutation operation
Step 8: Check the feasibility of the newly generated solution set
Step 9: Replace infeasible solutions by best feasible solutions
Step 10: Perform steps 2-9 until a stopping criterion be achieved.

5 OPPOSITION-BASED LEARNING

Opposition-based learning (OBL), introduced by Tizhoosh (Tizhoosh, 2005), has proven to be an effective method to accelerate PSO (Zhang, Ni, Wu & Gu, 2009), DE (Rahnamayan, Tizhoosh & Salama, 2007), ACO (Haiping, Xieyon & Baogen, 2010), TLBO (Mandal, & Roy, 2013), GSA (Shaw, Mukherjee, & Ghoshal, 2012), HSA (Chatterjee, Ghosal, & Mukherjee, 2012). Like other evolutionary algorithms, BBO starts with an initial population string, which is randomly generated when no preliminary knowledge about the solution space is known. The process of evolution terminates when predefined criteria are satisfied. The computation time is directly related to distance of the guess from optimal solution. The chances of getting optimal solutions can be improved by starting with a closer (fitter) solution by checking the opposite solution simultaneously. So the closer solution between the guess solution and the opposite guess solution can be chosen as initial solution. In fact, according to the theory of probability, 50% of the time a guess is further from the solution than its opposite guess. Therefore, staring with the opposite guess may accelerate convergence rate. This approach can be applied not only to initial solutions but also to each iterative solution in the current population.

It has been proved that, in general, opposite numbers are more likely to be closer to the optimal solution than purely random ones (Rahnamayan, Tizhoosh, & Salama, 2008). It has also been proved that the probability of getting better solution using quasi-opposite points is more compare to opposite points (Rahnamayan, Tizhoosh, & Salama, 2007).

A Novel Evolutionary Optimization Technique

- **Opposite Number:** Let $x \in [a,b]$ be a real number. Its opposite number x^o is defined by:

$$x^o = a + b - x \tag{16}$$

Similarly, the definition may be generalized to higher dimensions as follows:

- **Opposite Point:** Let $P(x_1, x_2, ..., x_n)$ be a point in n-dimensional space, where $x_i \in [a,b]$; $i = \{1, 2, ..., n\}$. The opposite point $OP(x_1^o, x_2^o, ..., x_n^o)$ is defined by:

$$x_i^o = a_i + b_i - x_i \tag{17}$$

- **Quasi-Oppositite Number and Quasi-Opposite Point:** In (Rahnamayan, Tizhoosh & Salama, 2007), Rahnamayan introduces quasi-opposition-based learning and proves that a quasi-opposite point is more likely to be closer to the solution than the opposite point.
- **Quasi-Opposite Number:** Let $x \in [a,b]$ be a real number. Its quasi-opposite number x^{qo} is defined by:

$$x^{qo} = rand(c, x^o) \tag{18}$$

where $rand(c, x^o)$ is a random number uniformly distributed between c and x^o; c is the center of the interval [a, b] and is given by:

$$c = \frac{a+b}{2} \tag{19}$$

- **Quasi-Oppositional Point:** Let $P(x_1, x_2, ..., x_n)$ be a point in n-dimensional space, where $x_i \in [a,b]$; $i = \{1, 2, ..., n\}$. The quasi-oppositional point $QOP(x_1^{qo}, x_2^{qo}, ..., x_n^{qo})$ is defined by:

$$x_i^{qo} = rand(c_i, x_i^o), \; i = \{1, 2, ..., n\} \tag{20}$$

The flow chart of the QOBBO algorithm is illustrated in Figure 1.

6 ALGORITHM OF QOBBO APPLIED TO ORPD

The proposed QOBBO algorithm integrated original BBO algorithm with QOBL concept to enhance the convergence rate of the standard BBO algorithm. The algorithmic steps of QOBBO are enumerated as follows:

Figure 1. Flow Chart of QOBBO Algorithm

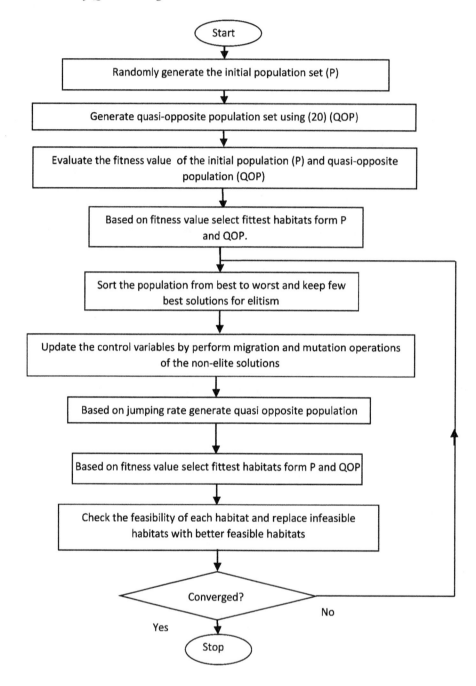

Step 1: The habitats' initial positions characterized by *SIV*s should be randomly selected while satisfying the equality and inequality constraints. The independent variables such as generators' voltages, tap settings of regulating transformers and reactive power injections of shunt capacitors of ORPD represents the SIVs of individual habitats.

Step 2: Calculate quasi-oppositional populations that is represented by

$$QOP_{i,j} = rand(c_j, OP_{i,j}); \; c_j = \frac{a_j + b_j}{2}; \; OP_{i,j} = a_j + b_j - P_{i,j} \tag{21}$$

where $i = 1, 2, ..., N_P$; $j = 1, 2, ..., N_D$; $rand(c_j, OP_{i,j})$ is a random point uniformly distributed between c_j and $OP_{i,j}$; $P_{i,j}$ is the j-th independent variables of the i-th vector of the population; $OP_{i,j}$ is the j-th independent variables of the i-th vector of the opposite population; N_P is the population size and N_D is the number of independent variables.

Step 3: Using Newton-Raphson based load flow analysis, the dependent variables of ORPD problem such as load voltages, active power of slack bus, generators' reactive powers, etc are evaluated.

Step 4: Calculate the fitness value (i.e. *HSI*) of each habitat of the population and quasi-oppositional population sets.

Step 5: Sort the population and quasi-oppositional population sets. From best to worst and identify the requisite number of fittest solutions from the entire solutions set.

Step 6: Based on the *HSI* values, identify the elite habitats.

Step 7: For each solution, map the *HSI* to the number of species, immigration rate λ_j and emigration rate μ_j.

Step 8: Perform immigration and emigration operation to modify the independent variables (*SIV*s) of each non-elite habitat.

Step 9: Perform mutation operation to update the independent variables (*SIV*s) of non-elite habitats.

Step 10: Use NR load flow analysis to update load voltages, generators' reactive powers, and calculate the fitness value of each habitat of the population set.

Step 11: Based on a jumping rate J_r (i.e. jumping probability), the quasi-opposite population is generated and fitness value of objective function (i.e. *HSI*) for each habitat of quasi-opposite population is calculated.

Step 12: Select requisite number of fittest solutions from the current population and the quasi-oppositional population.

Step 13: If the termination criterion (maximum iteration limit) is satisfied then stop the iteration otherwise go to step 6 for next generation.

7 INPUT PARAMETERS

The performance of the proposed algorithm depends on input parameters and they should be chosen carefully. After several runs, optimal input control parameters shown in Table 1 are found to be best for optimal performance of the proposed algorithm.

8 SIMULATION RESULTS

In this section, the performance of the QOBBO approach using three case studies of ORPD is evaluated. The optimization method is implemented in Matlab (MathWorks) using Microsoft Windows XP. All the programs are run on a 2.5 GHz core 2 duo processor with 1GB of random access memory. Due to

Table 1. Parameters for QOBBO in computation

Habitat modification probability	1	Elitism parameter	4
Mutation probability	0.005	Step size for numerical integration	1
Maximum immigration rate	1	Jumping rate	0.3
Maximum emigration rate	1		

the inherent randomness involved, the performance of heuristic search based optimization algorithms cannot be judged by the results of a single trial. Therefore, many trials with different initial populations are required to test the robustness of the proposed method. In each case study, 50 independent runs are made to generate 50 different initial trial solutions. In these case studies, the stopping criterion is 100 generations for the proposed QOBBO algorithm.

8.1 Case Study I (IEEE 30-Bus System)

The IEEE 30-bus system consists of 41 branches, 6 generator buses and 24 load buses. Four branches 6–9, 6–10, 4–12 and 27–28, are under tap setting transformer branches. In addition, buses 10, 12, 15, 17, 20, 21, 23, 24 and 29 have been selected as shunt Var compensation buses. Bus 1 is selected as slack bus and 2, 5, 8, 11 and 13 are the generator buses. The others are load buses. The total active and reactive load demand on this system are 2.834 *p.u.* and 1.260 *p.u.* respectively. The line data, bus data, generator data of (Lee, Park & Ortiz, 1985) are used for solving this problem. The system data are also given in Tables A1 and A2 in the appendix section. The voltage magnitudes limits of all buses are ($0.95 p.u.$, $1.1 p.u.$). Tap settings limits of regulating transformers are ($0.9 p.u.$, $1.1 p.u.$).The Var injection of the shunt capacitors are within the interval of *(0 MVar, 5 MVar)*. The upper and lower limits of reactive power generation are adopted from (Lee, Park & Ortiz, 1985). The optimal control variables and the transmission loss obtained for the IEEE 30-bus system using various techniques are given in Table 2 while satisfying all the equality and inequality constraints. Simulation results show that the transmission loss obtained by proposed algorithm is the best as compared to PSO (Mahadevann & Kannan, 2010), CLPSO (Mahadevann & Kannan, 2010) and BBO (Bhattacharya & Chattopadhyay, 2010).The convergence of optimal solution using QOBBO is shown in Figure 1.

Owing to the randomness, the proposed algorithm is executed 50 times when applied to the test system. The best, worst and average power losses found by the different methods are tabulated in Table 3. QOBBO shows good consistency by keeping the difference between the best and worst solutions within 0.035%. In addition, the average execution times summarized in Table 3 show that QOBBO is faster than PSO, CLPSO and BBO. The consistency of proposed QOBBO approach in attaining the near optimal value can also be noticed from Table 3. Out of 50 trails, the success rate of QOBBO is higher (98%) compare to PSO (43%), CLPSO (80%) and BBO (96%).

8.2 Case Study II (IEEE 57-Bus System)

In order to evaluate the applicability of the proposed method to medium size test system, the proposed QOBBO algorithm is also applied to the reactive power dispatch problem of the IEEE 57-bus system. This system consists of 57 buses, 80 transmission lines, 7 generators connected at bus numbers 1, 2, 3,

Table 2. Simulation results of ORPD using different techniques for the IEEE 30-bus system

Techniques	Base Case	PSO	CLPSO	BBO	QOBBO
V_1 (p.u.)	1.0500	1.1000	1.1000	1.1000	1.0999
V_2 (p.u.)	1.0400	1.1000	1.1000	1.0944	1.0942
V_5 (p.u.)	1.0100	1.0867	1.0795	1.0749	1.0747
V_8 (p.u.)	1.0100	1.1000	1.1000	1.0768	1.0766
V_{11} (p.u.)	1.0500	1.1000	1.1000	1.0999	1.1000
V_{13} (p.u.)	1.0500	1.1000	1.1000	1.0999	1.0999
TC_{6-9}	1.0780	0.9587	0.9197	1.0435	1.03873
TC_{6-10}	1.0690	1.0543	0.9000	0.90117	0.90000
TC_{4-12}	1.0320	1.0024	0.9000	0.98244	0.97244
TC_{27-28}	1.0680	0.9755	0.9397	0.96918	0.96275
Q_{C10} (Mvar)	0.0000	0.042803	0.049265	0.049998	0.500000
Q_{C12} (Mvar)	0.0000	0.500000	0.500000	0.04987	0.500000
Q_{C15} (Mvar)	0.0000	0.030288	0.500000	0.049906	0.500000
Q_{C17} (Mvar)	0.0000	0.040365	0.500000	0.04997	0.500000
Q_{C20} (Mvar)	0.0000	0.026697	0.500000	0.049901	0.041578
Q_{C21} (Mvar)	0.0000	0.038894	0.500000	0.049946	0.500000
Q_{C23} (Mvar)	0.0000	0.000000	0.500000	0.038753	0.025843
Q_{C24} (Mvar)	0.0000	0.035879	0.500000	0.049867	0.049926
Q_{C29} (Mvar)	0.0000	0.028415	0.500000	0.029098	0.022213
Transmission loss (p.u.)	0.05812	0.046282	0.045615	0.045511	0.045312
$P_{Loss}^{Save}\%$	-	20.36820	21.51583	21.69477	22.03716

Table 3. Comparison of the result obtained by different methods for the IEEE 30-bus system

Algorithms	Transmission Loss (p.u.)			Average Computational Time (sec.)	Success Rate (%)
	Maximum	Minimum	Average		
PSO	0.047986	0.046282	0.047363	130	43
CLPSO	0.046833	0.045615	0.046397	138	80
BBO	0.045522	0.045511	0.045515	110	96
QOBBO	0.045328	0.045312	0.045319	108	98

6, 8, 9 and 12; 15 regulating transformers connected between the line numbers 4-18, 4-18, 21-20, 24-26, 7-29, 34-32, 11-41, 15-45, 14-46, 10-51, 13-49, 11-43, 40-56, 39-57 and 9-55; 3 shunt compensators connected at bus numbers 18, 25 and 53. The total load demand on this system is 12.508 *p.u.* The line data, bus data, generator data of (IEEE 57-bus test system) are used for solving this problem. The system data are also listed in Tables A3 and A4 in the appendix section. The voltage magnitudes of all buses are considered within the range of ($0.94p.u, 1.06p.u.$). Tap settings of regulating transformers are within the range of ($0.9p.u., 1.1p.u.$). The Var injection of the shunt capacitors connected at bus numbers 18, 25 and 53 are taken within the interval of *(0 Mvar-10 MVar), (0 Mvar-5.9 MVar)* and *(0 Mvar-6.3 MVar)* respectively. The upper and lower limits of the reactive power generations are adopted from (Dai, Chen, Zhun & Zhang, 2009).

(I) Without Reactive Power Generation Constraints

The best, worst and average transmission losses in 50 trial runs using CGA (Dai, Chen, Zhun & Zhang, 2009), CLPSO (Dai, Chen, Zhun & Zhang, 2009), SPSO-07(Dai, Chen, Zhun & Zhang, 2009), L-SADE (Dai, Chen, Zhun & Zhang, 2009), PSO-cf (Dai, Chen, Zhun & Zhang, 2009), PSO-w (Dai, Chen, Zhun & Zhang, 2009), SOA (Dai, Chen, Zhun & Zhang, 2009), BBO (Bhattacharya & Chattopadhyay, 2010) methods are shown in Table 4. Table 5 shows the base case results, optimum results using different techniques for the IEEE 57-bus system ignoring reactive power generation constraints. The simulation results show that the QOBBO method can obtain lower generation cost than the other mentioned methods. Table 4 also shows that the average cost produced by QOBBO is least compared with other methods emphasizing its better solution quality. For comparing the computational speed, the CPU time requirement for the other recently published methods along with QOBBO are represented in Table 4. It is noted that the QOBBO method is computationally most efficient as time requirement is minimum amongst all the methods. To illustrate the convergence of the proposed algorithm, values of transmission loss over 100 trials are plotted in Figure 2.

Table 4. Comparison of the result obtained by different methods for the IEEE 57-bus system without reactive power generation constraints

Algorithms	Transmission Loss (*p.u.*)			Average Computational Time (sec.)	Success Rate (%)
	Maximum	Minimum	Average		
CGA	0.2750772	0.2524411	0.2629356	411.38	NA
CLPSO	0.2478083	0.2451520	0.2467307	426.85	NA
SPSO-07	0.2545745	0.2443043	0.2475227	137.35	NA
PSO-cf	0.2603275	0.2428022	0.2469805	408.19	NA
PSO-w	0.2615279	0.2427052	0.2472596	408.48	NA
L-SADE	0.2439142	0.2426739	0.2431129	410.14	NA
SOA	0.2428046	0.2426548	0.2427078	391.32	NA
BBO	0.242621	0.242616	0.242619	232.15	90.00
QOBBO	0.2426142	0.2426096	0.2426106	230.08	97.00

A Novel Evolutionary Optimization Technique

Figure 2. Convergence characteristic of transmission loss of the IEEE 30-bus system using QOBBO

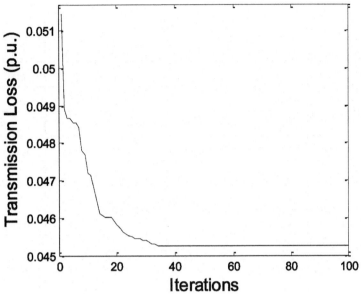

Table 5. Simulation results without reactive power generation constraints of IEEE 57-bus system

Techniques	Base Case+	CGA	CLPSO	SPSO-07	PSO-cf	PSO-w	L-SADE	SOA	BBO	QOBBO
V_1 (p.u.)	1.0400	0.9686	1.0541	1.0596	1.0600	1.0600	1.0600	1.0600	1.0600	1.0600
V_2 (p.u.)	1.0100	1.0493	1.0529	1.0580	1.0586	1.0578	1.0574	1.0580	1.0580	1.0585
V_3 (p.u.)	0.9850	1.0567	1.0337	1.0488	1.0464	1.04378	1.0438	1.0437	1.0442	1.0460
V_6 (p.u.)	0.9800	0.9877	1.0313	1.0362	1.0415	1.0356	1.0364	1.0352	1.0364	1.0373
V_8 (p.u.)	1.005	1.0223	1.0496	1.0600	1.0600	1.0546	1.0537	1.0548	1.0567	1.0567
V_9 (p.u.)	0.9800	0.9918	1.0302	1.0433	1.0423	1.0369	1.0366	1.0369	1.0377	1.0379
V_{12} (p.u.)	1.015	1.0044	1.0342	1.0356	1.0371	1.0334	1.0323	1.0336	1.0351	1.0351
TC_{4-18}	0.9700	0.92	0.99	0.95	0.98	0.90	0.94	1.00	0.99165	0.98001
TC_{4-18}	0.9780	0.92	0.98	0.99	0.98	1.02	1.00	0.96	0.96447	0.97429
TC_{21-20}	1.0430	0.97	0.99	0.99	1.01	1.01	1.01	1.01	1.0122	1.01460
TC_{24-26}	1.0430	0.90	1.01	1.02	1.01	1.01	1.01	1.01	1.0110	1.00890
TC_{7-29}	0.9670	0.91	0.99	0.97	0.98	0.97	0.97	0.97	0.97127	0.97204
TC_{34-32}	0.9750	1.10	0.93	0.96	0.97	0.97	0.97	0.97	0.97227	0.97450
TC_{11-41}	0.9550	0.94	0.91	0.92	0.90	0.90	0.90	0.90	0.90095	0.90000
TC_{15-45}	0.9550	0.95	0.97	0.96	0.97	0.97	0.97	0.97	0.97063	0.97174
TC_{14-46}	0.9000	1.03	0.95	0.95	0.96	0.95	0.96	0.95	0.95153	0.95010

continued on following page

Table 5. Continued

Techniques	Base Case+	CGA	CLPSO	SPSO-07	PSO-cf	PSO-w	L-SADE	SOA	BBO	QOBBO
TC_{10-51}	0.9300	1.09	0.98	0.97	0.97	0.96	0.96	0.96	0.96252	0.96267
TC_{13-49}	0.8950	0.90	0.95	0.92	0.93	0.92	0.92	0.92	0.92227	0.92193
TC_{11-43}	0.9580	0.90	0.95	1.00	0.97	0.96	0.96	0.96	0.95988	0.95986
TC_{40-56}	0.9850	1.00	1.00	1.00	0.99	1.00	1.00	1.00	1.0018	1.00930
TC_{39-57}	0.9800	0.96	0.96	0.95	0.96	0.96	0.96	0.96	0.96567	0.96601
TC_{9-55}	0.9400	1.00	0.97	0.98	0.98	0.97	0.97	0.97	0.97199	0.97096
Q_{C18} (Mvar)	0.0000	0.084	0.09888	0.03936	0.09984	0.05136	0.08112	0.09984	0.09640	0.09425
Q_{C25} (Mvar)	0.0000	0.00816	0.05424	0.05664	0.05904	0.05904	0.05808	0.05904	0.05897	0.05900
Q_{C53} (Mvar)	0.0000	0.05376	0.06288	0.03552	0.06288	0.06288	0.06192	0.06288	0.062948	0.06300
Transmission loss (p.u.)	0.28462	0.2524411	0.245152	0.2443043	0.2428022	0.2427052	0.2426739	0.2426548	0.242616	0.2426096
$P_{Loss}^{Save}\%$	-	11.3059	13.8669	14.1647	14.6925	14.7266	14.7376	14.7443	14.7579	14.7602

Table 6. Simulation results with reactive power generation constraints for the IEEE 57-bus system

Control variables	BBO	QOBBO	Control variables	BBO	QOBBO
V_1 (p.u.)	1.0600	1.06	TC_{15-45}	0.96602	0.96612
V_2 (p.u.)	1.0504	1.0504	TC_{14-46}	0.95079	0.95079
V_3 (p.u.)	1.0440	1.0436	TC_{10-51}	0.96414	0.96414
V_6 (p.u.)	1.0376	1.0381	TC_{13-49}	0.92462	0.92462
V_8 (p.u.)	1.0550	1.0574	TC_{11-43}	0.95022	0.95042
V_9 (p.u.)	1.0229	1.0242	TC_{40-56}	0.99666	0.99666
V_{12} (p.u.)	1.0323	1.0322	TC_{39-57}	0.96289	0.96289
TC_{4-18}	0.96693	0.96725	TC_{9-55}	0.96001	0.95931
TC_{4-18}	0.99022	0.98393	Q_{C18} (Mvar)	0.09782	0.0996
TC_{21-20}	1.0120	1.0124	Q_{C25} (Mvar)	0.058991	0.059
TC_{24-26}	1.0087	1.0068	Q_{C53} (Mvar)	0.06289	0.063
TC_{7-29}	0.97074	0.97228	Transmission loss (p.u.)	0.24544	0.24529
TC_{34-32}	0.96869	0.96869	$P_{Loss}^{Save}\%$	13.7657	13.8184
TC_{11-41}	0.90082	0.90082			

(II) With Reactive Power Generation Constraints

The optimal settings of control variables along with transmission loss obtained using the proposed QOBBO method and BBO method reported in the literature is presented in Table 6. The numerical results indicate that the transmission loss determined by QOBO is more advanced than that found by BBO with all control variables remained within their permissible limits. The best, worst and average active transmission loss obtained using BBO (Bhattacharya & Chattopadhyay, 2010) and QOBBO in 50 independent runs are summarized in Table 7, which show that the proposed QOBBO algorithm obviously performs better than the BBO algorithm. Also, the average CPU time for this case using QOBBO for a population size of 50 and iteration cycles of 100 is 230.21 Sec. whereas the CPU time for the BBO algorithm for the same population size and number of iteration cycles is 232.32 Sec. This clearly shows that the new proposed algorithm is capable of giving a better optimum solution with less computational time. Figure 3 shows the active power loss convergence obtained by QOBBO.

Table 7. Comparison of the result obtained by different methods for the IEEE 57-bus system with reactive power generation constraints

Algorithms	Transmission Loss (p.u.)			Average Computational Time (sec.)	Success Rate (%)
	Maximum	Minimum	Average		
BBO	0.245452	0.24544	0.245445	232.32	96.67
QOBBO	0.24541	0.24529	0.24534	230.21	97.84

Figure 3. Convergence characteristic of transmission loss of the IEEE 57-bus system without reactive power generation constraints using QOBBO

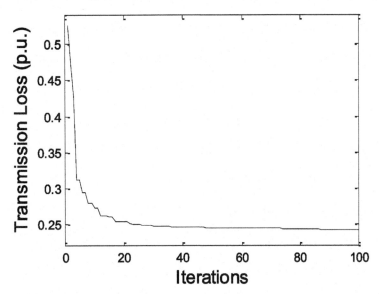

8.3 Case Study III (IEEE 118-Bus System)

To test the proposed QOBBO technique in solving ORPD problems of larger power systems, a standard IEEE 118-bus test system is considered. This test system consists of 54 generators, 9 tap-changing transformers, 186 transmission lines, 12 shunt capacitors, and two shunt reactors, so the system has seventy-five variables and the total active and reactive loads are 4 242 MW and 1438 Mvar, respectively. The bus data, branch data and generation limits are taken from (IEEE 118-bus test system). The system data are also illustrated in Tables A5 and A6 in the appendix section. The maximum and minimum limits of reactive power sources, bus voltage and tap-setting limits are taken from [29]. The initial active power generations and power losses are 4375.36 MW and 133.357 MW, respectively. To validate the superiority of the proposed QOBBO based approach, its results are compared with those obtained by PSO (Tehzeeb, Zafar, Mohsin, & Latee, 2012), FIPS (Tehzeeb, Zafar, Mohsin, & Latee, 2012), QEA (Vlachogiannis, & Lee, 2008), ACS (Vlachogiannis, & Lee, 2008) and DE (Varadarajan, & Swarup, 2008). approaches available in the literature.

Table 8 summarizes the results of the optimal settings as obtained by the proposed method. It is clearly observed from the simulation results that the proposed method converges to the optimal solution without violating any operating constraints. The statistical analysis of the proposed methodology is based on the information obtained by 50 independent runs with different initial populations to check the robustness of the proposed method. For each run, the final solutions are observed and important statistical parameters i.e. best, average and worst active power loss are listed in Table 9. Comparing the results given in Table 9, it is concluded that QOBBO has the best performance among all rival approaches. The simulation study demonstrates the superiority of the proposed method in terms of robustness, computational efficiency and solution quality than the PSO, FIPS, QEA, ACS, and DE methods. Figure 4 shows the convergence characteristics of real power loss by the number of iterations. It is clearly observed that the QOBBO obtained solution converges to high quality solutions at initial stage and finally converges to the optimal solution without any abrupt oscillation. This signifies the convergence superiority of the proposed method.

9 CONCLUSION

This study demonstrates the feasibility of employing QOBBO approach for efficient solving of ORPD problems. To enrich the searching behavior and to avoid being trapped into local optimum, quasi-opposition-based population initialization and quasi-opposition-based generation jumping are incorporated in conventional BBO approach.

Table 8. Comparison of test results of IEEE 118-bus system using different methods

Techniques	QOBBO	PSO	FIPS	QEA	ACS	DE
Best	114.5628	118.0	120.6	122.22	131.90	128.31
Worst	116.1824	122.3	120.7	NA	NA	NA
Mean	114.9327	120.6	120.6	NA	NA	NA

A Novel Evolutionary Optimization Technique

Table 9. Simulation resultobtained by QOBBO for loss minimization(IEEE 118-bus system)

V_{g1} (p.u.)	1.0218	V_{g49} (p.u.)	1.0582	V_{g90} (p.u.)	1.0424	QC_{48} (p.u.)	0.0626
V_{g4} (p.u.)	1.0439	V_{g54} (p.u.)	1.0295	V_{g91} (p.u.)	1.0393	QC_{74} (p.u.)	0.0447
V_{g6} (p.u.)	1.0311	V_{g55} (p.u.)	1.0298	V_{g92} (p.u.)	1.0528	QC_{79} (p.u.)	0.1929
V_{g8} (p.u.)	1.0997	V_{g56} (p.u.)	1.0271	V_{g99} (p.u.)	1.0266	QC_{82} (p.u.)	0.1792
V_{g10} (p.u.)	1.0955	V_{g59} (p.u.)	1.0551	V_{g100} (p.u.)	1.0396	QC_{83} (p.u.)	0.0809
V_{g12} (p.u.)	1.0247	V_{g61} (p.u.)	1.0541	V_{g103} (p.u.)	1.0251	QC_{105} (p.u.)	0.1029
V_{g15} (p.u.)	1.0215	V_{g62} (p.u.)	1.0563	V_{g104} (p.u.)	1.0176	QC_{107} (p.u.)	0.031
V_{g18} (p.u.)	1.0349	V_{g65} (p.u.)	1.0995	V_{g105} (p.u.)	1.0155	QC_{110} (p.u.)	0.0155
V_{g19} (p.u.)	1.0258	V_{g66} (p.u.)	1.0805	V_{g107} (p.u.)	1.0053	T_{8-5} T_{6-9}	1.0413
V_{g24} (p.u.)	1.0601	V_{g69} (p.u.)	1.0504	V_{g110} (p.u.)	1.026	T_{26-25}	1.0983
V_{g25} (p.u.)	1.0926	V_{g70} (p.u.)	1.0110	V_{g111} (p.u.)	1.0285	T_{30-17}	1.0392
V_{g26} (p.u.)	1.0991	V_{g72} (p.u.)	1.0351	V_{g112} (p.u.)	1.009	T_{38-37}	0.9926
V_{g27} (p.u.)	1.0506	V_{g73} (p.u.)	1.0129	V_{g113} (p.u.)	1.0382	T_{63-59} T_{6-9}	1.0313
V_{g31} (p.u.)	1.0318	V_{g74} (p.u.)	0.9993	V_{g116} (p.u.)	1.0786	T_{64-61}	1.0316
V_{g32} (p.u.)	1.0445	V_{g76} (p.u.)	0.9860	QC_{5} (p.u.)	-0.0933	T_{65-66}	0.9295
V_{g34} (p.u.)	1.0541	V_{g77} (p.u.)	1.0060	QC_{34} (p.u.)	0.0384	T_{68-69}	0.9815
V_{g36} (p.u.)	1.0517	V_{g80} (p.u.)	1.0221	QC_{37} (p.u.)	-0.2293	T_{81-80}	1.0297
V_{g40} (p.u.)	1.0370	V_{g85} (p.u.)	1.0512	QC_{44} (p.u.)	0.0848		
V_{g42} (p.u.)	1.0419	V_{g87} (p.u.)	1.0440	QC_{45} (p.u.)	0.099		
V_{g46} (p.u.)	1.0394	V_{g89} (p.u.)	1.0658	QC_{46} (p.u.)	0.0015		
Loss (MW)				114.5628			

Figure 4. Convergence characteristic of transmission loss of the IEEE 57-bus system with reactive power generation constraints using QOBBO

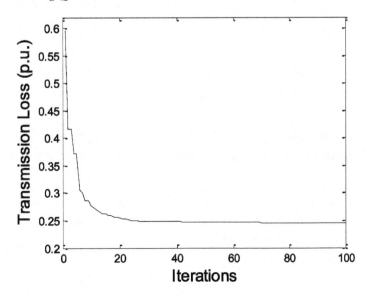

Figure 5. Convergence characteristic of transmission loss of the IEEE 118-bus system using QOBBO

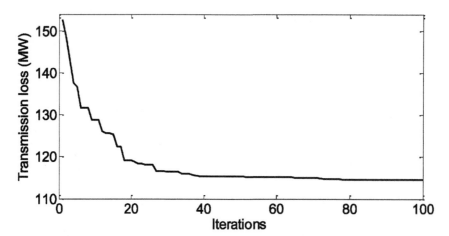

This study presents a close comparison of standard BBO and QOBBO strategies for solving the ORPD problem with complex constraints. The proposed QOBBO algorithm is implemented on standard IEEE 30-bus, 57-bus and 118- bus test systems with different number of generators. It is observed from the simulation study that BBO suffers from premature convergence while the quasi-oppositional based BBO is capable of achieving global best solutions efficiently for all the discussed test systems. Moreover, in order to prove the validity of the proposed method, its results are compared with those of other reputed algorithms available in the literature. The comparison of the numerical results and the convergence profiles of the transmission loss confirm the effectiveness and the superiority of the proposed approach.

It is also observed from the repeated trail runs that QOBBO converges to near optimal solution with high success rate, which demonstrates the robustness and effectiveness of the proposed method. It is also found that the proposed method not only produces minimum transmission loss, but requires lesser computational time compared to other reported results in the literature. This new algorithm may be applied to more application field to prove its efficiency in future.

REFERENCES

Aoki, K., Fan, M., & Nishikori, A. (1988). Optimal VAR planning by approximation method for recursive mixed-integer linear programming. *IEEE Transactions on Power Systems*, *3*(4), 1741–1747. doi:10.1109/59.192990

Bhattacharya, A., & Chattopadhyay, P. K. (2010). Solution of optimal reactive power flow using biogeography-based optimization. *International Journal of Electrical and Electronics Engineering.*, *4*(8), 580–588.

Bhattacharya, A., & Chattopadhyay, P. K. (2010). Hybrid differential evolution with biogeography based optimization for solution of economic load dispatch. *IEEE Transactions on Power Systems*, *25*(4), 1955–1964. doi:10.1109/TPWRS.2010.2043270

Bjelogrlic, M., Calovic, M. S., Babic, B. S., & Ristanovic, P. (1990). Application of Newton's optimal power flow in voltage/reactive power control. *IEEE Transactions on Power Systems*, *5*(4), 1447–1454. doi:10.1109/59.99399

Chatterjee, A., Ghosal, S. P., & Mukherjee, V. (2012). Solution of combined economic and emission dispatch problems of power systems by an opposition based harmony search algorithm. *International Journal of Electrical Power & Energy Systems*, *39*(1), 9–20. doi:10.1016/j.ijepes.2011.12.004

Chebbo, A. M., & Irving, M. R. (1995). Combined active and reactive power dispatch. Part I, Problem formulation and solution algorithm. *IEE Proceedings. Generation, Transmission and Distribution*, *142*(4), 393–400. doi:10.1049/ip-gtd:19951976

Dai, C., Chen, W., Zhu, Y., & Zhang, X. (2009). Reactive power dispatch considering voltage stability with seeker optimization algorithm. *Electric Power Systems Research*, *79*(10), 1462–1471. doi:10.1016/j.epsr.2009.04.020

Dai, C., Chen, W., Zhu, Y., & Zhang, X. (2009). Seeker optimization algorithm for optimal reactive power dispatch. *IEEE Transactions on Power Systems*, *24*(3), 1218–1231. doi:10.1109/TPWRS.2009.2021226

Deb, N., & Shahidehpour, S. M. (1990). Linear reactive power optimization in large power network using the decomposition approach. *IEEE Transactions on Power Systems*, *5*(2), 393–400.

Ghasemi, M., Ghavidel, S., Ghanbarian, M. M., & Habibi, A. (2014). A new hybrid algorithm for optimal reactive power dispatch problem with discrete and continuous control variables. *Applied Soft Computing*, *22*, 126–140. doi:10.1016/j.asoc.2014.05.006

Granada, M., Marcos, B., Rider, J., Mantovani, J. R. S., & Shahidehpour, M. (2012). A decentralized approach for optimal reactive power dispatch using Lagrangian decomposition method. *Electric Power Systems Research*, *89*, 148–156. doi:10.1016/j.epsr.2012.02.015

Granville, S. (1994). Optimal reactive dispatch through interior point methods. *IEEE Transactions on Power Systems*, *9*(1), 136–146. doi:10.1109/59.317548

Haiping, M., Xieyong, R., & Baogen, J. (2010). *Oppositional ant colony. optimization algorithm and its application to fault monitoring.* 2010 29th Chinese Control Conference (CCC).

Huang, C. M., & Huang, Y. C. (2012). Combined differential evolution algorithm and ant system for optimal reactive power dispatch. *Energy Procedia*, *14*, 1238–1243. doi:10.1016/j.egypro.2011.12.1082

Iba, K. (1994). Reactive power optimization by genetic algorithm. *IEEE Transactions on Power Systems*, *9*(2), 685–692. doi:10.1109/59.317674

Kawata, H. K., Fukuyama, Y., Takayama, S., & Nakanishi, Y. (2000). A particle swarm optimization for reactive power and voltage control considering voltage security assessment. *IEEE Transactions on Power Systems*, *15*(4), 1232–1239. doi:10.1109/59.898095

Khazali, A. H., & Kalantar, M. (2011). Optimal reactive power dispatch based on harmony search algorithm. *International Journal of Electrical Power & Energy Systems*, *33*(3), 684–692. doi:10.1016/j.ijepes.2010.11.018

Khorsandi, A., Alimardani, A., Vahidi, B., & Hosseinian, S. H. (2010). Hybrid shuffled frog leaping algorithm and Nelder–Mead simplex search for optimal reactive power dispatch. *IET Generation Transmission Distribution*, *5*(2), 249–256. doi:10.1049/iet-gtd.2010.0256

Lee, K., Park, Y., & Ortiz, J. (1985). A united approach to optimal real and reactive power dispatch. *IEEE Transactions on Power Apparatus and Systems*, *104*(5), 1147–1153. doi:10.1109/TPAS.1985.323466

Lee, K. Y., & Park, Y. M. (1995). Optimization method for reactive power planning by using a modified simple genetic algorithm. *IEEE Transactions on Power Systems*, *10*(4), 1843–1850. doi:10.1109/59.476049

Lee, K. Y., Park, Y. M., & Ortiz, J. L. (1984). Optimal real and reactive dispatch. *Electric Power Systems Research*, *7*(3), 201–212. doi:10.1016/0378-7796(84)90004-X

Li, Y., Cao, Y., Liu, Z., Liu, Y., & Jiang, Q. (2009). Dynamic optimal reactive power dispatch based on parallel particle swarm optimization algorithm. *Computers & Mathematics with Applications (Oxford, England)*, *57*(11-12), 1835–1842. doi:10.1016/j.camwa.2008.10.049

Liang, C. H., Chung, C. Y., Wong, K. P., & Dual, X. Z. (2007). Parallel optimal reactive power flow based on cooperative co-evolutionary differential evolution and power system decomposition. *IEEE Transactions on Power Systems*, *22*(1), 249–257. doi:10.1109/TPWRS.2006.887889

Lu, F. C., & Hsu, Y. Y. (1995). Reactive power/voltage control in a distribution substation using dynamic programming. *IEE Proceedings. Generation, Transmission and Distribution*, *142*(6), 639–645. doi:10.1049/ip-gtd:19952210

Ma, J. T., & Lai, L. L. (1996). Evolutionary programming approach to reactive power planning. *IEE Proceedings. Generation, Transmission and Distribution*, *143*(4), 365–370. doi:10.1049/ip-gtd:19960296

Mahadevan, K., & Kannan, P. S. (2010). Comprehensive learning particle swarm optimization for reactive power dispatch. *International Journal of Applied Soft Computing*, *10*(2), 641–652. doi:10.1016/j.asoc.2009.08.038

Mandal, B., & Roy, P. K. (2013). Optimal reactive power dispatch using quasi-oppositional teaching learning based optimization. *International Journal of Electrical Power & Energy Systems*, *53*, 123–134. doi:10.1016/j.ijepes.2013.04.011

Momeh, J. A., Guo, S. X., Oghuobiri, E. C., & Adapa, R. (1994). The quadratic interior point method solving the power system optimization problems. *IEEE Transactions on Power Systems*, *9*(3), 1327–1336. doi:10.1109/59.336133

Rahnamayan, S., Tizhoosh, H. R., & Salama, M. M. A. (2007). Quasi oppositional differential evolution. In *Proceeding of IEEE Congress on Evolutionary Computation CEC 2007*, (pp. 2229–2236). doi:10.1109/CEC.2007.4424748

Rahnamayan, S., Tizhoosh, H. R., & Salama, M. M. A. (2008). Opposition versus randomness in soft computing techniques. *Applied Soft Computing*, *8*(2), 906–918. doi:10.1016/j.asoc.2007.07.010

Ramesh, S., Kannan, S., & Baskar, S. (2012). An improved generalized differential evolution algorithm for multi-objective reactive power dispatch. *Engineering Optimization*, *44*(4), 391–405. doi:10.1080/0305215X.2011.576761

Rang-Mow, J., & Nanming, C. (1995). Application to the fast Newton Raphson economic dispatch and reactive power/voltage dispatch by sensitivity factors to optimal power flow. *IEEE Transactions on Energy Conversion*, *10*(2), 293–301. doi:10.1109/60.391895

Roy, P. K., Ghoshal, S. P., & Thakur, S. S. (2010). Biogeography-based Optimization for economic load dispatch problems. *Electric Power Components and Systems*, *38*(2), 166–181. doi:10.1080/15325000903273379

Roy, P. K., Ghoshal, S. P., & Thakur, S. S. (2010). Biogeography based optimization for multi-constraints optimal power flow with emission and non-smooth cost function. *Expert Systems with Applications*, *37*(12), 8221–8228. doi:10.1016/j.eswa.2010.05.064

Roy, P. K., Ghoshal, S. P., & Thakur, S. S. (2010). Multi-objective optimal power flow using biogeography-based optimization. *Electric Power Components and Systems*, *38*(12), 1406–1426. doi:10.1080/15325001003735176

Roy, P. K., Ghoshal, S. P., & Thakur, S. S. (2011). Optimal reactive power dispatch considering flexible AC transmission system devices using biogeography based optimization. *Electric Power Components and Systems*, *39*(8), 733–750. doi:10.1080/15325008.2010.541410

Roy, P. K., Ghoshal, S. P., & Thakur, S. S. (2012). Optimal VAR control for improvements in voltage profiles and for real power loss minimization using Biogeography Based Optimization. *International Journal of Electrical Power & Energy Systems*, *43*(1), 830–838. doi:10.1016/j.ijepes.2012.05.032

Shaw, B., Mukherjee, V., & Ghoshal, S. P. (2012). A novel opposition-based gravitational search algorithm for combined economic and emission dispatch problems of power systems. *International Journal of Electrical Power & Energy Systems*, *35*(1), 21–33. doi:10.1016/j.ijepes.2011.08.012

Simon, D. (2008). Biogeography-based optimization. *IEEE Transactions on Evolutionary Computation*, *12*(6), 702–713. doi:10.1109/TEVC.2008.919004

Swarup, K. S., Yoshimi, M., & Izui, Y. (1994). Genetic algorithm approach to reactive power planning in power systems. In *Proceedings of the 5th Annual Conference of Power and Energy Society*. IEEE. Available online: http://www.ee.washington.edu /research/pstca/

Tehzeeb, U. H. H., Zafar, R., Mohsin, S. A., & Latee, O. (2012). Reduction in power transmission loss using fully informed particle swarm optimization. *International Journal of Electrical Power & Energy Systems*, *43*(1), 364–368. doi:10.1016/j.ijepes.2012.05.028

The IEEE 118-Bus Test System. (n.d.). Available at: http://www.ee.washington.edu/ research/pstca/pf118/pg_tca118bus.htm

Titare, L. S., Singh, P., Arya, L. D., & Choube, S. C. (2014). Optimal reactive power rescheduling based on EPSED algorithm to enhance static voltage stability. *International Journal of Electrical Power & Energy Systems*, *63*, 588–599. doi:10.1016/j.ijepes.2014.05.078

Tizhoosh, H. (2005). Opposition-based learning: A new scheme for machine intelligence. In *Proceedings of the International Conference on Computational Intelligence for Modelling Control and Automation (CIMCA-2005)*. doi:10.1109/CIMCA.2005.1631345

Varadarajan, M., & Swarup, K. S. (2008). Differential evolution approach for optimal reactive power dispatch. *Applied Soft Computing, 8*(4), 1549–1561. doi:10.1016/j.asoc.2007.12.002

Varadarajan, M., & Swarup, K. S. (2008). Differential evolutionary algorithm for optimal reactive power dispatch. *International Journal of Electrical Power & Energy Systems, 30*(8), 435–441. doi:10.1016/j.ijepes.2008.03.003

Vasant, P. (2015). *Handbook of Research on Artificial Intelligence Techniques and Algorithms*. Hershey, PA: IGI Global; doi:10.4018/978-1-4666-7258-1

Vasant, P., Barsoum, N., & Webb, J. (2012). *Innovation in Power, Control, and Optimization: Emerging Energy Technologies*. Hershey, PA: IGI Global; doi:10.4018/978-1-61350-138-2

Vasant, P. M. (2013). *Meta-Heuristics Optimization Algorithms in Engineering, Business, Economics, and Finance*. Hershey, PA: IGI Global; doi:10.4018/978-1-4666-2086-5

Vasant, P. M. (2014). *Handbook of Research on Novel Soft Computing Intelligent Algorithms: Theory and Practical Applications*. Hershey, PA: IGI Global; doi:10.4018/978-1-4666-4450-2

Vlachogiannis, J. G., & Lee, K. Y. (2008). Quantum-inspired evolutionary algorithm for real and reactive power dispatch. *IEEE Transactions on Power Systems, 23*(4), 1627–1636. doi:10.1109/TPWRS.2008.2004743

Wenyin, G., & Zhihua, C. (2013). Differential evolution with ranking-based mutation operators. *IEEE Transaction on Cybernatics, 43*(6), 2066–2081. doi:10.1109/TCYB.2013.2239988 PMID:23757516

Wu, Q. H., & Ma, J. T. (1995). Power system optimal reactive power dispatch using evolutionary programming. *IEEE Transactions on Power Systems, 10*(3), 1243–1249. doi:10.1109/59.466531

Yan, W., Yu, J., Yu, D. C., & Bhattarai, K. (2006). A new optimal reactive power flow model in rectangular form and its solution by predictor corrector primal dual interior point method. *IEEE Transactions on Power Systems, 21*(1), 61–67. doi:10.1109/TPWRS.2005.861978

Yiqin, Z. (2010). *Optimal reactive power planning based on improved Tabu search algorithm. 2010 International Conference on Electrical and Control Engineering*. doi:10.1109/iCECE.2010.961

Zhang, C., Ni, Z., Wu, Z., & Gu, L. (2009). *A novel swarm model with quasi-oppositional particle*. 2009 International Forum on Information Technology and Applications. doi:10.1109/IFITA.2009.525

Zhang, X., Chen, W., Dai, C., & Cai, W. (2010). Dynamic multi-group self-adaptive differential evolution algorithm for reactive power optimization. *International Journal of Electrical Power & Energy Systems, 32*(5), 351–357. doi:10.1016/j.ijepes.2009.11.009

Zhao, B., Guo, C. X., & Cao, Y. J. (2005). A Multiagent-based particle swarm optimization approach for optimal reactive power dispatch. *IEEE Transactions on Power Systems, 20*(2), 1070–1078. doi:10.1109/TPWRS.2005.846064

ADDITIONAL READING

Abido, M. A., & Bakhashwain, J. M. (2005). Optimal VAR dispatch using a multiobjective evolutionary algorithm. *International Journal of Electrical Power & Energy Systems, 27*(1), 13–20. doi:10.1016/j.ijepes.2004.07.006

Dommel, H. W., & Tinney, W. F. (1968). Optimal power flow solutions. *IEEE Transactions on Power Systems and Apparatus., 87*(10), 1866–1876. doi:10.1109/TPAS.1968.292150

Huneault, M., & Galiana, F. D. (1991, May). A survey of the optimal power flow literature. *IEEE Transactions on Power Systems, 6*(2), 762–770. doi:10.1109/59.76723

Mansour, M. O., & Abdel-Rahman, T. M. (1984). Non-linear VAR optimization using decomposition and coordination. *IEEE Transactions on Power Apparatus and Systems, 103*(2), 246–255. doi:10.1109/TPAS.1984.318223

Momoh, J. A., El-Hawary, M. E., & Adapa, R. (1999). A review of selected optimal power flow literature to 1993. II. Newton, linear programming and interior point methods. *IEEE Transactions on Power Systems, 14*(1), 105–111. doi:10.1109/59.744495

KEY TERMS AND DEFINITIONS

Evolutionary Algorithm: It is a generic population based metaheuristic optimization algorithm.
Migration: The physical movement of species from one area to another.
Mutation: It is a permanent change of an organism, species etc.
Optimization: It is the selection of a best element from some set of available alternatives.

APPENDIX

A1 IEEE 30-Bus System

Table A1. Load data of IEEE-30 bus system

Bus no.	PD (MW)	QD (MVAR)	Bus no.	PD (MW)	QD (MVAR)	Bus no.	PD (MW)	QD (MVAR)
1	00.00	00.00	11	00.00	00.00	21	17.50	11.20
2	21.70	12.70	12	11.20	07.50	22	00.00	0.00
3	02.40	01.20	13	00.00	00.00	23	03.20	01.60
4	07.60	01.60	14	06.20	01.60	24	08.70	06.70
5	94.20	19.00	15	08.20	02.50	25	00.00	00.00
6	00.00	00.00	16	3.50	01.80	26	03.50	02.30
7	22.80	10.90	17	9.00	05.80	27	00.00	00.00
8	30.00	30.00	18	3.20	0.90	28	00.00	00.00
9	00.00	00.00	19	9.50	03.40	29	02.40	00.90
10	05.80	02.00	20	2.20	00.70	30	10.60	01.90

Table A2. Line data of IEEE-30 bus system

Line	From Bus	To Bus	Resistance (p.u.)	Reactance (p.u.)	Susceptance (p.u.)	Line	From Bus	To Bus	Resistance (p.u.)	Reactance (p.u.)	Susceptance (p.u.)
1	1	2	0.0192	0.0575	0.0264	22	15	18	0.1070	0.2185	0.0000
2	1	3	0.0452	0.1852	0.0204	23	18	19	0.0639	0.1292	0.0000
3	2	4	0.057	0.1737	0.0184	24	19	20	0.0340	0.0680	0.0000
4	3	4	0.0132	0.0379	0.0042	25	10	20	0.0936	0.2090	0.0000
5	2	5	0.0472	0.1983	0.0209	26	10	17	0.0324	0.0845	0.0000
6	2	6	0.0581	0.1763	0.0187	27	10	21	0.0348	0.0749	0.0000
7	4	6	0.0119	0.0414	0.0045	28	10	22	0.0727	0.1499	0.0000
8	5	7	0.0460	0.1160	0.0102	29	21	22	0.0116	0.0236	0.0000
9	6	7	0.0267	0.0820	0.0085	30	15	23	0.1000	0.2020	0.0000
10	6	8	0.0120	0.0420	0.0045	31	22	24	0.1150	0.1790	0.0000
11	6	9	0.0000	0.2080	0.0000	32	23	24	0.1320	0.2700	0.0000
12	6	10	0.0000	0.5560	0.0000	33	24	25	0.1885	0.3292	0.0000
13	9	11	0.0000	0.2080	0.0000	34	25	26	0.2544	0.3800	0.0000
14	9	10	0.0000	0.1100	0.0000	35	25	27	0.1093	0.2087	0.0000
15	4	12	0.0000	0.2560	0.0000	36	28	27	0.0000	0.3960	0.0000
16	12	13	0.0000	0.1400	0.0000	37	27	29	0.2198	0.4153	0.0000
17	12	14	0.1231	0.2559	0.0000	38	27	30	0.3202	0.6027	0.0000
18	12	15	0.0662	0.1304	0.0000	39	29	30	0.2399	0.4533	0.0000
19	12	16	0.0945	0.1987	0.0000	40	8	28	0.0636	0.2000	0.0214
20	14	15	0.2210	0.1997	0.0000	41	6	28	0.0169	0.0599	0.0065
21	16	17	0.0824	0.1923	0.0000						

A2 IEEE 57-Bus System

Table A3. Line data of IEEE-57 bus system

Line	From Bus	To Bus	Resistance (p.u.)	Reactance (p.u.)	Susceptance (p.u.)	Line	From Bus	To Bus	Resistance (p.u.)	Reactance (p.u.)	Susceptance (p.u.)
1	1	2	0.0083	0.028	0.129	41	7	29	0	0.0648	0
2	2	3	0.0298	0.085	0.0818	42	25	30	0.135	0.202	0
3	3	4	0.0112	0.0366	0.038	43	30	31	0.326	0.497	0
4	4	5	0.0625	0.132	0.0258	44	31	32	0.507	0.755	0
5	4	6	0.043	0.148	0.0348	45	32	33	0.0392	0.036	0
6	6	7	0.02	0.102	0.0276	46	34	32	0	0.953	0
7	6	8	0.0339	0.173	0.047	47	34	35	0.052	0.078	0.0032
8	8	9	0.0099	0.0505	0.0548	48	35	36	0.043	0.0537	0.0016
9	9	10	0.0369	0.1679	0.044	49	36	37	0.029	0.0366	0
10	9	11	0.0258	0.0848	0.0218	50	37	38	0.0651	0.1009	0.002
11	9	12	0.0648	0.295	0.0772	51	37	39	0.0239	0.0379	0
12	9	13	0.0481	0.158	0.0406	52	36	40	0.03	0.0466	0
13	13	14	0.0132	0.0434	0.011	53	22	38	0.0192	0.0295	0
14	13	15	0.0269	0.0869	0.023	54	11	41	0	0.749	0
15	1	15	0.0178	0.091	0.0988	55	41	42	0.207	0.352	0
16	1	16	0.0454	0.206	0.0546	56	41	43	0	0.412	0
17	1	17	0.0238	0.108	0.0286	57	38	44	0.0289	0.0585	0.002
18	3	15	0.0162	0.053	0.0544	58	15	45	0	0.1042	0
19	4	18	0	0.555	0	59	14	46	0	0.0735	0
20	4	18	0	0.43	0	60	46	47	0.023	0.068	0.0032
21	5	6	0.0302	0.0641	0.0124	61	47	48	0.0182	0.0233	0
22	7	8	0.0139	0.712	0.0194	62	48	49	0.0834	0.129	0.0048
23	10	12	0.0277	0.1262	0.0328	63	49	50	0.0801	0.128	0
24	11	13	0.0223	0.0732	0.0188	64	50	51	0.1386	0.22	0
25	12	13	0.0178	0.058	0.0604	65	10	51	0	0.0712	0
26	12	16	0.018	0.0813	0.0216	66	13	49	0	0.191	0
27	12	17	0.0397	0.179	0.0476	67	29	52	0.1442	0.187	0
28	14	15	0.0171	0.0547	0.0148	68	52	53	0.0762	0.0984	0
29	18	19	0.461	0.685	0	69	53	54	0.1878	0.232	0
30	19	20	0.283	0.434	0	70	54	55	0.1732	0.2265	0
31	21	20	0	0.7767	0	71	11	43	0	0.153	0
32	21	22	0.0736	0.117	0	72	44	45	0.0624	0.1242	0.004
33	22	23	0.009	0.0152	0	73	40	56	0	1.195	0
34	23	24	0.166	0.256	0.0084	74	56	41	0.553	0549	0
35	24	25	0	1.182	0	75	56	42	0.2125	0.354	0
36	24	25	0	1.23	0	76	39	57	0	1.355	0
37	24	26	0	0.0473	0	77	57	56	0.174	0.26	0
38	26	27	0.165	0.254	0	78	38	49	0.115	0.177	0.003
39	27	28	0.0618	0.0954	0	79	38	48	0.0312	0.0482	0
40	28	29	0.0418	0.0587	0	80	9	55	0	0.1205	0

Table A4. Load data of IEEE-57 bus system

Bus no.	PD (MW)	QD (MVA)	Bus no.	PD (MW)	QD (MVAR)	Bus no.	PD (MW)	QD (MVAR)	Bus no.	PD (MW)	QD (MVAR)
1	55	17	16	43	3	31	5.8	2.9	46	0	0
2	3	88	17	42	8	32	1.6	0.8	47	29.7	11.6
3	41	21	18	27.2	9.8	33	3.8	1.9	48	0	0
4	0	0	19	3.3	0.6	34	0	0	49	18	8.5
5	13	4	20	2.3	1	35	6	3	50	21	10.5
6	75	2	21	0	0	36	0	0	51	18	5.3
7	0	0	22	0	0	37	0	0	52	4.9	2.2
8	150	22	23	6.3	2.1	38	14	7	53	20	10
9	121	26	24	0	0	39	0	0	54	4.1	1.4
10	5	2	25	6.3	3.2	40	0	0	55	6.8	3.4
11	0	0	26	0	0	41	6.3	3	56	7.6	2.2
12	377	24	27	9.3	0.5	42	7.1	4.4	57	6.7	2
13	18	2.3	28	4.6	2.3	43	2	1			
14	10.5	5.3	29	17	2.6	44	12	1.8			
15	2	5	30	3.6	1.8	45	0	0			

A3 IEEE 118-Bus System

Table A5. Line data of IEEE-118 bus system

Line No.	From Bus	To Bus	R (pu)	X (pu)	B (pu)	Flow Limit (MW)	Line No.	From Bus	To Bus	R (pu)	X (pu)	B (pu)	Flow Limit (MW)
1	1	2	0.0303	0.0999	0.0254	175	56	40	41	0.0145	0.0487	0.01222	175
2	1	3	0.0129	0.0424	0.01082	175	57	40	42	0.0555	0.183	0.0466	175
3	4	5	0.00176	0.00798	0.0021	500	58	41	42	0.041	0.135	0.0344	175
4	3	5	0.0241	0.108	0.0284	175	59	43	44	0.0608	0.2454	0.06068	175
5	5	6	0.0119	0.054	0.01426	175	60	34	43	0.0413	0.1681	0.04226	175
6	6	7	0.00459	0.0208	0.0055	175	61	44	45	0.0224	0.0901	0.0224	175
7	8	9	0.00244	0.0305	1.162	500	62	45	46	0.04	0.1356	0.0332	175
8	8	5	0	0.0267	0	500	63	46	47	0.038	0.127	0.0316	175
9	9	10	0.00258	0.0322	1.23	500	64	46	48	0.0601	0.189	0.0472	175
10	4	11	0.0209	0.0688	0.01748	175	65	47	49	0.0191	0.0625	0.01604	175
11	5	11	0.0203	0.0682	0.01738	175	66	42	49	0.0715	0.323	0.086	175
12	11	12	0.00595	0.0196	0.00502	175	67	42	49	0.0715	0.323	0.086	175
13	2	12	0.0187	0.0616	0.01572	175	68	45	49	0.0684	0.186	0.0444	175
14	3	12	0.0484	0.16	0.0406	175	69	48	49	0.0179	0.0505	0.01258	175
15	7	12	0.00862	0.034	0.00874	175	70	49	50	0.0267	0.0752	0.01874	175
16	11	13	0.02225	0.0731	0.01876	175	71	49	51	0.0486	0.137	0.0342	175

continued on following page

Table A5. Continued

Line No.	From Bus	To Bus	R (pu)	X (pu)	B (pu)	Flow Limit (MW)	Line No.	From Bus	To Bus	R (pu)	X (pu)	B (pu)	Flow Limit (MW)
17	12	14	0.0215	0.0707	0.01816	175	72	51	52	0.0203	0.0588	0.01396	175
18	13	15	0.0744	0.2444	0.06268	175	73	52	53	0.0405	0.1635	0.04058	175
19	14	15	0.0595	0.195	0.0502	175	74	53	54	0.0263	0.122	0.031	175
20	12	16	0.0212	0.0834	0.0214	175	75	49	54	0.073	0.289	0.0738	175
21	15	17	0.0132	0.0437	0.0444	500	76	49	54	0.0869	0.291	0.073	175
22	16	17	0.0454	0.1801	0.0466	175	77	54	55	0.0169	0.0707	0.0202	175
23	17	18	0.0123	0.0505	0.01298	175	78	54	56	0.00275	0.00955	0.00732	175
24	18	19	0.01119	0.0493	0.01142	175	79	55	56	0.00488	0.0151	0.00374	175
25	19	20	0.0252	0.117	0.0298	175	80	56	57	0.0343	0.0966	0.0242	175
26	15	19	0.012	0.0394	0.0101	175	81	50	57	0.0474	0.134	0.0332	175
27	20	21	0.0183	0.0849	0.0216	175	82	56	58	0.0343	0.0966	0.0242	175
28	21	22	0.0209	0.097	0.0246	175	83	51	58	0.0255	0.0719	0.01788	175
29	22	23	0.0342	0.159	0.0404	175	84	54	59	0.0503	0.2293	0.0598	175
30	23	24	0.0135	0.0492	0.0498	175	85	56	59	0.0825	0.251	0.0569	175
31	23	25	0.0156	0.08	0.0864	500	86	56	59	0.0803	0.239	0.0536	175
32	26	25	0	0.0382	0	500	87	55	59	0.04739	0.2158	0.05646	175
33	25	27	0.0318	0.163	0.1764	500	88	59	60	0.0317	0.145	0.0376	175
34	27	28	0.01913	0.0855	0.0216	175	89	59	61	0.0328	0.15	0.0388	175
35	28	29	0.0237	0.0943	0.0238	175	90	60	61	0.00264	0.0135	0.01456	500
36	30	17	0	0.0388	0	500	91	60	62	0.0123	0.0561	0.01468	175
37	8	30	0.00431	0.0504	0.514	175	92	61	62	0.00824	0.0376	0.0098	175
38	26	30	0.00799	0.086	0.908	500	93	63	59	0	0.0386	0	500
39	17	31	0.0474	0.1563	0.0399	175	94	63	64	0.00172	0.02	0.216	500
40	29	31	0.0108	0.0331	0.0083	175	95	64	61	0	0.0268	0	500
41	23	32	0.0317	0.1153	0.1173	140	96	38	65	0.00901	0.0986	1.046	500
42	31	32	0.0298	0.0985	0.0251	175	97	64	65	0.00269	0.0302	0.38	500
43	27	32	0.0229	0.0755	0.01926	175	98	49	66	0.018	0.0919	0.0248	500
44	15	33	0.038	0.1244	0.03194	175	99	49	66	0.018	0.0919	0.0248	500
45	19	34	0.0752	0.247	0.0632	175	100	62	66	0.0482	0.218	0.0578	175
46	35	36	0.00224	0.0102	0.00268	175	101	62	67	0.0258	0.117	0.031	175
47	35	37	0.011	0.0497	0.01318	175	102	65	66	0	0.037	0	500
48	33	37	0.0415	0.142	0.0366	175	103	66	67	0.0224	0.1015	0.02682	175
49	34	36	0.00871	0.0268	0.00568	175	104	65	68	0.00138	0.016	0.638	500
50	34	37	0.00256	0.0094	0.00984	500	105	47	69	0.0844	0.2778	0.07092	175
51	38	37	0	0.0375	0	500	106	49	69	0.0985	0.324	0.0828	175
52	37	39	0.0321	0.106	0.027	175	107	68	69	0	0.037	0	500
53	37	40	0.0593	0.168	0.042	175	108	69	70	0.03	0.127	0.122	500
54	30	38	0.00464	0.054	0.422	175	109	24	70	0.00221	0.4115	0.10198	175
55	39	40	0.0184	0.0605	0.01552	175	110	70	71	0.00882	0.0355	0.00878	175
111	24	72	0.0488	0.196	0.0488	175	149	82	96	0.0162	0.053	0.0544	175

continued on following page

Table A5. Continued

Line No.	From Bus	To Bus	R (pu)	X (pu)	B (pu)	Flow Limit (MW)	Line No.	From Bus	To Bus	R (pu)	X (pu)	B (pu)	Flow Limit (MW)
112	71	72	0.0446	0.18	0.04444	175	150	94	96	0.0269	0.0869	0.023	175
113	71	73	0.00866	0.0454	0.01178	175	151	80	97	0.0183	0.0934	0.0254	175
114	70	74	0.0401	0.1323	0.03368	175	152	80	98	0.0238	0.108	0.0286	175
115	70	75	0.0428	0.141	0.036	175	153	80	99	0.0454	0.206	0.0546	200
116	69	75	0.0405	0.122	0.124	500	154	92	100	0.0648	0.295	0.0472	175
117	74	75	0.0123	0.0406	0.01034	175	155	94	100	0.0178	0.058	0.0604	175
118	76	77	0.0444	0.148	0.0368	175	156	95	96	0.0171	0.0547	0.01474	175
119	69	77	0.0309	0.101	0.1038	175	157	96	97	0.0173	0.0885	0.024	175
120	75	77	0.0601	0.1999	0.04978	175	158	98	100	0.0397	0.179	0.0476	175
121	77	78	0.00376	0.0124	0.01264	175	159	99	100	0.018	0.0813	0.0216	175
122	78	79	0.00546	0.0244	0.00648	175	160	100	101	0.0277	0.1262	0.0328	175
123	77	80	0.017	0.0485	0.0472	500	161	92	102	0.0123	0.0559	0.01464	175
124	77	80	0.0294	0.105	0.0228	500	162	101	102	0.0246	0.112	0.0294	175
125	79	80	0.0156	0.0704	0.0187	175	163	100	103	0.016	0.0525	0.0536	500
126	68	81	0.00175	0.0202	0.808	500	164	100	104	0.0451	0.204	0.0541	175
127	81	80	0	0.037	0	500	165	103	104	0.0466	0.1584	0.0407	175
128	77	82	0.0298	0.0853	0.08174	200	166	103	105	0.0535	0.1625	0.0408	175
129	82	83	0.0112	0.03665	0.03796	200	167	100	106	0.0605	0.229	0.062	175
130	83	84	0.0625	0.132	0.0258	175	168	104	105	0.00994	0.0378	0.00986	175
131	83	85	0.043	0.148	0.0348	175	169	105	106	0.014	0.0547	0.01434	175
132	84	85	0.0302	0.0641	0.01234	175	170	105	107	0.053	0.183	0.0472	175
133	85	86	0.035	0.123	0.0276	500	171	105	108	0.0261	0.0703	0.01844	175
134	86	87	0.02828	0.2074	0.0445	500	172	106	107	0.053	0.183	0.0472	175
135	85	88	0.02	0.102	0.0276	175	173	108	109	0.0105	0.0288	0.0076	175
136	85	89	0.0239	0.173	0.047	175	174	103	110	0.03906	0.1813	0.0461	175
137	88	89	0.0139	0.0712	0.01934	500	175	109	110	0.0278	0.0762	0.0202	175
138	89	90	0.0518	0.188	0.0528	500	176	110	111	0.022	0.0755	0.02	175
139	89	90	0.0238	0.0997	0.106	500	177	110	112	0.0247	0.064	0.062	175
140	90	91	0.0254	0.0836	0.0214	175	178	17	113	0.00913	0.0301	0.00768	175
141	89	92	0.0099	0.0505	0.0548	500	179	32	113	0.0615	0.203	0.0518	500
142	89	92	0.0393	0.1581	0.0414	500	180	32	114	0.0135	0.0612	0.01628	175
143	91	92	0.0387	0.1272	0.03268	175	181	27	115	0.0164	0.0741	0.01972	175
144	92	93	0.0258	0.0848	0.0218	175	182	114	115	0.0023	0.0104	0.00276	175
145	92	94	0.0481	0.158	0.0406	175	183	68	116	0.00034	0.00405	0.164	500
146	93	94	0.0223	0.0732	0.01876	175	184	12	117	0.0329	0.14	0.0358	175
147	94	95	0.0132	0.0434	0.0111	175	185	75	118	0.0145	0.0481	0.01198	175
148	80	96	0.0356	0.182	0.0494	175	186	76	118	0.0164	0.0544	0.01356	175

Table A6. Load data of IEEE-118 bus system

Bus no.	PD (MW)	QD (MVAR)	Bus no.	PD (MW)	QD (MVAR)	Bus no.	PD (MW)	QD (MVAR)	Bus no.	PD (MW)	QD (MVAR)
1	54.14	8.66	31	45.65	28.66	61	0.00	0.00	90	78	42
2	21.23	9.55	32	62.63	24.42	62	77	14	91	0.00	0.00
3	41.4	10.62	33	24.42	9.55	63	0.00	0.00	92	65	10
4	31.85	12.74	34	62.63	27.6	64	0.00	0.00	93	12	7
5	0.00	0.00	35	35.03	9.55	65	0.00	0.00	94	30	16
6	55.2	23.35	36	32.91	18.05	66	39	18	95	42	31
7	20.17	2.12	37	0.00	0.00	67	28	7	96	38	15
8	0.00	0.00	38	0.00	0.00	68	0.00	0.00	97	15	9
9	0.00	0.00	39	27	11	69	0.00	0.00	98	34	8
10	0.00	0.00	40	20	23	70	66	20	99	0.00	0.00
11	74.31	24.42	41	37	10	71	0.00	0.00	100	37	18
12	49.89	10.62	42	37	23	72	0.00	0.00	101	22	15
13	36.09	16.99	43	18	7	73	0.00	0.00	102	5	3
14	14.86	1.06	44	16	8	74	68	27	103	23	16
15	95.54	31.85	45	53	22	75	47	11	104	38	25
16	26.54	10.62	46	28	10	76	68	36	105	31	26
17	11.68	3.18	47	34	0	77	61	28	106	43	16
18	63.69	36.09	48	20	11	78	71	26	107	28	12
19	47.77	26.54	49	87	30	79	39	32	108	2	1
20	19.11	3.18	50	17	4	80	130	26	109	8	3
21	14.86	8.49	51	17	8	81	0.00	0.00	110	39	30
22	10.62	5.31	52	18	5	82	54	27	111	0.00	0.00
23	7.43	3.18	53	23	11	83	20	10	112	25	13
24	0.00	0.00	54	113	32	84	11	7	113	0.00	0.00
25	0.00	0.00	55	63	22	85	24	15	114	8.49	3.18
26	0.00	0.00	56	84	18	86	21	10	115	23.35	7.43
27	65.82	13.8	57	12	3	87	0.00	0.00	116	0.00	0.00
28	18.05	7.43	58	12	3	88	48	10	117	21.23	8.49
29	25.48	4.25	59	277	113	89	0.00	0.00	118	33	15
30	0.00	0.00	60	78	3						

Chapter 12
Application of Adaptive Tabu Search Algorithm in Hybrid Power Filter and Shunt Active Power Filters:
Application of ATS Algorithm in HPF and APF

Saifullah Khalid
IEEE, India

ABSTRACT

A novel hybrid series active power filter to eliminate harmonics and compensate reactive power is presented and analyzed. The proposed active compensation technique is based on a hybrid series active filter using ATS algorithm in the conventional Sinusoidal Fryze voltage (SFV) control technique. This chapter discusses the comparative performances of conventional Sinusoidal Fryze voltage control strategy and ATS-optimized controllers. ATS algorithm has been used to obtain the optimum value of Kp and Ki. Analysis of the hybrid series active power filter system under non-linear load condition and its impact on the performance of the controllers is evaluated. MATLAB/Simulink results and Total harmonic distortion (THD) shows the practical viability of the controller for hybrid series active power filter to provide harmonic isolation of non-linear loads and to comply with IEEE 519 recommended harmonic standards. The ATS-optimized controller has been attempted for shunt active power filter too, and its performance has also been discussed in brief.

INTRODUCTION

Over the years, there has been an incessant proliferation of nonlinear type of loads due to the rigorous use of power electronic control in all branches of industry as well as by the ordinary consumers of electric energy (Aredes, 1991; Moran, 1995; Maniya, 2013; Ruminot, 2006). This solid state control of AC power using thyristors and other semiconductor switches is extensively used to feed controlled electric power

DOI: 10.4018/978-1-4666-9755-3.ch012

to electrical loads, for example adjustable speed drives (ASD's) furnaces, computer power supplies, etc. Such controllers are also employed in HVDC systems and renewable electrical power generation (Vijay, 1996). These days, the power electronic converters are able of processing vast amount of power, and due to their advantages such as improved efficiency and ease of control, have caused a spectacular increase in the number of power electronic loads in the industry/system.

Problem Statement, Definition and Its Description

Unfortunately, power electronic loads have an intrinsically nonlinear nature, and they, therefore, draw a distorted current from the mains supply. Specifically, they draw non-sinusoidal current, which is not in proportion to the sinusoidal voltage. Consequently, the utility supplying these loads has to offer large reactive volt-amperes. Also, the harmonics produced by the load pollutes it. As nonlinear loads, these solid-state converters draw harmonic and reactive power part of the current from AC mains. Figure 1 shows the distorted currents passing through the linear, series impedance of the power distribution system.

The problem, which has been seen while using non-linear loads, is injected harmonics, reactive power burden, unbalance and excessive neutral current. Due to them, also to poor power factor, system's efficiency has also been reduced drastically. They also cause disturbance to other consumers and interference in nearby communication networks, excessive heating in transmission and distribution equipment, errors in metering and malfunctioning of utility relays. The inflict able tariffs levied by utilities against excessive VARs, and the threat of stricter harmonics standards have led to extensive surveys to quantify the problems associated with electric power networks having nonlinear loads. i.e. the load compensation techniques for power quality improvement.

To remove these problems, passive filters has been formulated, but they are designed to compensate selected harmonic components, so some harmonics are always there. The passive filter not only affects inverter harmonic injection but forces on the harmonics created by a joined nonlinear load. There are numerous techniques for controlling harmonic current flow, for example, DC ripple injection, harmonic current injection, series and parallel active filter systems, magnetic flux compensation. Passive harmonic filters are frequently employed to decrease current distortion and voltage harmonics in distributed generation systems.

The shunt passive filters expose lower impedance at tuned harmonic frequency than the source impedance so that reduced harmonic currents flow into the source (Rubén, 2005). However, the filtering

Figure 1. Harmonic voltages at the PCC due harmonic currents

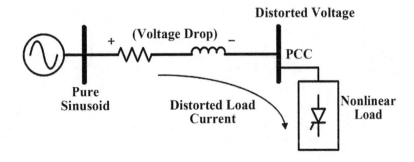

characteristics of shunt passive filter are decided by the impedance ratio of the source and the shunt passive filter. However, in practical application these passive second order filters show the following disadvantages:

- The source impedance powerfully affects filtering characteristics.
- Since both the harmonic and the fundamental current components flow into the filter, the capacity of the filter must be rated by taking into account both currents.
- When the harmonic current components increase, the filter can be overloaded.
- The parallel resonance between the passive filter and the power system causes amplification of harmonic currents on the source side at a particular frequency.
- The passive filter may fall into series resonance with the power system so that voltage distortion produces excessive harmonic currents flowing into the passive filter. Figure 2 presents common types of passive filters and their configurations.

Compensation methods based on active and passive filters to remove current harmonics, voltage harmonics and to compensate reactive power have already been presented. Shunt, Series, and hybrid active filter topologies have been conferred and demonstrated to be a feasible alternative for industrial compensation. Although passive filters LC are most frequently used to compensate current harmonics, it is well recognized that they are not the most excellent solution, since they generate resonance problems, affect voltage regulation, and produce high inrush currents. Shunt active filter is a better option for current harmonic and reactive power compensation; however its application in high power load compensation is due to power semiconductors restrictions (Khalid, 2013; Khalid; 2012). In this chapter, these problems have been resolved using series, hybrid and shunt active power filter using Sinusoidal Fryze, sinusoidal current control strategy along with the application of Adaptive Tabu search algorithm for different non-linear loads.

Related Works

Hybrid active power filter has been a favorite issue for research and application due to its low cost as compared to universal power quality conditioner. There have been many researchers, who have worked in the area of shunt active, series active, passive, and hybrid power filters. Few of them, related to this chapter have been discussed in brief. Maurício Aredes (2003) has applied SFV control for series and

Figure 2. Common types of different passive filters and their configurations

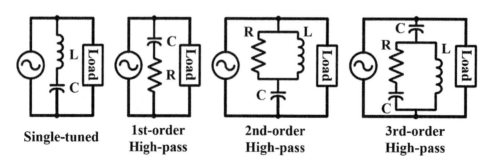

shunt both filters, in which the control strategy judges the presence of harmonics and unbalances concurrently in the system voltage and load current. The series control strategy discussed offers compensated voltages in such a way that the voltage distributed to the most critical load is sinusoidal as well balanced. Bhim Singh (2001) has proposed indirect current control approach in terms of its effortlessness and effectiveness for the operation of a series APF. They have shown perfectly that with the indirect current control adequately compensates harmonics for the voltage fed type harmonics producing the load. It is that the series active filter is capable to self-support its DC bus through the control under varying load connected to the rectifier (Pan, 2005). D. Puansdownreong (2011) presents a new Adaptive Tabu Search (ATS) method. In which they have added back-tracking and adaptive search radius mechanisms as a contrast to conventional Tabu Search (TS) method to attain faster and more efficient search. The primary application of the ATS method discussed in this article is the identification of system models.

The series active power filter works as a voltage controlled source while the shunt active approach acts as a current controlled source (le Roux, 2002; Abraham, 2003; Habrouk, 2002; Marco, 2004; Niraj, 2013; Sangsun Kim, 2002).The series approach compensates for voltage distortion, unbalances, and regulation (sags and swells). Another well-known topology is the hybrid filter, which employs a combination of active and passive filters and is an excellent and successful alternative for current harmonics compensation.

Now days, advance, soft computing techniques are used widely in the automatic control system or for optimization of the system applied. Some of them are such as adaptive tabu search, optimization of active power filter using Genetic algorithm, power loss minimization using particle swarm optimization, neural network control applied in both machinery and filter devices (Avik, 2011; Babak, 2010; Bhim, 2007; Cebi, 2012; Gaby, 2010; Elamvazuthi, 2011; Jiang, 2009; Vasant, 2012; Vasant, 2010; Polprasert, 2013; Pramod, 2009; Radha, 2010; Dehini, 2009; Saifullah, 2012;Turan,2012; Vasant, 2012; Vasant, 2013; Lin, 2011, Vasant, 2012; Zelinka, 2013).

Nomenclature

ATS: Adaptive Tabu Search algorithm.
TS: Tabu Search.
APF: Active Power Filter.
SFV: Sinusoidal Fryze voltage control.
THD: Total Harmonic Distortion.
HPF: Hybrid-Series Power filter.
PWM: Pulse width modulation.
HVDC: High-Voltage DC Transmission System.
PI: Proportional Integral Controller.
OF: objective function.
ATS-SFV: Adaptive Tabu Search algorithm optimized Sinusoidal Fryze Voltage Control Strategy.
PCC: Point of common coupling.
SI: Swarm Intelligence.
GA: Genetic Algorithm.
ANN: Artificial Neural Network.
SCC: Sinusoidal Current Control Strategy.
ATS-SCC: Adaptive Tabu Search algorithm optimized Sinusoidal Current Control Strategy.

V_{sa}, V_{sb}, V_{sc}: Source voltages.
V_a, V_b, V_c: Load Voltages.
V_{Ca}, V_{Cb}, V_{Cc}: Active filter Voltages.
$\dot{V}_c$: Active filter voltage as Controlled voltage source.
z_F: Equivalent impedance of passive filter.
$\dot{I}_{Lh}$: Load harmonic current.
$\dot{V}_{Sh}$: Source harmonic voltage.
z_s: Source impedance.

BACKGROUND

In 1983, when one of the first prototypes based on instantaneous reactive power theory was reported, active filters had been developed, The use of this method permits compensating independently the average or oscillating portions of the active (real) and reactive (imaginary) powers. One of the difficulties of the controllers based on the famous PQ Theory is the use of low-pass filters to separate the average and oscillating portions of powers. This power theory had been also challenged by Leszek S. Czarnecki for misinterpretation of properties of the power in three phase three-wire power systems using own theory "Currents' Physical Components" power theory or simply said "CPC power theory". His analysis shows that the result of the IRP p-q theory does not match with some regular interpretations of properties in the three-phase circuits. According to the defined p-q theory, the instantaneous reactive current is present in the supply current without checking that an unbalanced though purely resistive load. It also clearly shows that the values of the instantaneous powers p and q does not allow us to relate to the power phenomena in a three-phase unbalanced system with sinusoidal voltage. Likewise, the instantaneous identification of the power properties of the three-phase power system, as suggested by the IRP p-q theory is not possible.

Other control strategies also show similar drawbacks. Some compensation methods involve a variation on the proportional-integral compensation method described by N. Mendalek and K. Al-Haddad and a sliding mode compensation method described by N. Mendalek, K. Al-Haddad, and F. Fnaiech, and L.A. Dessaint. In the 30's of the last century Fryze (1932) suggested a set of active and non-active (reactive) power definitions in the time domain. From these concepts, Tenti (1986) built up a control strategy for shunt active filters that assurances compensated currents in the network that are sinusoidal even if the system voltage at the point of common coupling (PCC) already haves harmonics. Though, this control strategy does not guarantee balanced compensated currents if the system voltage itself is unbalanced (i.e. it contains a fundamental negative sequence component). This problem may be defeated by adding a positive-sequence voltage detector in the shunt active filter controller. This positive-sequence voltage detector will be accountable to determine control voltages that have the same magnitude and phase angle of the fundamental positive sequence component in the measured system voltages (Akagi, 1983; Akagi, 1984; Khalid, 2012; Watanabe, 1993). K. Chaijarurnudomrung (2011) has presented a controller design for a three-phase controlled rectifier using an adaptive tabu search. The averaging model of the power electronic system is utilized as an objective function as opposed to the model from software packages. Besides, this approach has been found suitable and flexible for electrical engineering to design the controller ofpower electronic systems with superior performances.

Recently, there is an increasing concern about the environment. The basic need to generate pollution-free energy has triggered considerable effort toward renewable energy (RE). Bull, S. R., Hassmann, K., Hammons, T. J. *et al.* discusses RE sources such as sunlight, the wind, flowing water and biomass offer the promise of clean and abundant energy. Maricar, N. M. *et al.* explains how they do not generate any greenhouse gasses and are inexhaustible. Solar energy, in particular, is especially attractive in a sunshine country like Malaysia. This energy is in DC form from photovoltaic (PV) arrays. It is converted into a more convenient alternating current (AC) power through an inverter system. Kim, S. et al., Komatsu, Y, and Wu, T. F. discusses the efforts that have been made to combine the APF with PV array. However, it appears that no attempt has been made to combine a hybrid APF with PV array.

Literature Survey

The key chapters that play a major role for the implementing HPF are described in the following paragraphs.

C.G.Terbobri, M.F. Saidon, M.S.Khanniche (2000): Primarily non-linear loads, which are mainly power electronics loads, cause power system harmonics. Injected current harmonics into the power system lead to distorted voltages at the point of common coupling (PCC). These distorted voltages result in poor power factor operation, and unacceptable current/voltage distortion levels. Harmonic filtering is one of the potential solutions, either by passive filters or by active filters. In this chapter trends in real-time APFs were discussed and experimental results of active power filtering using two real- time controlled, low costs, high costs, high-performance topologies were presented. In this chapter, the supply current detection method was used for the control of the APF converter.

Leobardo Cartes B., Sergio Horta M., Abraham Claudio S., Victor M., Cardenas G. (1998): This chapter presents design consideration of active power filter and a simple method to calculate reactive power components and current harmonics for single-phase non-linear loads. This allows compensation harmonic currents into the unity power. Simulation and experimental results are for non-linear loads.

Bimal K. Bose (1990): A hysteresis-band instantaneous current control PWM technique is popularity used because of its simplicity of implementation, inherent peak current limiting capability, and high-speed current control response. However, a current controller with a fixed hysteresis band has the disadvantage that the modulation frequency varies in a band and, as a result, generates anon-optimum current ripple in the load. This chapter describes an adaptive hysteresis band current control method where the band is by the system parameters to maintain the modulation frequency to be nearly constant. Systematic, analytical expressions of the hysteresis band are as a function of system parameters.

T. Sopapirm, K-N. Areerak, K-L. Areerak, and A. Srikaew (2011): A new artificial intelligence technique i.e. adaptive tabu search has been used to design the controller of a buck converter. The averaging model outcompeting from the DQ and generalized state-space averaging methods is employed to simulate the system during a searching process.

Maurício Aredes, Luís. F. C. Monteiro Jaime M. Miguel (2003): Sinusoidal Fryze control strategies, for series and shunt active filters have been described. The fundamentals of the PQ Theory was browbeaten and introduced into a minimization method, which jointly a vigorous synchronizing circuit allowed an improvement of a control strategy. It has compensation characteristics identical to the control strategy based on the PQ Theory; with lesser computational efforts to build it up. The controller computes the compensating currents that comprise all components that vary from the active portion of the fundamental positive sequence current of the load.

Hao Yi, Fang Zhuo, Yanjun Zhang, Yu Li, Wenda Zhan, Wenjie Chen, and Jinjun Liu (2014): A shunt active power filter (APF) with current detection on the source side is considered as a closed-loop system from the view of the whole power distribution system, which is expected with better harmonics filtering performance compared with the one with load current detection has been discussed.

Mohammed Qasim, Parag Kanjiya, and Vinod Khadkikar (2014): A phase-locking control scheme based on artificial neural networks (ANNs) for active power filters (APFs) has been discussed. The proposed phase locking has been achieved by estimating the fundamental supply frequency and by generating a phase-locking signal. The nonlinear-least-squares-based approach is modified to estimate the supply frequency. Finally, an adaptive-linear neuron- based scheme is proposed to extract the phase information of the supply voltage.

Mohammed Qasim, Vinod Khadkikar (2014): ADALINE and feed-forward MNN-based control algorithms realization for shunt APF has beendiscussed. A step-by-step procedure to implement these ANN-based techniques in MATLAB/Simulink environment is. Furthermore, a detailed analysis of the performance, limitation, and advantages of both methods has been described.

Deepak Somayajula, Mariesa L. Crow (2014): The concept of providing active/reactive power support and renewable intermittency smoothing to the distribution grid is by UCAPs. The UCAP is into the dc-link of the active power filter (APF) through a dc–dc converter. The dc–dc converter provides a stiff dc-link voltage that improves the performance of the grid tied inverter. Design and control of both the dc–AC inverter and the dc–dc converter are imperative in this regard and so discussed in detail.

Parag Kanjiya, Vinod Khadkikar and Hatem H. Zeineldin, (2015): An optimal algorithm to control a three-phase four-wire shunt APF under non-ideal supply conditions has been discussed. The optimization problem is aiming at maximizing the power factor subject to current harmonic constraints as per IEEE Std. 519 has been formulated and solved mathematically using the Lagrangian formulation. The proposed algorithm avoids the use of complex iterative optimization techniques and thus is simple to implement and has fast dynamic response.

Zhi-Xiang Zou, Keliang Zhou, Zheng Wang, and Ming Cheng (2015): A fractional- order RC (FORC) strategy at a fixed sampling rate is proposed to deal with any periodic signal of variable frequency, where a Lagrange-interpolation-based fractional delay (FD) filter is used to approximate the fractional delay items.The proposed FORC offers fast online tuning of the FD and the quick update of the coefficients, and then provides APFs with a simple but very accurate real-time frequency-adaptive control solution to the elimination of harmonic distortions under grid frequency variations.

An Algorithm that is used to optimize the switching angle by solving the non-linear equation is Metaheuristic Algorithm. Metaheuristic is a scientific method to sort out problems and to pinpoint it. The Metaheuristic algorithms are classified as Genetic Algorithm, bacterial foraging algorithm, Particle Swarm Optimization, Ant Colony Optimization, Bee Algorithm, etc.

Bee Algorithm is a meta-heuristic optimization algorithm based on the natural foraging behavior of honeybees to find the optimal solution. The Bees Algorithm is a new population-based search algorithm, first developed in 2005 by Pham DT and Karaboga. D independently. A bee colony consists of three kinds of basic bees: employed bees, onlooker bees, and scout bees. Employed bees carry information about the place and amount of nectar in a particular food source. They transfer the information to onlooker bees with dance in the hive.

We can also find interesting Swarm Intelligence (SI) based continuous optimization approaches. Socha, K. et al. suggested. Ant colony optimization for continuous domains, whereas Kennedy, J., Eberhart, R.C. have discussed Particle swarm optimization. Artificial Bee Colony Optimization has been done and explained by D. Karaboga, B. Basturk (2007) for numerical function optimization. Y. Lui, K.M. Passino (2002), have applied Social Foraging Bacteria for Distributed Optimization. Later on, other Metaheuristic models have also been considered to deal with continuous optimization problems. Chang Wook Ahn, R. S. Ramakrishna (2008) have done the research on the scalability of the real-codedBayesian optimization algorithm. Other different metaheuristics approaches like Simulated Annealing (Vanderbilt, D., Louie, S.G.), Variable Neighborhood Search (N. Mladenovic, M. Drazic, V. Kovacevic-Vujcic, M. Cangalovi), Scatter Search (F. Herrera, M. Lozano, D. Molina), Continuous GRASP (M.J. Hirsch, P.M. Pardalos, M.G.C. Resende), Central Force Optimization (R.A. Formato) has also been used for optimizing the system.

KEY TERMS (KT)

Following are the key terms, which will help the readers for understanding this chapter and related issues.

Power Quality: The term power quality is rather a general concept. Broadly, it may be defined as the provision of voltages and System design so that the user of electric power can utilized electric energy from the distribution system successfully, without interference on interruption.

Power quality is defined in the IEEE 100 Authoritative Dictionary of IEEE Standard Terms as the concept of powering and grounding electronic equipment in a manner that is suitable to the operation of that equipment and compatible with the premise wiring system and other connected equipment Utilities may want to define power quality as reliability.

Power Quality Problems: A recent survey of Power Quality experts has shown the concern that 50% of all Power Quality problems are related to grounding, ground current, ground bonds, ground loops, and neutral to ground voltages, or other ground associated issues. The following indications are pointers of Power Quality Problems:

- Piece of equipment misoperates at the same time of day.
- A circuit breakers trip deprived of being overloaded.
- Equipment fails in a thunderstorm.
- Automated systems halt for no apparent reason.
- Electronic systems fail or fail to operate on a common basis.
- Electronic systems work in one location but not in another location.

The commonly used terms those describe the parameters of electrical power that describe or measure power quality are Voltage sags, Voltage variations, Interruptions Swells, Brownouts, Blackouts, Voltage imbalance, Distortion, Harmonics, Harmonic resonance, Inter-harmonics, Noise, Notching, Spikes (Voltage), Impulse, Ground noise, Common mode noise, Crest factor, Critical load, Dropout, Electromagnetic compatibility, Fault, Flicker, Ground, Raw power, Ground loops, Clean ground, Transient, Voltage fluctuations, Dirty power, Under voltage, Momentary interruption, Over voltage, THD, Nonlinear load, Triplens, Voltage regulation, Voltage dip, Blink, Oscillatory Transient etc.

Power Quality Standard: Power quality is a worldwide issue, and keeping related standards current is a never-ending task. It typically takes years to push changes through the process.

Most of the ongoing work by the IEEE in harmonic standards development has shifted to modifying Standard 519-1992.

IEEE 519: IEEE 519-1992, Recommended Practices and Requirements for Harmonic Control in Electric Power Systems, established limits on harmonic currents and voltages at the point of common coupling (PCC), or point of metering. The limits of IEEE 519 are intended to:

1. Assure that the electric utility can deliver relatively clean power to all of its customers;
2. Assure that the electric utility can protect its electrical equipment from overheating, loss of life from excessive harmonic currents, and excessive voltage stress due to excessive harmonic voltage. Each point from IEEE 519 lists the limits for harmonic distortion at the point of common coupling (PCC) or metering point with the utility. The voltage distortion limits are 3% for individual harmonics and 5% THD.

In this chapter, the system has been polluted due to the non-linear load connected and for solving this problem; a passive filter has been connected, which was unable to remove the harmonics effectively. However, series active filter has also been found unable to compensate current harmonics effectively. So, for complete compensation, we have to reach a solution that employs both series and passive filter i.e. hybrid power filter.

In this chapter, we have presented an active compensation scheme that is based on Sinusoidal Fryze voltage control & optimized using adaptive tabu search algorithm, has been implemented in a hybrid series active power filter. In this chapter, ATS algorithm searches the PI controller parameters and then with these values, sinusoidal Fryze voltage control (SFV) strategy based voltage controller, gives the best output performance under non-linear load connected. Its results have been compared with the conventional sinusoidal Fryze voltage control strategy for the series filter. Results justified their effectiveness. Analysis of the results has been done on the basis of Total harmonic distortion (THD).It has been tried to prove that this new scheme gives better results and improvise the conventional technique with the application ARS algorithm.

Last but not the least, there has been a brief discussion of the future aspect of using ATS algorithm with shunt active power filter using some simulation results. The simulation has been done on a computer of configurations of AMD Athlon™ X2 Dual-core QL-60 processor with 3GB RAMusing MATLAB/ Simulink.

The chapter has been organized in the following manner. The HPF configuration and the load under consideration are in Section II. Adaptive Tabu search algorithm and other optimization algorithms have been discussed in Section III. The control algorithm for HPF is in Section IV. Optimization using adaptive Tabu search has been presented in Section V. Comparative evaluation using MATLAB/ Simulink results are discussed in Section V. Discussions has been done in Section VI and finally Section VII concludes the chapter.

SYSTEM DESCRIPTION

As shown in Figure 3, Hybrid Series Power Filter improves the power quality and compensates the harmonics in the system (Fujita, 2000). The series active filter ideally behaves as a controlled voltage source in such a way that the load voltage will have only positive-sequence at the fundamental frequency component.

Figure 3. Hybrid series active power filter

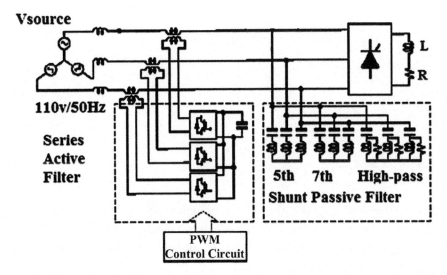

The voltages on the sources are given by V_{sa}, V_{sb}, and V_{sc}. Conversely, the relation among the source voltage, the load voltage, and the active filter voltage is given by

$$\begin{bmatrix} V_{Sa} \\ V_{Sb} \\ V_{Sc} \end{bmatrix} = \begin{bmatrix} V_a \\ V_b \\ V_c \end{bmatrix} \begin{bmatrix} V_{Ca} \\ V_{Cb} \\ V_{Cc} \end{bmatrix} \quad (1)$$

The fundamental series active filter voltages are synthesized by three single-phase converters with a common dc capacitor. The reference voltage for these converters is calculated by the "PWM Control circuit block which has active filter controller" shown in Figure 3.

We know that thyristor rectifier shown behaves like a current source; the voltage source can be split into its fundamental voltage source and a harmonic voltage source. The series active filter behaves like the controlled voltage source $\dot{V}_c$ and the shunt passive filter can be represented by equivalent impedance z_F. As per the basic compensation principle, the series active filter should synthesize the active impedance presenting zero impedance at the fundamental frequency and a high resistance k at the source or load harmonic frequencies.

The harmonic current running in the source is dependent on the both the load harmonic current $\dot{I}_{Lh}$ and the source harmonic voltage $\dot{V}_{Sh}$. z_s presents the source impedance of the system.

It is given by

$$i_{sh} = \frac{Z_F}{Z_F + Z_S + k} I_{Lh} + \frac{\dot{V}_{Sh}}{Z_F + Z_S + k} \quad (2)$$

where

$$i_{sh} \cong 0 \quad i_{fk \gg z_s, z_F} \tag{3}$$

The output voltage of the series active filter is given by;

$$\dot{V}_c = k\dot{I}_{sh} = k\frac{z_f \dot{I}_{Lh} + \dot{V}_{sh}}{Z_F + Z_s + k} \tag{4}$$

$$\dot{V}_c \cong z_F \dot{I}_{Lh} + \dot{V}_{sh} \quad if \quad k \gg z_s, z_F \tag{5}$$

Equation (5) shows undoubtedly that the voltage rating of the series active filter $\dot{V}_c$ is given by two factors; the first term in the right hand side of the equation, which is inversely proportional to the shunt passive filter's quality factors, and the second term are equal to the source harmonic voltage.

The harmonic voltage on the shunt passive filter is given by

$$\dot{V}_{Fh} = \frac{Z_s + k}{Z_F + Z_s + k} Z_F \dot{I}_{Lh} + \frac{Z_F}{Z_F + Z_s + k} \dot{V}_{sh} \tag{6}$$

$$\dot{V}_{Fh} \cong -z_F \dot{I}_{Lh} \quad if \quad k \gg z_s, z_F \tag{7}$$

The above equations show that by choosing k >>zS, z_F, no source harmonic voltage shows the shunt passive filter. Two harmonic current flows should be examined: one from the load to the source and the other from the source to the load.

ADAPTIVE TABU SEARCH ALGORITHM AND OTHER OPTIMIZATION ALGORITHMS

Since power electronics technology remains to develop, there is a growing need for automated synthesis that starts with a high-level statement of the anticipated behavior and optimizes the circuit component values for fulfilling the operational requirements. Numerous proposals for analog circuit design automation have been emerged in the early 1970s. These methods incorporate heuristics, knowledge bases, simulated annealing, and other algorithms. Traditional optimization techniques, such as the gradient methods, and hill-climbing techniques, have been applied. However, they might be subject to becoming stuck into local minima, leading to suboptimal parameter values, and therefore having a restraint of operating in large, multimodal, and noisy spaces.

Genetic algorithm (GA), which is one of the metaheuristic optimization methods, has been shown to be an actual way to find answers near to the overall optimum. It is less reliant on upon the initial guess, and has been applied to optimize detailed parts in the power electronic systems, for example system controllers, modulation schemes, optimization of component values, and specific applications, like battery chargers. Though, the values optimized by GAs for the circuits are occasionally not readily available but require postfabrication. For example, resistors are contrived with discrete values. It is from time to time necessary to associate several resistors in series and/or parallel so as to compose the values optimized by GAs. Furthermore, inevitable component tolerance would mark the actual component value vary from its nominal values to some amount. Designers can select the nearest discrete values accessible for the values optimized by the existing algorithms, which usually treat circuit optimization as finding the solution for a mathematical function and do not take component tolerance into consideration—an significant issue in practical circuit design. If the values used in the actual circuit are dissimilar from the optimized values, the tolerance range will be transformed, and the circuit performance will be different from the expected one. Thus, it is important to search the available values of the discrete-valued components.

Lately, a new metaheuristic optimization method named ant colony optimization (ACO) has been also been proposed. The technique simulates the behaviors of ants in finding pathsfrom the colony to food. It is a multi-agent approach for solving combinatorial optimization problems, like traveling salesman problems, data mining, network routing, and controller design. Recently, there is a growing application of ATS Algorithm to electrical engineering,

Adaptive Tabu Search (ATS) is a modified version of original tabu search formula for combinational optimization problem suggested by Glover. This method is very useful and quite different & simple for solving nonlinear continuous optimization problems as a comparison to other optimization algorithms. The modification that has been added to the new version is discretized continuous search space, backtracking and adaptive radius.

The Tabu Search algorithm is an iterative process that finds out for the best solution by moving from a current solution to finding a better solution again and again. One of the most significant constituents that make the TS method dissimilar from other searching methods is its Tabu list. It saves the history of paths. This list is utilized as information to find directions of a new action. This new action should guide the search to the improved local optimum solution and at last to the optimum global one. Features of a Tabu list are changed to go with each different problem. With the intention of get better the performance of the Tabu Search algorithm, it has been modified and called as adaptive Tabu search (ATS). This new ATS consists of two new steps, namely back- tracking and adaptive radius, different from the conventional Tabu Search algorithm. The back-tracking process permits the system to go back and look up the preceding solutions that have been already searched. The superior solution is then chosen from the current, and the previous solutions, and the adaptive radius process reduces the search area throughout the searching process. This added attributes continues until the close to the global solution is found.

In this chapter, the GA is applied to determine the value of inductor filter used in active shunt filter. GA will try to search the best value of the filter inductor. Supply side has been taken as input side for inductor filter. Inductor filter value used in this thesis is 0.25mH. Offline, computer simulation using MATLAB Simulink has been applied to find out the optimum value for inductor filter. For the program, the limits, inequality and bounds need to be defined. This work has attempted to develop a single GA code program for optimizing an objective function.

Boundary and limits of parameters data have been collected using MATLAB/Simulink. Finally, a program using genetic algorithm has been written to generate the best value of the filter inductor. After the calculation, GA generates the value of 0.187mH. After using this inductor value, total harmonic distortion of source current and voltage have been reduced so we can say that inductor value calculated is optimum.

CONTROL THEORY

The series active filter controller generates the reference voltage that will be synthesized by the PWM converter and positioned in series with the supply voltage, to force the load voltage, to become sinusoidal and balanced.

Based on Sinusoidal Fryze Voltages Control Strategy, the series active filter controller is composed of a circuit that senses the fundamental positive sequence component of the system voltage and produce compensating voltage references by performing the distinction between that fundamental element and the measured system voltages.

The combination of a shunt passive filter and a low-rated series active resulted in a very realistic and economical way to filter harmonic currents. The concept adopted in this approach differs from conventional shunt active filters or pure series active filters in terms of guaranteeing better filtering characteristics, as well as lower initial costs. The required rating of the series active filter is mainly decided by the quality existing in the power system.

Although the quality factor of the shunt passive filter used in the experiments was equal to 14, it may be in the range of 50 to 80 in real cases. Therefore, the rated power of the series active filter may be as little as 1% of the rated power of the nonlinear load (thyristor rectifier or cyclo converter) if no background harmonic voltage lives in the source. This shows the way to the necessary conclusion that this combined series active filter and shunt passive filter is one of the most suitable solutions to high-power thyristor rectifiers, like those used in high–voltage dc transmission system (HVDC) or cyclo converters.

In this approach, the active filter is connected in series with the passive filter, as shown in Figure 3. The composed branch of passive and active filters is connected in parallel, as close as achievable to the harmonic producing load. This new method provides almost hybrid filters be somewhat different in circuit configuration; they are almost the same in operating principle and filtering performance. Such a mixture with the passive filter makes it possible significantly to reduce the rating of the active filter. Two passive filters tuned to compensate 5th and seventh harmonics have been connected in parallel to the series filter.

The job of the active filter is not to compensate for harmonic currents produced by the thyristor rectifier, but to achieve harmonic isolation between the supply and the load. As a result, no harmonic resonance occurs, and harmonic current flows in the supply. In this chapter, ATS algorithm has been applied such that Sinusoidal Fryze Voltage control strategy can work optimum.

OPTIMIZATION USING ADAPTIVE TABU SEARCH ALGORITHM

In this chapter, the proposed ATS method searches the optimum value of the proportional integral controller parameters i.e. K_p and K_I and the objective function (OF) is decided such as to give their optimum value with the conditions of percent overshoot, rise time and settling time. The objective function has an equation that has three variables; percent overshoot, rise time and settling time. Initially, the Boundary of K_p and K_i, their upper limits and lower limits, then radius value, conditions for ATS back tracking, objective function and stop criteria has been defined. Maximum Searching iteration (500 rounds) for ATS has been set as stop criterion. Figure 4 shows the flow chart for the search of parameters using adaptive tabu search method. The values initially used for K_p and K_i were 0.1 and 50 respectively. After the calculation, ATS algorithm gives the value 0.199 and 15.36. It has been observed that while using

Figure 4. Flowchart for the search of K_p and K_i values using Adaptive Tabu Search algorithm

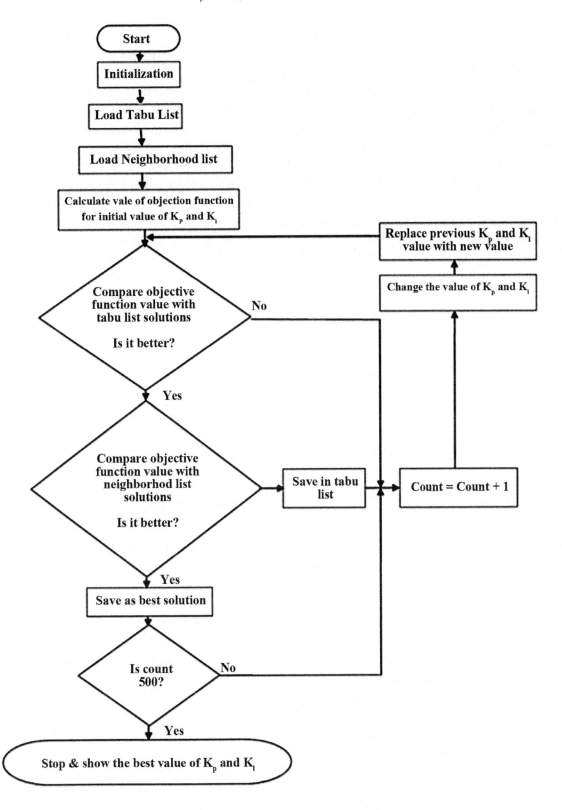

these ATS calculated values of K_p and K_i, the THD of source current and voltage have been reduced tremendously which proves that the values are optimum.

Function Evaluations: since this chapter is based on the critical analysis based on THD of the source it has been seen that the objective function taken has shown its effectiveness, which can be seen from the reduction of THD.

The computational time has also been seen very less i.e. within seconds; all iterations are over, and optimum values of Kp and Ki can be seen on MATLAB/Simulink complier. We can see that this process is very stable since it has been calculated offline and then can be used to replace the existing values. Robustness of this algorithm can be understood by the good results and less computational time. The Number of iterations with variation in K_p and K_i values has been taken to prove the flexibility of the algorithm. Figure4 shows a flow chart for the search of K_p and K_i values using Adaptive Tabu Search algorithm.

This algorithm is very convenient to use due to the programming and less computational time. The feasibility and benefit of the algorithm have been proved by the simulation results. It is very fast. The parameters i.e. K_p and K_i have been set randomly initially, and then it has been tuned by using this algorithm offline.

There has been a counter used, which will count the number of iterations, and the program will stop automatically when the count is equal to 500 i.e. stopping criteria is 500 iterations.

Objective function (O.F.) is defined by

$$O.F.(T_{Rise}, T_{Settling}, P.O.) = A(T_{Rise}) + B(T_{Settling}) + C(P.O.) \tag{8}$$

$$A + B + C = 1 \tag{9}$$

- P.O. is the percent overshoot.
- T_{Rise} is the rise time.
- $T_{Settling}$ is the setting time.
- A, B and C are the priority coefficients of T_{Rise}, $T_{Settling}$, P.O. respectively.

In this chapter, the values of (A, B, and C are set to 0.33, 0.33, and 0.34, respectively. The ATS search will try to find the best controller parameters to achieve the minimum O.F. value.

Step 1: Tabu list and neighborhood list having values of K_p and K_i have been loaded, and the counter has been made zero, which will check the number of iteration.
Step 2: Thevalue of the objective function has been calculated for initial values of K_p and K_i.
Step 3: Resultant of step 2 has been compared from the Tabu list and if it is better than it has been compared from neighborhood list. If it is not better than Tabu list solutions, then these values will be changed by varying the values of K_p and K_i.
Step 4: If the results are not better than neighborhood list solutions, it will be saved in tabu list, and then counter will automatically increased and there will be variation in K_p and K_i value and these value will replace the previous value and again go to step 3.
Step 5: If the results are better than neighborhood list solutions, it will be saved as the best solution.

Step 6: If number of iteration i.e. count value is 500, the results with the optimum value of K_p and K_i will be shown otherwise it will check the counter value and vary the K_p and K_i values and go to step 3.

It has been seen that the objective function actually based on percent overshoot, rise time and settling time, but the THD results outcome are excellent and since, this chapter does the critical analysis of the system based on THD of source, detail equations and values for percent overshoot, rise time and settling time have not been given, only THD details have been discussed.

COMPARATIVE EVALUATION USING SIMULATION RESULTS

The proposed scheme of Hybrid-Series Power filter (HPF) is simulated in MATLAB environment to estimate its performance. The load consists of three-phase thyristor rectifier connected with inductance and a pure resistance directly. The proposed control scheme has been simulated to compute the performance of HPF and analysis through THD of source voltage and current. The simulation results clearly demonstrate that the scheme can successfully reduce the significant amount of THD in source voltage and current within limits of IEEE 519-1992 (Khalid, 2012; Khalid, 2011). Simulation results have been analyzed on the basis of THD obtained. A simulation has been done for 15 cycles. Comparative evaluation of simulation using optimized SFV by ATS algorithm for series and hybrid power filter have been done.

Uncompensated System

After doing simulation in MATLAB/Simulink without using any filter (Figure 5) i.e. for Uncompensated System, it has been observed that the THD of source voltage and source current found when load is connected with the system is 31.6% and 34.1% respectively. By observing these data, we can easily understand supply has been polluted when the load has been connected.

Performance of Series APF

During the analysis of simulation results based on THD, this has been observed (Figure 5) that while doing simulation of Series power filter based on Sinusoidal Fryze Voltages Control Strategy that the THD of source Voltage and source current were 3.19% and 32.53%; whereas when model has been optimized using ATS algorithm has been used, it has been observed that the THD of source voltage reduces to 1.55% that is absolutely the improvement from conventional one. However, it has been observed that THD for source current has been reduced to 20.96%, but it was more than specified limit in IEEE 519-1992 standard.

Performance of Hybrid APF

During the analysis of simulation results based on THD, this has been observed (Figure 6) that while doing Simulation of Hybrid Series power filter optimized using ATS-SFV control has been used, it has been observed that the THD of source voltage reduces to 0.20% and THD of source current decreases to 0.0049%; which is excellent improvement from both the series filter results. It has also been observed that THD for source current was within the specified limit of IEEE 519-1992 standard.

Figure 5. Source voltage waveforms of uncompensated system, using series filter utilizing Sin Fryze voltage technique and using series filter utilizing ATS- Sin Fryze voltage technique

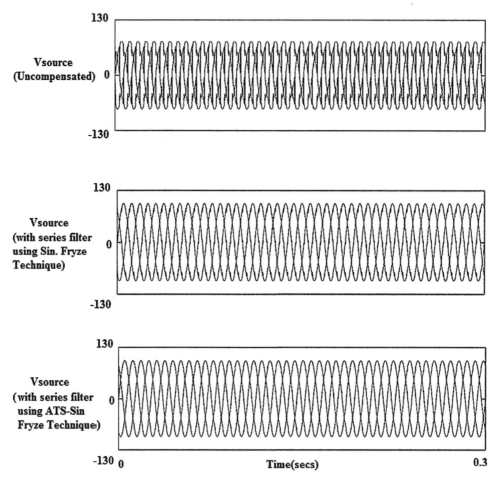

Figure 7 and Figure 8 present the graphical representation for THD-V and THD-I for different control schemes respectively. We can see that the hybrid power filter is the best as a comparison to other schemes.

DISCUSSIONS

We can see in Figure 7 and Figure 8 that Hybrid series filter using ATS-SFV control has least THD for voltage as well as for current as compared to series filter and uncompensated system. For source voltage THD, we can see that for the hybrid filter it has been reduced tremendously to 0.20% that is much better than the series filter. The results are also in the limit of the IEEE-519 standard defined for harmonics.

Excellent results have been observed for the consideration of current harmonics, and THD-I have been reduced to 0.0049%, which is extremely small. In other words, we can simply conclude that the source waveforms have become completely sinusoidal. Table 1 presents the statistical view of THD response for different filter and schemes.

Figure 6. Waveforms of source voltage and source current for hybrid power filter

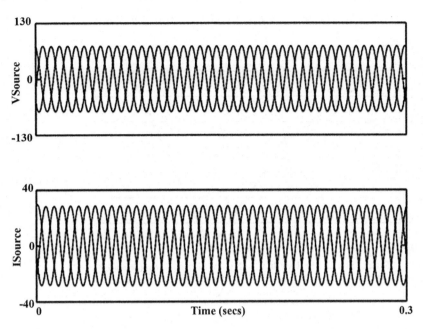

Figure 7. THD of source voltage for Compensated and Uncompensated system

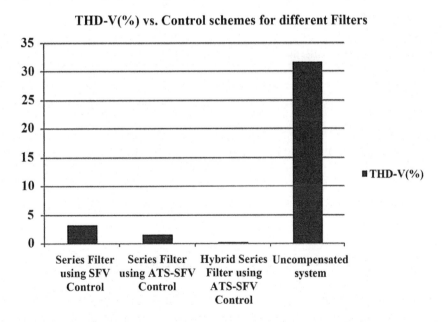

The major advantages of this approach are adequate compensation of the voltage and current harmonics and its sound speed. Unlike, Fuzzy logic system, which has very long run time, ATS algorithm works very fast and after obtaining the values of PI controllers' parameters, we can directly use them in the controller.

Figure 8. THD of source current for Compensated and Uncompensated system

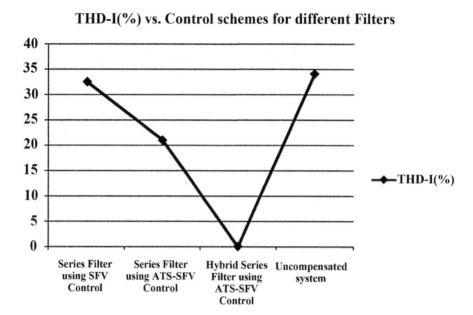

Table 1. Comparisons of THD for series filter

Strategy Applied	THD-V (%)	THD-I (%)
None (Uncompensated System)	31.6	34.1
Sinusoidal Fryze Voltage control (SFV) for series filter	3.19	32.53
ATS- SFV control for series filter	1.55	20.96
ATS- SFV control for Hybrid filter	0.20	0.0049

The Major disadvantage of this approach is the limitations of using same parameters. We cannot use same optimized parameter value for every load conditions. For each load condition, we have to find out different optimum values as well there is also the need for the design of new passive filter for each load conditions. In the case of load change, it does not work so optimum as in individual one.

RECENT FINDINGS AND FUTURE RESEARCH DIRECTIONS

ATS algorithm has been applied to shunt active power filter, and it has been observed that it gave wonderful results. We have considered sinusoidal current control strategy as the conventional control strategy and it has been optimized using ATS algorithm. After that, the results obtained from both the strategy have been compared. Three loads have been used for the simulation. The first load (Load 1) uses three phase 6-pulse current source-bridge converter. The second one (Load 2) is the three-phase diode bridge rectifier with the inductance of 300mH, and the third one (Load 3) is three-phase diode bridge rectifier with a capacitance of 1000uF (Load 3).To check the effectiveness of the filter, all three loads have been connected at different time interval. In this case, Load 1 is always connected, Load 2 is initially connected and is disconnected after every 2.5 cycles, and Load 3 is connected and disconnected after every half cycle. All the simulations have been done for 15 cycles.

Uncompensated System

After doing the simulation in MATLAB/Simulink, it has been observed in figure 9 that THD of the source current is 25.63%, and THD of source Voltage were 4.1%. By observing these data, we can quickly understand that they are out of the limit of IEEE 519-1992 limit. We have seen that supply has been polluted when different nonlinear loads are connected.

Performance of Shunt APF Based on Sinusoidal Current Control Strategy

From the simulation results shown in Figure 10, it has been observed that that the THD of source current & source voltage was 1.6% and 2.93% respectively. The compensation time was 0.059 sec. At t=0.059sec, we can see that the waveforms for source voltage and source current have become sinusoidal.

From Figure 10, we can see the waveforms of compensation current, dc capacitor voltage and load current. The variation in dc voltage can be clearly seen in the waveforms. As per requirement for increasing the compensation current for fulfilling the load current demand, it releases the energy and after that it charges and tries to regain its set value. If we carefully observe, we can find out that the compensation current is fulfilling the demand of load current and after the active filtering the source current and voltage is forced to be sinusoidal. This canbe easily observed that the results are within the limit of IEEE 519- 1992 standard defined for voltage and current harmonics.

Figure 9. Source voltage and source current waveforms of uncompensated system

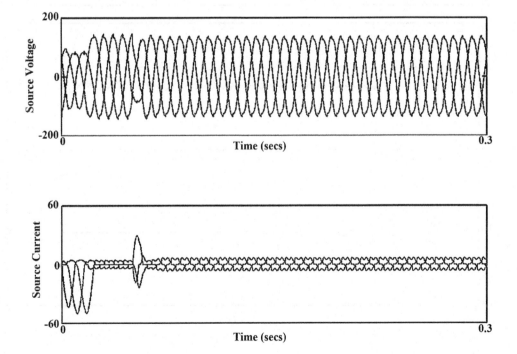

Figure 10. Source Voltage, source current, compensation current (phase b), DC link Voltage and load current waveforms of Active power filter using Sinusoidal Current Control strategy with all three loads connected at different time interval

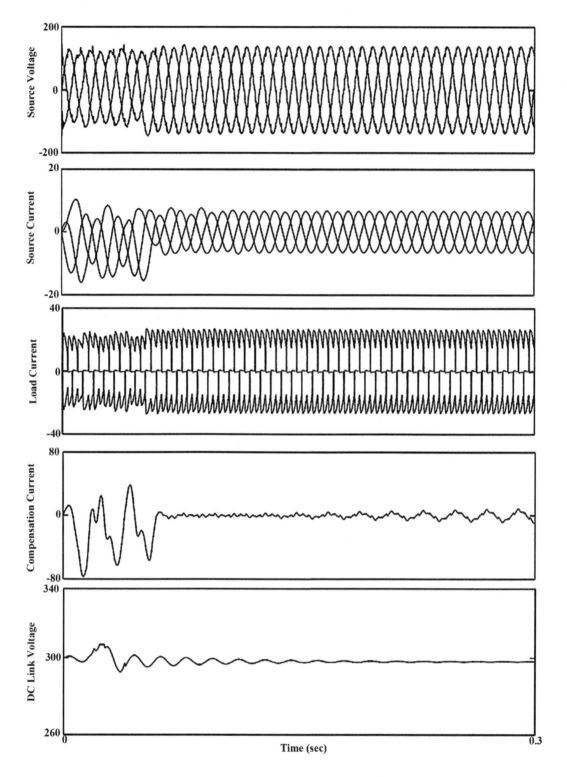

Performance of Shunt APF Optimized Using ATS Algorithm

The results from the simulation are shown in Figure 11. From the results, it is found that the THDs of source current & source voltage were 1.53% and 2.93% respectively. At t=0.057 sec, we can see that the waveforms for source voltage and source current have become sinusoidal. The observed compensation time was 0.057 sec. The waveforms of compensation current, dc capacitor voltage and load current can be seen from figure 11. There is variation in dc voltage which can be seen clearly in the waveforms. Due to that, the source voltage variations can also be clearly seen in the waveforms. If there is need of increasing the compensation current for fulfilling the demand of load current, it releases the energy and after that it charges and tries to regain its set value.

Performance of Shunt APF Optimized Using Genetic Algorithm

The results from the simulationare shown in Figure 12. From the results, it is found that the THDs of source current & source voltage were 1.55% and 3.87% respectively. At t=0.055 sec, we can see that the waveforms for source voltage and source current have become sinusoidal. The observed compensation time was 0.055 sec.

From Table 2, we can observe the amount of THD that the ATS-SCC has given better THD results as compare to GA-SCC, but GA-SCC has been observed fast as compare to ATS-SCC.

In future, this same scheme can be applied to shunt active power filter in high-frequency aircraft system with some modifications ATS algorithm along with some artificial intelligent techniques can be used to optimize the present conventional control methods. The artificial intelligent techniques like ANN, Fuzzy Logic, Genetic algorithm, particle swarm optimization, etc. can be applied to the optimization of the conventional schemes for 50 Hz as well as 400 Hz aircraft systems.

Some more soft computing techniques i.e. Bee algorithm, Ant Algorithm, etc. can be deployed in this field for optimization of controllers. The research work in the field of the active filter can also be extended by using the photo-voltaic cell in place of the capacitor of the active power filter unit. Harmonic compensation of variable frequency aircraft system can be explored. Application of Space Vector Modulation based hysteresis band controller can be explored as further research work.

Table 2. Comparisons of THD for shunt filter

Strategy Applied	THD-I (%)	THD-V (%)	Compensation Time (sec)
None (Uncompensated System)	4.1	25.63	NA
Sinusoidal current control (SCC)	1.6	2.93	0.059
ATS- SCC	1.53	2.93	0.057
GA-SCC	1.55	3.87	0.055

Figure 11. Source Voltage, source current, compensation current (phase b), DC link Voltage and load current waveforms of Active power filter using Sinusoidal Current Control strategy using ATS Algorithm with all three loads connected together at different time interval

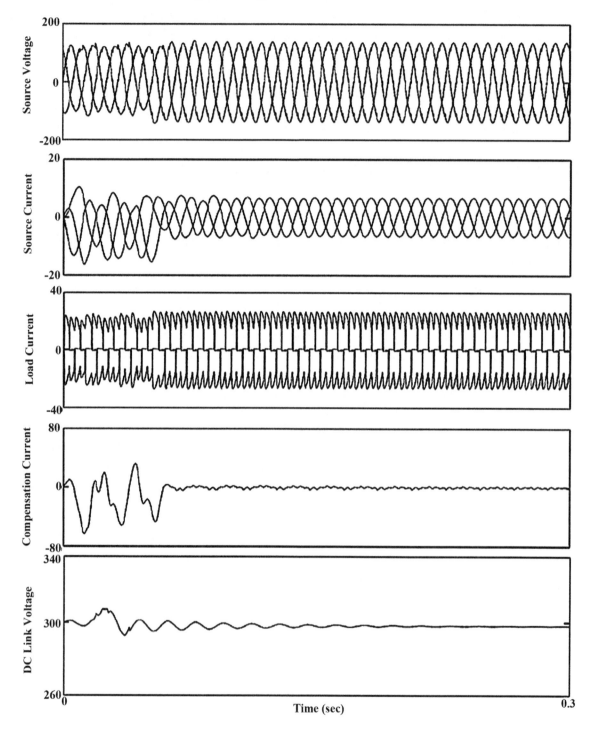

Figure 12. Source Voltage, source current, compensation current (phase b), DC link Voltage and load current waveforms of Active power filter using Sinusoidal Current Control strategy using Genetic Algorithm with all three loads connected together at different time interval

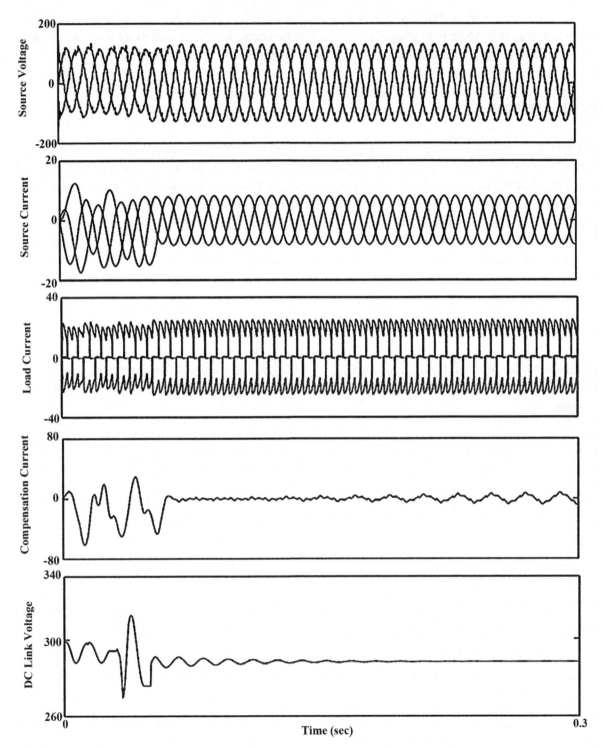

MERITS AND DEMERITS

The proposed Adaptive Tabu search algorithm optimizes the existing conventional schemes without any other complex circuitry addition, which makes it simple and low costly. Whereas other artificial intelligent schemes like ANN, Fuzzy logic, etc. optimize the complete system and make it robust for other disturbances. However, the additional circuitry makes them costly.

CONCLUSION

A novel ATS-Sin Fryze control techniques for hybrid series power filter and ATS-Sinusoidal current control method have been reported which clearly demonstrates its compensation ability. This also has been observed that adaptive Tabu search algorithm has well optimized the model and increased the ability of conventional models. It has been monitored that the system has a fast, vibrant response with the suggested control scheme and is capable to keep the THD of the source current and source voltage within the limits of IEEE 519 standard. The scheme has the advantage of simplicity and can provide harmonic compensation has also been observed with but balanced supply system. From the simulation results, this can be easily seen that the proposed novel active filter is effectively better than the conventional filters.

REFERENCES

Abraham, D. R., Mouton, H. D. T., & Akagi, H. (2002). Digital Control of an integrated series active filter and passive rectifier with voltage regulation. *Proceedings of the Power Conversion Conference*, (pp. 68-73).

Abraham, D. R., Mouton, H. D. T., & Akagi, H. (2003). Digital Control of an Integrated Series Active Filter and Diode Rectifier with Voltage regulation. *IEEE Transactions on Industry Applications*, *39*(6), 1–14.

Akagi, H., Kanazawa, Y., & Nabae, A. (1983). Generalized Theory of the Instantaneous Reactive Power in Three-Phase Circuits. *Proc. IPEC Tokyo 93 Int. Conf. Power Electronics*, (pp. 1375-1386).

Akagi, H., Kanazawa, Y., & Nabae, A. (1984). Instantaneous Reactive Power Compensator Comprising Switching Devices without Energy Storage Components. *IEEE Transactions on Industry Applications*, *IA-20*(3), 625–630. doi:10.1109/TIA.1984.4504460

Aredes, M. (1991). *New Concepts of Power and its Application on Active Filters*. (M.Sc. Thesis). COPPE – Federal University of Rio de Janeiro, Brazil.

Aredes, M., Luís, F. C., Miguel, M., & Jaime, M. (2003). Control Strategies for Series and Shunt Active Filters. In Proceedings of IEEE Bologna Power Tech Conference, (pp. 1122-1126). IEEE.

Avik, B., & Chandan, C. (2011). A Shunt Active Power Filter With Enhanced Performance Using ANN-Based Predictive and Adaptive Controllers. *IEEE Transactions on Industrial Electronics*, *58*(2), 421–428. doi:10.1109/TIE.2010.2070770

Babak, K., & Abdolreza, R. (2010). Genetic Algorithm Application in Controlling Performance and Power Dissipation of Active Power Filters. *Canadian Journal on Electrical & Electronics Engineering*, *1*(1), 15–19.

Bhim, S., Vishal, V., & Jitendra, S. (2007). Neural Network-Based Selective Compensation of Current Quality Problems in Distribution System. *IEEE Transactions on Industrial Electronics*, *54*(1), 53–60. doi:10.1109/TIE.2006.888754

Bose Bimal, K. (1990). An adaptive Hysteresis-band current control technique of a voltage fed PWM inverter for machine drive system. *IEEE Transactions on Industrial Electronics*, *37*(5), 402–408. doi:10.1109/41.103436

Bull, S. R. (2001). Renewable Energy Today and Tomorrow. *Proceedings of the IEEE*, *89*(8), 1216–1226. doi:10.1109/5.940290

Cebi, S., Kahraman, C., & Kaya, I. (2012). Soft Computing and Computational Intelligent Techniques in the Evaluation of Emerging Energy Technologies. In P. Vasant, N. Barsoum, & J. Webb (Eds.), Innovation in Power, Control, and Optimization: Emerging Energy Technologies, (pp. 164-197). doi:10.4018/978-1-61350-138-2.ch005

Chaijarurnudomrung, K., Areerak, K.-N., Areerak, K.-L., & Srikaew, A. (2011). The Controller Design of Three-Phase Controlled Rectifier Using an Adaptive Tabu Search Algorithm. In Proceedings of 8th Electrical Engineering/Electronics. Computer. Telecommunications and Information Technology (ECTI).

Chan, R., Lee, Y., Sudhoff, S., & Zivi, E. (2008). Evolutionary optimization of power electronics based power systems. *IEEE Transactions on Power Electronics*, *23*(4), 1908–1917. doi:10.1109/TPEL.2008.925197

Dahidah, M., & Agelidis, V. (2008). Selective harmonic elimination PWM control for cascaded multilevel voltage source converters: A generalized formula. *IEEE Transactions on Power Electronics*, *23*(4), 1620–1630. doi:10.1109/TPEL.2008.925179

Deepak, S., & Mariesa, L. C. (2014). An Integrated Active Power Filter–Ultracapacitor Design to Provide Intermittency Smoothing and Reactive Power Support to the Distribution Grid. *IEEE Transactions on Sustainable Energy*, *5*(4), 1116–1125. doi:10.1109/TSTE.2014.2331355

Dehini, R., Bassou, A., & Ferdi, B. (2009). Artificial Neural Networks Application to Improve Shunt Active Power Filter. *International Journal of Computer and Information Engineering*, *3*(4), 247–254.

DelCasale, M., Femia, N., Lamberti, P., & Mainardi, V. (2004). Selection of optimal closed-loop controllers for dc–dc voltage regulators based on nominal and tolerance design. *IEEE Transactions on Industrial Electronics*, *51*(4), 840–849. doi:10.1109/TIE.2004.831737

Dorigo, M., & Gambardella, L. (1997). Ant colony system: A cooperative learning approach to the traveling salesman problem. *IEEE Transactions on Evolutionary Computation*, *1*(1), 53–66. doi:10.1109/4235.585892

Dorigo, M., Maniezzo,, V., &Colorni, A. (1996). Ant system: Optimization by a colony of cooperating agents. *IEEE Trans. Syst., Man, Cybern. B, Cybern*, *26*(1), 29–41.

Dorigo, M., & Stützle, T. (2004). *Ant Colony Optimization*. Cambridge, MA: MIT Press. doi:10.1007/b99492

El-Habrouk, M., & Darwish, M. K. (2002). A New Control Technique for Active Power Filters Using a Combined Genetic Algorithm/Conventional Analysis. *IEEE Transactions on Industrial Electronics*, *49*(1), 58–66. doi:10.1109/41.982249

Elamvazuthi, I., Ganesan, T., & Vasant, P. (2011). A comparative study of HNN and Hybrid HNN-PSO techniques in the optimization of distributed generation power systems. *Proceedings of the 2011 International Conference on Advanced Computer Science and Information Systems (ICACSIS'11)*.

Favuzza, S., Graditi, G., Ippolito, M., &Sanseverino, E. (2007). Optimal electrical distribution systems reinforcement planning using gas micro turbines by dynamic ant colony search algorithm. *IEEE Trans. Power Syst, 22*(2), 580–587.

Formato, R. (2007). Central Force Optimization: A New Metaheuristic with Applications in Applied Electromagnetics. *Progress in Electromagnetics Research*, *77*, 425–491. doi:10.2528/PIER07082403

Fryze, S., Wirk, & Blind, U. (1932). Scheinleistung in elektrischen Strom kainsenmitnicht-sinusfömigenVerlauf von Strom und Spannung. ETZArch. *Elektrotech, 53*, 596–599.

Fujita, H. T., Yamasaki, T., & Akagi, H. (2000). A Hybrid Active Filter for Damping of Harmonic Resonance in Industrial Power Systems. *IEEE Transactions on Power Electronics*, *15*(2), 2–22. doi:10.1109/63.838093

Gaby, P., Uriel, I., Inessa, A., & Gad, R. (2010). GA for the Resource Sharing and Scheduling Problem. *Global Journal Technology and Optimization, 1*, 1985–1994.

Glover, F. (1994). Tabu search for nonlinear and parametric optimization (with links to genetic algorithms). *Discrete Applied Mathematics*, *49*(1-3), 231–255. doi:10.1016/0166-218X(94)90211-9

Goldberg, D. (1989). *Genetic Algorithms in Search, Optimization and Machine Learning*. Addison-Wesley.

Gomez, J., Oliveira, P., Yusta, J., Villasana, R., & Urdaneta, A. (2004). Ant colony system algorithm for the planning of primary distribution circuits. *IEEE Transactions on Power Systems*, *19*(2), 996–1004. doi:10.1109/TPWRS.2004.825867

Hagh, M., Taghizadeh, H., &Razi, K. (2009). Harmonic minimization in multilevel inverters using modified species-based particle swarm optimization. *IEEE Trans. Power Electron, 24*(10), 2259–2267.

Hammons, T., Boyer, J. C., Conners, S. R., Davies, M., Ellis, M., Fraser, M., & Markard, J. et al. (2000). Renewable Energy Alternatives for Developed Countries. *IEEE Transactions on Energy Conversion*, *15*(4), 481–493. doi:10.1109/60.900511

Hao, Y., Fang, Z., & Yanjun, Z. (2014). A Source-Current-Detected Shunt Active Power Filter Control Scheme Based on Vector Resonant Controller. *IEEE Transactions on Industry Applications*, *50*(3), 1953–1965. doi:10.1109/TIA.2013.2289956

Hassmann, K. (1993). Electric Power Generation. *Proceedings of the IEEE*, *81*(3), 346–354. doi:10.1109/5.241493

Herrera, F., Lozano, M., & Molina, D. (2005). Continuous scatter search: An analysis of the integration of some combination methods and improvement strategies. *European Journal of Operational Research, 169*(2), 450–476. doi:10.1016/j.ejor.2004.08.009

Hirsch, M., Pardalos, P., & Resende, M. (2009). Solving systems of nonlinear equations with continuous GRASP. *Nonlinear Analysis Real World Applications, 10*(4), 2000–2006. doi:10.1016/j.nonrwa.2008.03.006

Jiang, Y.-H., & Chen, Y.-W. (2009). Neural Network Control Techniques of Hybrid Active Power Filter. *Proceedings of International Conference on Artificial Intelligence and Computational Intelligence*, (pp. 26-30). doi:10.1109/AICI.2009.296

Juang, C., Lu, C., Lo, C., & Wang, C. (2008). Ant colony optimization algorithm for fuzzy controller design and its FPGA implementation. *IEEE Transactions on Industrial Electronics, 55*(3), 1453–1462. doi:10.1109/TIE.2007.909762

Kavousi, A., Vahidi, B., Salehi, R., Kazem, M., Farokhnia, N., & Hamid Fathi, S. (2012). Application of the Bee Algorithm for Selective Harmonic Elimination Strategy in Multilevel Inverters. *IEEE Transactions on Power Electronics, 27*(4), 95–101. doi:10.1109/TPEL.2011.2166124

Kennedy, J., & Eberhart, R. (1995). Particle swarm optimization. *Proc. of the IEEE International Conference on Neural Networks*, (pp. 1942–1948). doi:10.1109/ICNN.1995.488968

Kim, S., Yoo, G., & Song, J. (1996). A Bifunctional Utility Connected Photovoltaic System with Power Factor Correction and U.P.S. Facility. *Proceedings of the IEEE Conference on Photovoltaic Specialist*, (pp. 1363-1368).

Komatsu, Y. (2002). Application of the Extension pq Theory to a Mains-Coupled Photovoltaic System. *Proceedings of the Power Conversion Conference (PCC)*, (pp. 816-821). doi:10.1109/PCC.2002.997625

Lee, Y., Wang, W., & Kuo, T. (2008). Soft computing for battery state-of-charge (BSOC) estimation in battery string systems. *IEEE Transactions on Industrial Electronics, 55*(1), 229–239. doi:10.1109/TIE.2007.896496

Leobardo, C. B., Sergio H.M., Abraham, C.S., Victor, M., & Cardenas, G. (1998). Single- Phase active power filter and harmonic compensation. Centro Nacional de investigacion y Desarrollo Technologico interior internado Palmira s/n, Cuernavaca, Mor., Mexico, Instituto Technologico y de Estodios Superiores de Monterrey, Campus Ciudad de Mexico Calle del Puente 222. *Esquina Periferico Sur, C.P. D.F. Mexico,* 184-187.

Liserre, M., Dell'Aquila, A., & Blaabjerg, F. (2004). Genetic algorithm-based design of the active damping for an LCL-filter three-phase active rectifier. *IEEE Transactions on Power Electronics, 19*(1), 76–86. doi:10.1109/TPEL.2003.820540

Lui, Y., & Passino, K. (2002). Biomimicry of Social Foraging Bacteria for Distributed Optimization: Models, Principles, and Emergent Behaviors. *Journal of Optimization Theory and Applications, 115*(3), 603–628. doi:10.1023/A:1021207331209

Luis, A. M., Juan, W. D., Rogel, R. W. (1995). A Three- Phase Active Power Filter Operating with Fixed Switching Frequency for Reactive Power and Current Harmonic Compensation. *IEEE Trans. on Ind. Electronics, 42*(4), 14-18.

Malesani, L., Rosseto, L., & Tenti. (1986). Active Filter for Reactive Power and Harmonics Compensation. *IEEE – PESC*, (321-330). IEEE.

Maniya, K. D., & Bhatt, M. G. (2013). A Selection of Optimal Electrical Energy Equipment Using Integrated Multi Criteria Decision Making Methodology. *International Journal of Energy Optimization and Engineering, 2*(1), 101–116. doi:10.4018/ijeoe.2013010107

Marco, L., Antonio, D., & Frede, B. (2004). Genetic Algorithm-Based Design of the Active Damping for an LCL-Filter Three-Phase Active Rectifier. *IEEE Transactions on Power Electronics, 19*(1), 76–86. doi:10.1109/TPEL.2003.820540

Maricar, N. (2003). Photovoltaic Solar Energy Technology Overview for Malaysia Scenario. *Proceedings of the IEEE National Conference on Power and Energy Conference (PECon)*, (pp. 300-305). doi:10.1109/PECON.2003.1437462

Mladenovic, N., Drazic, M., Kovacevic-Vujcic, V., & Cangalovi, M. (2008). General variable neighborhood search for the continuous optimization. *European Journal of Operational Research, 191*(3), 753–770. doi:10.1016/j.ejor.2006.12.064

Mohamed, Y., & Saadany, E. (2008). Hybrid variable-structure control with evolutionary optimum-tuning algorithm for fast grid-voltage regulation using inverter-based distributed generation. *IEEE Transactions on Power Electronics, 23*(3), 1334–1341. doi:10.1109/TPEL.2008.921106

Mohammed, Q., Parag, K., & Vinod, K. (2014). Artificial-Neural-Network-Based Phase-Locking Scheme for Active Power Filters. *IEEE Transactions on Industrial Informatics, 61*(8), 3857–3866. doi:10.1109/TIE.2013.2284132

Niraj, K., Khem, B. T., & Khalid, S. (2013). A Novel Series Active Power Filter Scheme using Adaptive Tabu search Algorithm for Harmonic Compensation. *International Journal of Application or Innovation in Engineering & Management, 2*(2), 155–159.

Ozpineci, B., Tolbert, L., & Chiasson, J. (2005). Harmonic optimization of multilevel converters using genetic algorithms. *IEEE Power Electronics Letters, 3*(3), 92–95. doi:10.1109/LPEL.2005.856713

Parag, K., Vinod, K., & Hatem, H. Z. (2015). Optimal Control of Shunt Active Power Filter to Meet IEEE Std. 519 Current Harmonic Constraints Under Non-ideal Supply Condition. *IEEE Transactions on Industrial Electronics, 62*(2), 724–734. doi:10.1109/TIE.2014.2341559

Parpinelli, R., Lopes, H., & Freitas, A. (2002). Data mining with an ant colony optimization algorithm. *IEEE Transactions on Evolutionary Computation, 6*(4), 321–332. doi:10.1109/TEVC.2002.802452

Polprasert, J., Ongsakul, W., & Dieu, V. N. (2013). A New Improved Particle Swarm Optimization for Solving Nonconvex Economic Dispatch Problems. *International Journal of Energy Optimization and Engineering, 2*(1), 60–77. doi:10.4018/ijeoe.2013010105

Pramod, K., & Alka, M. (2009). Soft Computing Techniques for the Control of an Active Power Filter. *IEEE Transactions on Power Delivery, 24*(1), 452–461. doi:10.1109/TPWRD.2008.2005881

Radha, T., Thanga, R. C., Millie, P. A., & Crina, G. (2010). Optimal gain tuning of PI speed controller in induction motor drives using particle swarm optimization. *Logic Journal of IGPL Advance Access, 2*(2), 1-14.

Ramakrishna, R. S. (2008). On the scalability of real-coded bayesian optimization algorithm. *IEEE Transactions on Evolutionary Computation, 12*(3), 307–322. doi:10.1109/TEVC.2007.902856

Rubén, I., & Hirofumi, A. (2005). A 6.6-kV Transformerless Shunt Hybrid Active Filter for Installation on a Power Distribution System. *IEEE Transactions on Power Electronics, 20*(4), 454–459.

Ruminot, P., Morán, L., Aeloiza, E., Enjeti, P., & Dixon, J. (2006). A New Compensation Method for High Current Non-Linear Loads *IEEE International Symposium on Industrial Electronics*, (vol. 2, pp. 1480-1485). doi:10.1109/ISIE.2006.295690

Saifullah, K., & Anurag, T. (2012). Control and Analysis Of Active Filter Based on Constant Instantaneous Power Control Strategy in High Frequency (400 Hz) Aircraft System with Balanced and Unbalanced Load Conditions. *International Journal of Scientific Computing, 6*(2), 342–345.

Saifullah, K., & Anurag, T. (2012). Comparison of Constant Source Instantaneous Power & Sinusoidal Current Control Strategy for Total Harmonic Reduction for Power Electronic Converters in High Frequency Aircraft System. *International Journal of Advanced Research in Electrical. Electronics and Instrumentation Engineering, l*(5), 314–322.

Saifullah, K., & Anurag, T. (2012). Comparison of Sinusoidal Current Control Strategy & Synchronous Rotating Frame Strategy for Total Harmonic Reduction for Power Electronic Converters in Aircraft System under Different Load Conditions. *International Journal of Advanced Research in Electrical. Electronics and Instrumentation Engineering, l*(5), 305–313.

Saifullah, K., & Anurag, T. (2013). Simulation of Shunt Active Filters based on Sinusoidal Current Control Strategy in High Frequency Aircraft System. *International Journal of Management. IT and Engineering, 3*(1), 218–230.

Saifullah, K., & Dwivedi, B. (2011). Power Quality Issues. Problems. Standards & their Effects in Industry with Corrective Means. *International Journal of Advances in Engineering & Technology, 1*(2), 1–11.

Saifullah, K, Dwivedi, B., & Singh, B. (2012). New Optimum Three-Phase Shunt Active Power Filter based on Adaptive Tabu Search and Genetic Algorithm using ANN control in unbalanced and distorted supply conditions. *Elektrika: Journal of Electrical Engineering, 14*(2), 9-14.

Sangsun, K., & Prasad, N. E. (2002). A New Hybrid Active Power Filter (APF) Topology. *IEEE Transactions on Power Electronics, 17*(1), 23–29.

Sim, K., & Sun, W. (2003). Ant colony optimization for routing and load balancing: Survey and new directions. *IEEE Transactions on Systems, Man, and Cybernetics. Part A, Systems and Humans, 33*(5), 560–572. doi:10.1109/TSMCA.2003.817391

Singh, B., & Verma, V. (2003). A New Control Scheme of Series Active Filter for Varying Rectifier Loads. *The Fifth International Conference on Power Electronics and Drive Systems*. doi:10.1109/PEDS.2003.1282900

Socha, K., & Dorigo, M. (2008). Ant colony optimization for Continuous domains. *European Journal of Operational Research*, *18*(3), 1155–1173. doi:10.1016/j.ejor.2006.06.046

Sopapirm, T., Areerak, K.-N., Areerak, K.-L., & Srikaew, A. (2011). The Application of Adaptive Tabu Search Algorithm and Averaging Z`Model to the Optimal Controller Design of Buck Converters. *World Academy of Science. Engineering and Technology*, *60*, 477–483.

Surmann, H. (1996). Genetic optimization of a fuzzy system for charging batteries. *IEEE Transactions on Industrial Electronics*, *43*(5), 541–548. doi:10.1109/41.538611

Terbobri, C. G., Saidon, M. F., & Khanniche, M. S. (2000). Trends of Real Time Controlled Active Power Filters. Power Electronics and variable Speed Drives. Conference publication No. 475 (c) IEE, 410-415. doi:10.1049/cp:20000282

Turan, P., Eren, Ö., Nimet, Y. P., & Gerhard-Wilhelm, W. (2012). Particle Swarm Optimization Approach for Estimation of Energy Demand of Turkey. *Global Journal Technology and Optimization*, *3*(1), 2229–8711.

Vanderbilt, D., & Louie, S. (1984). A monte-carlo simulated annealing approach to optimization over continuous-variables. *Journal of Computational Physics*, *56*(2), 259–271. doi:10.1016/0021-9991(84)90095-0

Vasant, P. (2012). Solving Fuzzy Optimization Problems of Uncertain Technological Coefficients with Genetic Algorithms and Hybrid Genetic Algorithms Pattern Search Approaches. *Innovation in Power, Control, and Optimization: Emerging Energy Technologies,* (pp. 344-368). doi:.ch01210.4018/978-1-61350-138-2

Vasant, P. (2013). *Meta-Heuristics Optimization Algorithms in Engineering, Business, Economics, and Finance*. doi:.10.4018/978-1-4666-2086-5

Vasant, P., Barsoum, N., & Webb, J. (2012*). Innovation in Power, Control, and Optimization: Emerging Energy Technologies*. doi:10.4018/978-1-61350-138-2

Vasant, P., Elamvazuthi, I., & Webb, J. F. (2010). Fuzzy technique for optimization of objective function with uncertain resource variables and technological coefficients. *International Journal of Modeling, Simulation, and Scientific Computing*, *1*(3), 349–367. doi:10.1142/S1793962310000225

Vasant, P., Ganesan, T., & Elamvazuthi, I. (2012). Elamvazuthi. Improved Tabu Search Recursive Fuzzy method for crude oil industry. *International Journal of Modeling, Simulation, and Scientific Computing*, *3*(1), 20. doi:10.1142/S1793962311500024

Vijay, B. B., & Prasad, N. E. (1996). An Active line Conditioner to Balance Voltages in a Three-Phase System. *IEEE Trans. on Ind. Application*, *32*(2), 14-19.

Wai, R., & Tu, C. (2007). Design of total sliding-mode-based genetic algorithm control for hybrid resonant-driven linear piezoelectric ceramic motor. *IEEE Transactions on Power Electronics*, *22*(2), 563–575. doi:10.1109/TPEL.2006.889988

Whei-Min, L., & Chih-Ming, H. (2011). A New Elman Neural Network-Based Control Algorithm for Adjustable-Pitch Variable-Speed Wind-Energy Conversion Systems. *IEEE Transactions on Power Electronics*, *26*(2), 473–481. doi:10.1109/TPEL.2010.2085454

Wu, T., Shen, C., Chang, C., & Chiu, J. (2003). 1/spl phi/ 3W Grid-Connection PV Power Inverter with Partial Active Power Filter. *IEEE Transactions on Aerospace and Electronic Systems*, *39*(2), 635–646. doi:10.1109/TAES.2003.1207271

Yang, Y., & Yu, X. (2007). Cooperative co evolutionary genetic algorithm for digital IIR filter design. *IEEE Transactions on Industrial Electronics*, *54*(3), 1311–1318. doi:10.1109/TIE.2007.893063

Zelinka, I., Vasant, P., & Barsoum, N. (2013). Power, Control and Optimization. *Series: Lecture Notes in Electrical Engineering*, *239*, 176.

Zhang, J., Chung, H., Lo, W., Hui, S., & Wu, A. (2001). Implementation of a decoupled optimization technique for design of switching regulators using genetic algorithm. *IEEE Trans. Power Electron*, 752–763.

Zhi-Xiang, Z., Keliang, Z., Zheng, W., & Ming, C. (2015). Frequency-Adaptive Fractional-Order Repetitive Control of Shunt Active Power Filters. *IEEE Transactions on Industrial Electronics*, *62*(3), 1659–1668. doi:10.1109/TIE.2014.2363442

Zhiguo, P., Fang, Z. P., & Suilin, W. (2005). Power Factor Correction Using a Series Active Filter. *IEEE Transactions on Power Electronics*, *20*(1), 12–19.

APPENDIX

The system parameters used are as follows:

Three-phase source voltage: 110V/50 Hz
Source impedance: R=0.1Ω, L=0.5 mH
Filter inductor=2.5 mH
Filter capacitor: 30 μF,
DC voltage reference: 600 V
DC capacitor: 3000 μF

Chapter 13
Recent Techniques to Identify the Stator Fault Diagnosis in Three Phase Induction Motor

K. Vinoth Kumar
Karunya Institute of Technology and Sciences University, India

A. Immanuel Selvakumar
Karunya Institute of Technology and Sciences University, India

S. Suresh Kumar
Dr. N. G. P. Institute of Technology, India

R. Saravana Kumar
Vellore Institute of Technology University, India

ABSTRACT

Induction motors have gained its popularity as most suitable industrial workhorse, due to its ruggedness and reliability. With the passage of time, these workhorses are susceptible to faults, some are incipient and some are major. Such fault can be catastrophic, if unattended and may develop serious problem that may lead to shut down the machine causing production and financial losses. To avoid such breakdown, an early stage prognosis can help in preparing the maintenance schedule, which will lead to improve its life span. Scientist and engineers worked with different scheme to diagnose the machine faults. In this paper, the authors diagnose the turn-to-turn faults condition of the stator through symmetrical component analysis. The results of the analysis also verified through Power Decomposition Technique (PDT) in Matlab /SIMULINK. The results are compatible with the published results for known faults.

1. INTRODUCTION

The Induction Motor, due to its ruggedness and simplicity in design is the most common electromechanical energy conversion device used as industrial drives. But these machines are prone to faults. The faults may be major like turn–to-turn fault or incipient which may ultimately lead to a major fault. To avoid the ultimate catastrophe of breakdown, the condition of the motor should be identified at an early stage. Since 80's engineers and scientists worked together to identify the condition of the motor at an early stage by sensing the magnetic flux, stator current, rotor current, shaft leakage flux, ground fault current, vibration, temperature and speed. It has been found that electrical and magnetic parameters used in analyzing the machine condition are much more accurate and reliable than mechanical parameters.

DOI: 10.4018/978-1-4666-9755-3.ch013

The electrical and magnetic sensor is inexpensive, reliable and can be installed easily. G.B.Kliman, W.J.Premerlani, R.A.Koegl and D.Hoeweler (1996) reported a method to determine the sequence component from the current phasor and explained the theory of the coupled circuit model to identify the turn to turn fault. M.Arkan and P.J.Unsworth (1999) reported a Power Decomposition technique that has been used to derive positive and negative sequence components of arbitrary three phase signals in the time domain, in order to detect the stator fault. Gojko M.Joksimvoic and Jim Penman (2000) proposed an induction motor model to detect the stator inter turn short circuit and the dynamic model of motor by using winding function method. Zhongming ye, Binwu and Navid Zargari (2000) proposed an induction motor model to identify the mechanical faults and the dynamic model of motor by using winding function method. Stefano P.C. and Stank Vic (2003) proposed an induction motor model to analyze the sudden loss of balance in the supply voltage and the dynamic model of motor by using dynamic phasor. Burak Ozpineci, Leon M.tolbert (2003) proposed an induction motor model to analyze the open loop V/f and indirect vector control operation and the dynamic model of motor by using Krause model. Marius Marcu, Ilie Utu, Leon Pana, Maria Orban (2004) proposed an induction motor model to identify the electrical and mechanical parameters and the dynamic model of motor can be developed by using stator reference frame theory . Arafat Siddique, G.S.Yadava and Bhim Singh (2005) presented a comprehensive survey on the subject of condition monitoring of stator faults of induction motors. Ali M.Eltamaly, A.L.Alola and R.M.Hamouda (2006) proposed an induction motor model to identify the performance parameter and the induction motor model by reducing its equivalent circuit to be an R-L load. Ramtin Hadidi Iman Mazhari Ahad Kazemi (2007) used an induction motor model to analyze and simulate a induction motor with normal, star /delta and soft starter and in this paper, the diagnosis of the motor turn-to-turn fault is carried out using sequence component analysis of the stator current. The hardware circuits are developed as shown in figures 3, 4 and 5 to obtain the sequence component of the stator current. These sequence components are which validates the simulation model results. A new dynamic model of the induction machine is developed, considering stator and rotor turns (Ns, Nr). This model is used for simulation in Matalb/simulink to obtain the performance parameters i.e., current, speed and torque of the motor. Direct and Quadrature axis voltage and current parameters obtained from this simulation model is used to develop another dynamic model to identify the sequence component by applying power decomposition technique (PDT). The paper is organized as follows. Hardware implementation is dealt in Section 2. Simulink implementation and results are discussed in Section 3. Finally the paper concludes with Section 4.

2. HARDWARE IMPLEMENTATION

It is well known that stator internal faults create an unbalanced operating condition. The analysis of any unbalanced system is complicated and cumbersome. Fortescue proves that an unbalanced phasor can be resolved into a balanced phasors called symmetrical components of the original phasor. Several authors had applied this technique to analyze the system operation under unbalanced conditions. The principal merit of symmetrical component analysis is that a relatively complicated problem can be solved by developing no more than three balanced network for three phase unbalance system. Due to the faults in the stator winding, the impedance parameter which effectively controls the current magnitude vis-à-vis flux and phase gets affected due to change in mutual coupling between the turns of the winding. Using symmetrical component analysis, the three unbalanced line currents of a three phase induction motor can be expressed by three balanced sequence components as shown below.

$$I_R = I_{a0} + I_{a1} + I_{a2}; \quad I_Y = I_{a0} + a*I_{a1} + a^2*I_{a2}; \quad I_B = I_{a0} + a^2*I_{a1} + a*I_{a2};$$

where, I_{a0} - Zero sequence component, I_{a1} - Positive sequence component, I_{a2} - Negative sequence component of currents. By analyzing the sequence component and observing the responses of the solution it is possible to identify the nature of fault current /flux waveform. Different authors have identified the stator faults through the symmetrical component analysis as listed in Table – 1. It is seen that no attempt is made so far to identify the turn-to-turn fault of the stator winding through symmetrical component analysis. The hardware circuits are developed to realize the sequence component current due to artificially created stator turn-to-turn fault and capture the response using a Power Analyzer. These results verify the PDT used in Matlab/Simulink. The results so obtained are presented in the section IV.

Hardware Implementation

An Op-amp based active all pass filter is designed with Resistance (R) =18KΩ and Capacitance (C) =0.1μF to obtain an accurate phase shift of 120° of the input signal as shown in Figure 1. In Figure 2 two such all pass filters are cascaded to obtain 240° phase shift. These hard-ware circuits are used to capture the symmetrical component of the line current as presented in Figures 3, 4 and 5. The symmetrical components are obtained in terms of voltage proportional to current through three identically rated Current transformers (CT).

The experiment is carried out using hard-ware circuits on a 2.2kW, 3 phase, 400V, 50 Hz, 4 pole, 2.5 Amp, squirrel cage induction motor shown in Figure 6. The sequence component of the machine is

Table 1. Fault and there sequence components

Sl.no	Faults	Creation of Fault	Fault Identification
1	High resistance fault	Connecting a high resistance across the phase 'R' of the supply and the earth	Zero sequence components of low value
2	Single phasing	Opening any one phase.	1. Negative sequence of almost full reading 2. Positive sequence of reasonable reading
3	High resistance phase to phase fault.	Connecting high value of resistance between winding 'R' and 'Y'.	1. Positive sequence of reasonable reading 2. Negative sequence of small value.
4	Unbalanced stator impedance.	Connecting low value resistance in series with the phase 'R' winding.	1. Positive sequence of high value 2. Negative sequence of small reading
5	Earth fault	Connecting resistance between phase winding 'R' and earth.	Zero sequence components

Figure 1. Typical 120 ° all pass filter

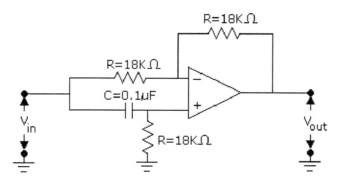

Figure 2. Typical 240 ° all pass filter

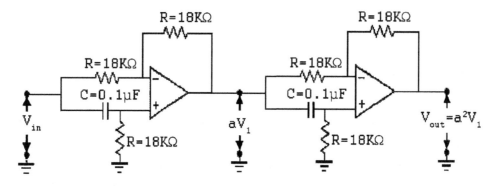

Figure 3. Measurement of zero sequence components

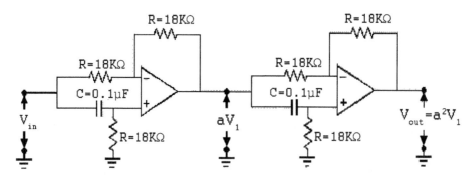

captured for different operating conditions like normal and artificially created stator turn-to- turn fault. The captured waveforms for normal condition are shown in Figure 7. The Figure 8 shows the measured sequence component waveforms under Inter turn fault. The results clearly show that for normal operating condition only positive sequence component is present. For stator turn-to-turn fault the positive and negative sequence component is present. This confirms with the theoretical analysis.

Figure 4. Measurement of positive sequence components

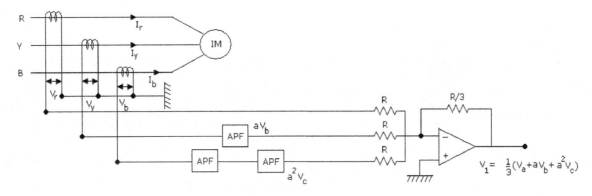

Figure 5. Measurement of negative sequence components

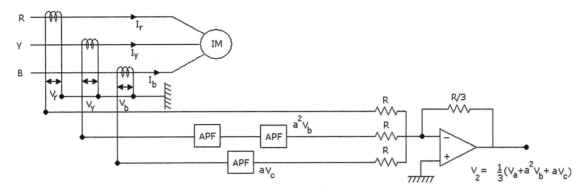

Figure 6. Experimental setup

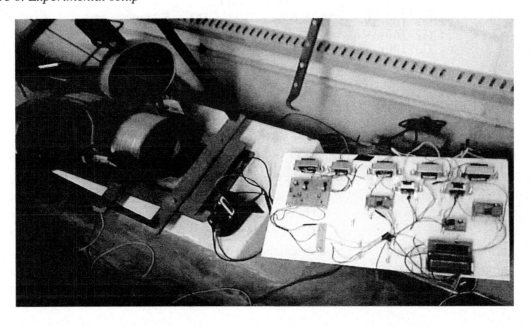

Figure 7. Measured positive sequence, negative sequence, zero sequence component waveform under healthy condition

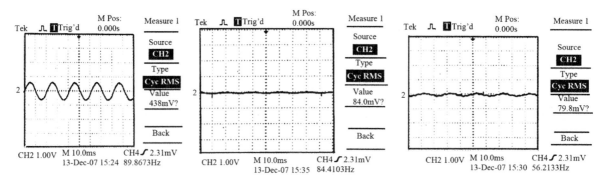

Figure 8. Measured sequence component waveform under inter turn fault

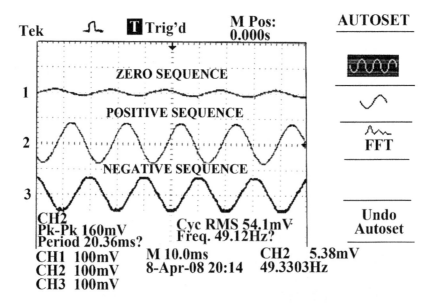

3. SIMULINK IMPLEMENTATION

The dynamic performance of an induction machine is complex because the coupling coefficients between the stator and rotor phases change continuously with the variation of rotor position 'θ_r'. This machine model can be described by differential equations with time-varying mutual inductance for dynamic analysis. But such model is complex. Different machine models are available for dynamic analysis of the machine operation. Two axis d-q model have been well tested and proven to be reliable and accurate for dynamic / transient study of the machines. The above theory is used to develop an equivalent two-phase machine of the original three phase machine. The development of the d-q model for three phase induction motor is shown through Equations 1 to 10. Since the stator turn-to-turn faults relates with number of turns it is necessary to develop the machine model in terms of number of turns. In this paper, a dynamic model considering the stator and rotor number of turns of the test machine is developed to analyze the performance of the machine in Matlab / Simulink, which is shown through Equations 11 to Equations 48. To identify the sequence components, another dynamic model is developed using Power Decomposition Technique described by Equations 49 to 64. The simulation is carried out to identify and analyze the machine performance under various conditions.

Induction Motor Dynamic Model for Healthy Case

Induction motor dynamic model for normal case can be obtained with assumption of $N_a = N_b = N_c = N_s = N_r$ i.e. number of turns in all windings are equal. The simulation model shown in Figure 9. Consists of six subsystems namely Supply block, sampling block, Inductance and resistance block, Flux block and PDT block. These blocks are developed using dynamic equations of the machine which is listed through Equations (11- to- 25).

Supply Block

$$V_{ao} = V_m \sin\omega t \tag{1}$$

$$V_{bo} = V_m \sin(\omega t - 120); \tag{2}$$

Figure 9. Induction motor model for healthy condition

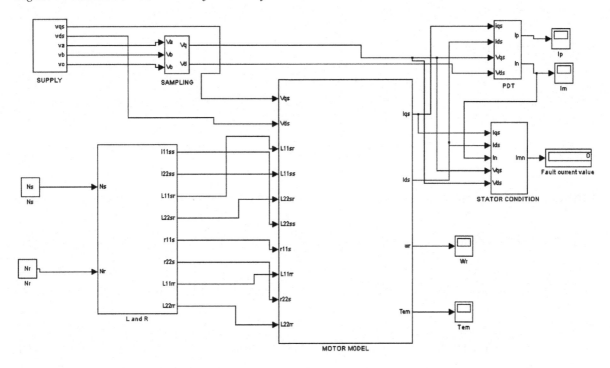

$$V_{co} = V_m \sin(\omega t - 240); \tag{3}$$

$$V_{an} = [(2/3 \times V_{ao}) - (1/3 \times V_{bo}) - (1/3 \times V_{co})]; \tag{4}$$

$$V_{bn} = [-(1/3 \times V_{ao}) + (2/3 \times V_{bo}) - (1/3 \times V_{co})]; \tag{5}$$

$$V_{cn} = [-(1/3 \times V_{ao}) - (1/3 \times V_{bo}) + (2/3 \times V_{co})]; \tag{6}$$

$$V_q^s = 2/3 \times [V_{an} - 1/2 \times (V_{bn} + V_{cn})]; \tag{7}$$

$$V_d^s = 1/\sqrt{3} \times [-V_{bn} + V_{cn}]; \tag{8}$$

Sampling Block

The input voltages V_{an}, V_{bn} and V_{cn} of the sampling block are sampled every 0.000515 seconds and 1000 samples of output voltages V_d and V_q is given by the block during this period.

$$V_d = 2/3 \times [(-0.5 \times V_{bn}) \times 0.866 + (0.5 \times V_{cn}) \times 0.866] \tag{9}$$

$$V_q = 2/3 \times [V_{an} - (0.5 \times V_{bn}) - (0.5 \times V_{cn})] \tag{10}$$

Inductance and Resistance Block

$$L_{11}^{ss} = L_{22}^{ss} = L_{ls} + L_m \tag{11}$$

$$L_{11}^{sr} = L_{22}^{sr} = (N_r / N_s) \times L_m \tag{12}$$

$$L_{12}^{sr} = L_{21}^{sr} = L_{31}^{sr} = L_{32}^{sr} = L_{13}^{sr} = L_{23}^{sr} = L_{33}^{sr} = 0 \tag{13}$$

$$L_{11}^{rr} = L_{22}^{rr} = L_{lr} + (N_r^2 / N_s^2) \times L_m \tag{14}$$

$$r_{11}^s = r_{22}^s = R_s \tag{15}$$

Motor Model Block

$$\lambda_q^s = \int (V_q^s - r_{11}^s \times I_q^s) \, dt \tag{16}$$

$$\lambda_d^s = \int (V_d^s - r_{22}^s \times I_d^s) \, dt \tag{17}$$

$$\lambda_q^r = \int (\omega_r \times \lambda_d^r - r_r^r \times I_q^r) \, dt \tag{18}$$

$$\lambda_d^r = -\int (\omega_r \times \lambda_q^r + r_r^r \times I_d^r) \, dt \tag{19}$$

$$I_q^s = (\lambda_q^s - L_{11}^{sr} \times I_q^r) / L_{11}^{ss} \tag{20}$$

$$I_d^s = (\lambda_d^s - L_{22}^{sr} \times I_d^r) / L_{22}^{ss} \qquad (21)$$

$$I_q^r = (\lambda_q^r - L_{11}^{sr} \times I_q^s) / L_{11}^{rr} \qquad (22)$$

$$I_d^r = (\lambda_d^r - L_{22}^{sr} \times I_d^s) / L_{22}^{rr} \qquad (23)$$

$$\omega_r = (P/2 \times J) \int (T_{em} + T_{mech} - T_{damp}) \qquad (24)$$

$$T_{em} = (3/2) \times (P/2) \times (\lambda_d^s \times I_q^s - \lambda_q^s \times I_d^s) \qquad (25)$$

Induction Motor Dynamic Model for Faulty Case

In order to develop an induction machine model with a stator turn-to-turn fault, it has been considered that stator phase a has two windings in series, comprising N_{us} un-shorted turns and N_{sh} shorted turns, where $N_{as} = N_{us} + N_{sh} = N_s$, the overall number of turns N_s. The stator phases b and c have $N_{bs} = N_{cs} = N_s$. The simulation model is shown in Figure 10. It also consist of six subsystem blocks – Supply block, sampling block, Inductance and resistance block, motor model block, PDT block. The dynamic equations involved in deriving the supply block, sampling block, speed and Torque block remains the same as healthy condition simulation block given in section III b. In this model the turn to turn fault is introduced by incorporating the changes in the Inductance, resistance and flux linkages Equations through 26 to 48.

Figure 10. Induction motor dynamic model for turn-to-turn fault

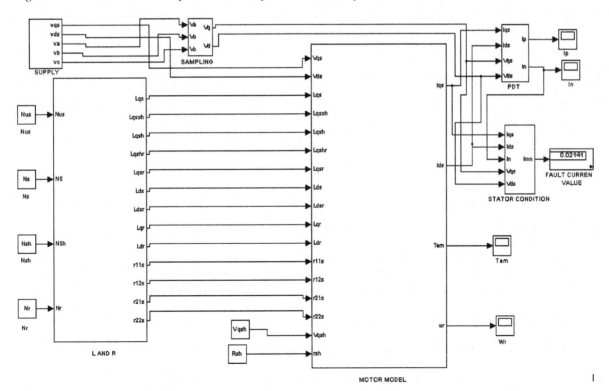

Inductance and Resistance Block

$$L_q^s = 2/3 \times [L_{asas} + 0.5 \times (L_{bsbs} + L_{bscs}) - 2 \times L_{asbs}]; \tag{26}$$

where

$$L_{asas} = N_{us}^2 / N_s^2 \times [L_{ls} + 2/3 \times L_m];$$

$$L_{bsbs} = N_s^2/N_s^2 [L_{ls} + 2/3 \times L_m];$$

$$L_{bscs} = 1/3 \times L_m;$$

$$L_{asbs} = -1/3 \times N_{us} / N_s \times L_m;$$

$$L_q^{ssh} = 2/3 \times L_{assh}; \tag{27}$$

where

$$L_{assh} = N_{us} \times N_{sh} \times [2/3 \times L_m/N_s^2];$$

$$L_q^{sh} = 2/3 \times [L_{shsh} - 2 \times L_{shbs}]; \tag{28}$$

where

$$L_{shsh} = N_{sh}^2 / N_s^2 \times [L_{ls} + 2/3 \times L_m];$$

$$L_{shbs} = -1/3 \times [N_{sh}/N_s] \times L_m;$$

$$L_q^{shr} = L_{shar}; \tag{29}$$

where

$$L_{shar} = L_m \times 2/3 \times N_r \times N_{sh} \times 1/N_s^2;$$

$$L_q^{sr} = L_{asr} + 0.5 \times L_{bsr}; \tag{30}$$

$$L_d^{sr} = 3/2 \times L_{bsr}; \tag{31}$$

where

$$L_{asr} = (2/3 \times N_{us} \times N_r \times L_m) / N_s^2;$$

$$L_{bsr} = [(2/3 \times N_s \times N_r) / N_s^2] \times L_m;$$

$$L_q^r = L_{11}^{rr} = L_{lm} + (N_r^2 / N_s^2 \times L_m); \tag{32}$$

$$L_d^r = L_{22}^{rr} = L_{lr} + (N_r^2/N_s^2) \times L_m; \tag{33}$$

$$L_d^s = L_{ls} + L_m; \tag{34}$$

$$r_{11}^s = 2/3 \times [r_{as} + rs]; \tag{35}$$

where

$$r_{as} = [(N_{us} / N_s) \times r_s + (N_{sh} / N_s) \times r_s];$$

$$r_{12}^s = \sqrt{3}/6 \times [r_{bs} - r_{cs}]; \tag{36}$$

$$r_{21}^s = r_{12}^s; \tag{37}$$

$$r_{22}^s = r_s; \tag{38}$$

Motor Model Block

$$\lambda qsh = \int (Vqsh - rsh \times Iqsh) \times dt; \tag{39}$$

$$\lambda qs = \int (Vqs - Vqsh - r11s \times Iqs - r12s \times Ids) \times dt; \tag{40}$$

$$\lambda ds = \int (Vds - r21s \times Iqs - r22s \times Ids) \times dt; \tag{41}$$

$$\lambda_q^r = \int (\omega_r \times \lambda_d^r - r_r^r \times I_q^r) \times dt; \tag{42}$$

$$\lambda_d^r = -\int (\omega_r \times \lambda_q^r + r_r^r \times I_d^r) \times dt; \tag{43}$$

$$I_q^{sh} = [\lambda_q^{sh} - L_q^{ssh} \times I_q^s - L_q^{shr} \times I_q^r] / L_q^{sh}; \tag{44}$$

$$I_q^s = [\lambda_q^s - L_q^{ssh} \times I_q^{sh} - L_q^{sr} \times I_q^r] / L_q^s; \tag{45}$$

$$I_d^s = [\lambda_d^s - L_d^{sr} \times I_d^r] / L_d^s; \tag{46}$$

$$I_q^r = [\lambda_q^r - L_q^{shr} \times I_q^{sh} - L_q^{sr} \times I_q^s] / L_q^r; \tag{47}$$

$$I_d^r = [\lambda_d^r - L_d^{sr} \times I_d^s] / L_d^r; \tag{48}$$

Power Decomposition (PDT) Block

A method to derive the positive and negative sequence components of arbitrary three-phase signals in the time domain has been proposed by describing their relevant powers in two dimensional ordinates by S.P. Huang. Arbitrary three phase signals can be divided into positive, negative and zero sequence components using time phasors as described in the conventional theory of symmetrical components. The PDT provides an alternative direct way to derive positive and negative sequence component in time domain. The values of positive sequence voltage V_{sp}, positive sequence current I_{sp}, negative sequence voltage V_{sn} and negative sequence current I_{sn} are derived from sampled voltage and current data by using PDT as shown in Equations (49 -64).

$$P_\alpha(t) = V_{qs} \times I_{qs} \tag{49}$$

$$P_\beta(t) = V_{ds} \times I_{ds} \tag{50}$$

$$Q_\alpha(t) = -V_{ds} \times I_{qs} \tag{51}$$

$$Q_\beta(t) = V_{qs} \times I_{ds} \tag{52}$$

$$P_\alpha = (1/T) \int_0^T (P_\alpha(t))dt \tag{53}$$

$$P_\beta = (1/T) \int_0^T (P_\beta(t))dt \tag{54}$$

$$Q_\alpha = (1/T) \int_0^T (Q_\alpha(t))dt \tag{55}$$

$$Q_\beta = (1/T) \int_0^T (Q_\beta(t))dt \tag{56}$$

$$P^i_{\alpha p} = P_\alpha + P_\beta \tag{57}$$

$$Q^i_{\alpha p} = Q_\alpha + Q_\beta \tag{58}$$

$$P^i_{\alpha n} = P_\alpha - P_\beta \tag{59}$$

$$Q^i_{\alpha n} = Q_\alpha - Q_\beta \tag{60}$$

$$I_{sn} = P^i_{\alpha n} + j Q^i_{\alpha n} \tag{61}$$

$$I_{sp} = P^i_{\alpha p} + j Q^i_{\alpha p} \tag{62}$$

$$V_{sn} = P^v_{\alpha n} + j Q^v_{\alpha n} \tag{63}$$

$$V_{sp} = P^v_{\alpha p} + j Q^v_{\alpha p} \tag{64}$$

Stator Condition Monitoring Block

From the sequence voltage V_{sn} and current I_{sn}, the sequence impedances Z_{sn} are calculated as

$$Z_{sn} = V_{sn} / I_{sn} \tag{65}$$

In the presence of a stator fault, the observed negative sequence current I_n is considered as the sum of a negative sequence component I_{sn}, caused by supply voltage unbalance V_{sn}, plus a component arising from the stator winding fault I_{mn},

i.e. $I_n = I_{sn} + I_{mn}$ (66)

Hence, the stator fault is associated with a fault current I_{mn}, given by

$I_{mn} = I_n - I_{sn}$ (67)

The magnitude and phase of this current component I_{mn} indicates the unbalance state of the stator due to winding fault.

Results and Conclusions

Simulation results of Induction Motor under healthy condition for speed, Torque, positive and negative sequence component is shown in Figure 12. These figures clearly indicate that there is no pulsating torque and speed attains its final steady state after definite time interval. For normal operation of Induction Motor, only the positive sequence component will persist under steady state. The simulation results also indicate the same. The three phase motor current also shows equal magnitude with 120° phase shifts, confirming the balance condition as shown in Figure 11. Simulation results of Induction Motor under Turn to turn faulty condition as indicated in Figure 13, has torque pulsations even under steady state condition. This torque pulsation is attributed to the effect of Turn to turn fault. The magnitude of positive and negative sequence components of current clearly indicates that there is an unbalanced operat-

Figure 11. Stator current Ia, Ib, Ic of motor under healthy conditions

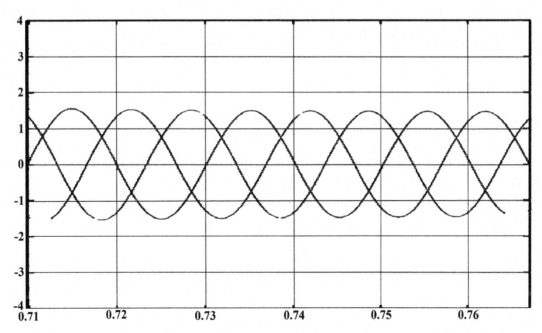

Figure 12. Torque, speed and positive and negative sequence component of motor under healthy conditions

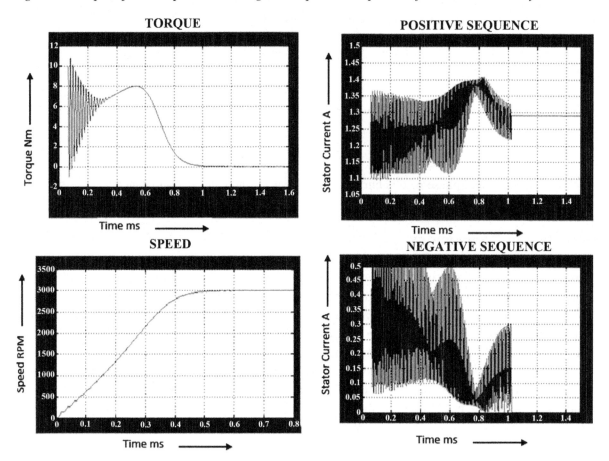

ing condition developed due to stator turn to turn fault. This unbalanced condition is also observable in three phase currents shown in Figure 14. The data for various case studies for turn-to-turn faults are listed in Table 2. The increase in I_{mn} clearly shows increase in short circuited number of turns. So the simulation results confirms that the model used for simulation is reliable and accurate in identifying the motor operating conditions.

Healthy Conditions

Machine Data: Squirrel cage induction motor 400V, 50 Hz, 2 Pole, and Nr = 1440 rpm, I=3A, delta connection.

Machine Parameter: The following machine parameters are found by conducting No load test, blocked rotor test and dc test.

N_s=560; N_r=560; V_m=375; f=50Hz; L_{ls}=L_{lr}=0.03157; L_m=0.71769; r_s=7.8; r_r=7.508642; T_{mech}=T_{damp}=0; P=2; J=0.013.

Figure 13. Torque, Speed, Positive and negative sequence component of motor under inter turn fault conditions

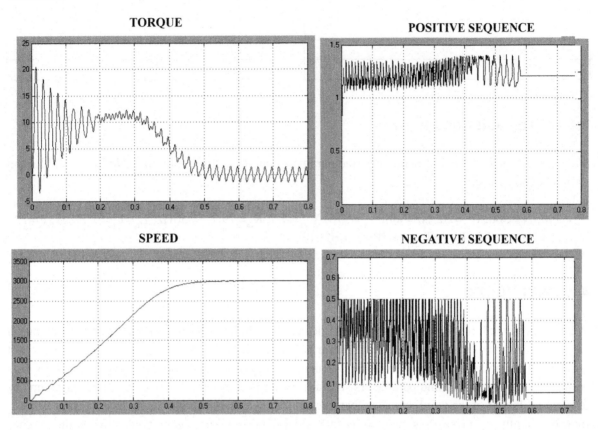

Figure 14. Stator current Ia,Ib,Ic of motor under inter turn fault conditions

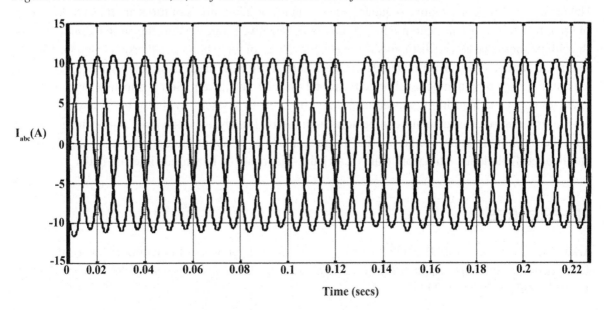

Table 2.

Nsh	In	Imn
45	161.3	76.64
50	170.3	169.8
55	254.6	254

Inter Turn Fault Conditions

Machine Parameter: The following machine parameters are found by conducting No load test, blocked rotor test and dc test.

N_{us}=540; N_s=560; N_{sh}=20; Nr=560; V_m=375; V_{qsh}=4.5; R_{sh}=79.7; f=50Hz; L_{ls}=L_{lr}=0.03157; r_s=7.8; L_m=0.71769; r_r=7.5; T_{mech}=T_{damp}=0; P=2; J=0.013.

4. CONCLUSION

We have presented the turn to turn stator fault detection technique for the induction motor. Simulation was done for various cases using PDT technique in Matlab /Simulink by changing the values of shorted turns N_{sh}. Two models have been developed for analyze an asymmetrical induction motor. These models are based on general machine parameters so that it is not necessary to know detailed motor geometry or physical layout of the windings. New parameters for asymmetrical conditions have been presented. Inter-turn faults can be easily simulated by the model. Fault severity can be controlled by the number of shorted turns and an optional current limiting resistance to short circuit the windings. These models have been used to study the relation between the number of shorted turns and negative sequence current. The results confirm that negative sequence current and shorted numbers of turns are linearly dependent and that resistive unbalance can produce comparable negative sequence current, so that fault detection algorithms should be able to distinguish between the effects of turn shorts and thermal unbalance. These results are obtained for short circuit under limited short circuit current (by external resistance) reveal that sensitivity depends on number of shorted turns and short circuit current. Self-repairing electrical drives based upon genetic-algorithm-assisted neural and fuzzy neural systems will also be widely used in the near future. Future enhancements can be done with the FFT analysis on the three phase currents taken from the machine to know the harmonics in the machine at different fault conditions using Artificial Intelligence Techniques. This work can be carried to diagnose the hazardous internal faults.

REFERENCES

Arkan, M., & Unsworth, P. J. (1999, October). Stator fault diagnosis in Induction motors using Power decomposition. In *Proceedings of the IEEE Industrial Application Conference 34th Annual Meeting.* doi:10.1109/IAS.1999.805999

Eltamaly, A. M., Alolah, A. L., & Hamouda, R. M. (2006). MATLAB simulation of three-phase SCR controller for three-phase induction motor. *Mansoura Engineering Journal, 31*(3), 213–234.

Hadidi, R., Mazhari, I., & Kazemi, A. (2007). Simulation Induction Motor with Svc. In *Proceedings of the Second IEEE Conference on Industrial Electronics and Application.*

Joksimvoic, G. M., & Penman, J. (2000). Detection of inter turn short circuits in the stator windings of operating motors. In *Proceedings of the 24th International Conference of the IEEE Industrial Electronics Society.* doi:10.1109/41.873216

Kliman, G. B., Premerlani, W. J., Koegl, R. A., & Hoewelerl, D. (1996). A new approach to on-line turn fault detection in AC motors. In *Proceedings of the IEEE Industrial Application Conference 34th Annual Meeting.* doi:10.1109/IAS.1996.557113

Marcu, M., Utu, I., Pana, L., & Orban, M. (2004). Computer simulation of real time identification for induction motor drives. In *Proceedings of the International Conference on Theory and Applications of Mathematics and Informatics.*

Moradi, M. H., & Khorasani, P. G. (2008). A new Matlab simulation of induction motor. In *Proceedings of the Australasian Universities Power Engineering Conference.*

Ozpineci, B., & Tolbert, L. M. (2003). Simulink implementation of induction motor model – A modular approach. In *Proceedings of the International Conference on Electric Machines and Drives.* doi:10.1109/IEMDC.2003.1210317

Siddique, A., Yadava, G. S., & Singh, B. (2005). A Review of stator fault monitoring techniques of induction motors. *IEEE Transactions on Energy Conversion, 20*(1), 215–223. doi:10.1109/TEC.2004.837304

Stefanov, P. C., & Stank Vic, A. M. (2002). Dynamic phasor in modeling and analysis of unbalanced polyphase ac machines. *IEEE Transactions on Energy Conversion, 17*(1), 107–113. doi:10.1109/60.986446

Ye, Z., Wu, B., & Zargari, N. (2000). Modelling and simulation of induction motor with mechanical fault based on winding function method. In *Proceedings of the 26th International Conference of the IEEE Industrial Electronics Society.*

Chapter 14
Optimal Reactive Power Dispatch Incorporating TCSC-TCPS Devices Using Different Evolutionary Optimization Techniques

Provas Kumar Roy
Jalpaiguri Government Engineering College, India

Susanta Dutta
Dr. B. C. Roy Engineering College, India

Debashis Nandi
National Institute of Technology, India

ABSTRACT

The chapter presents two effective evolutionary methods, namely, artificial bee colony optimization (ABC) and biogeography based optimization (BBO) for solving optimal reactive power dispatch (ORPD) problem using flexible AC transmission systems (FACTS) devices. The idea is to allocate two types of FACTS devices such as thyristor-controlled series capacitor (TCSC) and thyristor-controlled phase shifter (TCPS) in such a manner that the cost of operation is minimized. In this paper, IEEE 30-bus test system with multiple TCSC and TCPS devices is considered for investigations and the results clearly show that the proposed ABC and BBO methods are very competent in solving ORPD problem in comparison with other existing methods.

1. INTRODUCTION

The secure operation of power system has become an important and critical issue in today's large, complicated, and interconnected power systems. Security constraints such as thermal limits of transmission lines and bus voltage limits must be satisfied under any operating point. The best alternative solution of improvement of the security of power system is the use of Flexible AC Transmission Systems (FACTS)

DOI: 10.4018/978-1-4666-9755-3.ch014

devices. FACTS devices can be used to reduce the flow of power on the overloaded line and to increase the use of alternative paths to improve power transmission capacity. This allows existing transmission and distribution systems to operate under normal operating conditions and allows the load lines to operate much closer to their thermal limits. Now-a-days these power electronics based devices have become very popular in improving the overall performances of power system under both steady state and dynamic condition.

Optimal reactive power dispatch (ORPD) problem is one of the major issues in operation of power systems. Because of its significant influence on secure and economic operation of power systems, ORPD has received an ever-increasing interest from electric utilities. On the other hand, FACTS devices can reduce the power flows of heavily loaded lines, maintain the bus voltages at desired level, and consequently, they can reduce the transmission loss. Therefore, the optimal reactive power dispatch incorporating FACTS devices has become an important area of research in recent years and it motivates the authors to work on FACTS based ORPD problem whose main objective is to minimize the network losses at improve voltage level and maintain the power system under normal operating conditions.

In this study, flexible AC transmission system (FACTS) devices are considered as additional control parameters in the ORPD formulation. Static models of two FACTS devices consisting of thyristor controlled series compensator (TCSC) and thyristor controlled phase shifter (TCPS) are used in the present work. Minimization of the transmission loss and voltage deviation are achieved by finding suitable values of FACTS devices along with other control variables such as generator bus voltage, reactive power generation and transformer tap settings in the original ORPD problem.

A number of numerical optimization techniques have been proposed to solve the traditional ORPD problem, such as quadratic programming (Chung & Yun 1998), Newton's method (Ambriz-Pérez, Acha & Fuerte-Esquivel, 2006), linear programming (Sundar & Ravikumar, 2012), interior-point method (Rakpenthai, Premrudeepreechacharn & Uatrongjit, 2009), etc. However, these techniques have limitations in handling nonlinear, discontinuous functions and constraints, and often lead to a local minimum point (Chung & Yun, 1998). But due to non-continuous, non-linear and non-differential objective function and constraints of FACTS based ORPD, these methods are unable to locate the global optimum solution. Recently, some new heuristic methods like evolutionary programming (EP) "Ma (2003);Ongsakul and Jirapong (2005)", simulated annealing (SA) (Majumdar, Chakraborty, Chattopadhyay & Bhattacharjee, 2012), hybrid tabu search (Bhasaputra & Ongsakul, 2002), genetic algorithm (GA) (Cai & Erlich 2003; Ippolito, Cortiglia & Petrocelli, 2006; Mahdad, Srairia & Bouktir, 2010; Wirmond, Fernandes, & Tortelli, 2011), particle swarm optimization (PSO) (Benabid, Boudour & Abido, 2009; Kennedy & Eberhart, 1995; Mollazei, Farsangi & Nezamabadi-pour 2007; Mondal, Chakrabarti & Sengupta, 2012; Roy, Ghoshal & Thakur, 2010; Saravanan, Slochanal, Venkatesh & Abraham, 2007), differential evolution (DE) (Basu, 2008; Shaheen, Rashed & Cheng, 2011), ant colony optimization (ACO) (Sreejith, Chandrasekaran & Simon, 2009), bacteria foraging optimization (BFO) (Abd-Elazim & Ali, 2012), biogeography based optimization (BBO) (Roy, Ghoshal & Thakur, 2011; Roy, Ghoshal & Thakur, 2012), immune algorithm (Taher & Amooshahi, 2011), harmony search algorithm (HSA) (Sirjani, Mohamed & Shareef, 2012), gravitational search algorithm (GSA) (Bhattacharya & Roy, 2012; Roy, Mandal & Bhattacharya, 2012; Sonmez, Duman, Guvenc & Yorukeren, 2012), have been developed for the nonlinear optimization problem. These methods can generate high-quality solutions and have stable convergence characteristic when dealing with ORPD problem and other complex optimization problems in power systems.

In the literature, the optimal reactive power dispatch (ORPD) and setting of FACTS devices has retained the interest of worldwide researchers in power systems, where various methods and criteria are used in this field. Padhy *et al.* (Padhy, Abdel-Moamen & Kumar, 2004) used TCSC, to minimize an objective function that was composed of complex power lost in transmission lines (i.e. real and reactive line losses) and the generated reactive power. Benabid *et al.* (Benabid, Boudour & Abido, 2009) reported non-dominated sorting PSO to identify optimal location and settings of SVC and TCSC devices for multi-objective optimization problem. This method was used to improve voltage stability margin, to reduce real power losses, and to reduce load voltage deviation. The proposed method was tested on IEEE 30-bus system and Algerian 114-bus power system and showed excellent results. BFO technique was developed by Tripathy *et al.* (Tripathy & Mishra, 2007) to solve the multi-objective OPF problem that optimized the real power losses and voltage stability limit simultaneously. The control variables considered were the UPFC optimal placement and setting in addition to the existing transformer settings.

Marouani *et al.* (Marouani, Guesmi, Abdallah & Ouali, 2009) presented a multi-objective evolutionary algorithm to solve ORPD problem with FACTS devices. They used UPFC to regulate the power flow such that the real power losses and voltage deviation were minimized. IEEE 6-bus test system was considered for successful implementation of the proposed approach. Preedavichit *et al.* (Preedavichit & Srivastava, 1997) used three different FACTS devices, namely TCSC, TCPS and SVC to solve ORPD problem. A SQP was used to locate the optimal placements and settings of TCSC, TCPS and SVC devices. A technique based on LP to solve ORPD in AC–DC power systems using FACTS controllers with an objective of minimization of sum of the squares of the voltage deviations was proposed by Thukaram *et al.* (Thukaram & Yesuratnam, 2008). The proposed approach was tested on a real life equivalent 96-bus AC system and a two terminal DC system using UPFC. Sharifzadeh *et al.* (Sharifzadeh & Jazaeri, 2009) proposed a PSO based method to determine the optimal settings of different control variables, TCSC and SVC into the ORPD problem that minimizes active power loss. Their proposed approach was implemented on IEEE 30-bus and IEEE 118-bus systems and appeared to give better optimal results than DE and GA. An application employing GA in solving the ORPD problem was presented by Sadeghzadeh *et al.* (Sadeghzadeh, Khazali & Zare, 2009). In this case, GA algorithm was used to solve ORPD problem to minimize transmission loss and voltage deviation with the inclusion of additional control variables of UPFC and TCPAR settings. The feasibility of the proposed method was tested on IEEE 30-bus system. Fuzzy based evolutionary algorithm used to solve ORPD incorporating SVC and TCSC was developed by Bhattacharyya *et al.* (Bhattacharya and Gupta, 2014). Ghasemi *et al.* (Ghasemi, Valipour & Tohidi, 2014) proposed a chaotic parallel vector evaluated interactive honey bee mating optimization (CPVEIHBMO) approach to find the feasible solution of the multi objective ORPD problem with considering operational constraints of the generators. Moreover, Vasant et al proposed various new optimization techniques (Vasant, Barsoum, & Webb, 2012; Vasant, 2013; Vasant, 2014; Vasant, 2015) to solve nonlinear optimization problem.

It may be found from the literature that most of these population based optimization techniques have successfully been implemented to solve ORPD problem. However, many of them suffer from local optimality. This motivate the present authors to introduce simple, efficient and fast population based optimization techniques to solve FACTS based ORPD problem. Artificial bee colony (ABC) is a new population-based heuristic search algorithm, which describes how the foraging behavior of honeybees can be modeled to solve general optimization problems. The application of ABC to solve non-linear optimization problem has been first presented by Karaboga (Karaboga & Basturk, 2007a; Karaboga & Basturk, 2007b). Later, ABC algorithm has successfully been applied for solving OPF (Ayan & Kılıç, 2012), and ORPD (Khorsandi, Hosseinian & Ghazanfari, 2013), problems. However, this algorithm has never been applied in FACTS based ORPD problem. In this paper, two objective functions, namely minimization of real power losses in transmission lines and minimization of summation of voltage deviations of load busses are used as objectives and the proposed ABC method is implemented on IEEE 30-bus with multiple TCSC and TCPS devices. To show the effectiveness and superiority, the performance of the proposed technique is compared with other popular optimization techniques.

The remainder of the paper is organized as follows. In Section 2, the problem formulation is presented. In Section 3, brief introduction of artificial bee colony optimization is made. In Section 4, the various steps of three different computational intelligence techniques namely, (PSOIWA), (RGA) and ABC algorithms are described. In Section 5, the ABC algorithm based on ORPD incorporating TCSC and TCPS devices is developed. In Section 6, the studies of application cases are presented and demonstrate the potential of the presented algorithm. Finally, in Section 7, the conclusions are given.

Table 1. Nomenclature/abbreviation

FACTS	Flexible AC Transmission Systems
ORPD	Optimal reactive power dispatch
OPF	Optimal power flow
TCSC	Thyristor controlled series compensator
TCPS	Thyristor controlled phase shifter
UPFC	Unified power flow controller
SVC	Static var compensator
EP	Evolutionary programming
SA	Simulated annealing
GA	Genetic algorithm
PSO	Particle swarm optimization
DE	Differential evolution
GSA	Gravitational search algorithm
ACO	Ant colony optimization
ABC	Artificial bee colony
BFO	Bacteria foraging optimization
HSA	Harmony search algorithm
NG	Number of generator bus
NL	Number of load bus
NTL	Number of transmission lines
NT	Number of regulating transformers
NC	Number of shunt compensators

continued on following page

Table 1. Continued

Symbol	Description
V_{ref}	Reference voltage of the load buses
V_{L_i}	Load voltage of ith bus
P_{mn}, Q_{mn}	Real and reactive power flows from m^{th} bus to n^{th} bus
P_{nm}, Q_{nm}	Real and reactive power flows from n^{th} bus to m^{th} bus
V_m, V_n	Voltage magnitudes at m^{th} bus and n^{th} bus, respectively
∂_m, ∂_n	Phase angle of mth bus and n^{th} bus voltages
r_{mn}, x_{mn}	Resistance and reactance of transmission line that connected between m^{th} bus and n^{th} bus;
x_k	Reactance of TCSC placed in k^{th} line that connected between m^{th} bus and n^{th} bus
g_{mn}, b_{mn}	Conductance and susceptance of the line connected between m^{th} bus and n^{th} bus
$V_{Gi}^{min}, V_{Gi}^{max}$	Minimum and maximum voltage limits of ith generator bus
$P_{Gi}^{min}, P_{Gi}^{max}$	Minimum and maximum active power generation limits of ith bus
$Q_{Gi}^{min}, Q_{Gi}^{max}$	Minimum and maximum reactive power generation limits of ith bus
$V_{Li}^{min}, V_{Li}^{max}$	Minimum and maximum voltage limits of ith load bus
S_{Li}, S_{Li}^{max}	Actual and maximum apparent power flow limit of ith branch
T_i^{min}, T_i^{max}	Minimum and maximum tap setting limits of ith regulating transformer
$Q_{Ci}^{min}, Q_{Ci}^{max}$	Minimum and maximum var injection limits of ith shunt compensator
$x_{k_i}^{min}, x_{k_i}^{max}$	Minimum and maximum reactance limits of ith TCSC
$\alpha_i^{min}, \alpha_i^{max}$	Minimum and maximum phase shift angle limits of ith TCPS
$NTCSC$	Number of TCSC devices
$NTCPS$	Number of TCPS devices
E_{max}	Maximum emigration rate
$N_{S\,max}$	Maximum number of species

continued on following page

Table 1. Continued

N_{S0}	Equilibrium number of species
I_r	Immigration rate
E_r	Emigration rates
I_{max}	Maximum immigration rate
SIV	Suitability index variables
HSI	Habitat suitability index
v_{id}, x_{id}	Velocity and position of dth dimension of ith particle
rand	Uniform random number in the range [0, 1]
c1, c2	Cognitive and the social parameters
$pBest_{id}$	Position best of dth dimension of ith particle
$gBest_d$	Group best position of dth dimension of ith particle
w	Inertia weight factor
w_{max}, w_{min}	Initial inertia weight and final inertia weight
$iter_{max}$	Maximum iteration cycles
iter	Current iteration cycle

2. MATHEMATICAL PROBLEM FORMULATION

2.1 Static Model of Facts Devices

Although, there are several FACTS devices for controlling power flow and voltage profile in power system, for this study, only TCSC, TCPS devices have been considered to improve the voltage profile and to minimize the transmission loss. Static models of these two FACTS devices are described below.

2.1.1 Thyristor Controlled Series Compensator (TCSC)

The effect of TCSC on a network may be considered as a controllable reactance inserted in the related transmission line (Acha & Ambriz-Perez, 2000). The model of the network with (TCSC) is shown in Figure 1.

The power flow equations of the branch (connected between the mth bus and n^{th} bus) may be derived as follows:

$$P_{mn} = V_m^2 G_{mn} - V_m V_n \left[G_{mn} \cos(\partial_m - \partial_n) + B_{mn} \sin(\partial_m - \partial_n) \right] \quad (1)$$

$$Q_{mn} = -V_m^2 B_{mn} - V_m V_n \left[G_{mn} \sin(\partial_m - \partial_n) - B_{mn} \cos(\partial_m - \partial_n) \right] \quad (2)$$

where

$$G_{mn} = \frac{r_{mn}}{r_{mn}^2 + (x_{mn} - x_k)^2}, \quad B_{ij} = \frac{(x_{mn} - x_k)}{r_{mn}^2 + (x_{mn} - x_k)^2}; \quad P_{mn}, Q_{mn}$$

are the real and reactive power flows from m^{th} bus to n^{th} bus; V_m, V_n are the voltage magnitudes at m^{th} bus and n^{th} bus respectively; ∂_m, ∂_n are the phase angle of mth bus and n^{th} bus voltages; r_{mn}, x_{mn} are the resistance and reactance of transmission line that connected between m^{th} bus and n^{th} bus; x_k is the reactance of TCSC placed in k^{th} line that connected between m^{th} bus and n^{th} bus.

2.1.2 Thyristor Controlled Phase Shifter (TCPS)

A phase shifter model (Mathur & Basati, 1981) can be represented by an equivalent circuit, which is shown in Figure 2.

It consists of admittance connected in series with transformer having a complex turns ratio $k\angle\varphi$. The power flows of the line (connected between m^{th} and n^{th} bus) having a phase shifter can be written as:

$$P_{mn} = k^2 V_m^2 g_{mn} - k V_m V_n \left[g_{mn} \cos(\partial_m - \partial_n + \varphi) + b_{mn} \sin(\partial_m - \partial_n + \varphi) \right] \quad (3)$$

$$Q_{mn} = -k^2 V_m^2 b_{mn} - k V_m V_n \left[g_{mn} \sin(\partial_m - \partial_n + \varphi) - b_{mn} \cos(\partial_m - \partial_n + \varphi) \right] \quad (4)$$

Figure 1. Circuit model of TCSC connected between m^{th} bus and n^{th} bus

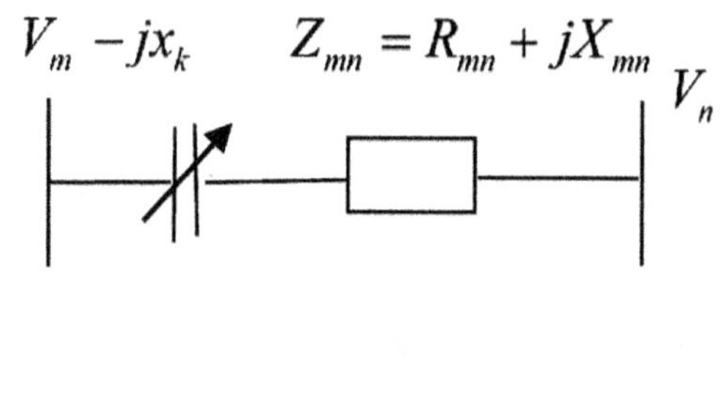

Figure 2. Circuit model of TCPS connected between m^{th} bus and n^{th} bus

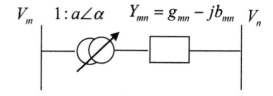

$$P_{nm} = V_n^2 g_{nm} - kV_n V_m \left[g_{mn} \cos(\partial_m - \partial_n + \varphi) - b_{mn} \sin(\partial_m - \partial_n + \varphi) \right] \tag{5}$$

$$Q_{nm} = -V_n^2 b_{mn} + kV_m V_n \left[g_{mn} \sin(\partial_m - \partial_n + \varphi) + b_{mn} \cos(\partial_m - \partial_n + \varphi) \right] \tag{6}$$

where P_{mn}, Q_{mn} are the real and reactive power flows from m^{th} bus to n^{th} bus; P_{nm}, Q_{nm} are the real and reactive power flows from n^{th} bus to m^{th} bus; g_{mn}, b_{mn} are the conductance and susceptance of the line connected between m^{th} bus and n^{th} bus.

2.2 Objective Function

The conventional formulation of ORPD problem determines the optimal setting of control variables such as generator terminal voltages, transformers tap setting, reactive power of shunt compensators, reactance values of TCSCs and phase shifting angles of TCPSs while minimizing two objective functions.

The first objective function is used to minimize the voltage deviation of all load buses and it may be expressed as follows: $f_1(x,y) = \sum_{i=1}^{NL} \left| \left(V_{l_i} - V_{ref} \right) \right|$ (7) where, $f_1(x,y)$ is the voltage deviation minimization objective function; x is the vector of dependent variable consisting of load voltages ($V_{l_1}, \cdots, V_{l_{NL}}$), generators' reactive powers ($Q_{g_1}, \cdots, Q_{g_{NG}}$), transmission lines' loadings ($S_{l_1}, \cdots, S_{l_{NTL}}$), reactance of TCSC devices ($x_{k_1}, \ldots, x_{k_{NTCSC}}$) and phase shifting angle of TCPS devices ($\alpha_1, \ldots, \alpha_{NTCPS}$); y is the vector of independent variables consisting of generators' voltages ($V_{g_1}, \cdots, V_{g_{NG}}$), transformers' tap settings ($T_1, \cdots, T_{NT}$), reactive power injections ($Q_{C_1}, \cdots, Q_{C_{NC}}$), active power flow through transmission line having TCSC devices ($PF_{TCSC_1}, \cdots, PF_{TCSC_{NTCSC}}$) and active power flow through transmission line having TCPS devices ($PF_{TCPS_1}, \cdots, PF_{TCPS_{NTCPS}}$).

Therefore x & y can be expressed as:

$$x = [V_{l_1}, \cdots, V_{l_{NL}}, Q_{g_1}, \cdots, Q_{g_{NG}}, S_{l_1}, \cdots, S_{l_{NTL}}, x_{k_1}, \ldots, x_{k_{NTCSC}}, \alpha_1, \ldots, \alpha_{NTCPS}] \tag{8}$$

$$y = [V_{g_1}, \cdots, V_{g_{NG}}, T_1, \cdots, T_{NT}, Q_{C_1}, \cdots, Q_{C_{NC}}, PF_{TCSC_1}, \cdots, PF_{TCSC_{NTCSC}}, PF_{TCPS_1}, \cdots, PF_{TCPS_{NTCPS}}] \tag{9}$$

where NG, NL are the number of generator and load buses; NTL, NT, NC are the number of transmission lines, regulating transformers and shunt compensators; V_{ref}, V_{L_i} are the reference voltage of the load buses and load voltage of ith bus.

The second objective function is used to minimize the transmission loss and is given by

$$f_2(x,y) = P_{loss} = \sum_{k=1}^{NTL} G_k \left[V_m^2 + V_n^2 - 2|V_m||V_n| \cos(\partial_m - \partial_n) \right] \tag{10}$$

where, $f_2(x,y)$ is the transmission loss minimization objective function; P_{loss} is the total power loss; G_k is the conductance of kth line connected between m^{th} and n^{th} buses; V_m, V_n are the voltage of mth and n^{th} buses respectively; ∂_m, ∂_n are the phase angle of mth and n^{th} bus voltages.

2.3 Constraints

The ORPD with TCSC and TCPS devices is subjected to the following constraints.

2.3.1 Equality Constraints

$$\begin{cases} \sum_{i=1}^{NB}(P_{Gi}-P_{Li}) = \sum_{i=1}^{NB}\sum_{j=1}^{NB} V_i V_j \left[G_{ij} \cos\theta_{ij} + B_{ij} \sin\theta_{ij} \right] \\ \sum_{i=1}^{NB}(Q_{Gi}-Q_{Li}) = -\sum_{i=1}^{NB}\sum_{j=1}^{NB} V_i V_j \left[G_{ij} \sin\theta_{ij} - B_{ij} \cos\theta_{ij} \right] \end{cases} \quad (11)$$

where P_{Li} is the active power demand of ith bus; Q_{Gi}, Q_{Li} are the reactive power generation and demand of ith bus; G_{ij}, B_{ij} are the conductance and susceptance of the line connected between i^{th} bus and j^{th} bus and NB is number of buses.

2.3.2 Inequality Constraints

$$\begin{cases} V_{Gi}^{\min} \leq V_{Gi} \leq V_{Gi}^{\max} \\ P_{Gi}^{\min} \leq P_{Gi} \leq P_{Gi}^{\max} \quad i \in NG \\ Q_{Gi}^{\min} \leq Q_{Gi} \leq Q_{Gi}^{\max} \end{cases} \quad (12)$$

$$V_{Li}^{\min} \leq V_{Li} \leq V_{Li}^{\max} \quad i \in NL \quad (13)$$

$$S_{Li} \leq S_{Li}^{\max} \quad i \in NTL \quad (14)$$

$$T_i^{\min} \leq T_i \leq T_i^{\max} \quad i \in NT \quad (15)$$

$$Q_{Ci}^{\min} \leq Q_{Ci} \leq Q_{Ci}^{\max} \quad i \in NC \quad (16)$$

$$x_{k_i}^{\min} \leq x_{k_i} \leq x_{k_i}^{\max} \quad i \in NTCSC \quad (17)$$

$$\alpha_i^{\min} \leq \alpha_i \leq \alpha_i^{\max} \quad i \in NTCPS \quad (18)$$

where $V_{Gi}^{\min}, V_{Gi}^{\max}$ are the minimum and maximum voltage limits of ith generator bus; $P_{Gi}^{\min}, P_{Gi}^{\max}$ are the minimum and maximum active power generation limits of ith bus; $Q_{Gi}^{\min}, Q_{Gi}^{\max}$ are the minimum and maximum reactive power generation limits of ith bus; $V_{Li}^{\min}, V_{Li}^{\max}$ are the minimum and maximum voltage limits of ith load bus; $S_{Li}, S_{Li}^{\max}$ are the apparent power flow and maximum apparent power flow limit of ith branch; $T_i^{\min}, T_i^{\max}$ are the minimum and maximum tap setting limits of ith regulating transformer; $Q_{Ci}^{\min}, Q_{Ci}^{\max}$ are the minimum and maximum reactive power injection limits of ith shunt compensator; $x_{k_i}^{\min}, x_{k_i}^{\max}$ are the minimum and maximum reactance limits of ith TCSC; $\alpha_i^{\min}, \alpha_i^{\max}$ are the minimum and maximum phase shift angle limits of ith TCPS; NG, NL, NTL, NT, NC are the number of generator bus, load bus, transmission line, regulating transformer and shunt compensator; respectively $NTCSC$, $NTCPS$ are the number of TCSC and TCPS devices.

3. OVERVIEW OF OPTIMIZATION

3.1 Artificial Bee Colony (ABC)

Artificial Bee Colony (ABC) is one of the most recently developed population based optimization technique developed by Karaboga and Basturk in 2006. ABC is based on adoptive process of natural system. In this method a artificial system is designed that retains the robustness of natural systems. It is a swarm intelligence approach inspired by the collective intelligent behavior of Bees colony. The intelligence behavior of honey bees consists of three essential components.

1. Food Sources which are evaluated by each bee's distances from the hive and amount of nectar in it. The nectar of the food source represents the quality or fitness of the solution.
2. Employed foragers those are currently associated with the food source, exploit it and carry the information about the source. They abandon worse sources after discovering better sources.
3. Unemployed foragers those look for the new food sources. The unemployed foragers are of two types: onlooker and scout.

Onlooker bees wait in the nest and establish a food source through the information shared by employed foragers and they employed themselves to the most profitable food source. Scouts bees search the new food source. They fly and choose the food sources randomly without using experience. Each food source chosen represents a possible solution to the problem under consideration. The number of scout bees is generally taken between 5% -10% of the population.

3.2 Biogeography Based Optimization (BBO)

3.2.1 Biogeography

Biogeography is nature's way of distributing species. Figure 3 describes a model of species abundance in a habitat. The immigration curve of Figure 3 shows that immigration rate (I_r) to the habitat is maximum ($I_{\max}$), when there are zero species in the habitat. As the number of species increases, the habitat becomes more crowded, fewer species are able to successfully survive immigration to the habitat, and the immigration rate decreases. The largest possible number of species that the habitat can support is $N_{S\max}$, at which the immigration rate becomes zero. The emigration curve of Figure 3 depicts that if there is no species in the habitat, the emigration rate will be zero. As the number of species increases, the habitat becomes more crowded, more species are able to leave the habitat to explore other possible residences and the emigration rate increases. The maximum emigration rate is $E_{\max}$, which occurs when the habitat contains the largest number of species ($N_{S\max}$). The equilibrium number of species is N_{S0}, at which the immigration rate (I_r) and emigration rates (E_r) are equal. The immigration and emigration curves in Figure 3 are shown as straight lines but, in general, they might be more complicated curves. Nevertheless, this simple model gives us a general description of the process of immigration and emigration.

From Figure 3, I_r and E_r for N_S number of species may be formulated as follows:

$$I_r = I_{\max} \cdot \left(1 - \frac{N_S}{N_{S\max}}\right) \tag{19}$$

$$E_r = \frac{E_{\max} \cdot N_S}{N_{S\max}} \tag{20}$$

Mathematically the concept of emigration and immigration can be represented by a probabilistic model. P_{N_S} is the probability of the habitat that contains exactly N_S species. P_{N_S} Changes from time t to time $t + \Delta t$ as follows:

Figure 3. Species model of a single habitat

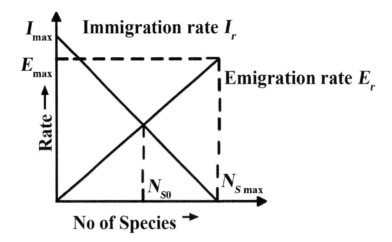

$$P_{N_S}(t+\Delta t) = P_{N_S}\left(1 - I_r\Delta t - E_r\Delta t\right) + P_{N_{s-1}} I_{r_{S-1}} \Delta t + N_{P_{S+1}} I_{r_{S+1}} \Delta t \qquad (21)$$

where, $I_{r_{S+1}}, I_{r_{S-1}}$ are the immigration rate when there are N_{S+1} and N_{S-1} species in the habitat; $E_{r_{S+1}}, E_{r_{S-1}}$ are the emigration rate when there are N_{S+1} and N_{S-1} species in the habitat.

This equation holds because in order to have N_S species at time $t + \Delta t$, one of the following conditions must hold:

1. There are N_S species at time t, and no immigration or emigration occurs between t and $t + \Delta t$;
2. There are N_{S-1} species at time t, and one species immigrates;
3. There are N_{S+1} species at time t, and one species emigrates.

If time t is small enough so that the probability of more than one immigration or emigration can be ignored then taking the limit of (21) as $\Delta t \to 0$ gives the following equation:

$$\dot{P}_{N_S} = \begin{cases} -(I_r + E_r)P_{N_S} + E_{rs+1}P_{Ns+1} & N_S = 0 \\ -(I_r + E_r)P_{Ns} + I_{rs-1}P_{Ns-1} + E_{rs+1}P_{Ns+1} & 1 \le N_S \le N_{S\max}-1 \\ -(I_r + E_r)P_{Ns} + I_{rs-1}P_{Ns-1} & N_S = N_{S\max} \end{cases} \qquad (22)$$

3.2.2 Overview of BBO Technique

BBO (Simon, 2008) has been developed based on the theory of biogeography. BBO concept is mainly based on migration and mutation. The concept and mathematical formulation of migration and mutation steps are given below:

A. Migration

This BBO technique is similar to other population based optimization techniques. In BBO, each possible solution is termed as habitat and their features that characterize habitability are called suitability index variables (SIV). The fitness of each solution is called its habitat suitability index (HSI). In BBO, each habitat H is a vector of N (number of control variables) integers. Each element of each vector is considered as one suitability index variable (SIV). Fitness of each candidate solution is evaluated using its $SIVs$. HSI represents the quality of each candidate solution. High HSI solution represents better quality solution and low HSI solution represents inferior solution in the optimization problem. In migration, the information is shared between habitats that depend on emigration rates and immigration rates of each solution. Immigration rate (I_r) and emigration rate (E_r) can be evaluated by the equations (19) and (20). Each solution is modified based on other solutions using habitat modification probability. Immigration rate, I_r of each solution is used to probabilistically decide whether or not to modify each SIV in that solution. If a SIV in a given solution is selected for modification, emigration rates, E_r of other solutions are used to probabilistically select which solution's SIV should migrate to that solution. Poor solutions accept more useful information from good solution, which improve the exploitation ability of the technique. The main difference between recombination approach of evolutionary strategies (ES) and migration process of BBO is that in ES, global recombination process is used to create a completely new solution, while in BBO; migration is used to bring changes within the existing solutions. In order to prevent the best solutions from being corrupted by the immigration process, few elite solutions are kept in BBO algorithm.

B. Mutation

Due to some natural calamities or other events HSI of a natural habitat can change suddenly and it may deviate from its equilibrium value. In BBO, this event is represented by the mutation of SIV and species count probabilities are used to determine the mutation rates. The species count probability can be calculated using the differential equation (22). Each population member has an associated probability, which indicates the likelihood that it exists as a solution for a given problem. Mutation rate of each set of solution can be calculated in terms of species count probability using the equation (23):

$$M_r = m_{max}\left(\frac{1-P_{N_S}}{P_{max}}\right) \quad (23)$$

where, M_r is the mutation rate for a habitat that contains N_s species; m_{max} is the maximum mutation rate; P_{max} is the maximum probability.

From (23), it can be inferred that if the probability (P_{N_S}) of a given solution is very low, then that solution is likely to mutate to some other solution. Similarly, if the probability (P_{N_S}) of a solution is high, then that solution has very little chance to mutate. Now, both habitats having very high HSI values and very low HSI values have low species count probability (P_{NS}) and therefore are equally probable for mutation i.e. they have high chances to produce much improved $SIVs$ in the later stage. But, medium HSI solutions have higher values of species count probability (P_{NS}) and less chance to create better solutions after mutation operation. This mutation scheme tends to increase diversity among the populations. Without this modification, the highly probable solutions will tend to be more dominant in the population. This mutation approach makes both low and high HSI solutions likely to mutate, which gives a chance of improving both types of solutions in comparison to their earlier values. Few elite solutions are kept in the mutation process to save the best solutions, so if a solution becomes inferior after mutation process then, previous solution (solution of that set before mutation) can revert back to that place again, if needed. So, mutation operation is a high-risk process. It is normally applied to both poor and better solutions. Since medium quality solutions are in improving stage, so it is better not to apply mutation on medium quality solutions.

Here, mutation of a selected solution is performed simply by replacing it with a randomly generated new solution. Other than this any other mutation scheme that has been implemented for GA can also be implemented for BBO.

4. ALGORITHMS

4.1 Particle Swarm Optimization (PSO)

PSO, developed by Eberhart and Kennedy (Kennedy & Eberhart, 1995) is one of the evolutionary computation techniques. PSO, like GA, is a population based optimization algorithm. Instead of the survival of the fittest, it is the simulation of the social behavior that motivates PSO. Here, the population is called 'swarm'. Each potential solution, called particle, is given a random velocity and is flown through the solution space (similar to the search process for food of a bird swarm) looking for the optimal position. The particles have memory and each particle keeps track of its previous best position, called pbest and corresponding fitness. The swarm remembers another value called gbest, which is the best position discovered by the swarm. If a particle discovers a promising new solution, all the other particles will move closer to it. Based on PSO concept, mathematical equations for the searching process may be written as follow (Kennedy & Eberhart, 1995):

$$v_{id}^{k+1} = w \times v_{id}^k + c_1 \times rand \times (pBest_{id} - x_{id}^k) + c_2 \times rand \times (gBest_d - x_{id}^k) \tag{24}$$

$$x_{id}^{k+1} = x_{id}^k + v_{id}^{k+1} \tag{25}$$

where v_{id} and x_{id} represent the velocity and position of dth dimension of ith particle respectively and 'rand' is a uniform random number in the range [0, 1]; $c1, c2$ are the cognitive and the social parameters; $pBest_{id}$ is the position best of dth dimension of ith particle, $gBest_d$ is the group best position of dth dimension of ith particle and w is the inertia weight factor.

4.1.1 Particle Swarm Optimization With Inertia Weight Approach (PSOIWA)

With some modifications in the basic PSO algorithm, better quality solution in terms of better objective function value, accelerated speed of convergence, lesser computational memory requirement and lesser execution time may be attributed. To get better solution and better convergence characteristics, a modified PSO with inertia weight approach (PSOIWA) is used. In PSOIWA, the inertia weight parameter w which controls the global and local exploration capabilities of the particle is modified. A large inertia weight factor is used at initial stages to enhance the global exploration and this value is gradually reduced as the search progresses to enhance local exploration. Linearly decreasing inertia weight factor is given by (26)

$$w = W_{max} - \frac{W_{max} - W_{min}}{iter_{max}} \times iter \tag{26}$$

where w_{max}, w_{min} are the initial inertia weight and final inertia weight respectively; $iter_{max}$ is the maximum iteration cycles and $iter$ is the current iteration cycle.

4.2 Real Coded Genetic Algorithm (RGA)

Genetic algorithms (GA) (Cai & Erlich, 2003; Ippolito, Cortiglia & Petrocelli, 2006; Mahdad, Srairia & Bouktir, 2010; Wirmond, Fernandes & Tortelli, 2011) are essentially search algorithms based on the mechanics of nature (e.g. natural selection, survival of the fittest) and natural genetics.

However, traditional binary coded GA (BCGA) suffers from few drawbacks when applied to multi-dimensional and high-precision numerical problems. The situation can be improved if GA is used with real number data (called as RGA) because the real number representation seems particularly natural when optimization problems with variables in continuous search spaces are tackled. Each chromosome is coded as a vector of floating point numbers that has the same length as the solution vector. Since real-parameters are used directly (without any bit-coding), solving real-parameter optimization problems is a step easier when compared to the BCGA. Unlike in the BCGA, decision variables can be directly used to compute the fitness values. Two commonly used genetic operators namely crossover and mutation are described below.

4.2.1 Crossover

Crossover refers to the mixing of information from both parents to create the children. Thus, in the crossover process, more fit individuals are allowed to produce more off-springs than less fit individuals, which tend to homogenize the population and improve the average result as the algorithm progresses. The child chromosome is generated by performing the crossover mechanism between two parent chromosomes as described below:

```
Let parent chromosomes  P₁ and P₂ are selected to be crossed and r₁ be a random number between [0, 1].
        if  r₁ < C_R
```
$$\text{if } f(P_1) < f(P_2)$$
$$P_1^0 = P_1 + rand \times (P_1 - P_2)$$
$$P_2^0 = P_2 + rand \times (P_1 - P_2)$$
$$\text{elseif } f(P_1) > f(P_2)$$
$$P_1^0 = P_1 + rand \times (P_2 - P_1)$$
$$P_2^0 = P_2 + rand \times (P_2 - P_1)$$
```
            end
        else
```
$$P_1^0 = P_1$$
$$P_2^0 = P_2$$
```
        end
```

4.2.1 Mutation

In crossover operation, individuals of the population resemble each other as the iteration process continues and if individuals become similar to the earlier iterations, the possible solutions will be trapped by local optima due to lack of diversity. To avoid local optima, the mutation operation is performed to increase the diversity in the population and explore new areas of the parameter search space. The probability that a particular element is mutated is governed by the mutation probability.

The mutation operation used in this article may be mathematically expressed as follows:

```
for i=1: N_p
        if rand < M_p
                Select a chromosome C (x_{i1}, x_{i2}, ....., x_{ij}, ..., x_{id}) from the entire popula-
tion set;
                Set the j^{th} mutation site randomly between [1, N_d];
                Replace x_{ij} by x'_{ij}, where x'_{ij} is given by
                x'_{ij} = a_j + rand × (b_j - a_j) where a_j and b_j are the limiting value of
x_j
        end
end
```

4.3 Algorithm Steps

Steps of RGA as implemented for optimization are as given below:

Step 1: Initialization of floating point chromosome strings of n_p population, each consisting of the parameters to be optimized.
Step 2: Decoding of strings and evaluation of fitness of each string.
Step 3: Selection of the elite strings in order of increasing fitness from minimum value.
Step 4: Apply crossover and mutation as explain in Section 4.1 and 4.2, respectively, to generate offspring.
Step 5: Calculate the objective function value (fitness value) of the specified problem using (7) and (10) for voltage deviation and transmission loss minimization objective, respectively.
Step 6: Adopt the elitist strategy i.e. replace the worst offspring by the elite string.
Step 7: If the generation cycle number (k) reaches the maximum limit, go to Step 8. Otherwise, set generation cycle number, k = k + 1, and go back to Step 3.
Step 8: Terminate the GA operation.

4.4. Artificial Bee Colony Algorithm

Main steps of ABC technique is given below:

Step 1: Generate the initial population (P) of the solutions (X_i) for the employed bees and calculate their fitness values where

$$X_i^e = \left[x_{i1}^e, x_{i2}^e, ..., x_{id}^e \right]$$

is d- dimensional optimization parameters.

Step 2: Modify each solution of the employed bees using following equation:

$$v_{ij}^e(t+1) = x_{ij}^e(t) + \phi_{ij}\left[x_{ij}^e(t) - x_{kj}^e(t)\right] \qquad (27)$$

where k is the randomly chosen index from population (P), t is the iteration number, ϕ_{ij} is a random number between -1 to 1, $x_{ij}^e(t)$ is the jth dimension (variable) of the ith employed bee at t^{th} iteration and $v_{ij}^e(t+1)$ is the new position of the jth dimension of the ith employed bee.

Step 3: Calculate the objective function values using the modified positions of the employed bees and compare these values with the objective function values using the old position of the employed bees. If the new objective function values are better than the old values then the new positions are retained otherwise the old positions are retained in its memory. This may be described briefly as follows:

```
if
```
$f\left[v_{i1}^e(t+1), v_{i2}^e(t+1), ..., v_{id}^e(t+1)\right] > f\left[x_{i1}^e(t), x_{i2}^e(t), ..., x_{id}^e(t)\right]$
$\left[x_{i1}^e(t+1), x_{i2}^e(t+1), ..., x_{id}^e(t+1)\right] = \left[v_{i1}^e(t+1), v_{i2}^e(t+1), ..., v_{id}^e(t+1)\right]$;

```
else
```
$\left[x_{i1}^e(t+1), x_{i2}^e(t+1), ..., x_{id}^e(t+1)\right] = \left[x_{i1}^e(t), x_{i2}^e(t), ..., x_{id}^e(t)\right]$;

```
end
```

where $f\left[v_{i1}^e(t+1), v_{i2}^e(t+1), ..., v_{id}^e(t+1)\right]$ is the modified objective function value of the ith employed bee, $f\left[x_{i1}^e(t), x_{i2}^e(t), ..., x_{id}^e(t)\right]$ is the current objective function value of ith employed bee, $\left[x_{i1}^e(t+1), x_{i2}^e(t+1), ..., x_{id}^e(t+1)\right]$ is the modified solution parameters of the ith employed bee.

Step 4: The new solutions of the employed bees share the information with onlooker Bees (i.e. the solution of employed bee becomes the solution of onlooker bee).

Step 5: Calculate the probability of each solution by following equation:

$$p_i = \frac{f_i}{\sum f_i} \qquad (28)$$

where p_i, f_i are the probability and objective function value of ith onlooker bee. The objective function value (fitness value) is evaluated using (7) and (10) for voltage deviation and transmission loss minimization objectives, respectively.

Step 6: Depending on the probability, modify the solution of onlooker bees using following equation:

Optimal Reactive Power Dispatch Incorporating TCSC-TCPS Devices

$$v_{ij}^o(t+1) = x_{ij}^o(t) + \phi_{ij}\left[x_{ij}^o(t) - x_{kj}^o(t)\right] \tag{29}$$

where $x_{ij}^o(t)$ is the jth dimension of the ith onlooker bee at t^{th} iteration and $v_{ij}^o(t+1)$ is the new position of the jth dimension of the ith onlooker bee.

Step 7: Calculate the nectar amount (objective function value) of the onlooker bee. If the new food has equal or better nectar than the old source, it is replaced with the old one in the memory. Otherwise, the old one is retained in the memory. This may be described briefly as follows:

```
if
```
$$f\left[v_{i1}^o(t+1), v_{i2}^o(t+1), \ldots, v_{id}^o(t+1)\right] > f\left[x_{i1}^o(t), x_{i2}^o(t), \ldots, x_{id}^o(t)\right]$$
$$\left[x_{i1}^o(t+1), x_{i2}^o(t+1), \ldots, x_{id}^o(t+1)\right] = \left[v_{i1}^o(t+1), v_{i2}^o(t+1), \ldots, v_{id}^o(t+1)\right]$$
```
else
```
$$\left[x_{i1}^o(t+1), x_{i2}^o(t+1), \ldots, x_{id}^o(t+1)\right] = \left[x_{i1}^o(t), x_{i2}^o(t), \ldots, x_{id}^o(t)\right]$$
```
end
```

where $f\left[v_{i1}^o(t+1), v_{i2}^o(t+1), \ldots, v_{id}^o(t+1)\right]$ is the modified objective function value of the ith onlooker bee; $f\left[x_{i1}^o(t), x_{i2}^o(t), \ldots, x_{id}^o(t)\right]$ is the current objective function value of ith employed bee, $\left[x_{i1}^o(t+1), x_{i2}^o(t+1), \ldots, x_{id}^o(t+1)\right]$ is the modified solution parameters of the ith employed bee.

Step 8: Replace the worst solutions by the new solutions. Number of replacement solutions are controlled by scout limit (SL). New Solutions are generated using the following equation:

$$x_{ij} = x_j^{min} + r_1\left[x_j^{max} - x_j^{min}\right] \tag{30}$$

where x_j^{max}, x_j^{min} are the maximum and minimum value of jth independent variable, r_1 is a random number between 0 to 1.

Step 9: Calculate the objective function value of new solutions.
Step 10: Go to step 2 for the next iteration.
Step 11: Stop iterations after a predefined number of iterations.

4.4 BBO Algorithm

The algorithm of BBO technique may be briefly described by the following steps:

Step 1: Initialize the parameters of BBO like habitat modification probability P_{mod}, mutation probability, maximum mutation rate m_{max}, maximum immigration rate I_{max}, maximum emigration rate E_{max}, step size for numerical integration dt etc. Also set maximum generation, maximum species count S_{max} and an elitism parameter. Generate the *SIV*s of the given problem within their feasible region. This means deriving a method of mapping problem solutions to *SIV*s. A complete solution consisting of *SIV*s is known as habitat H. All these are problem dependent.

Step 2: Initialize several numbers of habitats depending upon the population size. Each habitat represents a potential solution to the given problem.

Step 3: Calculate the *HSI* for each habitat of the population set for given emigration rate E_r, immigration rate I_r and Species N_S.

Step 4: Based on the *HSI* value elite habitats are identified.

Step 5: Probabilistically immigration rate and emigration rate are used to modify each non-elite habitat using migration operation. The probability that a habitat H_i is modified is proportional to its immigration rate I_{ri} and the probability that the source of the modification comes from a habitat H_j is proportional to the emigration rate E_{rj}. After modification of each non-elite habitat using migration operation, each *HSI* is recomputed.

Step 6: For each habitat, the species count probability is updated using (22). Then, mutation operation is performed on each non-elite habitat as per discussed in section 3.2.2.B and compute each *HSI* value again.

Step 7: Go to step (3) for the next generation. This loop can be terminated after a predefined number of generations or after an acceptable problem solution has been found.

After each habitat is modified (steps 2, 5 and 6), its feasibility as a problem solution should be verified. If it does not represent a feasible solution, then some method needs to be implemented in order to map it to the set of feasible solutions.

5. ALGORITHM APPLIED TO ORPD WITH TCSC AND TCPS DEVICES

5.1 ABC Algorithm Applied to ORPD with TCSC and TCPS Devices

In this chapter ABC algorithm has been employed to solve constrained ORPD problem with FACTS devices like TCSC and TCPS and to find the optimal solutions satisfying both equality and inequality constraints.

In ABC algorithm, applied to the ORPD incorporating TCSC and TCPS, each solution of the bee is defined by a vector with d independent variables such generators' voltages, tap settings of regulating transformers, reactive power injections of shunt compensator, power flow through the transmission lines where TCSC & TCPS devices are placed and power injections of TCPS devices. If there are d independent variables, then the ith employed bee E^i can be defined as follows:

$$E^i = [V^i_{g_1},...,V^i_{g_{NG}}, T^i_1,...,T^i_{NT}, Q^i_{c_1},...,Q^i_{c_{NC}}, PF^i_{TCSC_1},...,PF^i_{TCSC_{NTCSC}}, PF^i_{TCPS_1},...,PF^i_{TCPS_{NTCPS}}] \qquad (31)$$

where n is number of employed bees and m is the number of independent variables. The algorithm of the proposed method is as enumerated below.

Optimal Reactive Power Dispatch Incorporating TCSC-TCPS Devices

Step 1: The independent parameters (i.e. generators' voltages, tap settings of regulating transformers, reactive power injections, power flow through the transmission lines where TCSC & TCPS devices are placed and power injections of TCPS devices) of each population of the employed bees should be randomly selected while satisfying different inequality constraints of ORPD problems. Several numbers of employed bees depending upon the population size are being generated.

Step 2: To satisfy the equality constraint (11), perform load flow of the system having TCSC and TCPS devices using any classical technique and determine all dependent variables such as load voltages, active power of slack bus, generators' reactive powers, reactance of TCSC and phase shifting angle of TCPS etc. If any value of dependent variables do not satisfy the inequality constraints; discard that population (student) set and re-initialize the corresponding set. In this paper load flow is performed by Newton-Raphson method.

Step 3: Calculate objective function value for each employed bee.

Step 4: Each independent parameter of the employed bees is modified using (22). Then, the objective function values (voltage deviation and transmission loss) of the new food source position (new solution) are evaluated. The objective function values of the modified position are compared to the fitness of the old position computed in step 3. If the new objective function values are better than the old values then the new position is retained otherwise the old one is retained in its memory. This is described briefly in step 3 of section 4.3.

Step 5: The employed bees share the nectar information of the food sources and their position information with the onlooker bees. The probability each solution is calculated using (23). Then it is used to probabilistically decide whether or not to modify the solution of onlooker bee. If the solution of a onlooker bee is selected for modification, then it is modified using (24) and checks the nectar amount of the onlooker bee. If the new food has equal or better nectar than the old source, it is replaced with the old one in the memory. Otherwise, the old one is retained in the memory. This is described briefly in step 7 of section 4.3.

Step 6: Depending upon the scout limit, some worst solutions are replaced by the randomly generated new solutions. The independent variables of the new solutions are generated using (25).

Step 7: Go to step 3 for the next iteration.

Step 8: Stop iterations after a predefined number of iterations.

5.2 BBO Algorithm Applied to ORPD with TCSC and TCPS Devices

In this chapter BBO algorithm is employed to solve constrained ORPD problem incorporating multiple FACTS devices like TCSC and TCPS and to find the optimal solutions satisfying both equality and inequality constraints.

The algorithm of the proposed method is as enumerated below.

Step 1: Initialize the population randomly in the search space, i.e. the initial positions of *SIVs* (independent variables such as generators' voltages, tap settings of regulating transformers, reactive power injections, power flow through the transmission lines where TCSC & TCPS devices are placed and power injections of TCPS devices) of each habitat should be randomly selected while satisfying different operational constraints. Several numbers of habitats depending upon the population size are being generated. Each habitat represents a potential solution.

Step 2: Perform TCSC and TCPS based load flow using these *SIVs* and determine all dependent variables such as load voltages, active power of slack bus, generators' reactive powers, reactance of TCSC and phase shifting angle of TCPS etc. In this paper load flow is performed by Newton-Raphson method.
Step 3: Evaluate the voltage deviation and transmission loss (i.e. *HSI*) for each habitat of the population set using (7).
Step 4: Sort the habitats from the best to the worst using *HSI* values.
Step 5: Identify few elite habitats based on the *HSI* values. These best habitat sets are kept as it is without any modification.
Step 6: Assign number of species to each habitat based on *HSI*
Step 7: Update species count probability of each habitat using (28). Based on habitat modification probability, modify the independent variables (*SIVs*) of non-elite habitats by probabilistic mutation operation using (27).
Step 8: Go to Step 2 for the next iteration.
Step 9: Stop iterations after a predefined number of iterations.

6. SIMULATION RESULTS AND DISCUSSIONS

The feasibility of the proposed method has been tested on IEEE 30 bus system, which consists of six generators and 41 transmission lines. The generator and transmission-line data relevant to the system are taken from (Basu, 2008). The upper and lower generation limits of the generators are listed in Table 2. The upper and lower voltage limits at all the bus bars are taken as 1.10 p.u and 0.95 p.u respectively. The upper and lower tap settings limits of regulating transformers are taken as 1.10 p.u and 0.9 p.u respectively. The Var injection of the shunt capacitors are within the interval of [0 MVar, 30 MVar]. The reactance limits of TCSC in p.u. is $0 \leq x_t \leq 0.15$. The limits of phase shift angle (radian) of TCPS are taken as $-0.075 \leq \alpha \leq 0.075$. The simulation results shown in different Tables are the best solutions obtained in fifty trial runs. All the algorithms are implemented in Matlab 7.0 and run on a PC with core 2 duo processor, 2.9GHz, 1GB RAM.

Table 2. Generating units capacity limits of IEEE 30-bus system

Bus No	P_{gi}^{min} (MW)	P_{gi}^{max} (MW)	Q_{gi}^{min} (MW)	Q_{gi}^{max} (MW)
1	50	250	-20	10
2	20	80	-20	50
5	15	50	-15	40
8	10	35	-15	40
11	10	30	-10	24
13	12	40	-15	24

Since the performance of the algorithms depends on input parameters, they should be carefully chosen. After several runs, the input control parameters shown in Table 3 are found to be the best for the optimal performance of the different algorithms. A population size of 50 and iteration cycles of 100 are taken for all the algorithms. Owing to the randomness in these heuristic algorithms, the whole optimization procedures are executed 50 times when applied to the test system.

In order to show the power flow control capability of the artificial bee colony ORPD algorithm four cases for two different objective functions are carried out.

Case 1: ORPD without FACTS Devices

In this case PSOIWA, RGA, BBO and ABC algorithms are applied to ORPD problem with no FACTS devices in IEEE 30-bus system.

1. Minimization of Voltage Deviation

The results obtained for voltage deviation minimization objective function by PSOIWA, RGA, BBO and ABC are reported in Table 4. In this case, the sum of voltage deviations has been improved from (1.725779 p.u) to (0.127187 p.u.) by PSOIWA, from (1.519971 p.u.) to (0.126877 p.u.) by RGA and from (1.959679 p.u.) to (0.124406 p.u). by ABC method compare to transmission loss minimization objective function. The simulation results indicate that the optimal solutions (best) determined by ABC algorithm are more advanced than that found by PSOIWA, RGA, BBO methods.

2. Minimization of Transmission Loss

The simulation results of controlled variables like voltages of generator buses, tap settings of the regulating transformers, Var injections of the shunt capacitors, transmission loss and summation of voltage deviation obtained with power loss minimization objective function are given in Table 4. It shows that the loss reductions of 38.15% (from 8.9295 MW to 6.4638 MW), 56.45% (from 10.0937 MW to 6.4516 MW), (from 9.9101 MW to 6.3858 MW) and 59.09% (from 10.1080 to 6.3535 MW) from loss values obtained by voltage deviation minimization objective function are achieved by PSOIWA, RGA, BBO and ABC methods respectively. The results also show that the loss found by the proposed ABC method is lower than PSOIWA and RGA. Thus ABC gives the best results amongst the algorithms.

Case 2: ORPD with Three TCSC

In this case, TCSC devices are placed in the transmission lines 4-12, 16-17 & 27-28.

Table 3. Input parameters setting of different algorithms

RGA	PSOIWA	ABC	BBO
mutation probability:0.001 crossover: 0.8	$c1, c2 : 2.05$; $w_{max} : 0.9; w_{min} : 0.4$	Scout Limit (SL): 8% of the population	$P^{mod} =1; \mu= 0.005, I =1, E =1, dt =1$

Table 4. Comparison of control variable settings, transmission loss and voltage deviation obtained by PSOIWA, RGA, ABC and BBO (Case 1)

Algorithms	Minimization of Voltage Deviation				Minimization of Transmission Loss			
	PSOIWA	**RGA**	**ABC**	**BBO**	**PSOIWA**	**RGA**	**ABC**	**BBO**
V_1 (p.u.)	1.0011	1.0151	1.0207	1.0174	1.0936	1.0925	1.0999	1.0977
V_2 (p.u.)	0.9786	0.9694	0.9728	0.9710	1.0782	1.0835	1.0905	1.0891
V_5 (p.u.)	1.0085	1.0154	1.0183	1.0182	1.0590	1.0600	1.0622	1.0600
V_8 (p.u.)	1.0060	1.0163	1.0193	1.0077	1.0601	1.0691	1.0747	1.0699
V_{11} (p.u.)	1.0661	1.0077	0.9943	1.0574	1.0937	1.0956	1.0993	1.0986
V_{13} (p.u.)	1.0618	1.0551	1.0329	1.0411	1.0993	1.0936	1.0993	1.0981
QC_{10} (MVAR)	22.58	15.52	17.92	0.2157	24.85	17.92	28.16	0.2301
QC_{24} (MVAR)	15.92	15.09	13.86	0.1309	10.73	10.53	10.73	0.1060
TC_{6-9}	1.0877	1.0244	1.0074	1.0761	0.9865	1.0488	0.9918	1.0308
TC_{6-10}	0.9788	0.9402	0.9477	0.9632	0.9957	0.9388	1.0102	0.9488
TC_{4-12}	1.0449	1.0153	0.9813	0.9962	0.9680	1.0031	0.9666	0.9800
TC_{27-28}	0.9553	0.9551	0.9545	0.9514	0.9574	0.9693	0.9549	0.9616
Voltage deviation (p.u.)	0.127187	0.126877	0.124406	0.126033	1.725779	1.519971	1.959679	1.7820
Transmission Loss (MW)	8.9295	10.0937	10.1080	9.9101	6.4638	6.4516	6.3535	6.3858

1. Minimization of Voltage Deviation

The optimal settings of controlled variables (i.e. voltage of generator buses, tap settings of the regulating transformers, Var injections of the shunt capacitors, reactances of TCSC), transmission loss and voltage deviation that are obtained by the different methods are shown in the left half portion of Table 5. The results clearly show that the voltage deviation has improved from (0.127187 p.u.) to (0.124194 p.u.) using PSOIWA, from (0.126877 p.u.) to (0.123788 p.u.) using RGA and from (0.124406 p.u.) to (0.119576 p.u.) using ABC compare to case 1. It may also be noticed in Table 5 that the objective function value (voltage deviation) obtained by ABC is better than PSOIWA and RGA. The convergence performances of the methods are illustrated in Figure 4.

2. Minimization of Transmission Loss

The results obtained for transmission loss minimization objective function by PSOIWA, RGA and ABC are reported in the right half portion Table 5. It may be seen that PSOIWA, RGA and ABC give less transmission loss compared to that produces by the same algorithms in case 1. It is also evident that the ABC approach outperforms the PSOIWA and RGA algorithm.

Case 3: ORPD with One TCSC and One TCPS

In this case, IEEE-30 bus system with one TCSC devices (connected at transmission line 4-12) and one TCPS (connected at transmission line 2-5) is used to test the effectiveness of the proposed method.

1. Minimization of Voltage Deviation

Table 6 summarizes the optimal value of control variables (i.e. generators' voltages, tap settings of regulating transformers, reactive power injections, reactance value of TCSC devices & phase shifting angle of TCPS devices), objective function value (voltage deviation), transmission loss obtained by PSOIWA, RGA and ABC algorithms. Table 6 indicates that ABC method has the smallest value of voltage deviation than all the other listed algorithms.

Table 5. Comparison of control variable settings, transmission loss and voltage deviation obtained by PSOIWA, RGA, ABC and BBO (Case 2)

Algorithms	Minimization of Voltage Deviation				Minimization of Transmission Loss			
	PSOIWA	RGA	ABC	BBO	PSOIWA	RGA	ABC	BBO
V_1 (p.u.)	1.0082	0.9986	1.0012	1.0155	1.0992	1.0999	1.1000	1.0990
V_2 (p.u.)	0.9607	0.9875	0.9534	0.9970	1.0861	1.0890	1.0904	1.0825
V_5 (p.u.)	1.0181	1.0164	1.0183	1.0204	1.0608	1.0677	1.0659	1.0476
V_8 (p.u.)	1.0117	1.0091	1.0207	1.0093	1.0715	1.0762	1.0767	1.0665
V_{11} (p.u.)	1.0469	0.9929	0.9627	0.9943	1.0997	1.0999	1.1000	1.0860
V_{13} (p.u.)	1.0655	1.0582	1.0917	1.0234	1.0994	1.0999	1.1000	1.0964
QC_{10} (MVAR)	10.23	30.00	30.00	0.2109	29.59	29.95	27.30	0.2342
QC_{24} (MVAR)	21.09	18.94	12.26	0.1640	10.37	10.03	10.01	0.1259
TC_{6-9}	1.0614	1.0051	0.9769	1.0070	0.9774	1.0070	1.0939	1.0407
TC_{6-10}	0.9374	1.0406	1.0028	0.9672	1.0114	0.9796	0.9000	0.9609
TC_{4-12}	1.0510	1.0369	1.0983	0.9729	0.9611	0.9692	0.9619	0.9893
TC_{27-28}	0.9594	0.9580	0.9530	0.9518	0.9557	0.9698	0.9543	0.9868
$TCSC_{4-12}$ reactance (p.u.)	0.009613	0.0277290	0.050852	0.031462	0.015375	0.022142	0.000118	0.013421
$TCSC_{16-17}$ reactance (p.u.)	0.208361	0.107366	0.128915	0.139621	0.053869	0.024689	0.000199	0.009643
$TCSC_{27-28}$ reactance (p.u.)	0.005639	0.010772	0.005753	0.009763	0.066721	0.103246	0.065141	0.034217
Voltage deviation (p.u.)	0.124194	0.123788	0.119576	0.118214	1.948844	1.924453	1.951781	1.915437
Transmission Loss (MW)	10.8450	8.7725	11.6822	8,2846	6.3706	6.3668	6.3405	6.3592

Figure 4. Convergence characteristics of voltage deviation using different algorithms (Case 2)

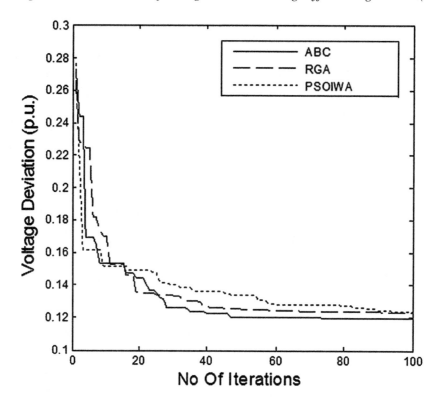

Table 6. Comparison of simulation results obtained by PSOIWA, RGA, ABC and BBO (Case 3)

Algorithms	Minimization of Voltage Deviation				Minimization of Transmission Loss			
	PSOIWA	RGA	ABC	BBO	PSOIWA	RGA	ABC	BBO
V_1 (p.u.)	0.9981	1.0003	0.9981	1.0278	1.0998	1.0997	1.0997	1.0991
V_2 (p.u.)	1.0137	1.0167	1.0077	0.9660	1.0899	1.0891	1.0892	1.0880
V_5 (p.u.)	1.0159	1.0165	1.0169	1.0197	1.0649	1.0645	1.0637	1.0626
V_8 (p.u.)	1.0120	1.0063	1.0167	1.0157	1.0760	1.0747	1.0734	1.0701
V_{11} (p.u.)	0.9587	0.9943	0.9797	1.0073	1.0957	1.0996	1.0973	1.0977
V_{13} (p.u.)	1.0360	1.0299	1.0465	1.0502	1.0999	1.0997	1.0998	1.0978
QC_{10} (MVAR)	29.90	19.43	07.25	0.1549	28.84	29.99	28.40	0.2877
QC_{24} (MVAR)	15.64	19.04	12.72	0.1347	10.00	09.98	10.07	0.1026
TC_{6-9}	0.9688	1.0074	0.9936	1.0234	1.0385	1.0251	1.0416	1.0133
TC_{6-10}	1.0260	0.9768	0.9000	0.9328	0.9571	0.9750	0.9486	0.9920
TC_{4-12}	0.9918	0.9856	1.0030	1.0084	0.9567	0.9599	0.9599	0.9653
TC_{27-28}	0.9555	0.9571	0.9537	0.9570	0.9589	0.9553	0.9564	0.9562
$TCSC_{4-12}$ reactance (p.u.)	0.018972	0.035989	0.002458	0.013628	0.007391	0.022805	0.015162	0.012134
$TCPS_{2-5}$ angle (radian)	0.026501	0.056525	0.027985	0.009983	0.061915	0.058594	0.028360	0.009328
Voltage deviation (p.u.)	0.124673	0.124275	0.121218	0.120399	1.958035	1.973084	1.942069	1.841135
Transmission Loss (MW)	8.4926	8.4613	8.5117	8.6743	6.3690	6.3590	6.2944	6.3096

2. Minimization of Transmission Loss

Table 6 compares the results obtained by different algorithms. The simulation results shows that the proposed ABC method outperforms the other evolutionary computation methods like PSOIWA and RGA. It is also seen that a loss reduction of nearly 0.07MW is achieved by ABC over PSOIWA and RGA. The convergence of minimal transmission loss with evolution generations (Figure 5) certifies the results of Table 6 vividly. Especially, ABC algorithm can not only maintain the diversity of the objective function solutions at the beginning of searching but also converge in the best solution at the later searching.

Case 4: ORPD with Two TCSC and Three TCPS

To test the potential of the algorithms in solving complicated systems, a standard IEEE 30-bus test system with two TCSC devices (connected in transmission lines 4-12 and 27-28) and three TCPS devices (connected in transmission lines 2-5, 6-10 and 14-15) is considered in this case.

1. Minimization of Voltage Deviation

The comparison results are given in left half portion of Table 7. From the results, it is noticed that the voltage deviation has been improved significantly (3.64% using PSOIWA, 3.57% using RGA and 5.81% using ABC) for all the algorithms compare to case 1. Table 7 also shows the objective function value obtained by the proposed ABC method is best (0.117573 p.u.) amongst PSOIWA (0.122725 p.u.), RGA (0.122500 p.u.) and ABC. The corresponding reactance of TCSC devices and phase angle (radian) of

Figure 5. Convergence characteristics of transmission loss using different algorithms (Case 3)

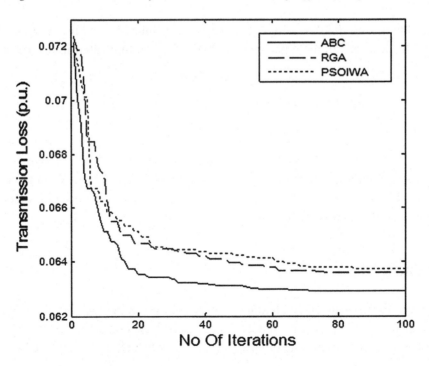

Table 7. Comparison of control variable settings, transmission loss and voltage deviation obtained by PSOIWA, RGA, ABC and BBO (Case 4)

Algorithms	Minimization of Voltage Deviation				Minimization of Transmission Loss			
	PSOIWA	RGA	ABC	BBO	PSOIWA	RGA	ABC	BBO
V_1 (p.u.)	0.9982	0.9985	0.9987	1.0064	1.0998	1.0999	1.1000	1.0994
V_2 (p.u.)	1.0676	1.0056	0.9583	0.9801	1.0887	1.0906	1.0901	1.0884
V_5 (p.u.)	1.0097	1.0143	1.0171	1.0213	1.0654	1.0659	1.0663	1.0632
V_8 (p.u.)	0.9980	1.0104	1.0163	1.0058	1.0673	1.0756	1.0760	1.0733
V_{11} (p.u.)	1.0745	1.0477	0.9868	1.0293	1.0995	1.0982	1.0819	1.0863
V_{13} (p.u.)	0.9818	1.0387	1.0894	1.0511	1.0996	1.0986	1.1000	1.0966
QC_{10} (MVAR)	13.18	07.55	29.88	0.2276	27.61	27.32	27.83	0.2895
QC_{24} (MVAR)	16.54	13.57	11.87	0.1746	09.98	09.97	10.07	0.1101
TC_{6-9}	1.1000	1.0699	1.0042	1.0450	1.0567	1.0492	1.0953	1.0277
TC_{6-10}	0.9477	0.9000	1.0062	0.9873	0.9205	0.9208	0.9000	0.9812
TC_{4-12}	0.9064	1.0028	1.0999	1.0215	0.9613	0.9697	0.9617	0.9774
TC_{27-28}	0.9537	0.9547	0.9523	0.9451	0.9570	0.9593	0.9581	0.9644
$TCSC_{4-12}$ reactance (p.u.)	0.077750	0.058789	0.031298	0.014521	0.024838	0.031732	0.005723	0.017633
$TCSC_{27-28}$ reactance (p.u.)	0.104720	0.075864	0.118562	0.009832	0.082103	0.066180	0.023526	0.005482
$TCPS_{2-5}$ angle (radian)	-0.107569	-0.081400	-0.089359	-0.015237	0.027457	0.028380	0.028730	-0.01324
$TCPS_{6-10}$ angle (radian)	-0.038022	-0.047322	-0.047066	0.019421	-0.018626	-0.018280	-0.021575	0.009932
$TCPS_{14-15}$ angle (radian)	-0.018768	-0.015715	-0.013745	0.008507	0.001949	0.002970	-0.005447	0.021659
Voltage deviation (p.u.)	0.122725	0.122500	0.117573	0.117396	1.894654	1.957837	1.913077	1.757314
Transmission Loss (MW)	14.2708	9.1870	11.8005	8.7962	6.3304	6.3185	6.2891	6.2933

TCPS devices obtained by PSOIWA, RGA and ABC algorithms are also listed in left half portion of Table 7. It is quite clear that all these parameters obtained by all three algorithms are within their permissible limits.

2. Minimization of Transmission Loss

The simulation results of transmission loss voltage deviation and optimal control variable those are obtained by PSOIWA, RGA, ABC are shown in the right half portion of Table 7. The simulation results show that the proposed ABC algorithm gives more reduction in loss (6.2891 MW) compared to PSOIWA (6.3304 MW) and RGA (6.3185 MW). Figure 6 shows the variation of real power loss against the number of iterations for the different algorithms.

7. CONCLUSION

This chapter develops two efficient evolutionary approach for solving the FACTS based optimal reactive power dispatch (ORPD) problem using two algorithms based on artificial bee colony (ABC) and biogeography based optimization. The proposed approach is tested on the modified 6 generator IEEE

Figure 6. Convergence characteristics of transmission loss using different algorithms (Case 4)

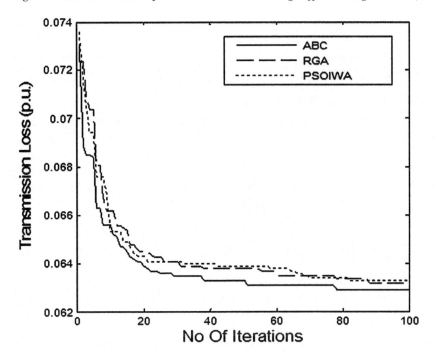

30-bus test system for four different cases for voltage deviation minimization and transmission loss minimization objective functions. It is observed that the FACTS devices (TCSC & TCPS) can reduce the transmission loss and can improve the voltage profile compare to normal system without FACTS devices. It is shown through different test cases that the ABC outperforms other methods, in terms of solution quality, computational efficiency, dynamic convergence, robustness and stability.

8. FUTURE RESEARCH DIRECTIONS

As the proposed methods, especially BBO is proved as an efficient optimization technique for the proposed areas, it may be recommended as a promising algorithm for solving other complex engineering optimization problems. The proposed algorithms may also be applied to other power system optimization problems like, unit commitment, state estimation, power system stabilizer etc. Moreover, the feasibility of the proposed methods may be verified by introducing more practical constraints like ramp rate limit and spinning reserve. Furthermore, the use of any new novel evolutionary technique may also be adopted to obtain better results of the proposed areas.

ACKNOWLEDGMENT

The authors are grateful to the anonymous reviewers for their valuable comments and helpful suggestions which greatly improved the paper's quality.

REFERENCES

Abd-Elazim, S. M., & Ali, E. S. (2012). Bacteria foraging optimization algorithm based SVC damping controller design for power system stability enhancement. *International Journal of Electrical Power & Energy Systems, 43*(1), 933–940. doi:10.1016/j.ijepes.2012.06.048

Acha, E., & Ambriz-Perez, H. (2000). A thyristor controlled series compensator model for the power flow solution of practical power networks. *IEEE Transactions on Power Systems, 15*(1), 58–64. doi:10.1109/59.852101

Ambriz-Perez, H., Acha, E., & Fuerte-Esquivel, C. R. (2006). TCSC-firing angle model for optimal power flow solutions using Newton's method. *International Journal of Electrical Power & Energy Systems, 28*(2), 77–85. doi:10.1016/j.ijepes.2005.10.003

Ayan, K., & Kılıç, U. (2012). Artificial bee colony algorithm solution for optimal reactive power flow. *Applied Soft Computing, 12*(5), 1477–1482. doi:10.1016/j.asoc.2012.01.006

Basu, M. (2008). Optimal power flow with FACTS devices using differential evolution. *International Journal of Electrical Power & Energy Systems, 30*(2), 150–156. doi:10.1016/j.ijepes.2007.06.011

Benabid, R., Boudour, M., & Abido, M. A. (2009). Optimal location and setting of SVC and TCSC devices using non-dominated sorting particle swarm optimization. *Electric Power Systems Research, 79*(12), 1668–1677. doi:10.1016/j.epsr.2009.07.004

Bhasaputra, P., & Ongsakul, W. (2002). *Optimal power flow with multi-type of FACTS devices by hybrid TS/SA approach.* IEEE ICIT'02, Bangkok, Thailand. doi: 10.1109/ICIT.2002.1189908

Bhattacharya, A., & Roy, P. K. (2012). Solution of multi-objective optimal power flow using gravitational search algorithm. *IET Generation Transmission Distribution, 6*(8), 751–763. doi:10.1049/iet-gtd.2011.0593

Cai, L. J., & Erlich, I. (2003). Optimal choice and allocation of FACTS devices using genetic algorithms. In *Proceedings on Twelfth Intelligent Systems Application to Power Systems Conference*, (pp. 1–6).

Chung, T. S., & Yun, G. S. (1998). Optimal power flow incorporating FACTS devices and power flow control constraints. *Power System Technology, Proceedings. POWERCON '98. 1998 International Conference on*, (pp. 415 – 419). doi:10.1109/ICPST.1998.728997

Ghasemi, A., Valipour, K., & Tohidi, A. (2014). Multi objective optimal reactive power dispatch using a new multi objective strategy. *International Journal of Electrical Power & Energy Systems, 57*, 318–334. doi:10.1016/j.ijepes.2013.11.049

Ippolito, L., Cortiglia, A. L., & Petrocelli, M. (2006). Optimal allocation of FACTS devices by using multi-objective optimal power flow and genetic algorithms. *International Journal of Emerging Electric Power Systems, 7*(2). doi:10.2202/1553-779X.1099

Karaboga, D., & Basturk, B. (2007). Artificial bee colony (ABC) optimization algorithm for solving constrained optimization problems, Springer-Verlag, IFSA 2007. *LNAI, 4529*, 789–798. doi:10.1007/978-3-540-72950-1_77

Karaboga, D., & Basturk, B. (2008). On the performance of artificial bee colony (ABC) algorithm. *Applied Soft Computing*, *8*(1), 687–697. doi:10.1016/j.asoc.2007.05.007

Kennedy, J., & Eberhart, R. (1995). Particle swarm optimization. In *Proceedings of IEEE International Conference on Neural Networks*, (vol. 4, pp. 1942–1948). doi:10.1109/ICNN.1995.488968

Khorsandi, A., Hosseinian, S. H., & Ghazanfari, A. (2013). Modified artificial bee colony algorithm based on fuzzy multi-objective technique for optimal power flow problem. *Electric Power Systems Research*, *95*, 206–213. doi:10.1016/j.epsr.2012.09.002

Ma, T. T. (2003). Enhancement of power transmission systems by using multiple UPFC on evolutionary programming. IEEE Bologna Power Tech Conference. doi:10.1109/PTC.2003.1304760

Mahdad, B., Srairia, K., & Bouktir, T. (2008). Optimal power flow for large-scale power system with shunt FACTS using efficient parallel GA. *International Journal of Electrical Power & Energy Systems*, *32*, 867–872. doi:10.1109/IECON.2008.4758067

Majumdar, S., Chakraborty, A. K., Chattopadhyay, P. K., & Bhattacharjee, T. (2012). Modified simulated annealing technique based optimal power flow with FACTS devices. *International Journal of Artificial Intelligence and Soft Computing*, *3*(1), 81–93. doi:10.1504/IJAISC.2012.048181

Marouani, I., Guesmi, T., Abdallah, H. H., & Ouali, A. (2009). *Application of a multi-objective evolutionary algorithm for optimal location and parameters of FACTS devices considering the real power loss in transmission lines and voltage deviation buses*. In 6th International Multi-conference on Systems (pp. 1–6). Djerba: Signals and Devices; doi:10.1109/SSD.2009.4956757

Mathur, R. M., & Basati, R. S. (1981). A thyristor controlled static phase-shifter for AC power transmission. *IEEE Transactions on Power Apparatus and Systems*, *100*(5), 2650–2655. doi:10.1109/TPAS.1981.316780

Mollazei, S., Farsangi, M. M., Nezamabadi-pour, H., & Lee, K. Y. (2007). Multi-objective optimization of power system performance with TCSC using the MOPSO algorithm. IEEE Power Engineering Society General Meeting. doi:10.1109/PES.2007.385878

Mondal, D., Chakrabarti, A., & Sengupta, A. (2012). Optimal placement and parameter setting of SVC and TCSC using PSO to mitigate small signal stability problem. *International Journal of Electrical Power & Energy Systems*, *42*(1), 334–340. doi:10.1016/j.ijepes.2012.04.017

Ongsakul, W., & Jirapong, P. (2005). Optimal allocation of FACTS devices to enhance total transfer capability using evolutionary programming. In *Proceedings of IEEE Int. Symposium on Circuits and Systems*. doi:10.1109/ISCAS.2005.1465551

Padhy, N. P., Abdel-Moamen, M. A., & Kumar, B. J. (2004). *Optimal location and initial parameter settings of multiple TCSCs for reactive power planning using genetic algorithms*. IEEE Power Engineering Society General Meeting. doi:10.1109/PES.2004.1373013

Preedavichit, P., & Srivastava, S. C. (1997) Optimal reactive power dispatch considering FACTS devices. *Proceedings of the 4th International Conference on Advances in Power System Control, Operation and Management, APSCOM-97*. doi:10.1049/cp:19971906

Rakpenthai, C., Premrudeepreechacharn, S., & Uatrongjit, S. (2009). Power system with multi-type FACTS devices states estimation based on predictor–corrector interior point algorithm. *International Journal of Electrical Power & Energy Systems, 31*(4), 160–166. doi:10.1016/j.ijepes.2008.10.010

Roy, P. K., Ghoshal, S. P., & Thakur, S. S. (2010). Optimal power flow with TCSC and TCPS modeling using craziness and turbulent crazy particle swarm optimization. *International Journal of Swarm Intelligence Research, 1*(3), 34–50. doi:10.4018/jsir.2010070103

Roy, P. K., Ghoshal, S.P., & Thakur, S. S. (2011). Optimal reactive power dispatch considering flexible AC transmission system devices using biogeography based optimization. *Electric Power Components and Systems, 39*(8), 733–750. doi:10.1080/15325008.2010.541410

Roy, P. K., Ghoshal, S. P., & Thakur, S. S. (2012). Optimal VAR control for improvements in voltage profiles and for real power loss minimization using biogeography based optimization. *International Journal of Electrical Power & Energy Systems, 43*(1), 830–838. doi:10.1016/j.ijepes.2012.05.032

Roy, P. K., Mandal, B., & Bhattacharya, K. (2012). Gravitational search algorithm based optimal reactive power dispatch for voltage stability enhancement. *Electric Power Components and Systems, 40*(9), 956–976. doi:10.1080/15325008.2012.675405

Sadeghzadeh, S. M., Khazali, A. H., & Zare, S., (2009). *Optimal reactive power dispatch considering TCPAR and UPFC*. EUROCON, St-Petersburg. Doi: 10.1109/EURCON.2009.5167690

Saravanan, M., Slochanal, S. M. R., Venkatesh, P., & Abraham, J. P. S. (2007). Application of particle swarm optimization technique for optimal location of FACTS devices considering cost of installation and system loadability. *Electric Power Systems Research, 77*(3-4), 276–283. doi:10.1016/j.epsr.2006.03.006

Shaheen, H. I., Rashed, G. I., & Cheng, S. J. (2011). Optimal location and parameter setting of UPFC for enhancing power system security based on differential evolution algorithm. *International Journal of Electrical Power & Energy Systems, 33*(1), 94–105. doi:10.1016/j.ijepes.2010.06.023

Sharifzadeh, H., & Jazaeri, M. (2009). Optimal reactive power dispatch based on particle swarm optimization considering FACTS devices. *International Conference on Sustainable Power Generation and Supply*. doi:10.1109/SUPERGEN.2009.5347956

Simon, D. (2008). Biogeography-Based Optimization. *IEEE Transactions on Evolutionary Computation, 12*(6), 702–713. doi:10.1109/TEVC.2008.919004

Sirjani, R., Mohamed, A., & Shareef, H. (2012). optimal allocation of shunt Var compensators in power systems using a novel global harmony search algorithm. *International Journal of Electrical Power & Energy Systems, 43*(1), 562–572. doi:10.1016/j.ijepes.2012.05.068

Sonmez, Y., Duman, S., Guvenc, U., & Yorukeren, N. (2012). *Optimal power flow incorporating FACTS devices using gravitational search algorithm*. International Symposium on Innovations in Intelligent Systems and Applications (INISTA- 2012). doi:10.1109/INISTA.2012.6246993

Sreejith, S., Chandrasekaran, K., & Simon, S. P. (2009). *Application of touring ant colony optimization technique for optimal power flow incorporating thyristor controlled series compensator*. IEEE Region 10 Conference-TENCON 2009. doi:.10.1109/TENCON.2009.5396226

Sundar, K. S., & Ravikumar, H. M. (2012). Selection of TCSC location for secured optimal power flow under normal and network contingencies. *International Journal of Electrical Power & Energy Systems*, *34*(1), 29–37. doi:10.1016/j.ijepes.2011.09.002

Taher, S. A., & Amooshahi, M. K. (2011). Optimal placement of UPFC in power systems using immune algorithm. *Simulation Modelling Practice and Theory*, *19*(5), 1399–1412. doi:10.1016/j.simpat.2011.03.001

Thukaram, D., & Yesuratnam, G. (2008). Optimal reactive power dispatch in a large power system with AC–DC and FACTS controllers. *IET Generation Transmission and Distribution*, *2*(1), 71–81. doi:10.1049/iet-gtd:20070163

Tripathy, M., & Mishra, S. (2007). Bacteria foraging-based solution to optimize both real power loss and voltage stability limit. *IEEE Transactions on Power Systems*, *22*(1), 240–248. doi:10.1109/TPWRS.2006.887968

Vasant, P. (2015). *Handbook of Research on Artificial Intelligence Techniques and Algorithms*. Hershey, PA: IGI Global; doi:10.4018/978-1-4666-7258-1

Vasant, P., Barsoum, N., & Webb, J. (2012). *Innovation in Power, Control, and Optimization: Emerging Energy Technologies*. Hershey, PA: IGI Global; doi:10.4018/978-1-61350-138-2

Vasant, P. M. (2013). *Meta-Heuristics Optimization Algorithms in Engineering, Business, Economics, and Finance*. Hershey, PA: IGI Global; doi:10.4018/978-1-4666-2086-5

Vasant, P. M. (2014). *Handbook of Research on Novel Soft Computing Intelligent Algorithms: Theory and Practical Applications*. Hershey, PA: IGI Global; doi:10.4018/978-1-4666-4450-2

Wirmond, V. E., Fernandes, T. S. P., & Tortelli, O. L. (2011). TCPST allocation using optimal power flow and genetic algorithms. *International Journal of Electrical Power & Energy Systems*, *33*(4), 880–886. doi:10.1016/j.ijepes.2010.11.017

APPENDIX

The network data of IEEE 30 bus system with base capacity of 100 Mva are given in Tables 8 and 9.

Table 8. Transmission line data

Bus No.	Bus No.	R (p.u.)	X (p.u.)	B/2 (p.u.)	Bus No.	Bus No.	R (p.u.)	X (p.u.)	B/2 (p.u.)
1	2	0.0192	0.0575	0.0264	15	18	0.1073	0.2185	0.0000
1	3	0.0452	0.1652	0.0204	18	19	0.0639	0.1292	0.0000
2	4	0.057	0.1737	0.0184	19	20	0.0340	0.0680	0.0000
3	4	0.0132	0.0379	0.0042	10	20	0.0936	0.2090	0.0000
2	5	0.0472	0.1983	0.0209	10	17	0.0324	0.0845	0.0000
2	6	0.0581	0.1763	0.0187	10	21	0.0348	0.0749	0.0000
4	6	0.0119	0.0414	0.0045	10	22	0.0727	0.1499	0.0000
5	7	0.0460	0.1160	0.0102	21	22	0.0116	0.0236	0.0000
6	7	0.0267	0.0820	0.0085	15	23	0.1000	0.2020	0.0000
6	8	0.0120	0.0420	0.0045	22	24	0.1150	0.1790	0.0000
6	9	0.0000	0.2080	0.000	23	24	0.1320	0.2700	0.0000
6	10	0.0000	0.5560	0.000	24	25	0.1885	0.3292	0.0000
9	11	0.0000	0.2080	0.000	25	26	0.2544	0.3800	0.0000
9	10	0.0000	0.1100	0.000	25	27	0.1093	0.2087	0.0000
4	12	0.0000	0.2560	0.000	28	27	0.0000	0.3960	0.0000
12	13	0.0000	0.1400	0.000	27	29	0.2198	0.4153	0.0000
12	14	0.1231	0.2559	0.000	27	30	0.3202	0.6027	0.0000
12	15	0.0662	0.1304	0.000	29	30	0.2399	0.4533	0.0000
12	16	0.0945	0.1987	0.000	8	28	0.0636	0.2000	0.0214
14	15	0.2210	0.1997	0.000	6	28	0.0169	0.0599	0.0065
16	17	0.0524	0.1923	0.000					

Table 9. Load data

Bus No.	Load (p.u.)		Bus No.	Load (p.u.)	
	Active Load (p.u.)	Reactive Load (p.u.)		Active Load (p.u.)	Reactive Load (p.u.)
1	0	0	16	0.0350	0.0180
2	0.2170	0.1270	17	0.0900	0.0580
3	0.0240	0.0120	18	0.0320	0.0090
4	0.0760	0.0160	19	0.0950	0.0340
5	0.9420	0.1900	20	0.0220	0.0070
6	0.0000	0.0000	21	0.1750	0.1120
7	0.2280	0.1090	22	0.0000	0.0000
8	0.3000	0.3000	23	0.0320	0.0160
9	0.0000	0.0000	24	0.0870	0.0670
10	0.0580	0.0200	25	0.0000	0.0000
11	0.0000	0.0000	26	0.0350	0.0230
12	0.1120	0.0750	27	0.0000	0.0000
13	0.0000	0.0000	28	0.0000	0.0000
14	0.0620	0.0160	29	0.0240	0.0090
15	0.0820	0.0250	30	0.1060	0.0190

Chapter 15
Scope of Biogeography-Based Optimization for Economic Load Dispatch and Multi-Objective Unit Commitment Problem

Vikram Kumar Kamboj
I. K. Gujral Punjab Technical University, India

S. K. Bath
Giani Zail Singh Campus College of Engineering & Technology, Bathinda, India

ABSTRACT

Biogeography Based Optimization (BBO) algorithm is a population-based algorithm based on biogeography concept, which uses the idea of the migration strategy of animals or other spices for solving optimization problems. Biogeography Based Optimization algorithm has a simple procedure to find the optimal solution for the non-smooth and non-convex problems through the steps of migration and mutation. This research chapter presents the solution to Economic Load Dispatch Problem for IEEE 3, 4, 6 and 10-unit generating model using Biogeography Based Optimization algorithm. It also presents the mathematical formulation of scalar and multi-objective unit commitment problem, which is a further extension of economic load dispatch problem.

INTRODUCTION

In the modern power system networks, there are various generating resources like thermal, hydro, nuclear etc. Also, the load demand varies during a day and attains different peak values. Thus, it is required to decide which generating unit to turn on and at what time it is needed in the power system network and also the sequence in which the units must be shut down keeping in mind the cost effectiveness of turning on and shutting down of respective units. The entire process of computing and making these decisions is known as unit commitment (UC). The unit which is decided or scheduled to be connected to the power system network, as and when required, is known to be committed unit. Unit commitment

DOI: 10.4018/978-1-4666-9755-3.ch015

in power systems refers to the problem of determining the on/off states of generating units that minimize the operating cost for a given time horizon (Kumar & Bath, 2013). Economic dispatch is the short-term determination of the optimal output of a number of electricity generation facilities, to meet the system load, at the lowest possible cost, subject to transmission and operational constraints. The economic load dispatch means that the generator's real and reactive power are allowed to vary within certain limits so as to meet a particular load demand within minimum fuel cost.

ECONOMIC LOAD DISPATCH

The scheduling of the units together with the allocation of the generation quantities which must be scheduled to meet the demand for a specific period represents the Unit Commitment Problem (Kamboj & Bath, 2013). Economic Load Dispatch (ELD) seeks the best generation schedule for the generating plants to supply the required demand plus transmission loss with the minimum generation cost. Significant economical benefits can be achieved by finding a better solution to the ELD problem. The economic dispatch problem is defined so as to minimize the total operating cost of a power system while meeting the total load plus transmission losses within generator limits (Dhillon & Kothari, 2010).

ECONOMIC LOAD DISPATCH PROBLEM FORMULATION

The objective of economic load dispatch of electric power generation is to schedule the committed generating unit outputs so as to meet the load demand at minimum operating cost while satisfying all units and operational constraints of the power system. The economic dispatch problem is a constrained optimization problem and it can be mathematically expressed as follows (Dhillon & Kothari, 2010):

Minimize

$$F(P_i) = \sum_{i=1}^{NG} (\alpha_i P_i^2 + \beta_i P_i + \gamma_i) \quad \$/h .$$

(27)

subject to:

The energy balance equation:

$$\sum_{i=1}^{NG} P_i = P_D + P_L .$$

(28)

The inequality constraints:

$$P_i^{\min} \leq P_i \leq P_i^{\max} \quad (i = 1, 2, 3, \dots\dots\dots, NG) .$$

(29)

where, α_i, β_i and γ_i are cost coefficients.

P_D is Load Demand.
P_L is power transmission Loss.
NG is the number of generation buses.
P_i is real power generation and will act as decision variable.

The most simple and approximate method of expressing power transmission loss, P_L as a function of generator powers is through George's Formula using B-coefficients and mathematically can be expressed as:

$$P_L = \sum_{i=1}^{NG} \sum_{j=1}^{NG} P_{g_i} B_{ij} P_{g_j} \text{ MW}. \tag{30}$$

where, P_{g_i} and P_{g_j} are the real power generations at the ith and jth buses respectively.

B_{ij} is the loss coefficients which are constant under certain assumed conditions and NG is the number of generation buses.

BIOGEOGRAPHY BASED OPTIMIZATION FOR ECONOMIC LOAD DISPATCH

Introduction

Biogeography based Optimization (BBO) algorithm is based on the concept of Biogeography. In Biogeography Migration and Mutation are carried out. The population of individuals or candidate solutions can be represented as a solution vector having integers. Every integer in the solution vector is equal to one SIV. The quality of the solutions is evaluated by SIVs. The good solutions are considered as high HSI habitats, where as others are known as low HSI habitats. The habitats HSI is the fitness function to a given problem. By using the migration operation the information is shared between habitats probabilistically. A sudden change can occur in the HSI of a natural habitat due to some natural calamities or other events known as mutation. The diversity of the populations is increased by the mutation (Vanitha & Thanushkodi, 2012).

Mathematical Model for BBO Algorithm

Mathematical models (Simon, 2008) of BBO describe the migration species from one island to another, how species arise and become extinct. Island in Biogeography Based Optimization is defined as any habitat isolated geographically from other habitat. Well suited habitats for species are said to have high habitat suitability index (HSI) while the habitat that is not well suited is said to have low HSI. Each habitat consists of features that decide the HSI for the habitat. These features were considered as independent variable and called suitability index variable (SIV) which map the value of the HSI of the habitat. High HSI habitat has a large number of species while low HSI habitat has small number of species. High HSI habitat has large no. of species, high emigration rate μ_2 and low immigration rate λ_2 as shown in Fig.-2, which represents the model of species abundance in a single habitat. Low HSI habitat has small number of species, low emigration rate μ_1 and high immigration rate λ_1 as in Fig.-2. Emigration of species from one habitat to other habitat does not mean that all the species will disappear from its home, only a few reprehensive emigrate. In BBO there are two main operations migration and mutation. Migration process may raise the HSI of the habitat however if the HSI does not rise, the species in that habitat will tend to go extinct due to mutation process and replaced by new species. Probability P_S that habitat contains exactly S species.

$$P_S(t + \Delta t) = P_S(t)(1 - \lambda_S \Delta t - \mu_S \Delta t) + P_{S-1}\lambda_{S-1}\Delta t + P_{S+1}\mu_{S+1}\Delta t. \tag{31}$$

In order to have S species at time $(t + \Delta t)$ we should have one of the following:

1. S species at time t and no immigration or emigration occurred between t and $(t + \Delta t)$
2. S-1 species at time t and one species immigrated.
3. S+1 species at time t and one species emigrated.

Migration

The process of probabilistically sharing information between habitat using the immigration and emigration rate is known as migration.

In migration process modify each solution based on other solutions. In Fig.-1 solutions S1 represent a low HSI solution since the number of species is small and considered as bad solution while S2 repre-

Figure 1. Illustration of candidate solution for Poor solution and good solution

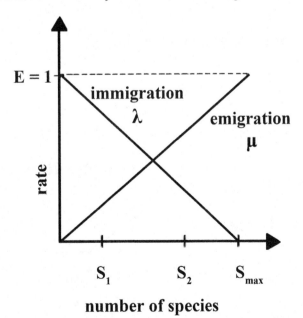

sent a High HSI solution, since the number of species is large and considered as good solution. S1 has high immigration rate λ and low emigration rate μ. S2 has low immigration rate λ and high emigration rate μ. Hence high HSI habitat can be selected based on probability which proportional to the emigration rate (high value) and low HSI habitat can be selected based on probability which proportional to the immigration rate. Consider H_i as habitat with low HSI and need to be modified using the migration process and has λ_x and H_x as habitat with high HSI as source of modification with μ_x. In migration operation first H_i With low HSI is selected to modify it. The selection should be with probability proportional to λ_i. After that H is with high HSI is selected to be the source of modification. The selection should be with probability proportional to μ_x. Then randomly select SIV_m from Hx. Finally randomly replace SIV in H_i with SIV_m. Emigration to neighboring habitat does not mean that all the species will disappear from its home only a few reprehensive species emigrate. Useful way to calculate P_s in programs is using the differentiation of P_s as shown in (32).

$$P_S(t + \Delta t) \approx P_S(t) + \frac{dP_S}{dt} \Delta t. \tag{32}$$

Mutation

Severely destructive events can extremely change the HSI of a habitat and can cause the species count deviate from its equilibrium point. In BBO the mutation is modeled as SIV mutation using species count probabilities to determine mutation rate. Very high HSI and very low HSI solutions are likely to mutated to a different solution using the mutation rate m that is calculated using (33).

$$m(S) = m_{max}(1 - \frac{P_S}{P_{max}}). \tag{33}$$

Figure 2. Emigration and Immigration in BBO

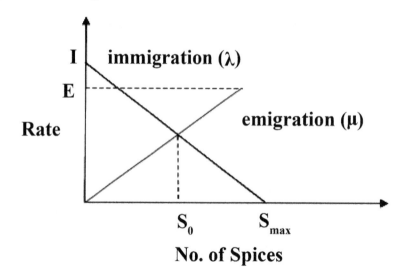

where, $m(S)$ is the mutation rate, m_{max} is the maximum mutation rate, P_s is the probability that S species in a habitat and P_{max} is the maximum probability that S species in a habitat. Mutation process is a problem dependent.

BBO Definitions and Algorithm

- **Habitat:** A habitat is a vector of integers that represents a feasible solution to some problem.
- **Suitability Index variable (SIV):** Suitability Index Variable is an integer that is allowed in a habitat, is the set of all integers that are allowed in a habitat.
- **Habitat Suitability Index (HSI):** Habitat Suitability Index is a measure of the goodness of the solution that is represented by the habitat.
- **Ecosystem:** An ecosystem is a group of habitats. The size of an ecosystem is constant.
- **Immigration Rate:** Immigration rate is a monotonically non increasing function of HIS is proportional to the likelihood that SIVs from neighboring habitats will migrate into habitat.
- **Emigration Rate:** Emigration rate is a monotonically non-decreasing function of HIS is proportional to the likelihood that SIVs from habitat will migrate into neighboring habitats.
- **Ecosystem Transition Function:** An ecosystem transition function is a 6-tuple that modifies the ecosystem from one optimization iteration to the next. An ecosystem transition function can be written as follows:

$$\Psi = \lambda^n \circ \mu^n \circ \Omega^n \circ HSI^n \circ M^n \circ HSI^n. \tag{34}$$

- **Habitat Modification:** It is a probabilistic operator that adjusts habitat based on the ecosystem. The probability that is modified is proportional to its immigration rate and the probability that the source of the modification comes from is proportional to the emigration rate. Habitat modification can loosely be described as follows:

```
Select H_i with probability ∝ λ_i
    If H_i is selected
      For j=1 to n
      Select H_j with probability ∝ μ_i
          If H_j is selected
              Randomly select an SIV σ from H_j
              Replace a random SIV in H_i with σ
          end
      end
    end
```

- **Mutation:** Mutation is a probabilistic operator that randomly modifies habitat SIVs based on the habitat's a priori probability of existence. Mutation can be described as follows:

```
for j=1 to m
    Use .. and μ_i to compute the probability P_i
        Select SIV H_i(j) with probability ∝ P_i
            If H_i(j) is selected
                Replace H_i(j) with a randomly generated SIV
            end
end
```

With habitat modification, elitism can be implemented by setting the probability of mutation selection to zero for the best habitats.

ECONOMIC LOAD DISPATCH USING BBO

The stepwise procedure to implement the Biogeography based optimization algorithm for economic load dispatch problem is elaborated below:

Step-1: Initialize the BBO Parameters like: popsize, Maxgen, numVar, p_{mutate}, p_{modify}, P_{Max}, P_{Min}, A_i, B_i, C_i, Load Demand, $\lambda_{lower}, \lambda_{upper}$, maximum Spices, E, I etc.

Step-2: Set the BBO parameters with the following values: generation=450, island =150, integration probability step size dt =1 and initial mutation probability =0.1, Maximum emigration rate for each island: E=1, Maximum immigration rate for each island, I=1, $\lambda_{lower} = 0.0$ and $\lambda_{upper} = 1$, elitism parameter=2.

Step-3: Generate a random set of habitats that consists of SIV's which represent a feasible solution.

Step-4: Calculate the λ, μ, HSI and the fuel cost for each habitat.

Step-5: Identify the best solutions based on the HSI value and save the best solutions.

Step-6: Probabilistically use λ and μ to modify the non elite habitat using the migration process.

Step-7: Based on species count probability of each habitat mutate the non-elite habitat then go to Step-4.

Step-8: Terminate the loop after specified number of generation i.e. generation=450.

Step-9: After the modification of each habitat, test the feasibility of the habitat as a candidate solution and if it is not feasible then variables are tuned to convert it to a feasible solution.

TEST SYSTEMS

In order to show the effectiveness of the BBO Algorithm for Economic Load Dispatch Problem, four different types of standard IEEE test systems have been taken into consideration:

- The first test system consists of 3-Generating units with a load demand of 350 MW.
- The second test system consists of 4-Generating units with a load demand of 660 MW.
- The third test system consist of 6-Generating Units Model with a load demand of 650 MW.
- The fourth test system consists of 10-Generating units with a load demand of 660 MW.

RESULTS AND DISCUSSION

The corresponding results of 3, 4, 6 and 10-generating unit model has been obtained using Biogeography Based Optimization Algorithm and shown in Table 2, 4, 7 and 8. The MATLAB Simulation software 7.12.0 (R2011a) is used to obtain the corresponding results. Transmission losses are neglected to obtain the corresponding results. Table 2 and 4 shows that total fuel cost for 3-unit generating model for 350MW load demand using BBO algorithm is 18316.0944 Rs./Hour. For 4-unit generating model, it is 1756.0148 Rs./Hour for the load demand of 660MW, which shows that performance of BBO algorithm is better than Fire-Fly Algorithm and Evolutionary Search Method.

Table 1. Generator characteristics of 3-unit test system (Reddy & Reddy, 2012)

No. of Generating Units	Real Powers(MW)		Cost Coefficients		
	P_{max}	P_{min}	A	B	C
1	210	35	1243.53	38.30533	0.003546
2	325	130	1658.57	36.32782	0.0211
3	315	125	1356.66	38.27041	0.01799

Table 2. Results of 3-unit system for 350 MW load demand

Method	P1(MW)	P2(MW)	P3(MW)	Fuel Cost(Rs./h)
Firefly Algorithm	70.3012	156.267	129.208	18564.5
BBO[Proposed Method]	65.90909	159.0909	125	18316.0944

Table 3. Generator characteristics for test system-2 (Dhillon & Kothari, 2010)

No. of Generating Units	Real Powers(MW)		Cost Coefficients		
	P_{max}	P_{min}	A	B	C
1	300	50	40.0	1.8	0.0015
2	125	20	60.0	1.8	0.003
3	175	30	100.0	2.1	0.0012
4	250	40	120.0	2.0	0.001

Table 4. Results of 4-unit system for 660 MW load demand

Method	P1(MW)	P2(MW)	P3(MW)	P4(MW)	Fuel Cost(Rs./h)
Evolutionary Search Method	201.6681	118.8477	141.1979	217.1224	1802.135
BBO[Proposed Method]	198	110	132	220	1756.0148

Table 5. Generator characteristics of 6-Unit test system (Reddy & Reddy, 2012)

No. of Generating Units	Real Powers(MW)		Cost Coefficients		
	P_{max}	P_{min}	A	B	C
1	125	10	756.79890	38.53000	0.15240
2	150	10	451.32510	46.15916	0.10587
3	225	35	1049.99800	40.39655	0.02803
4	210	35	1243.53100	38.30553	0.03546
5	325	130	1658.57000	36.32782	0.02111
6	315	125	1356.65900	38.27041	0.01799

Table 6. Generator characteristics of 10-Unit test system (Surekha, Archana, & Sumathi, 2012)

No. of Generating Units	Real Powers(MW)		Cost Coefficients		
	P_{max}	P_{min}	A	B	C
1	470	150	958.20000	21.60000	0.00043
2	460	135	1313.60000	21.05000	0.00063
3	340	73	604.97000	20.81000	0.00039
4	200	60	471.60000	23.90000	0.00070
5	243	73	480.29000	21.62000	0.00079
6	160	57	601.75000	17.87000	0.00056
7	130	20	502.70000	16.51000	0.00210
8	120	47	639.40000	23.23000	0.00480
9	80	20	455.60000	19.58000	0.10908
10	55	55	692.40000	22.54000	0.00951

Table 7. Results of 6-unit system for 650 MW load demand

Method	P1(MW)	P2(MW)	P3(MW)	P4(MW)	P5(MW)	P6(MW)	Fuel Cost(Rs./h)
Firefly Algorithm[6]	26.0279	10	107.264	109.668	216.775	196.954	34482.6
BBO[Proposed Method]	10	10	128.6842	171.0526	205.2632	125	34053.0355

Table 8. Results of 10-unit system for 1036 MW load demand

Method	P1(MW)	P2(MW)	P3(MW)	P4(MW)	P5(MW)	P6(MW)	P7(MW)	P8(MW)	P9(MW)	P10(MW)	Fuel Cost(Rs./h)
SADE	462.5504	0	0	228.4496	0	160	130	0	0	55	25135
BBO[Proposed Method]	227.9298	158.9825	109.0526	72.70175	73	145.1053	127.2281	47	20	55	28172.0279

SCOPE OF BIOGEOGRAPHY BASED OPTIMIZATION FOR MULTI-OBJECTIVE UNIT COMMITMENT PROBLEM

Unit Commitment Problem

The Unit Commitment (UC) is an important research challenge and vital optimization task in the daily operational planning of modern power systems due to its combinatorial nature. Because the total load of the power system varies throughout the day and reaches a different peak value from one day to another, the electric utility has to decide in advance which generators to start up and when to connect them to the network and the sequence in which the operating units should be shut down and for how long. The computational procedure for making such decisions is called unit commitment and a unit when scheduled for connection to the system is said to be committed (Wood & Wollenberg, 1996; Zhu, 2009). The objective of classical unit commitment is to find the optimal schedule for operating the available generating units in order to minimize the total operating cost of the power generation. The total operating cost includes fuel cost, start up cost and shut down cost. The fuel costs are calculated using the data of unit heat rate & fuel price information which is normally a quadratic equation of power output of each generator at each hour determined by Economic Load Dispatch (ELD) as per equation (1):

$$FC_i(P_{ih}) = \sum_{i=1}^{NG} \alpha_i P_{ih}^2 + \beta_i P_{ih} + \gamma_i \tag{1}$$

where, α_i (\$/MW²h), β_i (\$/MWh) and γ_i (\$/h) are fuel consumption coefficients of ith unit.

Eqn. (1) presents the fuel cost for various units without valve point effects. To take the effects of valve points, a sinusoidal function is added to the convex cost function and can be represented as:

$$FC_i(P_{ih}) = \sum_{i=1}^{NG} [\alpha_i P_{ih}^2 + \beta_i P_{ih} + \gamma_i + |\delta_i \sin\{\varepsilon_i(P_{ih}^{\min} - P_{ih})\}|]. \tag{2}$$

The fuel cost for each thermal generating unit in the system is usually determined by a second order function of the active power generation. During opening of each admission valve in a steam turbine, the so called drawing effect occurs. This rippling effect is modeled as sinusoidal function of the active power generation. Therefore, the fuel cost function is written as a non-linear and non-convex function (Blaze & Marko, 2013) as follows:

$$FC_i(P_{ih}) = \sum_{h=1}^{H} \sum_{i=1}^{NG} [\alpha_i P_{ih}^2 + \beta_i P_{ih} + \gamma_i + |\delta_i \sin\{\varepsilon_i(P_{ih}^{\min} - P_{ih})\}|]. \tag{3}$$

where, α_i=\$/MW², β_i=\$/MWh, γ_i=\$/h, δ_i=\$/h, ε_i=rad/MW are constants unique for each generating units, P_{ih} is the power output of the ith thermal unit, $P_{ih}^{\min}$ is the minimum power output of the ith thermal unit and H is the duration of the time interval in hours.

Multi-Objective Unit Commitment

The multi-objective unit commitment is a non-convex, non-linear, multidimensional and highly constrained problem (Kamboj & Bath, 2014). The problem comprises multiple and often conflicting optimization criteria for which no unique optimal solution can be determined with respect to all criteria. The objective function of multi-objective unit commitment problem depends upon fuel cost, Start-up cost, Shut Down cost, emissions of gaseous pollutants (sulfur oxides and nitrogen oxides) and unavailability of power generation.

Objective1: Minimization of *Fuel Cost* (including Start-Up and Shut-Down Cost).

The Total Cost (Chandrasekaran & Simon, 2012) for Unit commitment problem including fuel cost and start-up cost over the time period 'H' is given by

$$Cost_{NH} = \sum_{h=1}^{H}\sum_{i=1}^{NG}[FC_i(P_{ih}) + STC_{ih}]. \quad (4)$$

$$Cost_{NH} = \sum_{h=1}^{H}\sum_{i=1}^{NG}[FC_i(P_{ih}) + STC_i(1 - U_{i(h-1)})]U_{ih}. \quad (5)$$

$$Cost_{NH} = \sum_{h=1}^{H}\sum_{i=1}^{NG}[FC_i(P_{ih})*U_{ih} + STC_{ih}*(1 - U_{i(h-1)})*U_{ih}]. \quad (6)$$

where, U_{ih} is the status of ith unit at hth hour.

Start up cost is the cost involved in bringing the thermal unit online. Start up cost is expressed as a function of the number of hours the units has been shut down (exponential when cooling and linear when banking). Shut down costs are defined as a fixed amount for each unit/shutdown. A simplified start up cost model is used as follows:

$$STC_{ih} = \begin{cases} HSC_i, & if \quad MDT_i \leq DT_i < (MDT_i + CSH_i) \\ CSC_i, & if \; DT_i > (MDT_i + CSH_i) \end{cases}. \quad (7)$$

where, DT_i is shut down duration, MDT_i is Minimum down time, HSC_i is Hot start up cost, CSC_i is Cold start up cost and CSH_i is Cold start hour of ith generating unit.

Unit Commitment problem with Fuel cost, Start-up cost and Shut-down cost over the time period 'H' can be represented as:

$$Cost_{NH} = \sum_{h=1}^{H}\sum_{i=1}^{NG}[FC_i(P_{ih}) + STC_{ih} + SDC_{ih}]. \qquad (8)$$

$$Cost_{NH} = \sum_{h=1}^{H}\sum_{i=1}^{NG}[FC_i(P_{ih})*U_{ih} + STC_{ih}*(1-U_{i(h-1)})*U_{ih} + SDC_{ih}*(1-U_{ih})*U_{i(h-1)}]. \qquad (9)$$

Objective2: *Environmental objective:* Atmospheric pollutants emission.

Second objective of unit commitment problem is to minimize the total quantity of atmospheric pollutants emission (Blaze & Marko, 2013). The emissions are generally expressed as a quadratic function:

$$EC_i(P_{ih}) = a_i(P_{ih})^2 + b_i P_{ih} + c_i. \qquad (10)$$

where, a_i, b_i and c_i are the emission coefficients of unit i. So, the total emission of atmospheric pollutants cost over the time period 'H' is expressed as follows:

$$\min imize \sum_{h=1}^{H}\left(\sum_{i=1}^{NG} EC_i(P_{ih})*U_{ih} + STE_i*(1-U_{i(h-1)})*U_{ih}\right). \qquad (11)$$

$$TCost_{NH} = \min imize \sum_{h=1}^{H}\left(\sum_{i=1}^{NG}\{a_i(P_{ih})^2 + b_i P_{ih} + c_i\}*U_{ih} + STE_i*(1-U_{i(h-1)})*U_{ih}\right). \qquad (12)$$

The total emissions of sulfur oxides (SO_X) and nitrogen oxides (NO_X) emitted by thermal units can be modelled together and described by one function as follows:

$$F_i(P_{ih}) = \sum_{h=1}^{H}\sum_{i=1}^{NG} 10^{-2} \times [\alpha_i P_{ih}^2 + \beta_i P_{ih} + \gamma_i + \eta_i e^{\lambda_i P_{ih}}]. \qquad (13)$$

Objective3: *Reliability Function, LOLP.*

In UCP, the system reliability level (Blaze & Marko, 2013; Chandrasekaran & Simon, 2012) is dependent on the allocation of spinning reserve. Hence, to minimize *LOLP*, enough spinning reserve has to be scheduled for each hour. Reliability level of the power system at each hour is calculated using (14).

$$LOLP(h) = \sum_{j=1}^{LC} p(j), \quad h \in [1, H]. \qquad (14)$$

Implementation of Multi-Objective Function

Real time optimization problems involve simultaneous optimization of different conflicting objectives. Hence a multi-objective optimization problem consists of multiple objectives to be optimized simultaneously with the various equality and inequality constraints. This can be generally formulated as:

$$Min \quad f_l(x) \quad ,l = 1,2,3,\ldots\ldots\ldots,NOF. \tag{15}$$

$$\text{Subject to:} \begin{cases} g_m(x) = 0, & m = 1,2,3,\ldots\ldots\ldots,M \\ h_n(x) \leq 0, & n = 1,2,3,\ldots\ldots\ldots,K \end{cases}. \tag{16}$$

where, f_l is the lth objective function, x is a decision vector that represents a solution, NOF is the number of objective functions, M and K are the number of equality and inequality constraints, respectively. Multi-objective optimization problems with such conflicting objectives, give raise to a set of optimal solution, rather than a single optimal solution, since, no solution can be considered to be better than other solutions, without adequate information.

System and Physical Constraints of Unit Commitment Problem

The UCP is subjected to many constraints that include (Bhardwaj, Tung, Shukla, & Kamboj, 2012; Kumar & Bath, 2012):

- The total power generated must meet the load demand.
- There must be enough spinning reserve to cover any shortfalls in generation.
- The loading of each unit must be within its minimum and maximum allowable rating (limits).
- The minimum up and down times of each unit must be observed.
- Unit availability constraint is either unit is available/ not available, out aged/Must out, Must run, and Fixed Output Power (F.O.P).
- Unit initial status +/- either already up or already down.

The above constraints can be classified into two groups: System constraints and Physical constraints or unit constraints.

There are many constraints that are applicable on the Unit Commitment Problem (UCP). The constraints are applied according to the units under consideration and the various rules of the utilities. The various constraints are as follows:

System Constraints or Coupling Constraints

Constraints that concern all the units of the system are called system or coupling constraints. These constraints have two categories: the system power balance and system spinning reserve constraints.

Load Balance or Power Balance Constraints

The load balance or system power balance constraint requires that the sum of generation of all the committed units at h^{th} hour must be greater than or equal to the demand at a particular hour 'h'.

$$\sum_{i=1}^{NG} P_{ih} U_{ih} = D_h. \tag{17}$$

Spinning Reserve Constraints

Considering the important aspect of reliability, there is a provision of excess capacity of generation which is required to act instantly when there is a failure of already running unit or sudden load demand. This excess capacity of generation is known as Spinning Reserve and mathematically given as:

$$\sum_{i=1}^{N} P_{i(\max)} U_{ih} \geq D_h + R_h. \tag{18}$$

Thermal Constraints

A thermal generation unit needs to undergo gradual temperature changes and thus it takes some period of time to bring a thermal unit online. Also, the operation of a thermal unit is manually controlled. So a crew is required to perform the operation and maintenance of any thermal unit. This leads to many restrictions in the operation of thermal unit and thus it gives rise to many constraints.

Minimum up Time

If the units have already been shut down, there will be a minimum time before they can be restarted. This constraint in given as:

$$X_i^{on}(t) \geq MU_i. \tag{19}$$

where, $X_i^{on}(t)$ is duration for which unit i is continuously ON (in hrs) and MU_i is unit i minimum up time (in hrs).

Minimum Down Time

Once the unit is decommitted, there is a minimum time before it can be recommitted. This constraint is given as:

$$X_i^{off}(t) \geq MDT_i. \tag{20}$$

where, $X_i^{off}(t)$ is duration for which unit i is continuously OFF (in hrs) and MDT_i minimum down time (in hrs).

Crew Constraints

If a plant consists of two or more units, they cannot be turned on at the same time since there are not enough crew members to attend both units while starting up.

Startup/Shutdown Costs

A startup cost is incurred when a generator is put into operation. The cost is dependent on how long the unit has been inactive. While the startup cost function is nonlinear, it can be discretized into hourly periods, giving a stepwise function. Similarly, shutdown cost is incurred during shutting down generating units. In general, it is neglected from unit commitment decision.

Operation and Maintenance Costs

Operation and maintenance cost is actually the labour cost of operating crews and the cost of plant maintenance. Typically, this cost depends on the amount of generation output.

$$OMC_i^t = nP_i^t. \tag{21}$$

Fuel Cost

The quadratic approximation is the most widely used by the researchers, which is basically a convex shaped function and is mathematically represented as:

$$FC_i(P_{ih}) = \sum_{i=1}^{NG} \alpha_i P_{ih}^2 + \beta_i P_{ih} + \gamma_i. \tag{22}$$

Start up Cost

Start up cost is warmth-dependent. Mathematically it is represented as a step function:

$$ST_{ih} = h - \cos t : T_i^{down} \leq X_i^{off}(h) \leq T_i^{down} + c - s - hour_i \quad \$/h. \tag{23}$$

Shut down Cost

The typical value of the shut down cost is zero in the standard systems. This cost is considered as a fixed cost.

Initial Operating Status of Generating Units

The initial operating status of every unit should take the last day's previous schedule into account, so that every unit satisfies it's minimum up/down time.

Maximum and Minimum Power Limits

Every unit has its own maximum/minimum power level of generation, beyond and below which it cannot generate.

$$P_{i(\min)} \leq P_{ih} \leq P_{i(\max)}. \tag{24}$$

Other Constraints

Hydro-Constraints

Unit commitment if separated from the scheduling of hydro-units as a separate hydrothermal scheduling or coordination problem may not result in an optimal solution.

Must Run Units

Some units must run during certain times of the year for voltage support on the transmission network or for supply of steam for uses outside the steam plant itself.

Must-Off Units

Some units are required to be off-line due to maintenance schedule or forced outage. These units can be excluded from the unit commitment decision.

Emission Constraints

There are emissions like sulphur dioxide (SO_2), nitrogen oxides (NO_x), carbon dioxide (CO_2) and mercury which are produced by fossil-fuelled thermal power plants. The amount of emission depends on various factors such as the type fuel used, level of generation output and the efficiency of the unit. The production cost minimization may need to be compromised in order to have the generation schedule that meets the emission constraints.

Fuel Constraints

Some units may have limited fuel, or else have constraints that require them to burn a specified amount of fuel in a given time, presenting a challenging unit commitment problem.

Unit Constraints or Local Constraints

Constraints that concern individual units are called unit constraints or local constraints are described as follows:

1. **Units minimum and maximum generation limits:** The generation limits represent the minimum loading limit below which it is not economical to load the unit, and the maximum loading limit above which the unit should not be loaded.

2. **Minimum up and down time limits:** Once the unit is running, it should not be turned off immediately. Once the unit is de-committed, there is a minimum time before it can be recommitted.
3. **Unit availability constraints:** The availability constraint specifies the unit to be in one of the following different situations: Available/ not available, out aged/Must out, Must run, Fixed Output (F.O.P).
    a. **Must run units:** Some units are given a must run status during certain times of the year for reasons of voltage support on the transmission network or for such purposes as supply of steam for uses outside the steam plant itself, and to increase the reliability or stability of the system.
    b. **Must out units:** Units which are on forced outages and maintenance are unavailable for commitment and these are the must out units.
    c. **Units on fixed generation: (F.O.P):** These are the units which have been prescheduled and have their generation specified for certain time period. A unit of fixed generation is automatically a must run unit for the designated time period.
4. **Unit initial status:** The initial status value, if it is positive indicates the number of hours the unit is already up, and if it is negative indicates the number of hours the unit has been already down. The status of the unit +/- before the first hour in the schedule is an important factor to determine whether its new state violates the MUT/MDT constraints. The initial status also affects the start up cost calculations.
5. **Unit Derating constraints:** During the life time of a unit its performance could be change due to aging and cause derating of the unit. Therefore, the unit minimum and maximum limits are changed.
6. **Ramp Rate Constraints:** The operating range of all online units is restricted by their ramp rate limits for forcing the units' operation continually between two adjacent specific operation periods. Inequality constraints due to ramp rate limits can be expressed as (Baldwin, Dale, & Dittrich, 1959):
    a. When Generation increases:

$$P_i - P_i^O \leq UR_i. \tag{25}$$

where, UR_i is Up ramp limit of ith generator (MW/h).

P_i^O is previous output power of ith generating unit.

    b. When Generation decreases:

$$P_i^O - P_i \leq DR_i. \tag{26}$$

where, DR_i is the ramp limit of ith generator (MW/h)

CONCLUSION

In this chapter, researchers have presented the Biogeography Based Optimization algorithm to solve the economic load dispatch problem of electric power system. The effectiveness of proposed BBO algorithm is tested with the standard IEEE bus system consisting of 3, 4, 6 and 10-generating units model. It have been observed that the performance of Biogeography Based Optimization algorithm is better than Fire-fly algorithm and Evolutionary search algorithm. Also, it is observed that SADE algorithm yields better results as compared to BBO algorithm.

Also, the paper presents the complete mathematical formulation of Multi-Objective Unit Commitment Problem considering various system and physical constraints. The Biogeography Based Optimization approach can be applied to solve non-convex, non-linear, multidimensional and highly constrained problem unit commitment problem, which is an extension of Economic Load Dispatch Problem.

An appendix for standard IEEE 14-Bus, 30-Bus and 10-Unit generating model along with 24-hours load demand is also given at the last, which can be used as quick reference by the other researchers working in the similar area.

Future Scope

From the above analysis, it has been observed that the performance of Biogeography Based Optimization Algorithm is better than Fire-fly algorithm and Evolutionary search algorithm. It can be used to solve the Single Area and Multi-Area Unit Commitment Problem of electric power system, which is non-convex, non-linear, multidimensional and highly constrained problem.

Abbreviations

i= index for units.
h=index for time.
X_i^{on} = duration that the ith unit is continuously ON.
X_i^{off} = duration that the ith unit is continuously OFF.
MDT_i =minimum downtime of the ith unit.
MU_i =minimum up-time of the ith unit.
D_h =load demand at the hth hours.
$P_{i(\max)}$ =maximum generation limit of the ith unit.
$P_{i(\min)}$ =minimum generation limit of the ith unit.
$P_{ih}^{\min}$ = minimum generation limit of the ith unit at the hth hour.
P_{ih} = power output of the ith unit at the hth hour.
R_h =spinning reserve at the hth hours; spinning reserve is the surplus power of total generation capacity after meeting load demand at the hth hour (usually 10% of load demand).
U_{ih} =On/Off status of the ith unit; U_{ih} =0 and U_{ih} =1 are for Off and On statuses respectively.
FC_i =Fuel cost of ith generating unit.
STC_{ih} =Start-up cost of ith generating unit at the hth hour.
SDC_{ih} =Shut-down cost of ith generating unit at the hth hour.

REFERENCES

Ademovic, A., Bisanovic, S., & Hajro, M. (2010). A genetic algorithm solution to the unit commitment problem based on real-coded chromosomes and fuzzy optimization. In *Proc. The 15th IEEE Mediterranean Electrotechnical Conference (MELECON 2010)*. doi:10.1109/MELCON.2010.5476238

Anita, J. M., Raglend, I. J., & Kothari, D. P. (2012). Solution of unit commitment problem using shuffled frog leaping algorithm. *IOSR Journal of Electrical and Electronics Engineering, 1*(4), 9-26.

Aruldoss, T., Victoire, A., Ebenezer, A., & Jeyakumar. (2005). A simulated annealing based hybrid optimization approach for a fuzzy modeled unit commitment problem. *International Journal of Emerging Electric Power Systems, 3*(1).

Balci, H. H., & Valenzuela, J. F. (2004). Scheduling electric power generators using particle swarm optimization combined with the Lagrangian relaxation method. *International Journal of Applied Mathematics and Computer Science, 14*(3), 411–421.

Baldwin, C. J., Dale, K. M., & Dittrich, R. F. (1959). A study of economic shutdown of generating units in daily dispatch. *AIEE Transaction of Power Apparatus and Systems, PAS-78*(4), 1272–1284. doi:10.1109/AIEEPAS.1959.4500539

Bhardwaj, A., Tung, N. S., Shukla, V. K., & Kamboj, V. K. (2012). The important impacts of unit commitment constraints in power system planning. *International Journal of Emerging Trends in Engineering and Development, 5*(2), 301–306.

Blaze, G., & Marko, C. (2013). Multi-objective powergeneration scheduling: Slovenian power system case study. *Elektrotehniski Vestnik, 80*(5), 222–228.

Chandrasekaran, K., & Simon, S. P. (2012). Multiobjective unit commitment problem using Cuckoo search Lagrangian method. *International Journal of Engineering Science and Technology, 4*(2), 89–105.

Chusanapiputt, S., Nualhong, D., Jantarang, S., & Phoomvuthisarn, S. (2005). *Relative velocity updating in parallel particle swarm optimization based Lagrangian relaxation for large-scale unit commitment problem*. IEEE Region TENCON; doi:10.1109/TENCON.2005.300991

Dasgupta, D., & McGregor, D. R. (1993). Short term unit-commitment using genetic algorithms. In *Proceedings of the Fifth International Conference Proceedings, on Tools with Artificial Intelligence* (pp. 240-247).

Dhillon, J. S., & Kothari, D. P. (2010). *Power system optimization* (2nd ed.). New Delhi, India: Prentice Hall India.

Du, D. (2009). *Biogeography-based optimization: Synergies with evolutionary strategies, immigration refusal, and Kalman filters*. (M. S. Thesis). Department Electrical and Computer Engineering, Cleveland State University.

Du, D., Simon, D., & Ergezer, M. (2009). Biogeography-based optimization combined with evolutionary strategy and immigration refusal. In *IEEE Proc. International Conference on Systems, Man and Cybernetics* (pp. 997- 1002). doi:10.1109/ICSMC.2009.5346055

Jain, J., & Singh, R. (2013). Biogeographic-based optimization algorithm for load dispatch in power system. *International Journal of Emerging Technology and Advanced Engineering, 3*(7), 549-553.

Jalilzadeh, S., & Pirhayati, Y. (2009). An improved genetic algorithm for unit commitment problem with lowest cost. In *Proc. 2009 IEEE International Conference on Intelligent Computing and Intelligent Systems (ICIS 2009)* (pp. 571-575). doi:10.1109/ICICISYS.2009.5357777

Kamboj, V. K., & Bath, S. K. (2013). Single area unit commitment using dynamic programming. In *Proceeding of 4th International Conference on Emerging Trends in Engineering and Technology* (IETET-2013) (pp. 930-936). DOI: 03.AETS.2013.3.260

Kamboj, V. K., & Bath, S. K. (2014). Mathematical formulation of scalar and multiobjective unit commitment problem considering system and physical constraints. In *Proceedings of the 2nd National Conference on Advances in Computing, Communication Network & Electrical Systems* (NCACCNES-2014) (pp. 245-250).

Khokhar, B., Parmar, K. P. S., & Dahiya, S. (2012). Application of biogeography-based optimization for economic dispatch problems. *International Journal of Computers and Applications, 47*(13), 25–30. doi:10.5120/7249-0309

Kumar, V., & Bath, S. K. (2012, March 27-28). Optimization techniques for unit commitment problem-a review. In *Proceeding of the National Conference at Maharishi Deyanand University, Rohtak* (NCACCNES-2012) (pp. 157.1-157.9).

Kumar, V., & Bath, S. K. (2013). Single area unit commitment problem by modern soft computing techniques. *International Journal of Enhanced Research in Science Technology and Engineering, 2*(3), 1–6.

Kundra, H., Kaur, S., Verma, A., & Bedi, R. P. S. (2011). PBBO: A new approach for underground water analysis. *International Journal of Computers and Applications, 30*(3), 37–40. doi:10.5120/3620-5055

Li, M., & Tiaosheng, T. (2001). A gene complementary genetic algorithm for unit commitment. In *Proc. IEEE 5th International Conference on Electrical Machines & Systems* (ICEMS-2001) (Vol. 1, pp. 648-651). IEEE.

Ma, X., El-Keib, A. A., Smith, R. E., & Ma, H. (1995). Genetic algorithm based approach to thermal unit commitment of electric power systems. *Electric Power Systems Research, 34*(1), 29–36. doi:10.1016/0378-7796(95)00954-G

Orero, S. O., & Irving, M. R. (1996). A genetic algorithm for generator scheduling in power systems. *Electric Power Energy Systems, 18*(1), 19–26. doi:10.1016/0142-0615(94)00017-4

Rajan, C. C. A., Mohan, M. R., & Manivannan, K. (2003). Neural based Tabu search method for solving unit commitment problem. *IEE Proceedings. Generation, Transmission and Distribution, 150*(4), 469–474. doi:10.1049/ip-gtd:20030244

Rarick, R., Simon, D., Villaseca, F. E., & Vyakaranam, B. (2009). Biogeography-based optimization and the solution of the power flow problem. In *Proceedings of the IEEE Proc. International Conference on Systems, Man and Cybernetics* (pp. 1003-1008). doi:10.1109/ICSMC.2009.5346046

Reddy, K. S., & Reddy, M. D. (2012). Economic load dispatch using firefly algorithm. *International Journal of Engineering Research and Applications, 2*(4), 2325-2330.

Scheidegger, C., Shah, A., & Simon, D. (2011). Distributed learning with biogeography-based optimization: markov modeling and robot control. In *Proc. 24th International Conference on Industrial Engineering and Other Applications of Applied Intelligent Systems Conference on Modern Approaches in Applied Intelligence* (IEA/ AIE'11) (pp. 203-215).

Shi, L., Haoa, J., Zhou, J., & Xu, G. (2004). Ant colony optimization algorithm with random perturbation behavior to the problem of optimal unit commitment with probabilistic spinning reserve determination. *Electric Power Systems Research, 69*(2-3), 295–303. doi:.epsr.2003.10.00810.1016/j

Simon, D. (2008). Biogeography-based optimization. *IEEE Transactions on Evolutionary Computation, 12*(6), 702–713. doi:10.1109/TEVC.2008.919004

Simon, D., Rarick, R., Ergezer, M., & Du, D. (2011). Analytical and numerical comparisons of biogeography-based optimization and genetic algorithms. *ELSEVIER Journal Information Sciences, 181*(7), 1224–1248. doi:10.1016/j.ins.2010.12.006

Srinivasan, D., & Chazelas, J. (2004). A priority list-based evolutionary algorithm to solve large scale unit commitment problem. In *Proceedings Int. Conf. on Power System Technology* (pp. 1746–1751). doi:10.1109/ICPST.2004.1460285

Sriyanyong, P., & Song, Y. H. (2005). Unit commitment using particle swarm optimization combined with lagrange relaxation. *IEEE Power Engineering Society General Meeting, 3*, 2752-2759. doi:10.1109/PES.2005.1489390

Surekha, P., Archana, N., & Sumathi, S. (2012). Unit commitment and economic load dispatch using self adaptive differential evolution. *WSEAS Transactions on Power Systems, 7*(4), 159-171.

Vanitha, M., & Thanushkodi, K. (2012). An effective biogeography based optimization algorithm to solve economic load dispatch problem. *Journal of Computer Science, 8*(9), 1482-1486.

Victoire, T. A. A., & Jeyakumar, A. E. (2005). Unit commitment by a tabu-search-based hybrid optimisation technique. *IEE Proceedings. Generation, Transmission and Distribution, 152*(4), 563–574. doi:10.1049/ip-gtd:20045190

Wood, A. J., & Wollenberg, B. F. (1996). *Power generation operation and control*. New York, NY: John Wiley and Sons.

Zakaryia, M., & Talaq, J. (2011). Economic dispatch by biogeography based optimization method. In *Proceedings of the 2011 International Conference on Signal, Image Processing and Applications with workshop of ICEEA-2011 IPCSIT* (Vol. 21, pp. 161-165).

Zhu, J. (2009). *Unit commitment, optimization of power system operation*. Hoboken, NJ: John Wiley & Sons, Inc. Publication. doi:10.1002/9780470466971.ch7

APPENDIX A

Table 9. 14-bus system data (Kamboj, 2015)

	P_{max}	P_{min}	A	B	C	MU_i	MD_i	H_{cost}	C_{cost}	C_{hour}	IniState
Unit1	250	10	0.00315	2	0	1	1	70	176	2	1
Unit2	140	20	0.0175	1.75	0	2	1	74	187	2	-3
Unit3	100	15	0.0625	1	0	1	1	50	113	1	-2
Unit4	120	10	0.00834	3.25	0	2	2	110	267	1	-3
Unit5	45	10	0.025	3	0	1	1	72	180	1	-2

Figure 3. Load demand pattern for 24-hours for 14-bus system

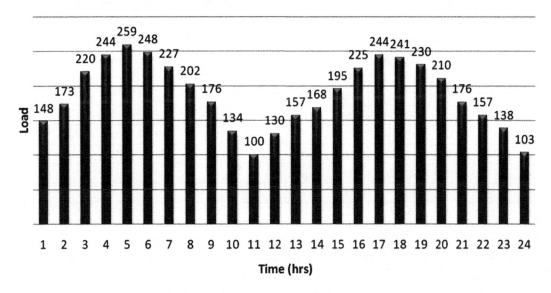

Table 10. 30-bus system data (Kamboj, 2015)

Unit No.	P_{max}	P_{min}	A	B	C	MU_i	MD_i	H_{cost}	C_{cost}	C_{hour}	IniState
Unit1	200	50	0.00375	2	0	1	1	70	176	2	1
Unit2	80	20	0.0175	1.7	0	2	2	74	187	1	-3
Unit3	50	15	0.0625	1	0	1	1	50	113	1	-2
Unit4	35	10	0.00834	3.25	0	1	2	110	267	1	-3
Unit5	30	10	0.025	3	0	2	1	72	180	1	-2
Unit6	40	12	0.025	3	0	1	1	40	113	1	-2

Figure 4. Load demand pattern for 24-hours for 30-bus system

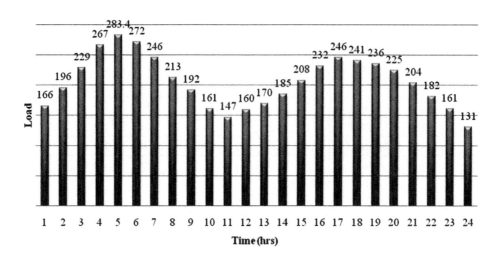

Table 11. 10-unit system data (Kamboj, 2015)

	P_{max}	P_{min}	A	B	C	MU_i	MD_i	H_{cost}	C_{cost}	C_{hour}	IniState
Unit1	455	150	0.00048	16.19	1000	8	8	4500	9000	5	8
Unit2	455	150	0.00031	17.26	970	8	8	5000	10000	5	8
Unit3	130	20	0.002	16.6	700	5	5	550	1100	4	-5
Unit4	130	20	0.00211	16.5	680	5	5	560	1120	4	-5
Unit5	162	25	0.00398	19.7	450	6	6	900	1800	4	-6
Unit6	80	20	0.00712	22.26	370	3	3	170	340	2	-3
Unit7	85	25	0.00079	27.74	480	3	3	260	520	2	-3
Unit8	55	10	0.00413	25.92	660	1	1	30	60	0	-1
Unit9	55	10	0.00222	27.27	665	1	1	30	60	0	-1
Unit10	55	10	0.00173	27.79	670	1	1	30	60	0	-1

Figure 5. Load demand pattern for 24-hours for 10-unit system

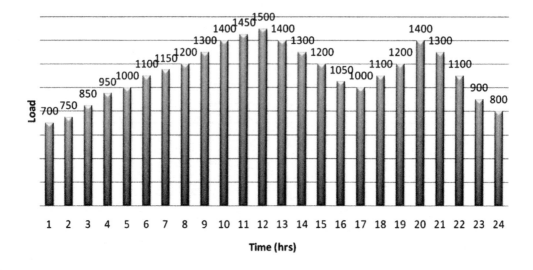

APPENDIX B

Data Sheet for Multi-Area Unit Commitment (Four Area Test System)

Table 12. Generating unit characteristics

Unit No.	Minimum Up Time [Hour]	Minimum Down Time[Hour]	Initial Condition [Hour]	Minimum Generation [MW]	Maximum Generation [MW]
1	0	0	-1	2.40	12
2	0	0	-1	2.40	12
3	0	0	-1	2.40	12
4	0	0	-1	2.40	12
5	0	0	-1	2.40	12
6	0	0	-1	4.00	20
7	0	0	-1	4.00	20
8	0	0	-1	4.00	20
9	0	0	-1	4.00	20
10	0	-2	3	15.20	76
11	3	-2	3	15.20	76
12	3	-2	3	15.20	76
13	3	-2	3	15.20	76
14	3	-2	-3	25.00	100
15	4	-2	-3	25.00	100
16	4	-2	-3	25.00	100
16	4	-3	5	54.25	155
18	4	-3	5	54.25	155
19	5	-3	5	54.25	155
20	5	-3	5	54.25	155
21	5	-4	-4	68.95	197
22	5	-4	-4	68.95	197
23	5	-4	-4	68.95	197
24	8	-5	10	140.00	350
25	8	-5	10	140.00	350
26	8	-5	10	140.00	350

Table 13. Cost functions for generating units in area 1

Unit No.	Generation Cost Coefficient a [$/MW²]	Generation Cost Coefficient b [$/MW]	Generation Cost Coefficient c [$]	Start Up Cost Coefficient A [$]	Start Up Cost Coefficient B [$]	Start Up Time Constant [tau]
1	24.36	25.237	0.012	0	0	1
2	24.379	25.255	0.0121	0	0	1
3	24.395	25.273	0.0125	0	0	1
4	24.42	25.299	0.0129	0	0	1
5	24.434	25.321	0.013	0	0	1
6	117.121	37	0.006	20	20	2
7	117.239	37.132	0.0062	20	20	2
8	117.358	37.307	0.0064	20	20	2
9	117.481	37.49	0.0066	20	20	2
10	81	13.322	0.0046	50	50	3
11	81.028	13.244	0.0047	50	50	3
12	81.104	13.3	0.0049	50	50	3
13	81.176	13.35	0.0052	50	50	3
14	217	18	0.0042	70	70	4
15	217.1	18.1	0.0044	70	70	4
16	217.2	18.2	0.0047	70	70	4
16	142.035	10.394	0.0043	150	150	6
18	142.229	10.515	0.0045	150	150	6
19	142.418	10.637	0.0047	150	150	6
20	143.497	10.708	0.0048	150	150	6
21	256.101	22	0.0025	200	200	8
22	257.649	22.1	0.0026	200	200	8
23	258.176	22.2	0.0026	200	200	8
24	175.057	10.462	0.0016	300	200	8
25	305.036	7.486	0.0019	500	500	10
26	306.91	7.493	0.0019	500	500	10

Table 14. Cost functions for generating units in area 2

Unit No.	Generation Cost Coefficient a [$/MW²]	Generation Cost Coefficient b [$/MW]	Generation Cost Coefficient c [$]	Start Up Cost Coefficient A [$]	Start Up Cost Coefficient B [$]	Start Up Time Constant [tau]
1	24.389	25.547	0.0123	0	0	1
2	24.411	25.675	0.0125	0	0	1
3	24.638	25.803	0.013	0	0	1
4	24.76	25.932	0.0134	0	0	1
5	24.488	26.061	0.0136	0	0	1
6	117.755	37.551	0.0059	20	20	2
7	118.108	37.664	0.0066	20	20	2
8	118.458	37.777	0.0066	20	20	2
9	118.821	37.89	0.0073	20	20	2
10	81.136	13.327	0.0047	50	50	3
11	81.298	13.354	0.0049	50	50	3
12	81.464	13.38	0.0051	50	50	3
13	81.626	13.407	0.0053	50	50	3
14	217.895	18	0.0043	70	70	4
15	218.355	18.1	0.0051	70	70	4
16	218.775	18.2	0.0049	70	70	4
16	142.735	10.695	0.0047	150	150	6
18	143.029	10.715	0.0047	150	150	6
19	143.318	10.737	0.0048	150	150	6
20	143.597	10.758	0.0049	150	150	6
21	259.131	23	0.0026	200	200	8
22	259.649	23.1	0.0026	200	200	8
23	260.176	23.2	0.0026	200	200	8
24	177.057	10.862	0.0015	300	200	8
25	310.002	7.492	0.0019	500	500	10
26	311.91	7.503	0.0019	500	500	10

Table 15. Cost functions for generating units in area 3

Unit No.	Generation Cost Coefficient a [$/MW²]	Generation Cost Coefficient b [$/MW]	Generation Cost Coefficient c [$]	Start Up Cost Coefficient A [$]	Start Up Cost Coefficient B [$]	Start Up Time Constant [tau]
1	24.451	26.547	0.0123	0	0	1
2	24.395	26.675	0.0125	0	0	1
3	24.738	26.803	0.013	0	0	1
4	24.861	26.932	0.0134	0	0	1
5	24.988	27.061	0.0136	0	0	1
6	118.755	38.551	0.0069	20	20	2
7	119.108	38.664	0.0076	20	20	2
8	119.458	38.777	0.0076	20	20	2
9	119.821	38.89	0.0083	20	20	2
10	82.136	14.327	0.0047	50	50	3
11	82.298	14.354	0.0059	50	50	3
12	82.464	14.481	0.0061	50	50	3
13	82.626	14.407	0.0063	50	50	3
14	218.895	19	0.0053	70	70	4
15	219.355	19.1	0.0061	70	70	4
16	219.775	19.2	0.0059	70	70	4
16	143.735	11.695	0.0056	150	150	6
18	144.029	11.715	0.0057	150	150	6
19	144.318	11.737	0.0058	150	150	6
20	144.597	11.758	0.0059	150	150	6
21	259.131	24	0.0036	200	200	8
22	259.649	24.1	0.0036	200	200	8
23	260.176	24.2	0.0036	200	200	8
24	177.057	11.862	0.0015	300	200	8
25	310.002	7.692	0.0019	500	500	10
26	311.91	7.703	0.0019	500	500	10

Table 16. Cost functions for generating units in area 4

Unit No.	Generation Cost Coefficient a [$/MW²]	Generation Cost Coefficient b [$/MW]	Generation Cost Coefficient c [$]	Start Up Cost Coefficient A [$]	Start Up Cost Coefficient B [$]	Start Up Time Constant [tau]
1	24.389	25.202	0.0123	0	0	1
2	24.411	25.255	0.0125	0	0	1
3	24.638	25.273	0.013	0	0	1
4	24.76	25.342	0.0134	0	0	1
5	24.888	25.366	0.0136	0	0	1
6	117.755	37.012	0.0059	20	20	2
7	118.108	37.055	0.0066	20	20	2
8	118.458	37.098	0.0066	20	20	2
9	118.821	37.156	0.0073	20	20	2
10	81.136	13.261	0.0047	50	50	3
11	81.298	13.278	0.0049	50	50	3
12	81.464	13.295	0.0051	50	50	3
13	81.626	13.309	0.0053	50	50	3
14	217.895	17.5	0.0043	70	70	4
15	218.355	17.6	0.0051	70	70	4
16	218.775	17.7	0.0049	70	70	4
16	142.735	10.21	0.0047	150	150	6
18	143.029	10.268	0.0047	150	150	6
19	143.318	10.307	0.0048	150	150	6
20	143.597	10.375	0.0049	150	150	6
21	259.131	22.5	0.0026	200	200	8
22	259.649	22.6	0.0026	200	200	8
23	260.176	22.7	0.0026	200	200	8
24	177.057	10.462	0.0015	300	200	8
25	310.002	7.492	0.0019	500	500	10
26	311.91	7.503	0.0019	500	500	10

APPENDIX C

Data Sheet for Multi Objective Multi-Area Unit Commitment (Three Area Test System)

Table 17. Generating unit characteristics for three area test system

Unit No.	Pmax	Pmin	Minimum Up-Down Time		Startup Cost		CSH_i	IS_i
			MUT_i	MDT_i	HSC_i	CSC_i		
U1	200	50	1	1	70	176	2	-1
U2	80	20	2	2	74	187	1	-3
U3	50	15	1	1	50	113	1	2
U4	35	10	1	2	110	267	1	3
U5	30	10	2	1	72	180	1	-2
U6	40	12	1	1	40	113	1	2

Table 18. Emission and fuel cost coefficients for three area test system

Unit No.	Fuel Cost Coefficients			Emission Coefficients		
	a($/MW²)	b ($/MWh)	c ($/h)	α	β	γ
U1	0.00375	2	0	22.983	-0.9	0.0126
U2	0.0175	1.75	0	25.313	-0.1	0.02
U3	0.0625	1	0	25.505	-0.01	0.027
U4	0.00834	3.25	0	24.9	-0.005	0.0291
U5	0.025	3	0	24.7	-0.004	0.029
U6	0.025	3	0	25.3	-0.0055	0.0271

Figure 6. Three area test system model for Multi Objective UCP

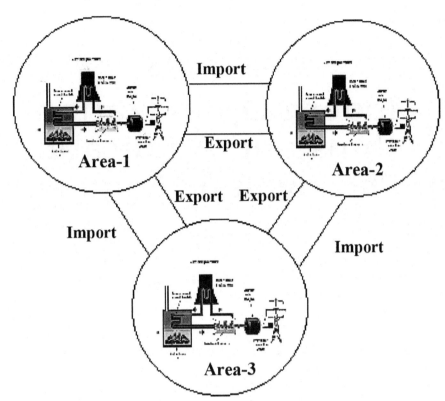

Figure 7. Load demand profile for three area test system

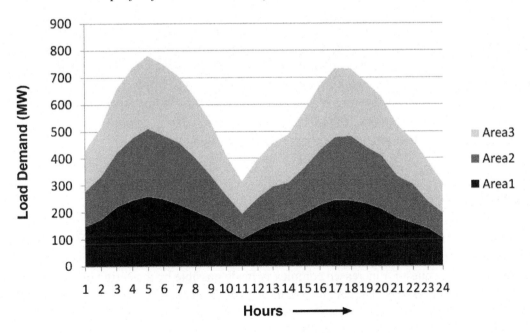

Chapter 16
Modern Optimization Algorithms and Applications in Solar Photovoltaic Engineering

Igor Tyukhov
All-Russian Research Institute for Electrification of Agriculture (VIESH), Russia

Hegazy Rezk
Minia University, Egypt

Pandian Vasant
Universiti Teknologi PETRONAS, Malaysia

ABSTRACT

This chapter is devoted to main tendencies of optimization in photovoltaic (PV) engineering showing the main trends in modern energy transition - the changes in the composition (structure) of primary energy supply, the gradual shift from a traditional (mainly based on fossil fuels) energy to a new stage based on renewable energy systems from history to current stage and to future. The concrete examples (case studies) of optimization PV systems in different concepts of using from power electronics (particularly maximum power point tracking optimization) to implementing geographic information system (GIS) are considered. The chapter shows the gradual shifting optimization from specific quite narrow areas to the new stages of optimization of the very complex energy systems (actually smart grids) based on photovoltaics and also other renewable energy sources and GIS.

INTRODUCTION

All processes in surrounding material world and life on Earth are driven by energy. Vaclav Smil, author of many books on energy, wrote: "…energy's role in world history seems to be a natural proposition, with history seen as a quest for increased complexity made possible by mastering higher energy flows" (Smil V., 2004). This may be because of an extraordinary difficulty and exceptional nature of the coming

DOI: 10.4018/978-1-4666-9755-3.ch016

energy transition—but, given the enormous challenges of ushering in a post-fossil world, it may also be because of the possibility of an unprecedented and persistent commitment to a rapid change (Smil V., 2010).

Any process can be analyzed in terms of its underlying energy conversions; any object, as well as any bit of information, can be valued for its energy content and for its potential contribution to future energy transformations (Smil V., 2008).

Energy and information are including in the infrastructure of whole activity of modern humankind. Energy and communicative revolutions are the main driving forces in the human development. Education and information are also connecting by very strong relationship. Very often these technologies are considered separately. It is totally incorrect and inappropriate. Table 1 shows roughly the main steps of appearing the new energy, communicational and educational technologies.

Before the industrial revolution, our energy needs were modest. The level of development civilization is characterized by the rate of liberation from nature influences. The history of civilization and liberation from nature influences is the history of new types of energy conquest.

This work is stimulated by the modern trends or diffusion in developing the most important sphere of humankind activity – energy and new computer technologies. Last decades traditional electrical power industry, which basic principles were formed more than hundred years ago, is changing and transforming very fast.

Table 1. Energy, communicative and educational revolutions

Years Ago (Very Approximately)	Communicative and Educational Revolutions	Energy Revolutions	Stages of Development
10^6	Oral communication through language	Conquer of fire, biomass	Forming human society based on additional to biological type energy sources
10^5	Picture and symbol communication	Solar energy, animal energy	Renewable energy – first stage Exchange information between people
10^4	Writing (from internal biological memory to external recorded memory)	Wind and hydro energy (from muscles to machines)	
10^3	Book printing, forming of scientific systems Forming schools and universities	Mechanical energy, Thermal energy (coal, oil, gas)	Using fossil fuels (accumulating solar energy) Human civilization, scientific and technology revolutions
10^2	System of education Mass media, radio, TV	Steam energy, electrical energy, atomic energy	
10^1	Computers, Internet, MOOC	Photovoltaic revolution	Current technology revolution (informatization, robotization, artificial intelligence) Extra planetary activity
Now	Practically total interconnectivity Satellite communication Mobile communication Mass open education	Smart grids, Energy mixt Decentralized energy Space technologies	
Near future	Global earth and near space communication Internet of things Big data technology	Mixt of traditional energy and renewable energy – second stage, new sources of energy Thermonuclear energy Global energy system	Energy and information everywhere Moon and Mars exploitation
Future	Trends of dominating distributed energy and information networks integrating by smart technologies Global technologies and Internet of everything including energy, artificial intelligence		Solar energy for the future of mankind
Far future	Widen areal of mankind from Earth to near space		

We are now in the period of a transition as a process of successful technical and infrastructural innovation. Energy transitions can be also studied as specific subsets of two more general processes of technical innovation and resource substitution. Transition from an energy supply dominated by nuclear and fossil fuels to a world relying mostly on non-fossil fuels and generating electricity by harnessing renewable energy flows is thus definitely desirable and, given the finite nature of fossil resources (Smil, V., 2010).

Any anthropogenic energy system—that is any arrangement whereby the humans use the Earth's resources to improve the human species chances of surviving and to enhancing their quality of life (and, less admirably, also to increase their individual and collective power and to dominate, and to conquer and destroy others biological species)—has three fundamental components: natural energy sources, their conversions, and a variety of specific uses of the available energy flows.

Technological and social developments of humankind have been well supported by the discovery of new sources of energy and its proper application in day-to-day life. Since the first installation of electricity grids more than 100 years ago, our society has been driven by this convenient source of energy. Most of this electric energy is produced from fossil-fired power generation plants using carbon-based fuels, such as gasoline in the transportation industry. Burning of fuel looks like burning fire on the dawn of humankind in spite we have already now many technologies direct converting fuel energy into electricity (fuel cells), heat into electricity (thermo-elements), solar radiation into electricity (solar cells) and others.

In today's modern era, however, demands development of new and more efficient technologies for next-generation electrical grids and shifting of the focus of power generation to cleaner and carbon-free sources. Both the exhausting of carbon-based fossil fuels and environmental pollution created by the burning of fossil fuels are responsible for this desirable shift power engineering to so called green energy sources. Possible sources of such clean and renewable energy are solar, nuclear, wind, geothermal energy and many others. It is worth mentioning here that the world consumes less energy in one year than the amount of energy Earth surface receive from sunlight in one hour (13 terawatts).

Solar energy represents one of the most significant energy solutions to avoid and even to eradicate our addiction to oil.

1. SOLAR ENERGY, PHOTOVOLTAICS

Solar cells are devices that directly convert solar radiation into electricity by photovoltaic effect. Let us start from the early beginning of photovoltaics. In 1839 French scientist Edmond Becquerel discovers the photovoltaic effect while experimenting with an electrolytic cell made up of two metal electrodes placed in an electricity-conducting solution—electricity-generation increased when exposed to light.

At the beginning of 1930th one of the greatest Russian scholar A. F. Ioffe concentrated on semiconductor physics and formulated many important practical tasks of direct converting solar energy. In result it was created the world level school of semiconductor physics, which further lead to elaborating the most efficient solar cells on heterostructures (Strebkov, D. S., 2005).

It seems obvious that capturing solar energy is going to be the main trend for the last few decades. In line with this expectation, research on solar cell science and technology has taken a huge jump in recent years. Since the first demonstration of the solar cell in 1954 by Bell Laboratories, much progress has been made in this field. The initial more or less practical solar cell by Bell was a silicon p-n junction cell.

In 1960, solar cells were essentially unique, handmade, cost approximately $1000 per peak Watt of power and were used almost exclusively for space applications. In 1970, cells were manufactured in batch processes, cost approximately $100 per peak Watt, were primarily used in space applications, but were making an entry into standalone applications.

In 1980, larger quantities of solar cells were produced in larger batch processes, cost approximately $10 per peak Watt, saw significant terrestrial applications (primarily stand-alone, remote devices), and began to be used for grid connected residential applications.

Today, different varieties of solar cells are being studied including amorphous silicon, microcrystalline silicon, polycrystalline silicon, cadmium telluride, copper indium selenide/sulfide, organic/polymer, dye-sensitized solar cells and many others. Actually three generations are considered in the book (M.A. Green, 2006) and many new concepts are appearing now

In spite the practically modern silicone cell dates back to 1954 and was developed at Bell Labs, it should be mentioned the first silicon solar cell was made entirely by accident in early 1940 when Russell Ohl, a researcher at Bell telephone Laboratories in New Jersey, made an important discovery. When he shone a flashlight onto a piece of silicon he was studying, the needle of the voltmeter connected across jumped to a surprisingly large reading.

Today, we are on the threshold of continuous-process manufacturing and photovoltaic modules are beginning to be purchased (in large orders) for approximately four dollars per peak Watt. In the United States, there are several thousand (mostly very small) residential systems and a few central power stations in operation or under development—primarily in California. In addition, there is a rapidly growing market for photovoltaic consumer products such as watches, calculators, radios, televisions, battery rechargers, and so on.

As our conventional source of energy—fossil fuel—rapidly depletes, the search for alternative energy sources has become an integral issue in modern industrialized society. Unavoidably, the world's energy demands and cost will rise. To keep pace with these demands, unlimited energy sources must be researched and optimized. The rapid depletion of non-renewable fossil resources will not continue, since it is now or soon will be technically and economically feasible to supply all of man's need from the most abundant energy source of all, the sun. [Green energies and the environment]

The predicted world energy demand will reach 28 TW by 2050 and 46 TW by 2100. The deployment of solar cells as a source of electricity will have to expand to a scale of tens of peak terawatts in order to become a noticeable source of energy in the future (Coby S.Tao, 2011), that is why even natural resource limitations to terawatt-scale solar cells are considered now and possible solutions are discussed (Tyukhov, 2012).

Solar photovoltaics (PV) have been growing exponentially since 2007, doubling in capacity every two years. This means that annual installation rates have mushroomed from less than 4GW in 2007 to 42GW in 2014 – a more than ten-fold increase in just seven years (Green, J, 2014).

2. DIFFERENT MPPT TECHNIQUES FOR OPTIMIZATION PHOTOVOLTAIC SYSTEMS

PV generation systems have two major problems: the conversion efficiency of electric power generation is low, and the amount of electric power generated by solar arrays changes continuously with weather conditions (Berrera et al., 2009; Sreekanth et al., 2012). Moreover, because of nonlinear I-V and P-V characteristics of PV systems, their output power is always changing with weather conditions, i.e., solar radiation, atmospheric temperature and also nature of load connected (Fangrui et al., 2008; Emad et al., 2010). Maximum power point tracking (MPPT) is essential as there is a probable mismatch between the load characteristics and the maximum power points (MPPs) of the PV module in order to ensure optimal utilization of solar cells (Xiao et al., 2004; Ashish et al., 2007). Using of MPPT leads to reduce the cost of energy generated by PV panels (Hohm et al., 2000).

2.1 Conventional MPPT under Uniform Insolation

A large number of techniques have been proposed for tracking the MPP of PV systems. There are many techniques available in the literature such as fractional open-circuit voltage and short-circuit current (Fangrui et al., 2008), the Artificial Neural Network (ANN) technique (El Sayed, 2013), and the Fuzzy Logic control (Eltamaly, 2010; Eltamaly et al., 2010). Also, it was demonstrated in (Ibrahim et al., 1999) technique to track the maximum power using look-up table in the microcomputer. Another common approach is to use the array power as the feedback. The popular tracking methods based on this approach are widely adopted in PV power systems (Faranda et al., 2008) which include but not limited to, perturb and observe method (P&O) (Koutroulis et al., 2001; Ioan et al., 2013), the incremental conductance method (INC) (Xiao et al., 2004; Esram et al., 2007) and the hill climbing method (HC) (Weidong et al., 2004). These techniques are widely applied in the MPPT controllers due to their simplicity and easy implementation.

2.1.1 Hill Climbing (HC) Technique

The advantage of the hill climbing MPPT technique is its simplicity. It uses the duty cycle of boost converter as the judging parameter when the task of the maximum power point tracking is implemented. When the condition dP/dD = 0 is accomplished, it represents that the maximum power point has been tracked (Xiao et al., 2004). The flow diagram of the HC algorithm is shown in Figure 1. The duty cycle in every sampling period is determined by the comparison of the power at present time and previous time. If the incremental power dP > 0, the duty cycle should be increased in order to make dD > 0. If dP < 0, the duty cycle is then reduced to make dD < 0. The main problem associated with this technique is because of the tradeoff between the stability of the system in constant radiation period and lack of fast response in rapidly changing radiation. The constant radiation period needs very small value of change in duty cycle, ΔD to prevent high oscillation of power around the maximum power point which reduces the energy captured from PV. On the other hand, the rapidly changing radiation needs higher value of duty cycle for fast tracking the maximum power. HC technique has been simulated in the PSIM software package as shown in Figure 2.

Where; ZOH is zero-order hold block; 1/Z is unit delay block (a unit delay block delays the input by one sampling period); and Sign is sign function block (The output of the sign function block is the sign of the input. When the input is positive, the output is 1. When the input is 0, the output is 0. When the input is negative, the output is -1.)

2.1.2 Incremental Conductance (INC) Technique

Among all the MPPT strategies, the incremental conductance technique is widely used due to the high tracking accuracy at steady state and good adaptability to the rapidly changing atmospheric conditions

Figure 1. State flow chart of HC MPPT technique

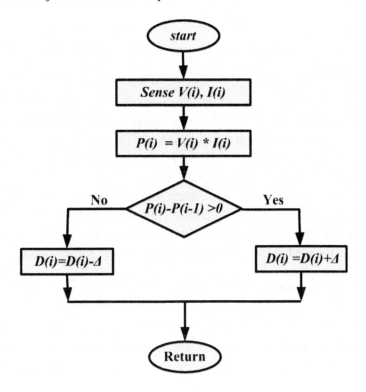

Figure 2. HC MPPT Control Technique using PSIM Software

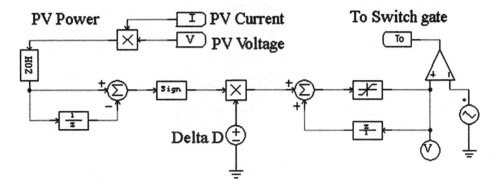

(Fangrui et al., 2008). This technique employs the slope of the PV array power characteristics to track MPP. The slope of the PV array power curve is zero at the MPP, positive for values of output voltage smaller than the voltage at MPP, and negative for values of the output voltage greater than the voltage at MPP. The derivative of the PV module power is given as in (1), and the resultant equation for the error e is as in (2) (Abouobaida et al., 2012; Emad et al., 2011; Chen et al., 2014).

$$\frac{dP}{dV} = \frac{d(V \times I)}{dV} = I + V\frac{dI}{dV} = 0 \qquad (1)$$

Also,

$$\frac{dI}{dV} + \frac{I}{V} = \frac{I(i) - I(i-1)}{V(i) - V(i-1)} + \frac{I(i)}{V(i)} = 0 \quad e = \frac{I(i) - I(i-1)}{V(i) - V(i-1)} + \frac{I(i)}{V(i)} \qquad (2)$$

Therefore tracking the MPP requires the following procedure as shown in Figure 3. It can be implemented by a simple discrete integrator with the error signal e as the input, and a scaling factor k. The function of the scaling factor k is to adapt the error signal e to a proper range before the integral compensator. As the operating point approaches the MPP, the error signal e becomes smaller, resulting in an adaptive and smooth tracking (Abouobaida et al., 2012).

To improve, both, the MPPT speed and accuracy simultaneously a modified dynamic change in step size for INC is introduced (Fangrui et al., 2008). This technique improves the performance of INC technique but at a cost of increased complexity of the control system.

INC technique was simulated in the PSIM software package as shown in Figure 4.

2.1.3. Perturb and Observe (P&O) Technique

P&O is the most frequently used technique to track the maximum power due to its simple structure (Ioan et al., 2013). This technique operates by periodically perturbing the PV module terminal voltage and

Figure 3. State-flow chart of INC MPPT technique

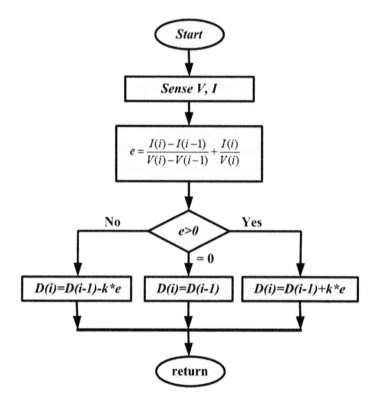

Figure 4. INC MPPT Control Technique using PSIM Software

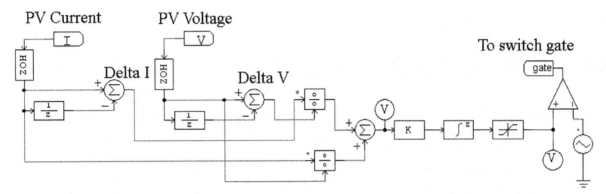

comparing the PV output power with that of the previous perturbation cycle (Faranda et al., 2008). As shown in Figure 5 if the PV module operating voltage changes and power increases the control system moves the operating point in that direction; otherwise the operating point is moved in the opposite direction. The flowchart of this technique is shown in Figure 5. PSIM software package has been used to simulate this technique as shown in Figure 6. A common problem in this technique is that the PV module terminal voltage is perturbed every MPPT cycle; therefore when the MPP is reached, the output power

Figure 5. State flowchart of P&O MPPT technique

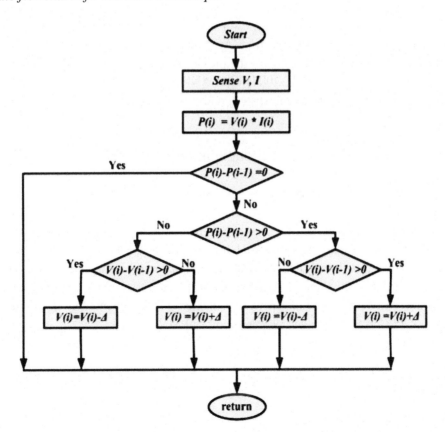

Figure 6. P&O MPPT control technique using PSIM software

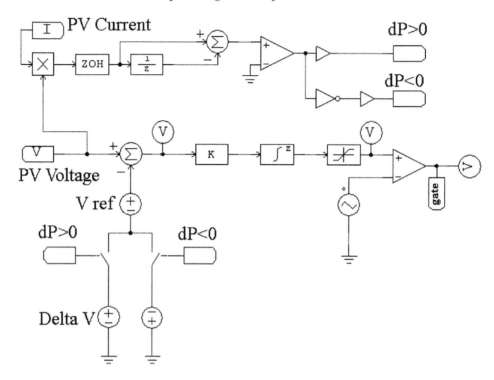

oscillates around the maximum, resulting in power loss in the PV system. To remedy this problem, a modified P&O technique has been introduced in (Al-Diab et al., 2010) by multiplying the change in the duty cycle by dynamic constant depending on the previous change in the extracted power as shown in (1). Another technique (Amrouch et al., 2007) used ANN to predict this multiplying constant. These techniques increase the complexity of the system and may cause more oscillations in stable weather conditions.

2.1.4. Fuzzy Logic Controller (FLC) Technique for PV MPPT

FLC has been introduced in many researches as in (Gounden et al., 2009; Ben Salah et al., 2008; Altasa et al., 2008; Khaehintung et al., 2004; Karlis et al., 2007; Veerachary et al., 2003; Rajesh et al., 2014) to force the PV to work around MPP. FLCs have the advantages of working with imprecise inputs, not needing an accurate mathematical model, and handling nonlinearity. The details of using FLC in MPPT of PV system is shown in (Eltamaly, 2010). The error signal can be calculated as shown in (3). The value of ΔE is calculated as shown in (4). The model of the proposed system has been simulated in co-simulation between PSIM and Simulink software packages. The idea behind using co-simulation is the easy simulation of FLC in Simulink and power circuit in PSIM.

$$E(n) = \frac{P(i) - p(i-1)}{V(i) - V(i-1)} \tag{3}$$

$$\Delta E(i) = E(i) - E(i-1) \tag{4}$$

The detailed logic, theory and implementation of this model can be found in (Eltamaly, 2010) and (Eltamaly et al., 2010). The output power from the PV system and the voltage are used to determine the E and ΔE based on (4) and (5). The inputs to a FLC are usually E and ΔE. The range of E and ΔE are fixed judiciously based on trial and error. These variables are expressed in terms of linguistic variables or labels such as PB (Positive Big), PM (Positive Medium), PS (Positive Small), ZE (Zero), NS (Negative Small), NM (Negative Medium), NB (Negative Big) using basic fuzzy subset. Each of these acronyms is described by mathematical membership functions, MF as shown in Figure 7. Once E and ΔE calculated and converted to the linguistic variables based on MF, the FLC output, which is typically a change in duty cycle, ΔD of the power converter can be looked up in a rule base given in Table 2. A triangular membership function can be used for both inputs and output variables, as it can easily be implemented on the digital control system. The linguistic variables assigned to *ΔD* for the different combinations of E and ΔE are based on the power converter being used and also on the knowledge of the user.

These linguistic variables of input and output MFs are then compared to a set of pre-designed values during aggregation stage. The proper choice of If-then rules or Fuzzy inference is essential for the appropriate response of the FLC system. The inference used in this work is tabulated in Table 2. Some researches proportionate these variables to only five fuzzy subset functions as in (Eltamaly et al., 2010). Table 2 can be translated into 49 fuzzy rules IF-THEN rules to describe the knowledge of control as follows;

Figure 7. A fuzzy system with two inputs, 1 output and 7 MFs each

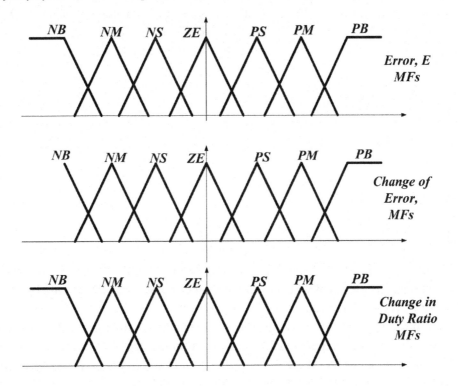

Table 2. Rules for a fuzzy system with 2-inputs and 1 output with 7-membership functions

E \ ΔE	NB	NM	NS	ZE	PS	PM	PB
NB	NB	NB	NB	NB	NM	NS	ZE
NM	NB	NB	NB	NM	NS	ZE	PS
NS	NB	NB	NM	NS	ZE	PS	PM
ZE	NB	NM	NS	ZE	PS	PM	PB
PS	NM	NS	ZE	PS	PM	PB	PB
PM	NS	ZE	PS	PM	PB	PB	PB
PB	ZE	PS	PM	PB	PB	PB	PB

R_{25} : if E is NM and ΔE is PS the ΔD is NS

R_{63} : if E is PM and ΔE is NS the ΔD is PS

……

R_{51} : if E is PS and ΔE is NB the ΔD is NM

Defuzzification is for converting the fuzzy subset of control form inference back to values. As the plant usually required a nonfuzzy value of control, a defuzzification stage is needed. Defuzzificaion for this system is the height method. The height method is both very simple and very fast method. The height defuzzification method in a system of rules by formally given by (5):

$$\Delta D = \left(\sum_{k=1}^{m} c(k) * W_k \right) / \sum_{k=1}^{n} W_k \qquad (5)$$

where

ΔD = change of duty cycle
$c(k)$ = peak value of each output
W_k = height of rule k.

In the defuzzification stage, FLC output is converted from a linguistic variable to a numerical variable. This provides an analog signal which is ΔD of the boost converter. This value is subtracted from previous value of D to get its new value.

PSIM model showing the calculating E and ΔE and the inputs to Simulink is shown in Figure 8. The simulation model showing the output signal from FLC in Simulink to PSIM and the switching circuit that control the switch is shown in Figure 9. The simulation model of the proposed system in Simulink is shown in Figure 10. Where; SLINK1 is the name that used in Simulink for the input to PSIM part of simulation

Figure 8. PSIM model showing the calculating E and ΔE and the inputs to Simulink

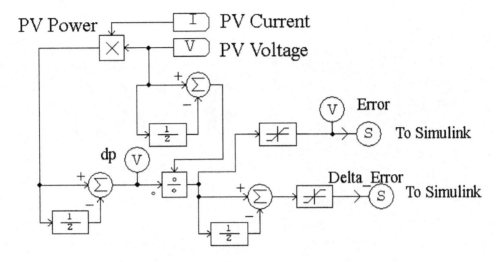

Figure 9. Simulation model of output signal from FLC in Simulink to PSIM and the switching circuit

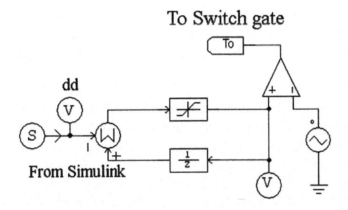

2.2. Soft Computing MPPT under Partial Shading Conditions

PV system comprises multiple modules interconnected in series and in parallel to create a system with the desired rating. Therefore, the probability for occurring partially shaded condition (PSC) is extremely high. PSC is a phenomenon which is taken place when a part or the whole PV module receives a non-uniform irradiance (Ahmed and Salam, 2014). Ideally, the PV system should be installed in a shade free location. However, it is usually installed in urban and sub-urban areas and PV modules are subjected to partially shadow which is caused by clouds, trees due to wind, neighboring buildings and utilities. Under this condition, the power against voltage (*P–V*) characteristics of PV system exhibits multiple local peaks and one global peak which is required to be tracked, this phenomena reduces the effectiveness of conventional maximum power point tracking (MPPT) techniques (Liu et al., 2012; Sarvi et al., 2015). Large efficient conventional MPPT techniques have been proposed to track maximum power point (MPP) to improve the conversion efficiency of a PV system under uniform solar irradiance (Ma J. et al., 2013). These techniques include hill climbing, incremental conductance, fuzzy logic control (FLC), perturb &

Figure 10. Simulink simulation model of FLC MPPT

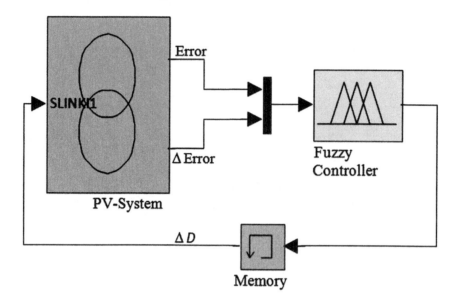

observe, and neural network (Liu et al., 2012; Tey et al., 2014). However, these techniques are highly efficient in tracking the MPP under uniform irradiance, fully illuminated system, where there is only single MPP that exists in the P–V curve, while they are not suitable for tracking the global MPP under PSC (Sarvi et al., 2015) and the efficiency of the PV system decreases (Tey et al., 2014). The idea of using global MPPT based on meta-heuristic optimization techniques algorithms such as genetic algorithms (GAs), Ant Colony Systems (ACS), Artificial Bee Colony (ABC), Artificial Immune Systems (AIS) and particle swarm optimization (PSO) have been presented in (Tey et al., 2014; Sundareswaran et al., 2014; Sundareswaran et al., 2015; Tajuddin et al., 2011). A global MPPT based on cuckoo search algorithm for PV system under PSC has been presented in (Ahmed and Salam, 2014); the authors exploit the significance of the Lévy flight to speed up the algorithm for reaching the global MPP. A modified PSO–MPPT has been presented in (Ishaqu et al., 2012) to minimize the steady state oscillation generated by unstable environmental conditions like change in solar irradiance, change in load and partial shading of PV system. Various extra coefficients have been added to conventional PSO to speed its performance in tracking the global power extracted from the PV system (Daraban et al., 2013); the disadvantage of adding these extra parameters is increasing the algorithm time consumption. A genetic optimization algorithm has been used to determine neuron numbers in multilayer perceptron neural network (ANN) (Kulaksız et al., 2012). A technique based on artificial bee colony has been proposed in (Kinattingal et al., 2015) for global MPP tracking in a PV power generation system. A new MPPT algorithm based on a colony of fireflies for quickly tracking global MPP in partially shaded A PV array has been presented in (Sundareswaran et al., 2014). The tracking procedure consists of positioning the fireflies in the possible solution space and based on the PV power output, the flies move to promised regions (Sundareswaran et al., 2014). An algorithm combines PSO and *P&O* for tracking the global MPP for the PV system has been proposed in (Sundareswaran et al., 2013). Varying step size decreased response time and oscillations around the MPP. Ideal size for ANN structure with five neurons in hidden layer was obtained. In (Paraskevadaki et al., 2011) the effect of partial shading on PV module has been analyzed; additionally, a mathematical formula to calculate the shadow effect on the main electrical characteristic of PV module and the MPP of voltage and current has been provided. Modified Differential evolution technique (Tajuddin et al., 2011) able to deal with large and rapid fluctuations of solar irradiance and reached the global MPP under various PS conditions. Average efficiency of 99.6% was attained during tests of partial-shading conditions. Proposed Ant colony optimization (ACO)-based MPPT technique (Besheer et al., 2012) uses heuristic algorithm to find global MPP with simpler control system and low cost. Simulations verified that the technique was successful and robust to various shading patterns. A method for tracking the global MPP of PV system based on the measurements of the PV array open-circuit voltage has been presented in (Lei et al., 2011) while the method described in (Kazmi et al., 2009) is dependent on the solar radiation or PV array short circuit current. A two stage method to track the global MPP of the shaded PV system has been developed in (Kobayashi et al., 2006). In the first one the point of operation moves nearest to the global MPP and in the second one it converges to the actual global MPP.

PV cell is an electrical device that converts the energy of light directly into electricity through the PV effect. They are connected in series and parallel to step up the voltage and current of PV module respectively. PV cell has a complex relationship between solar irradiance, temperature, and the load; so it exhibits a nonlinear output efficiency characteristic known as the P–V curve. Therefore, MPPT techniques should be developed in PV systems in order to maximize the output power (Chen et al., 2014; Rezk et al., 2015). Normally, numbers of PV modules are connected in series and parallel to form a PV array and the total output power is the combination of the power generated from each PV module. PSC is a common phenomenon occurred due to the shadows of buildings and clouds. If there is one or more of the PV modules are partially shaded, the shaded PV modules will consume the power generated by the rest of the PV modules and dissipates heat and cause hot spots (Tey et al., 2014; Chen et al., 2014). Under this condition, the current available in series-connected PV modules is limited by the current of the shaded PV module. This can be avoided by using bypass diodes which are placed in parallel with each PV module. The bypass diodes are reversing biased and have no impact under normal conditions (Tey et al., 2014). On the other hand, they are forward biased and carry the current instead of the PV module under shadowing effect. Figure 11 shows the power against voltage and the current against voltage curves of PV system, respectively, under different cases of PSC. This condition is given by four PV modules connected in series with solar irradiances as shown in Table 3.

From Figure 11 it can be seen that, the *I–V* curve under PSC have multiple steps, while the P–V curve has multiple peaks. Also, the number of peaks is equal to the number of different irradiance levels incident on the PV modules.

2.2.1. PSO–Based Tracker

PSO is a population–based evolutionary technique that has been recently used for global MPPT under PSC (Sarvi et al., 2015). It was inspired by the behavior of birds, which comprise a swarm of particles, and each particle suggests a candidate solution by exchange the information obtained in their respective search process to find the best solution (Boutasseta, 2012). These particles move in the search-space according to simple mathematical relation, exploiting their position and velocity. The position of the particle is affected by the agent position, which suggests the best solution through the current particles (local best), Pbest, as well as the best solution suggested through the entire population (global best), Gbest.

Figure 11. Power against voltage and current against voltage curves of PV system under uniform irradiance and different partial shading patterns

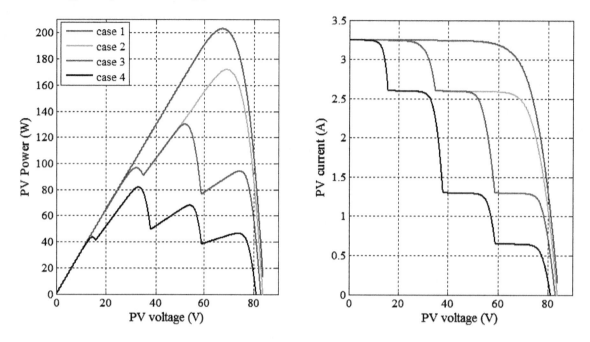

Table 3. Different partial shading patterns

Shadow Pattern	First PV Module G_1, w/m²	Second PV Module G_2, w/m²	Third PV Module G_3, w/m²	Fourth PV Module G_4, w/m²
Case 1	1000	1000	1000	1000
Case 2	1000	1000	800	800
Case 3	1000	1000	800	400
Case 4	1000	800	400	200

In this work, the particle swarm position is defined as the duty cycle of the DC/DC boost converter, and the fitness value evaluation function is taken as the PV system output power. The particle position di is updated by the following relation;

$$d_i^{k+1} = d_i^k + v_i^{k+1} \tag{6}$$

where the velocity component v_i represents the step size at iteration k+1 and is calculated by using the following expression:

$$v_i^{k+1} = wv_i^k + c_1 r_1 \left(P_{best} - d_i^k \right) + c_2 r_2 \left(G_{best} - d_i^k \right) \tag{7}$$

where *w* is the inertial weight, c1 and c2 are the acceleration coefficients, r1 and r2 are random values that belong to the interval of [0, 1], Pbest is the best position of particle i, and Gbest is the best position in the entire population (Mirhassaniet et al., 2015). A basic problem with the conventional PSO can be traced to its random nature. It can be seen that the last two terms in (2) is totally dependent on random numbers. The modified velocity equation by (Ishaque and Salam, 2013) can be written as the following;

$$v_i^{k+1} = wv_i^k + P_{best} + G_{best} - 2d_i^k \qquad (8)$$

Figure 12 shows the flowchart for searching mechanism by PSO–based tracker. Using a larger number of particles result in more accurate MPP tracking even under complicated shading patterns. However, a larger number of particles also lead to longer computation time. Therefore, a tradeoff should be made to ensure good tracking speed and accuracy. Firstly appropriate variables have to be selected for the search. The total number of particles is defined as n, in this work; the particles are defined as the duty cycles of boost converter. The particles can be placed on fixed position or be placed in the space randomly. The generated samples are applied to the PV system and the corresponding PV voltage and current are measured. These values used to calculate the PV power (fitness value) of particle i. If the current power of particle i is better than the best fitness value in history p_{best}, set current power as the new best fitness of particle i. Then, choose the particle with the best fitness value of all the particles as the G_{best}. After evaluation of all particles, the velocity and position of each particle in the swarm should be updated using (1) and (3). Finally, if the maximum number of iterations is reached, the algorithm will stop and give the obtained G_{best} solution, which represents the optimum value of the duty cycle corresponding to global MPP (Liu et al., 2012).

2.2.2. Teaching-Learning Based Optimization (TLBO)

TLBO is proposed by Reo and Patel in 2010 (Rao and Patel, 2013) It is a population based technique that uses a population of candidate solutions to proceed to the optimal solution (Rao et al., 2015). TLBO simulates the traditional teaching-learning process of a classical school. A group of n students is considered the population and different subjects offered to the students are analogous with the different design variable of the optimization problem. The results of students are analogous to the fitness values of the optimization problem. The best solution in the entire population is considered as the teacher (Rao and Patel, 2013). The teacher tries to improve the individuals by changing the during the teacher and student phases, where an individual is only replaced if his new solution is better than his previous one. The process of TELBO is divided into two stages. During the first stage, called teacher phase, a teacher imports knowledge directly to his students. During the second stage, called student stage, a student may learn with help of fellow students. Overall, how much knowledge is transferred to a student does not only depend on his teacher but also on interaction amongst students through peer teaching (Cerpinsek et al., 2012). The process of TLBO can be summarized as follow:

In the teacher phase, a teacher conveys knowledge among the students and makes an effort to increase the mean result if the class (Rao and Patel, 2013). A good teacher is one who brings his students up to his level in terms of knowledge. But in practice this is not possible and a teacher can only move the mean of the class up to some extent depending on the capability of the students (Rao et al., 2015).

Figure 12. Flowchart for searching mechanism by PSO–based tracker

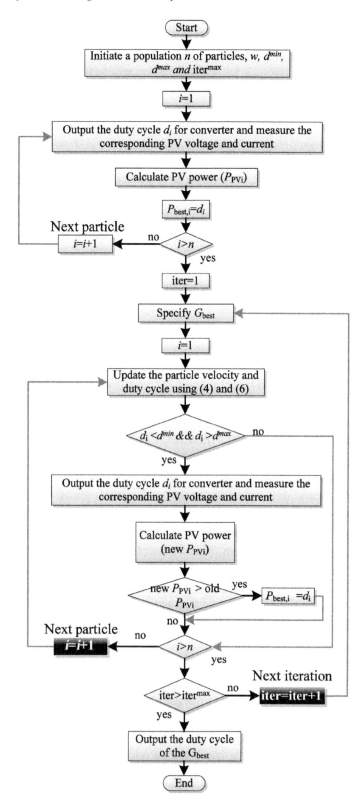

Suppose there is one subject (design variable) offered to n number of students. In this work the student represented by the duty cycle of boost converter di and the fitness value represented by the PV output power. At any iteration i, $d_{i,mean}$, is the mean value of the students. Since a teacher is the most experienced and knowledgeable person on a subject, the best student who is identified as a teacher for that iteration i. the teacher will put maximum effort into increasing the knowledge level of the whole class, but students will gain knowledge according to quality of teaching delivered by a teacher. The existing solution is updated in the teacher phase according to the difference between the value of the teacher di,best and the mean value of the students $d_{i,mean}$ as follows:

$$d_{i+1} = d_i + r\left(d_{i,best} - T_F d_{i,mean}\right), \tag{9}$$

where d_{i+1} is the updated value of d_i. Accept d_{i+1} if it gives a better function value. All the accepted function values at the end of the teacher phase are maintained, and these values become the input to the student phase (Rao and Patel, 2013). T_f is a teaching factor that decides the value of mean to be changed, and r is a random number in the range [0 1]. The value of T_f can be either 1 or 2.

In the student phase, TLBO simulates the learning of the students through interaction among themselves by discussing and interacting. A student will learn new information if the other students have more knowledge than him (Rao and Patel, 2013). During this phase, one student d_x tries to improve his knowledge by peer learning from arbitrary student d_y. In the case that d_y is better than d_x, d_x move toward d_y according to (10). Otherwise, it is moved away from d_y according (11).

$$d_{new} = d_x + r\left(d_y - d_x\right) \text{ if } P_{pv}(d_y) > P_{pv}(d_x) \tag{10}$$

$$d_{new} = d_x + r\left(d_x - d_y\right) \text{ if } P_{pv}(d_x) > P_{pv}(d_y) \tag{11}$$

If student d_{new} performs better by following Eq (10) and (11), he will be accepted into the next iteration. The algorithm will continue its iterations until reaching maximum number of iterations (Cerpinsek et al., 2012).

The searching mechanism by TLBO–based trackers as fellow; firstly appropriate variables have to be selected for the search. The total number of learners is defined as n, in the current work; the learners are defined as the duty cycles of boost converter. The initial duty cycles (learners) are applied to the PV system and the corresponding PV voltage and current are measured. These values used to calculate the PV power (fitness value) of each learner i. Then, choose the learner with the best fitness value of all the learners as the teacher. After evaluation of all learners, the duty cycle is updated in the teacher phase according to (9). Afterwards, select randomly any two learners and update the duty cycle according (10) and (11). These steps are executed for a certain number of iterations until the all nests reach at the best solution which represents the optimum duty cycle corresponding to global MPP.

2.2.3. Differential Evolution (DE) Based Tracker

DE, which is proposed by Storn and Price, is one of the best population based techniques used for global optimization because of its simplicity and efficiency. Basically, a population of n particle is required in

DE and little iteration needed to reach the optimal solution. The algorithm starts with an initial random population and implements mutation, crossover, and selection operations to update the population during the evolution toward the optimum solution (Tey et al., 2012; Salam and Saad, 2013). It implements mutation and crossover processes to yield a trial vector dui for each target vector. Thereafter, a selection operation is executed between the trial vector and the target vector. There are several mutation strategies, each of which is suited for obtaining particular results. The one used in this work is called DE/best/bin. The mutation, crossover and selection operations of this DE algorithm are explained as follow.

The duty cycle of boost converter is used as a target vector and the PV power as the fitness function. The target vectors are distributed on fixed positions with equal distance ranging over their corresponding feasible limits $[d_{min}, d_{max}]$. The initial population of duty cycles is send to the boost converter. Next, the fitness function values of population are evaluated by measuring PV system current and voltage correspond to each target vector. The measured current and voltage are then used to calculate the PV system power pi which is used as fitness function. After the initialization step finished, the highest PV power is selected as P_{best} and corresponding duty cycle di is saved as the best one D_{best}. Next, two different population members are randomly selected. The, a mutation factor F is used to weight the difference between the selected target vector and the weighted difference is added into the best duty cycle to generate mutated particle or so called donor vector d_{vi}. The mutation process is given by (Tey et al., 2012; Salam and Saad, 2013);

$$dv_i = D_{best} + F \times (d_{r_1} - d_{r_2}) \tag{12}$$

where indexes r_1, and r_2 represent mutually different integers that are randomly generated over [1, n], and F is the scaling factor.

In this work, if a one component of a mutant vector violates the boundary constraint, this component is reset as follows:

$$du_i = \begin{cases} du_i = d_{max} & \rightarrow \quad \text{if } du_i \text{ greater than } d_{max} \\ du_i = d_{min} & \rightarrow \quad \text{if } du_i \text{ smaller than } d_{min} \end{cases} \tag{13}$$

After the mutation process, the trail vector dui is generated by making use of a binomial crossover operation on the target vector di and the mutant vector d_{vi}. The condition used in crossover process is described as the following;

$$du_i = \begin{cases} dv_i & \text{if } rand \leq C_r \\ di \end{cases} \tag{14}$$

Cr is the crossover control parameter in rang of [0 1]

After crossover, the selection stage is reach. It is conducted by comparing the target vector di against the trial vector d_{ui}. Each trial vector dui is assessed through PV system and P_{ui} have to be measured. However, the value of dui may remain as d_i after crossover and P_{ui} is also equal to P_i. Therefore, only the P_{ui} which does not have the same duty cycle as P_i measured again from the input of boost converter.

Modern Optimization Algorithms and Applications in Solar Photovoltaic Engineering

This process helps to reduce the searching time. When all values of the P_{ui} are obtained, it is compared to P_i. From comparison, the duty cycle which corresponds to higher value of power is used as the next target vector according to the following;

$$d_{i+1} = \begin{cases} du_i & \text{if } f(du_i) \geq f(d_i) \\ di & \text{otherwise.} \end{cases} \qquad (15)$$

The iterations are carried out until a termination condition is met. The searching mechanism of DE-based tracker is shown in Figure 13.

2.2.4. Cuckoo Search Based MPPT Technique

Cuckoo Search represents a new optimization meta-heuristic technique recently developed by Yang and Deb (Yang and S. Deb, 2009). The key mechanism in cuckoo breeding is brood parasitism. This is the act of laying eggs in the nests owned by other birds, which may or may not be of the same species. There are three types of brood parasitism, which are termed intraspecific, cooperative and nest takeover (Payne et al., 2005). If the host bird discovers the cuckoo egg in her nest, it may destroy the egg or abandon the nest and build a new one elsewhere. In order to increase the probability of having a new cuckoo and reduce of abandoning eggs by host birds, cuckoos use several strategies such as imitating the colors and patterns and devolving the capacity of cuckoo to mimic the call of host bird to gain access to more feeding opportunity. The model described in (Yang and S. Deb, 2009) has been based on three proposed rules which are;

1. Each cuckoo lays one egg at a time and deposits it into a random nest,
2. The best nests contain the highest quality eggs pass to the next generation,
3. The number of host nests is fixed and there is always a probability that the cuckoo egg is discovered by the host. If an egg is discovered, the host bird throws it out of the nest.

One of the most powerful features of CS is the use of Lévy flights to generate new eggs. A Lévy flight is a random walk, characterized by a series of instantaneous laeps generated by a probability density function. Mathematically, a Lévy flight is a random walk where step sizes are extracted from Lévy distribution according to a power law as follows:

$$Lévy(\beta) = l^{-\beta} \qquad (16)$$

where l is the length of flight and $1 < \beta < 3$. New eggs are provided by performing a Lévy flight from a randomly selected egg. The new egg is then represented by the coordinates reached at the end of the flight. A coefficient α is used to control the size of the Lévy flight. When a Lévy step is generated, it is first multiplied by α before it is used to generate a new egg. There are important controlling parameter to be adjusted; the fraction of eggs to be abandoned in each generation. In this work, the fraction of eggs to be discarded $p_a = 0.25$ were used, as suggested by (Yang and S. Deb, 2009). The proposed strategy

Figure 13. Flowchart for searching mechanism by DE-based tracker

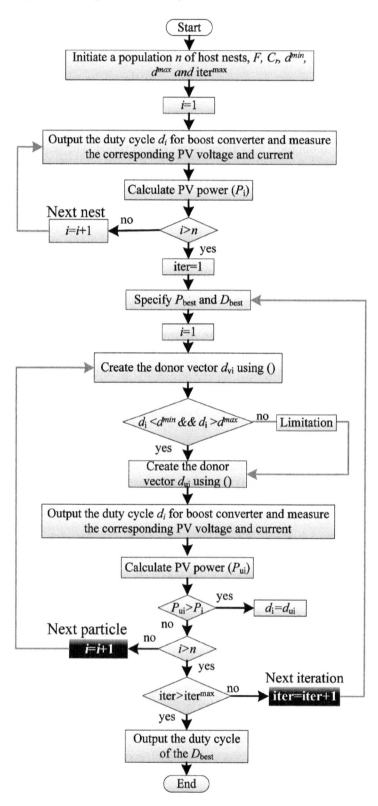

in this work is that finding the optimum duty cycle of boost converter corresponding to global MPP under different partial shading patterns. Therefore the duty cycle is used as a variable to be searched, independent variable. The PV output power is used as the fitness function, objective function. According the Lévy flight, the duty cycle is updated based on the following relation (Yang and S. Deb, 2009):

$$d_i^{k+1} = d_i^k + \alpha \oplus Lévy(\beta) \approx d_i^k + k_{Lévy} \times \left(\frac{u}{|v|^{1/\beta}}\right)(d_{best}^k - d_i^k) \tag{17}$$

where β=1.5, kLévy is the Lévy multiplying coefficient (selected as 0.1), u and v are determined from the normal distribution curves.

$$u \approx N(0, \sigma_u^2) \quad and \quad v \approx N(0, \sigma_v^2) \tag{18}$$

where the variable σu and σv are defined as follows:

$$\sigma_u = \left(\frac{\Gamma(1+\beta) \times \sin(\pi \times \beta/2)}{\Gamma\left(\frac{1+\beta}{2}\right) \times \beta \times 2^{\frac{(\beta-1)}{2}}}\right)^{1/\beta} \quad and \quad \sigma_v = 1 \tag{19}$$

Based on the above mentioned rules, the searching mechanism for CS–based tracker can be summarized in Figure 13. Considering fixed numbers of the host nests (n) initially population is generated as a duty cycle of boost converter $d_i=d_1, d_2,...,d_n$. The duty cycles are applied in the PV system and corresponding PV voltage and current are measured. The measured PV voltage and current are then used to calculate the PV power which used as the fitness function. The duty cycle corresponding to the maximum value among the fitnesses is considered as the current best nest (d_{best}). Then Lévy flight is performed and new nests are generated as in (9). New fitnesses values are tested through the PV system. Afterwards, the worst nest is randomly destroyed with a probability Pa, this process emulates the behavior of the host bird discovering the cuckoo's eggs and destroying them. The new nest is replaced the destroyed one via Lévy flight, then the PV power is measured and the current best the current best nest is selected. These steps are executed for a certain number of iterations until the all nests reach at the best solution which represents the optimum duty cycle corresponding to global MPP.

2.3. Summary

Judging by the amount of recent work, it can be concluded that the MPPT is still a very active research area. There are rooms for improvement, particularly when dealing with partial shading conditions. For uniform insolation, there is an increasing trend in combining the conventional methods with soft computing techniques.

Figure 14. Flowchart for searching mechanism by CS-based tracker

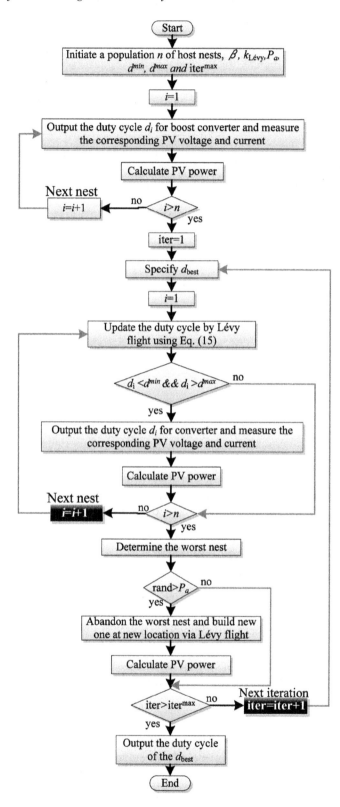

As for the partial shading, it appears that the EA methods are the most promising. The PSO, which has a very similar structure to HC has proven to be a viable option. However, there are still potential areas of concerns for EA, particularly the selection of appropriate evolutionary control parameters such as population size, mutation factor etc. This may be resolved with the assistance of other optimization technique such as ANN. There is also a need to consider the viability of EA methods when it comes to real time implementation. The way forward is to find an effective method such that the search space can be narrowed, so that the computational time is reduced. Furthermore, there are several more EA techniques, which are yet to be exploited fully, for example the differential evolution (DE) and TLBO.

3. SOLAR ENERGY AND RENEWABLE ENERGY

For reliable energy supply using only PV systems is unpractical because of needs too much accumulating equipment. More practical solution especially for decentralized electricity energy supply is using combination of renewable energy different types. Renewable energy (REN) is defined as energy that is produced by stable or intermittent natural energy currents - such as sunlight, wind, rain, waves, tides, and geothermal heat—that are naturally replenished within a time span of a few years (Lund H., 2010). Henrik Lund even asks: how can society convert to 100 percent renewable energy?

Hundred percent of renewable power is foundation for next step of human evolution. The time has come to abolish the combustion of coal, oil and gas for energy generation worldwide, along with the nuclear power threat (Peter Droege, 2009).

The main unpleasant features of renewable energy are low density, space distribution and time intermittence.

Decentralized electricity generation and fluctuating availability pose a challenge on grid stability. This section provides understanding of interaction between PV and other REN systems. Optimization of combined renewable energy sources (different types of converters) for standalone electrical power supply is an important practical task.

The use of renewable energy sources (RES) for autonomous power supply is characterized with a number of features that must be considered when specific systems are creating and designing. The low energy density of renewable energy flows, their variability over time, on the one hand, and the non-uniformity of the load curve the consumers on the other, requires the use of additional equipment - energy storage devices and "smart" devices with the appropriate software and hardware for control income and outcome energy. Energy storage compensates unevenness between energy generation and energy consumption by storing it in periods of surplus energy and energy from in deficit moments. However, accumulating system increases the cost of energy supplying, despite the lack of fuel in the total expenses for such systems. So, the effective way to solve this problem is the combined use of renewable energy sources.

The combined use of the different renewable energy sources allows compensating the intermittent character of RES, to reduce the capacity of the accumulating system units, and, consequently, reduce its cost. However, due to territorial inequity of energy currents from RES, different equipment performances, and variable demands of energy consumers, the optimal configuration of the autonomous power supply combined system (in terms of its economy) is not the same for different places (geographical positions) of operation. The developed methodology and software package allow determining the optimal structure and parameters of the combined power system taking into account all the above factors. The system of computer mathematical package MATLAB with the block of visual simulation Simulink has been developed and used for finding the optimal configuration of the system combination for specific geographic locations.

3.1. Outline of a Case Study

Simultaneous using of multiple energy sources allows to compensate uneven coming of renewable energy, which, in turn, can significantly reduce the capacity of the generating units, as well as storage capacity in such systems and, consequently, reduce their cost.

Planning and designing of such systems should also take into account the uneven territorial distribution of renewable energy resources. Thus, the problem of determining the optimal structure and parameters of the autonomous power supply combined system (APSCS) from renewable energy sources (APSCS-RES) with the real geographical and climatic conditions, characteristics of the equipment using and the characteristics of the energy consumer, including the expected variable graphics power consumption is arising. In this case, the optimality criterion must be the minimum cost of the system generating electricity.

In the example of a particular case, we consider the solution of the optimization problem for a system that can use only three renewable energy currents: solar, wind and small watercourse. Choice of data sources due to the possibility of increasing the reliability of uninterrupted power supply consumer as micro-hydro which is a relatively stable source of energy, and the installation of photovoltaic (PV) with a wind power unit (WPU) which help each other due to the synergistic effect. Moreover the existence of such sources as the sun and the wind is wide available and small streams flowing in abundance across the expanses of many countries. This result allows applying this methodology to solve optimization problems with a wide range of sources, as well as the relatively stable sources can be considered as geothermal, biomass and others.

3.2. Software Elaborating for Case Study

The solution of described problems is obtained by means of modern specialized software - computer systems Mathematics (CSM). The package MATLAB with visual simulation Simulink was chosen for this system, which is implemented software system that determines the optimal structure and parameters APSCS on the basis of three renewable energy: solar, wind, micro-hydro, taking into account the actual climatic and geographical conditions, characteristics of the equipment used and the characteristics of the consumer, including the expected variable graphics of electric power consumption, according to the criterion of the minimum system cost.

The developed complex consists of two interrelated parts:

- Sub-programs created in the m-file MATLAB;
- Mathematical models APSCS-RES created in Simulink.

The first part determines the optimal composition and preliminary parameters APSCS-RES, which are calculated on the basis of average values from the generalized expressions, and some dependencies are not taken into account, because the variability in the time is very difficult to express by the simple means. Accordingly, in the second stage there is a need to simulate the operation APSCS-RES throughout the calculating period: to analyze its operation, verify the optimal parameters, make corrections and adjustments. Moreover, the simulation can accurately determine the coefficients of power plants usage that characterize the climate conditions in selected areas and time characteristics of the equipment in order to solve the optimization problem.

The methodology developed in the first part of the software package implements the principle that determines the optimal structure and parameters APSCS-RES. The unknown parameters are the power of installed plants based on renewable energy sources N_{res}^{inst} and storage capacity C_{ab}^{inst} according to their type and their size range. From the available n unknown variables (corresponding to the standard size range) a certain amount of ξ (ξ <n) must enter in the optimal solution. In this case, each target variable (discrete value) corresponds to a binary variable δi. Thus, if as a result of solving the problem δi = 1, the variable and, therefore, appropriate that variable setting of the standard series, is part of the optimal solution, if δi = 0, then no.

Taking into account that the optimality criterion, i.e. extreme value of the objective function is a minimum cost, the objective function takes the following form:

$$\begin{aligned}
&c_{pv} \cdot N_{pv1}^{inst} \cdot \delta_1 + c_{pv} \cdot N_{pv2}^{inst} \cdot \delta_2 + \ldots + c_{pv} \cdot N_{pv\,l}^{inst} \cdot \delta_l \\
&+ c_{wind} \cdot N_{wind1}^{inst} \cdot \delta_{l+1} + c_{wind2} \cdot N_{wind2}^{inst} \cdot \delta_{l+2} + \ldots + c_{wind} \cdot N_{wind\,g}^{inst} \cdot \delta_{l+g} \\
&+ c_{hydro} \cdot N_{hydro1}^{inst} \cdot \delta_{l+g+1} + c_{hydro} \cdot N_{hydro\,2}^{inst} \cdot \delta_{l+g+2} + \ldots + c_{hydro} \cdot N_{hydro\,p}^{inst} \cdot \delta_{l+g+p} \\
&+ c_{ab} \cdot C_{ab1}^{inst} \cdot \delta_{l+g+p+1} + c_{ab} \cdot C_{ab\,2}^{inst} \cdot \delta_{l+g+p+2} + \ldots \\
&+ c_{ab} \cdot C_{abf}^{inst} \cdot \delta_{l+g+p+f=n} \rightarrow \min
\end{aligned} \qquad (20)$$

where,

c_{wind}, c_{hydro} : the specific cost for installed capacity in Watts for PV, wind turbines and micro hydro respectively, rub/W;

c_{ab} : unit cost of installed capacity accumulating battery at a voltage U_{ab}^{nom} for the entire life of the APSCS-RES, rub. / A • h;

$N_{pv1}^{inst}, N_{pv2}^{inst}, \ldots N_{wind\,l}^{inst}, N_{wind1}^{inst}, N_{wind2}^{inst}, \ldots N_{wind\,g}^{inst}, N_{hydro1}^{inst}, N_{hydro\,2}^{inst}, \ldots N_{hydrop}^{inst}$: installed capacity of PV, wind turbines and micro-hydro, respectively, in W;

$C_{ab\,1}^{inst}, C_{ab\,2}^{inst}, \ldots C_{ab\,f}^{inst}$: accumulator battery (AB) installed capacity at a nominal voltage U_{ab}^{nom}, A•h;

$\delta_1, \delta_2 \ldots \delta_l, \delta_{l+1}, \delta_{l+2}, \delta_{l+g}, \delta_{l+g+p+1}, \delta_{l+g+p+2}, \delta_{l+g+p+f=n}$: binary variables corresponding number of standard sizes PV, wind turbines, micro-hydro and AB, respectively.

For determining the unit cost of AB during the life time of APSCS-RES the number of substitutions is taking into account produced during this time, due to the limited number of charge-discharge cycles

$$k_{ab}^{sub} = \frac{T_{APSCS-RES}^{expl}}{(Q_{ab}^{u.r.p.} / Q_{A.B}^{u.r.})}, \qquad (21)$$

where

$T_{APSCS-RES}^{expl}$: the life APSCS-RES years (we took in calculations 20 years);

$Q_{ab}^{cdd.}$: the number of charge-discharge cycles at a given depth of discharge;

$Q_{ab}^{cd.}$: the number of charge-discharge cycles, the average for the year (in the calculations is equal to 365, in average one full cycle per day).

In this case, the unit cost of AB, determined during its life APSCS-RES is found by the following expression:

$$c_{ab} = c'_{ab} \cdot k_{ab}^{sub}, \qquad (22)$$

where c'_{ab}: the unit cost of battery at a voltage U_{ab}^{nom}, rub. / A • h.

The system of restrictions is a condition to ensure uninterrupted power supply to consumers. This is the equation of energy balance, made up for the average daily hours for each calendar month of the accounting period (years):

$$\begin{aligned}
&k_{pv}^{fac}(\Delta t_i^j) \cdot N_{pv\,1}^{inst} \cdot k_n \cdot \delta_1 \cdot \Delta t_i^j + k_{pv}^{inst}(\Delta t_i^j) \cdot N_{pv\,2}^{inst} \cdot k_n \cdot \delta_2 \cdot \Delta t_i^j + \ldots \\
&+ k_{pv}^{inst}(\Delta t_i^j) \cdot N_{pv\,l}^{inst} \cdot k_n \cdot \delta_l \cdot \Delta t_i^j + k_{wind}^{inst}(\Delta t_i^j) \cdot N_{wind\,1}^{inst} \cdot k_n \cdot \delta_{l+1} \cdot \Delta t_i^j + \\
&+ k_{wind}^{inst}(\Delta t_i^j) \cdot N_{wind\,2}^{inst} \cdot k_n \cdot \delta_{l+2} \cdot \Delta t_i^j + \ldots + k_{wind}^{inst}(\Delta t_i^j) \cdot N_{wind\,g}^{inst} \cdot k_n \cdot \delta_{l+g} \cdot \Delta t_i^j + \\
&+ k_{hydro}^{inst}(\Delta t_i^j) \cdot N_{hydro\,1}^{inst} \cdot k_n \cdot \delta_{l+g+1} \cdot \Delta t_i^j + k_{hydro}^{inst}(\Delta t_i^j) \cdot N_{hydro\,2}^{inst} \cdot k_n \cdot \delta_{l+g+2} \cdot \Delta t_i^j + \ldots, \\
&+ k_{hydro}^{inst}(\Delta t_i^j) \cdot N_{hydrop}^{inst} \cdot k_n \cdot \delta_{l+g+p} \cdot \Delta t_i^j + k_{ab}^{inst}(\Delta t_i^j) \cdot C_{ab\,1}^{inst} \cdot U_{ab}^{nom} \cdot k_n \cdot \delta_{l+g+p+1} + \\
&+ k_{ab}^{inst}(\Delta t_i^j) \cdot C_{ab\,2}^{inst} \cdot U_{ab}^{nom} \cdot k_n \cdot \delta_{l+g+p+2} + \ldots \\
&+ k_{ab}^{inst}(\Delta t_i^j) \cdot C_{ab\,f}^{inst} \cdot U_{ab}^{nom} \cdot k_n \cdot \delta_{l+g+p+f=n} \geq W_H(\Delta t_i^j)
\end{aligned} \qquad (23)$$

where

$k_{pv}^{fac}(\Delta t_i^j)$, $k_{wind}^{fac}(\Delta t_i^j)$, $k_{hydro}^{fac}(\Delta t_i^j)$: the capacity factors PV, wind turbines and micro-hydro, respectively, for an interval of time;

$k_{ab}^{fac}(\Delta t_i^j)$: coefficient of AB during the time interval Δt_i^j;

$k_l = 0,93$: the coefficient of losses of energy conversion system;

Δt_i^j: a time interval corresponding to the i-th hour diurnal (i = 1,2 ... 24) j-th month (j = 1,2 ... 12) Δt_i^j h;

$W_{gp}(\Delta t_i^j)$: the amount of electrical energy required for guaranteed power consumers during the interval Δt_i^j W • h:

$$W_{gp}(\Delta t_i^j) = N_{cons}^{av}(\Delta t_i^j)/k_n \cdot \Delta t_i^j + N_{on}^{av}(\Delta t_i^j) \cdot \Delta t_i^j + N_1^{av}(\Delta t_i^j) \cdot \Delta t_i^j \qquad (24)$$

where

$N_{\text{cons}}^{\text{av}}(\Delta t_i^j)$: the average amount of power consumers during the interval Δt_i^j, W (defined by schedule of consumer load);

$N_{\text{on}}^{\text{av}}(\Delta t_i^j)$: average value of the power of their own needs in a time interval Δt_i^j, W;

$N_{\text{l}}^{\text{av}}(\Delta t_i^j)$: the average power loss in the system during the interval Δt_i^j, W.

It should be noted that the utilization rate of energy discharge coefficient $k_{ab}^{fac}(\Delta t_i^j)$ corresponds to the $k_{ab}^{fac}(\Delta t_i^j)$ AB and it cannot be larger than the coefficient of the depth of discharge $k_{ab}^{dd.}$ (to take into account equal to 0.6), as in this case, during the hours of AB discharged more than adopted by the depth of discharge.

Total number of constraints from i = 24 and j = 12 is 12 • 24 = 288.

Utilization rates for each time interval determined on the basis of climatic and geographical conditions and the characteristics of the selected areas of the; $N_{\text{wind 1}}^{\text{inst}}, N_{\text{w inf 2}}^{\text{inst}}, \ldots N_{\text{wind } g}^{\text{inst}}$; $N_{\text{hydro 1}}^{\text{inst}}, N_{\text{hydro 2}}^{\text{inst}}, \ldots N_{\text{hydro } p}^{\text{inst}}$ equipment used by a mathematical model of the system.

$C_{\text{ab 1}}^{\text{inst}}, C_{\text{ab 2}}^{\text{inst}}, \ldots C_{\text{ab } f}^{\text{CAB}}$: limitations of the capability of the equipment used, the following:

$$\delta_1 + \delta_2 + \ldots + \delta_l \leq 1,$$

$$\delta_{l+1} + \delta_{l+2} + \ldots + \delta_{l+g} \leq 1,$$

$$\delta_{l+g+1} + \delta_{l+g+2} + \ldots + \delta_{l+g+p} \leq 1 \qquad (25)$$

$$\delta_{l+g+p+1} + \delta_{l+g+p+2} + \ldots + \delta_{l+g+p+f=n} \leq 1.$$

The first line corresponds to the standard size range of the PV, the second - of wind turbines, the third - the micro-hydro, and the fourth - the battery. This is possible quantity of facilities which can be selected from the corresponding range of power alone.

The boundary conditions are not recorded because the possible values of discrete variables are given, and the values of binary variables can only be 0 or 1.

With such a formalization of the optimization problem one cannot guarantee the accuracy of the results, because they do not take into account certain factors.

First, the ratio of depth of discharge (energy) AB depending on the desired energy outputted and determines the discharge capacity of the battery.

Second, inaccurate representation of AB in (23) as an independent energy source to consider introducing additional constraints arising from the fact that AB is a source of energy, the development of which is directly dependent on the production of generating units based on renewable energy. Thus, the average annual production power plants based on renewable energy must exceed the required annual average load factor taking into account the losses of electric energy for compensation. However, it is difficult to estimate accurately the amount of energy to be compensated. Thus, due to the above factors, the solution of the optimization problem by classical methods of linear programming results are not very accurate results, so the composition and parameters APSCS-RES determined by successive iterations.

The proposed method of calculation is to step through the energy balance equations created for the average daily hours for each calendar month of the accounting period (23) with the accumulation of ($-C_{ab}^{T} \cdot U_{ab}^{nom}$) and generation ($+C_{ab}^{T} \cdot U_{ab}^{nom}$) energy by AB:

$$k_{pv}^{fac}(\Delta t_i^j) \cdot N_{pv}^{inst} \cdot k_l \cdot \Delta t_i^j + k_{wind}^{fac}(\Delta t_i^j) \cdot N_{wind}^{inst} \cdot k_l \cdot \Delta t_i^j + \\ + k_{\Gamma}^{ucn}(\Delta t_i^j) \cdot N_{\Gamma}^{ycT} \cdot k_l \cdot \Delta t_i^j \pm C_{ab}^{B}(\Delta t_i^j) \cdot U_{ab}^{nom} \cdot k_l \geq W_H(\Delta t_i^j) \tag{26}$$

where

$$+C_{ab}^{B}(\Delta t_i^j) \cdot U_{ab}^{nom} = k_{ab}^{disch}(\Delta t_i^j) \cdot C_{ab}^{ac}(\Delta t_{i-1}^j) \cdot k_{ab}^{changC}(\Delta t_i^j) \cdot U_{ab}^{nom},$$

$$-C_{ab}^{B}(\Delta t_i^j) \cdot U_{ab}^{nom} = k_{ab}^{ch} \cdot C_{ab}^{inst} \cdot U_{ab}^{nom} \cdot \eta_{ab} \tag{27}$$

where $C_{ab}^{ac}(\Delta t_{i-1}^j)$: energy accumulated in the previous time by AB, defined in the previous step as:

$$C_{ab}^{ac}(\Delta t_{i-1}^j) = \pm C_{ab}^{B}(\Delta t_{i-1}^j) + C_{ab}^{ac}(\Delta t_{i-2}^j) \tag{28}$$

$k_{ab}^{changC}(\Delta t_i^j)$: coefficient characterizing changes in battery capacity based on the current discharge current is determined by the coefficient of discharge $k_{ab}^{disch}(\Delta t_i^j)$, which in turn is determined based on the required power compensation;

$k_{ab}^{ch}(\Delta t_i^j)$: coefficient characterizing the nominal charging current AB;

η_{ab}: efficiency of the batteries.

Since usually only efficiency of charge-discharge cycle is available it is counted only once during charging.

When checking each step takes into account the total amount of energy that can accumulate over the AB Δt_i^j:

$$\left| \sum_{i,j=1}^{288} \pm C_{ab}^{B} \cdot U_{ab}^{nom} \right| \leq C_{ab}^{inst} \cdot U_{ab}^{nom} \cdot k_{ab}^{dd} \tag{29}$$

where k_{ab}^{dd}: the coefficient of the depth of discharge of batteries.

Verification is required for all possible combinations of APSCS-RES with specified standard series equipment is used in it.

The second part of the software system, which determines the usage settings and adjusts the composition and parameters of APSCS-RES, a mathematical model and the modeling process is the means of APSCS-RES under different operating conditions. The model is based on the principle of hierarchy; it

Modern Optimization Algorithms and Applications in Solar Photovoltaic Engineering

is a basic block, in groups (subsystems) are functionally related units within the subsystems of the first level, second level are subsystems, etc.

Using the background information that characterizes the real climatic and geographical conditions of operation, the corresponding blocks in the model, the data are generated determining the flow of renewable energy (solar, wind and Small River). Credibility is defined as the quantity of data characterizing renewable resources, and data characterizing the unequal distribution of them in time, given how predictable patterns (day, night), and random factors (cloudy, windless days, hours). The other model blocks directly characterize the functioning process APSCS-renewable energy sources and define the characteristics of the equipment used, and especially the consumer, based on graphs of consumption of electrical energy.

The simulated time of working RES APSCS, due to the frequency of recurrence of renewable energy resources, is equal to one year and is divided into n = 8760 time slots (Δt_i) with a duration of one hour. Detailed description of the mathematical model is presented in paper by Simakin V.V. (Simakin V.V., 2011).

3.3. Calculation Procedure of Concrete Example and Results

The following describes the basic procedures and procedures developed complex.

1. The data relating to the renewable energy equipment and consumer are entered. The energy flows inherent to renewable energy sources are generated and the work APSCS-RES is simulated under these conditions with the help of the mathematical model of the system. Power generating devices are set equal to 1 kW, capacity AB - arbitrary. The simulation determines the values of the capacity factor of power plants APSCS-RES ($k_{pv}^{fac}(\Delta t_i^j)$, $k_{wind}^{fac}(\Delta t_i^j)$, $k_{hydro}^{fac}(\Delta t_i^j)$) and the amount of electrical energy needed to power consumer ($W_H(\Delta t_i^j)$) at different time intervals.
2. The variants of configurations of APSCS-RES are defined to meet the certain conditions of operation, distributed in order of increasing cost, and prefound the optimum composition and parameters of the system.
3. It checks received at the second stage of options APSCS-RES until a configuration satisfying the demand of 100% coverage of the load.

Once the final result is displayed that allows you to get a lot of additional information that characterizes the work of APSCS-RES, as well as a look at the overall picture of the system, evaluate its operation to a new qualitative level.

The effectiveness of the use of combined power from those in the example of renewable energy sources has been evaluated for several case study in the CIS with the help of the developed software system.

Fife geographic locations were taken for analysis: Sochi, Makhachkala, Bishkek (formerly Frunze, Kyrgyzstan), p. Ust-Kokshov (Altai), p South Kuril (Far East). Four rural houses modern buildings were selected as a consumer to which electricity from APSCS-RES is provided by Practical guidance (Practical guidance, 2008). The area of each house equals 300-360 m², daily consumption - 8,15 kW • h. Respectively a graph of the total daily load equal to, 32.6 kW • h is shown at Figure 15. According to the guidelines for the design of power supply agriculture (Guidance, 1981) the load factor was used: winter - 1, spring - 0.8; summer - 0.7; autumn - 0.9 in the seasonal distribution. Type of equipment used in APSCS-RES PV consisting of fixed PV modules optimally oriented to the south, vertical-axis wind turbine type, micro-hydro sleeve-type. Two types of AB considered: with lead-acid and lithium-iron-phosphate.

Specific costs of power units generated by renewable-energy installations APSCS were assumed: cpv = 110 rub/W; cwind = 100 rub/W; cwind = 75 rub/W, cAB la = 206 rub/A • h, cAB LiFePh = 280 rub/A • h (UAB = 12 V), which corresponds to the real market values presently. Parameters of power electronics: $k_p = 0,93$ The parameters for lead-acid AB: $k_{ab}^{ch} = 0,2 \times C$, $\eta_{ab} = 0,8$, $k_{ab}^{dd} = 0,6$ for the lithium-iron-phosphate $k_{ab}^{ch} = 0,5 \times C$, $\eta_{ab} = 0,94$, $k_{ab}^{dd} = 0,77$.

Optimization parameters APSCS-RES defined for the five selected geographical locations are shown in Figure 16. Contingent resources of water flow were selected based on the fact that the available head and flow of water can get through micro-hydro power of 1 kW. Because the energy generated by the micro-hydro is the cheapest, its capacity is determined in all cases, the maximum available. However, the energy generated by the micro-hydro is not enough to cover the load of the consumer and requires additional sources, which in the case of the city of Makhachkala and part of the South Kuril were wind turbines, in the case of Bishkek - PV, while in Sochi with. Ust-Kokshov are optimal, given the micro-hydro, and wind turbines, and PV.

The optimal configuration study was carried out while the limiting sources were used on the example of the city Sochi (Figure 17). It is evidently that the combined use of renewable energy allows to reduce the installed power plants. If to look at the total cost of power plants (Figure 17 column 1, 6, 7, 8), the most expensive it will be for electric power through wind turbines and significantly less if use the PV. Combination PV allows reducing costs by a factor of 2.2 compared with only wind turbine and 1.6 as compared with PV. Their combined use with a micro-hydro can have a 2-fold more reduction in the costs of power plants for electric power consumers in the district of Sochi.

Figure 15. Daily energy consumption diagram (consumer load)

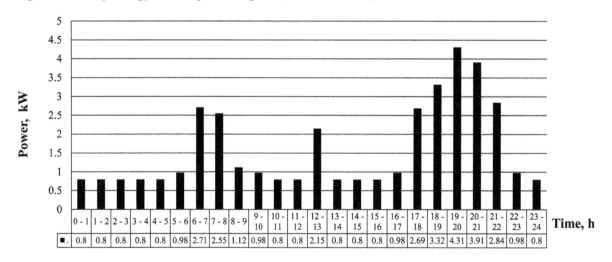

Figure 16. Results of optimization parameters APSCS-RES defined for the five selected geographical locations

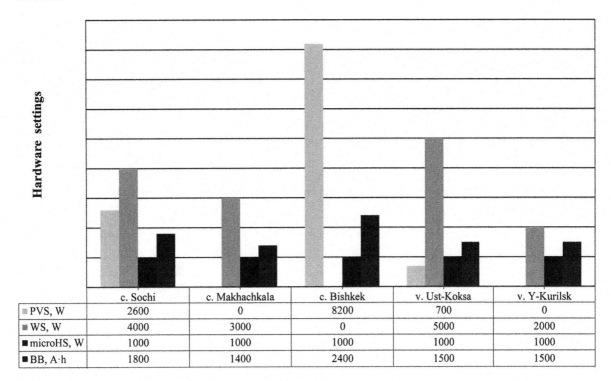

	c. Sochi	c. Makhachkala	c. Bishkek	v. Ust-Koksa	v. Y-Kurilsk
■ PVS, W	2600	0	8200	700	0
■ WS, W	4000	3000	0	5000	2000
■ microHS, W	1000	1000	1000	1000	1000
■ BB, A·h	1800	1400	2400	1500	1500

If to analyze the total cost of power plants in the five selected points (Figure 18 column 1 - 5) the most cost-effective electricity will be to consumers in Makhachkala and in the South Kuril village. And in the second case, power supply, despite a slightly lower utilization rate of wind turbines, their more profitable because of the best time distribution according to production of electrical energy by wind turbines and consumption generated energy.

The effect of seasonal factor of the load on the optimal configuration APSCS-RES was investigated on the example of Ust-Koksha village (Figure 19). For given the seasonal load factors in Ust-Koksha, along with a micro-hydro, the predominance of the wind turbines power is optimal, due to the coincidence of the seasonal increase in the load and generation of wind turbines. In the case where the load is not changing by season and throughout the year is constant, the optimal capacity of wind turbines and PV are comparable.

All results are presented for APSCS-RES using lithium-iron-phosphate AB, as the total cost of renewable energy power plants and AB in this case were 34% less in comparison with the use of more traditional for lead-acid AB. Thus, the superiority of lithium AB is already significantly in the field of power engineering, and without regard to their weight and size.

Figure 17. Economical estimations of systems with different share of RES in total electricity generation for Sochi district

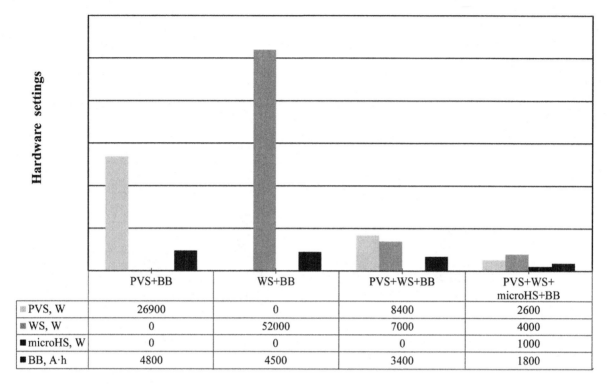

Figure 18. The total cost of power plants based on APSCS-RES under different operating conditions

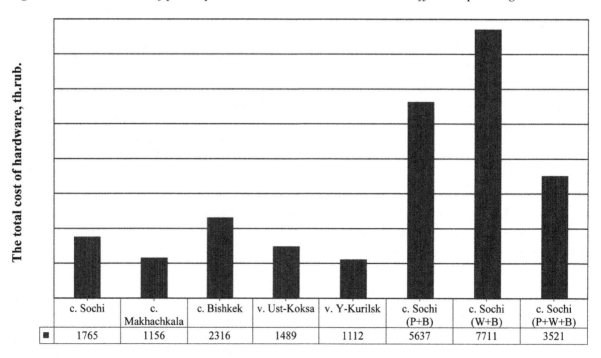

Figure 19. Influence of seasonal factor on optimal parameters of stand-alone power supply systems with RES

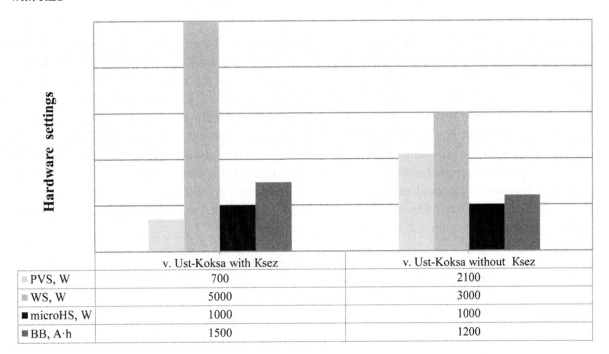

	v. Ust-Koksa with Ksez	v. Ust-Koksa without Ksez
PVS, W	700	2100
WS, W	5000	3000
microHS, W	1000	1000
BB, A·h	1500	1200

3.4. Summary

The technique and software package that allows any user to determine the optimal parameters of the combined systems of autonomous power supply using only renewable energy sources such as solar, wind and water. Results of optimization studies, conducted with the help of the program complex showed a significant dependence of the optimal structure and parameters APSCS-RES on many factors, which should certainly be taken into account when designing such systems.

For example, the Sochi shows how, under certain conditions the combined use of power plants based on renewable energy profitable. Combination PV multiplier to reduce costs by a factor of 2.2 compared with only wind turbine and 1.6 times as compared with a photomultiplier. Their combined use with a micro-hydro can have a 2-fold reduction in the costs of power plants for electric power consumers in the district of Sochi.

So, fluctuating and intermittent renewable energy production should be coordinated with the combination of RES, and also the size of the energy demand must be adjusted to the realistic amount of potential renewable sources. The implementation of 100 percent renewable energy systems is challenging task with the next step to integrate RES into existing energy systems on the large scale.

4. OPTIMIZATION OF GLOBAL-LOCAL MONITORING AND IMPLEMENTING CLIMATE-WEATHER CONDITIONS FOR SOLAR ENERGY

The main objective of this section is to describe activity associated with high technologies of real-time monitoring of the Earth surface and solar energy conversion. Geographic Information Systems (GIS) are currently being used to analyze the potential for producing electricity from renewable energy sources. Making decisions based on meteorological-geographical data is basic to human thinking nowadays. A geographic information system is a technological tool for comprehending geography and making intelligent decisions. Faced with grim predictions of energy supply and consumption, humankind is responding with tremendous efforts to capture and cultivate renewable resources (GIS for Renewable Energy, 2010).

Solar energy systems are becoming more widespread, cheaper and more efficient each year. At the same time, it is not very evident and clear what kind of solar energy system is most suitable for the concrete geographical location and weather-climatic conditions, what output parameters and characteristics concrete solar energy system will perform in real situations.

4.1. A Case Study: Global-Local Monitoring of Climate-Weather Conditions for Solar Energy

A concept of global-local monitoring of climate-weather conditions for solar energy was developed firstly in (Tyukhov, 2012). A data acquisition system has designed and implemented with facilities for monitoring meteorological data and solar radiation. The system, using photovoltaic monitoring equipment and hardware–software complex "Kosmos–3M", was developed for taking images of the Earth from National Oceanic and Atmospheric Administration (NOAA) satellites and for handling these images, analyzing many important geographical and meteorological parameters for forecasting of incoming solar radiation. With the help of the proposed monitoring system it is possible to study not only weather but also climatic aspects using solar energy in case of long-term collecting data at least during 10–30 years.

According to Renewables 2013 Global Status Report, renewable energy markets, industries and policy frameworks have evolved rapidly in recent years (Renewables 2013 Global Status Report, 2013). Solar photovoltaic (PV) stands out as having undergone one of the greatest technological developments. It offers the greatest potential for being competitive in the market (Trillo-Montero D., 2014, European Photovoltaic Industry Association (EPIA), 2013). PV market saw strong year, with total global operating capacity reaching the 100 GW in 2012. On the other hand the concentrating solar thermal power (CSP) market continued to advance in 2012, with total global capacity up more than 60% to about 2550 MW. The Middle East and North Africa region saw significant new developments in 2012 as the Egyptian Solar Plan, approved in July 2012, setting a target for 2800 MW of thermal CSP and 700 MW of solar PV by 2027 (Renewables 2013 Global Status Report, 2013). Europe is still the world leader in PV electricity generation, and for the second year running it has been the largest new source of installed generation (European Photovoltaic Industry Association (EPIA), 2013). In markets outside Europe, such as China, Japan or the USA, the percentage of installed power is still very small compared with their enormous potential. Moreover, there are large parts of Africa, Middle East, Southeast Asia or Latin America where this type of installations is only beginning now to be developed (Trillo-Montero D., 2014).

PV applications in developing countries have the potential to become a major force for social and economic development. Many isolated sites are not yet connected to the conventional electrical grid and face severe problems of water for domestic consumption and irrigation purposes (Benghanema M,

1999). Egypt and Russia are situated among the countries, which are better placed for the utilization of solar energy.

Egypt has a high solar radiation intensity—the annual global solar radiation of over 2000 kWh/m^2: areas appear in red or pink (see Figure 20, including Egypt) (Rezk, H, 2013, Dousoky, G.M., 2010). Moreover, Egypt has large areas of low-cost land. It has been estimated that Russia's gross potential for solar energy is 2.3 trillion tce. The regions with the best solar radiation potential are the North Caucasus, the Black Sea and the Caspian Sea areas, and southern parts of Siberia and the Far East. This potential is largely unused, although the possibilities for off-grid solar energy or hybrid applications in remote areas are huge (Strebkov D., 2000, Energy in Russia, 2013). Prediction of the solar radiation coming onto the earth's surface at the given geographical location, depending on the cloudiness, is particularly very important when large solar power plants are used as energy sources for power supply, giving a significant contribution to the power grid system based on conventional sources (Tyukhov, I, 2009, Tyukhov, I., Vignola. F., 2008). Knowing the arrival of solar radiation will allow for planning of system operators of grids. Many factors like cloudiness, atmospheric transmissivity, latitude and orientation of the Earth relative to the Sun, time of day, slope and the surface orientation determine the spatial and temporal distribution of the aspect of solar radiation on a surface. At the same time, the availability of solar radiation data from meteorological stations is more restricted, for example, than for precipitation and temperature. Solar resource assessment is a significant input for a wide range of different solar system applications. Solar radiation assessment is the most important for operational purposes of the big solar plants working in combination with the utilities. The aim of this paper was to design and develop a software application that enables users to perform an automated analysis of data from the monitoring of photovoltaic systems. This application integrates data from all devices which are already in operation such as environmental sensors and meters, and which record information on typical PV installations (Trillo-Montero D., 2014).

Solar radiation and other environmental parameters are needed in the design of solar systems; it is difficult to obtain these data for remote areas (Mukaro, R, 1999). Developing new technologies aimed at solving the world's energy and climate change challenges, and environmental problem demand young people with a good background in solar energy (Tyukhov, I, Schakhramanyan M, 2008). Satellite imagery is widely used to study and solve problems in many industries, including the assessment of natural resources of the Earth, mapping, forestry and agriculture, etc. Currently, analysis and collection of meteorological data mainly benefited from surface, upper-air, radar and satellite observations. Remote sensing by satellites allows quickly covering large areas of capturing Earth images, and provides acceptable accuracy and high visibility materials (Kasyanov, O.V., 2014). Analysis and short-term forecasting of atmospheric processes on successive series of satellite images of the clouds to identify the characteristic features of the evolution of cloud structures and the direction of their movement were considered in (Bespalov, D.P., 2011). Authors (Chubarova N.Y., 2008) measured the arrival of natural ultraviolet radiation under actual cloud conditions and the absence of clouds. Study of theoretical and experimental methods of forecasting and evaluation of solar radiation as a function of cloud resource space monitoring the earth's surface are also studied in (Hammer, A, 1999, Lorenz, E, 2012, Vignola, 2012).

Figure 20. Total global annual solar irradiance (Wh/m^2)

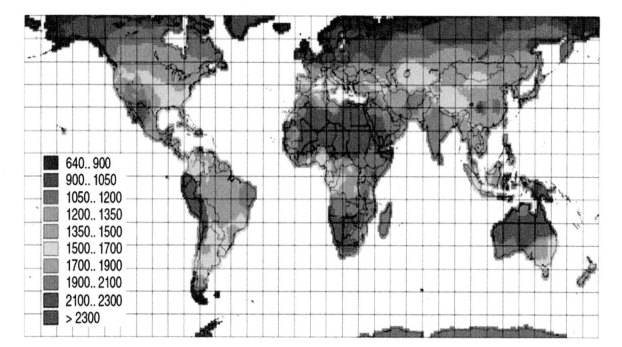

4.2. Components and Principle of Operation of Space Monitoring System

In the current manuscript, a computer-based data-acquisition system for monitoring both meteorological data and photovoltaic solar radiation is discussed. Data-acquisition system allows observing current incoming radiation and observing cloud motion using space technology; with this, one is able to predict cloudy or sunny sky and to make forecast for any solar application.

A block diagram of the proposed system is shown in Figure 21. The collected data are processed, displayed on the monitor and stored in the personal computer using the Lab-VIEW software and program APT Viewer 3.5. Lab-VIEW provides an easy-to-use graphical environment that permits the system operators to process easily the collected data, using complex data-processing algorithms, without detailed knowledge of the data acquisition system design like in (Koutroulis, E, 2003). The system shown in Figure 21 consists of solar photovoltaic module "FSM-25" installed on the roof of a University of Mechanical Engineering, Moscow (Umech), data acquisition device USB-6009 with output on a computer and software, developed in the graphical programming environment Lab-VIEW. The structure of devices and software for receiving and processing satellite images includes hardware-software complex "Kosmos-M3", which includes an antenna mounted on the roof of an Umech university, a receiver and program APT Viewer 3.5.

PV module FSM-25 is made of 35 multi-crystalline silicon solar cells in series and provides 26.7W of nominal maximum power at 1000W/m^2 and 25 °C (standard conditions). The installed PV module in the monitoring system is based on the silicone gel lamination technology. A new encapsulant developed at the All Russian Research Institute for Electrification of Agriculture (VIESH) was used in an attempt

Figure 21. Schematic diagram of the hardware proposed data acquisition system

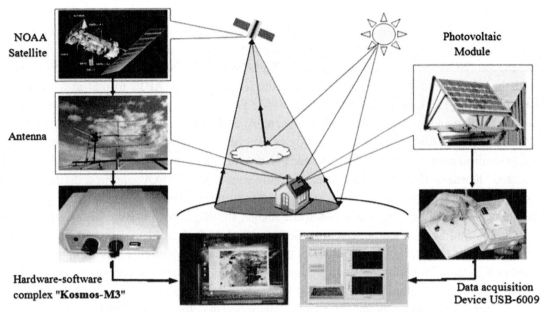

Display of data by using program APT-Viewer 3.5 and monitoring of
PV module on the programming environment Lab-VIEW

to increase the service life time of the module. The main factor determining the operational life of PV modules is the encapsulation material (encapsulant). Encapsulant is a protective material completely enveloping and isolating the photovoltaic cells from moisture, heat and mechanical damage and providing good optical contact between the surface of the PV cells and the protective outer coating. Silicone gel encapsulant allows better stress relaxation during thermal cycling of PV panels, compared to traditional solid EVA encapsulant. Silicone gel laminated PV panel with power reduction about 15% over 50 years service life is possible because of strongly reduced corrosiveness of solar cells in modules by silicone gel. Corrosiveness of solar cells is the main source of failures in EVA laminated PV panel (Poulek, V., 2012).

Typically, fixed (nonadjustable) PV arrays should be tilted toward south by an angle equal to the latitude of the array's location to capture the most year round solar energy. If the PV array is mounted with a tilt angle equal to the site latitude, it is perpendicular to the sun twice a year (on each equinox date) and very close to perpendicular for the weeks before and after the equinox; this makes the array perpendicular to the sun's position in the sky for the greatest number of hours throughout the year (Goswami, D.Y., 2000).

A satellite receiver and software were elaborated for demonstrating how GIS systems can be used, particularly for solar engineering projects. Complex consists of satellite antenna mounted at the roof of Umech, receiver of satellite signal and software of processing aerospace images of the Earth (Figure 21, left part). Software of the receiving complex "Kosmos–M3" is able to determine the temperature of underlying surface at any point of the image obtained; to measure distance from one point to another with regards the Earth's geometry; to determine surface area; selection of map layer and so on, to have real time images within interval 2–4-h interval. With the help of this complex it is possible to study the climatic aspects of solar energy if to collect data at least during 10–30 years.

In general A hardware/software complex, Kosmos–M3 is designed under the supervision of Prof. M. Schakhramanyan for receiving Earth images transmitted from the National Oceanic and Atmospheric Administration (NOAA) series polar orbiting satellites (NOAA-12, NOAA-14, NOAA-15 and NOAA-17), and the Meteor satellites, in Automatic Picture Transmission (APT) format at a frequency of 137MHz (Tyukhov, I., Vignola. F., 2008). Kosmos–M3 is a portable, relatively inexpensive tool. It was designed to receive and process images of Earth sent by satellites in real time. These systems are quite useful for introducing certain modern technologies. Images can be received without payment of a fee, helping to minimize the expense involved in research cost in this subject matter. Basic capabilities of the multifaceted "Kosmos–3M" system include: receiving signal from NOAA satellites; digital processing of space images with geographical fixing, superposition of maps of cities and coordinate grid; finding of geographical coordinates at any point of space image; finding of temperature of underlying surface at given points; finding of albedo (reflection coefficient) at any point of space image; finding of upper boundary of clouds (cloudiness); forecasting of dangerous weather phenomena; defining wind fields in cyclones; precipitations forecast; measuring distances between given points; measuring surfaces (areas) and forming of electronic library of images of the Earth. Work is underway to use the "Kosmos–3M" cloudiness images to estimate the incident solar radiation values for evaluating terrestrial solar energy performance in real time. Such kind of system would have a wide variety of uses from the classroom to the field (Tyukhov, I, Schakhramanyan, M, Strebkov, D, 2009).

Solar radiation is scattered and absorbed by Earth's atmosphere and is reflected from the Earth's surface and clouds. Reflected light carries information about the properties of the underlying surface area and, particularly, about underlying clouds. This radiation is collected by the satellite receiver system and is transformed into an electrical signal and transmitted to an antenna for further processing (Figure 21 left part). Swath width is about 3000 km, and the spatial resolution of the image 4 km. There are tradeoffs related to data volume and spatial resolution. Our working experience showed that the swath width is quite big to see the whole picture (all cloudy and non-cloudy areas in a region) and quite small spatial resolution to see how clouds move on a specific place with enough accuracy. The higher resolution will complicate saving and analyze big numerical data for climatic investigations during long time of collecting information. Images of good quality with a single satellite can be captured 3–4 times a day. During receiving the image, the trajectory of the satellite in the web portal can be watched as shown in Figure 22 (Real time satellite tracking and prediction, 2012).

Table 4 shows an example of more detailed parameters on the image pixel by pixel. Thus, this system allows for local monitoring of incoming solar radiation and global monitoring of climatic conditions in real time. APT Viewer 3.5 program allows to monitor meteorological phenomena such as the appearance or disappearance of cyclones and anticyclones, wind speed and direction, temperature change of the land and water surfaces, temperature and height of cloud top forecast rainfall, cloud species, etc. as shown in Figure 23 (Shakhramanyan, M.A., 2009).

4.3. Monitoring Weather-Climatic Conditions and Data Processing

Solar radiation was monitored in real time, by measuring electrical energy generated by the photovoltaic solar module. Then this information compares with information received from satellites.

Figure 22. Satellite trajectory observation

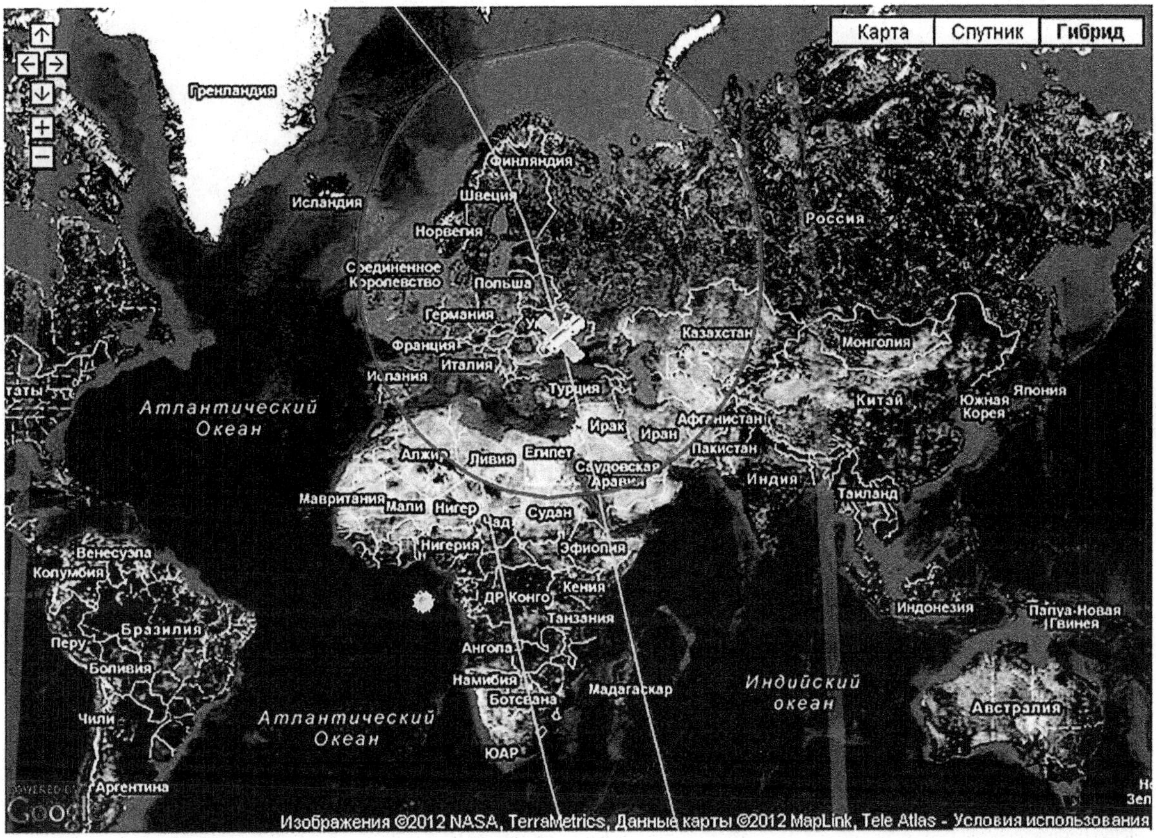

Figure 23. Computer interface to display the PV module short circuit current and solar radiation

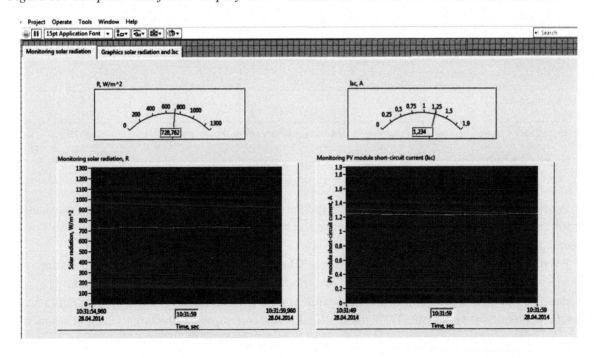

Table 4. The pixels of different underlying surfaces in the picture and their parameters

The image from the satellite NOAA 18	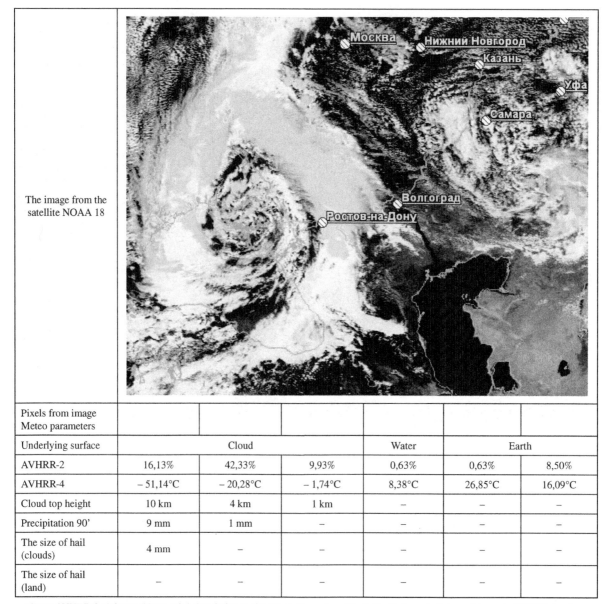					
Pixels from image Meteo parameters						
Underlying surface	Cloud			Water	Earth	
AVHRR-2	16,13%	42,33%	9,93%	0,63%	0,63%	8,50%
AVHRR-4	– 51,14°C	– 20,28°C	– 1,74°C	8,38°C	26,85°C	16,09°C
Cloud top height	10 km	4 km	1 km	–	–	–
Precipitation 90'	9 mm	1 mm	–	–	–	–
The size of hail (clouds)	4 mm	–	–	–	–	–
The size of hail (land)	–	–	–	–	–	–

Notes: AVHRR-2 (Advanced Very High Resolution Radiometer): for surface albedo the 2nd spectral channel is used; AVHRR-4 —for surface temperature the 4th spectral channel is used; "precipitation 90" is an indicator of the average total amount of convective precipitation that may fall out in 90 min.

Predicting the incident solar radiation on the earth's surface in cloudy conditions is particularly important when large solar power plants are used as sources for power generating in order to allow for planning of power energy system operation.

PV module short-circuit current is the current when the voltage across the PV module is zero (i.e. when PV module is short circuited). This current is due to the generation and collection of light-generated carriers. For an ideal PV module at most moderate resistive loss mechanisms, the short-circuit current and the photocurrent are identical. Therefore, the short-circuit current is the largest current which may be drawn from the PV module. At normal levels of solar radiation, the short-circuit current can be considered equivalent to photocurrent, i.e. proportional to the solar radiation G (W/m²). But this may result in some deviation from the experimental result, so a power law having exponent α to account for the non-linear affect that the photocurrent depends on. PV module short-circuit current is not strongly temperature dependent. It tends to increase slightly with the increase of the module temperature.

For the purposes of PV module performance, modeling this variation can be considered negligible. Then, the short-circuit current can be simply calculated by (Zhou, W., 2007):

$$I_{sc} = I_{sc0} \left(\frac{G}{G_0} \right)^\alpha \quad (30)$$

where I_{sc0} is the short-circuit current of the PV module under the standard solar radiation G_0; while I_{sc} is the short-circuit current of the PV module under the solar irradiance G; α is the exponent responsible for all the non-linear effects that the photocurrent depends on.

From (1) it can be concluded that the solar radiation can be estimated by the following equation;

$$G = G_0 \left(\frac{I_{sc}}{I_{sc0}} \right)^{\frac{1}{\alpha}} \quad (31)$$

The proposed monitoring system allows us to measure PV short circuit current and solar radiation. The data are transmitted and stored in a computer through data acquisition device USB-6009. The monitoring variables are processed and displayed in the computer screen using virtual instruments developed with Lab-VIEW in the real-time interface system (Forero, N, 2006). Figure 23 shows a computer interface used to display monitoring data of short circuit current and solar radiation. A print screen for monitored solar radiation is shown in Figure 24.

Simultaneously, the space image receiver allows to record information from satellite about weather conditions on the earth surface. Program APT Viewer allows monitoring the actual state of the weather situation, including the state of the atmosphere, the presence or absence of clouds on radio brightness parameters laid down in the program and settlement functions. In addition, it can be used "Cloud Atlas" to study the more properties of cloud cover (Bespalov, D.P., 2011).

4.3.1. Study the Effect of Clouds on Incident Solar Radiation

The recorded solar radiation from monitoring system can be compared to the pixel value of satellite image. This can provide ground true data for testing solar radiation derived from the satellite data (Tyukhov, I, Schakhramanyan, M, Strebkov, D, 2009). Figure 25 shows incident solar radiation and NOAA 15 satellite image taken at 16:46 in clear weather.

The measurements carried out in one of the sunny days. Due to the fact that in the absence of clouds demonstrable change between multiple images is not significant, there is shown only one satellite image recorded on the afternoon of the same day. From the shown figure it can be clearly seen that, solar radiation curve varies smoothly, reaching its peak at noon. On the other hand it can be seen some fluctuations

Figure 24. Monitoring solar radiation in specific day

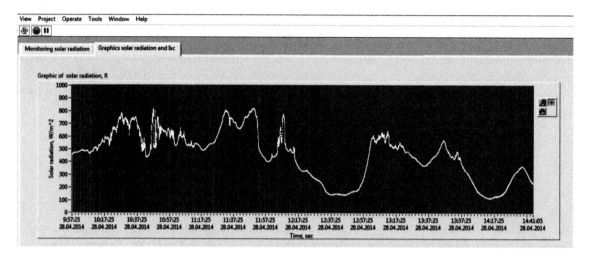

on the curve, and these fluctuations are caused by various factors: the absorption and the scattering of dust, dry air molecules, a selective absorption of water vapor, carbon dioxide, etc.

Figure 26 shows the experimental measurement of solar radiation at cloudy weather. From the showed figure it can be seen that the values of incident solar radiation decrease with the time. There are also three satellite images at different intervals in time. First half of the day due to the variability of thickness of cloud layers there are sudden changes in availability of solar radiation; it changes from 278 to 784W/m^2. Furthermore due to the appearance of dense layers of cloud, the solar radiation decreased and changed from 114 to 84 W/m^2.

4.3.2. Observing and Forecasting the Movement of Clouds

Predicting the incident solar radiation on the earth's surface in cloudy conditions is particularly important when large solar power plants are used as sources for power in order to allow for planning of system operation. Considering Figure 26b it can be seen that, there is a clear cloudless weather until 13:00 at afternoon and then the weather changes to cloudy skies, which clearly illustrate the solar radiation curve advent of periodic dips. The process of cloud formations in the current site and the gradual movement of other clouds from different sides are shown in the following figure.

Figure 27 shows that four satellites images were taken during one specific day from two meteorological satellites at different time intervals and elevation angles of the satellites relative to the horizon.

The results of measurement are shown in Table 5. Considering Figure 27 it can be seen the movement of the clouds, also directed towards the movement of wind, as forecast.

4.3.3. Three-Dimensional Modeling and Visualization of Clouds

Constructing a three-dimensional model of clouds and modeling of incident direct solar radiation in three-dimensional modeling software package allows to analyze not only the actual state, but also to obtain predictive information about cloud cover impact on solar power plants. The prediction and study of cloud cover impact on the incident direct solar radiation are particularly needed for concentrated

Figure 25. a) Incident solar radiation and b) NOAA 15 satellite image taken at 16:46 in clear weather in Moscow

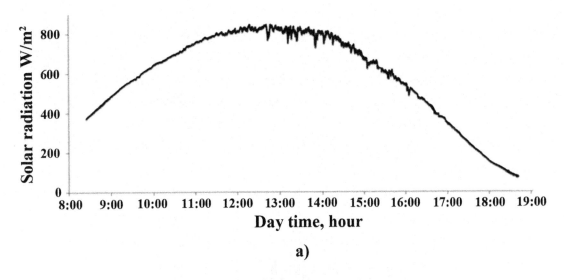

a)

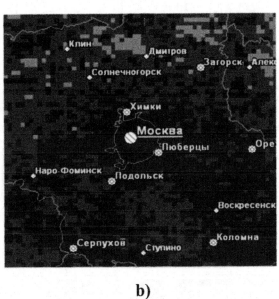

b)

Table 5. The experimental results of clouds movements by proposed monitoring system

N° Image	The Time of Receiving	Distance from Moscow to the Nearest Edge of the Clouds, km	Change in the Distance, km	Speed, m/s	Average Speed, m/s
1	11:50	54.3	-	-	6.25
2	12:25	63.4	9.1	4.3	
3	13:30	77.6	14.2	3.6	
4	14:05	105	27.4	13.1	

Figure 26. Measured solar radiation and satellite images, taken in: a) clear cloudy overcast and b) variable cloudy weather in Moscow

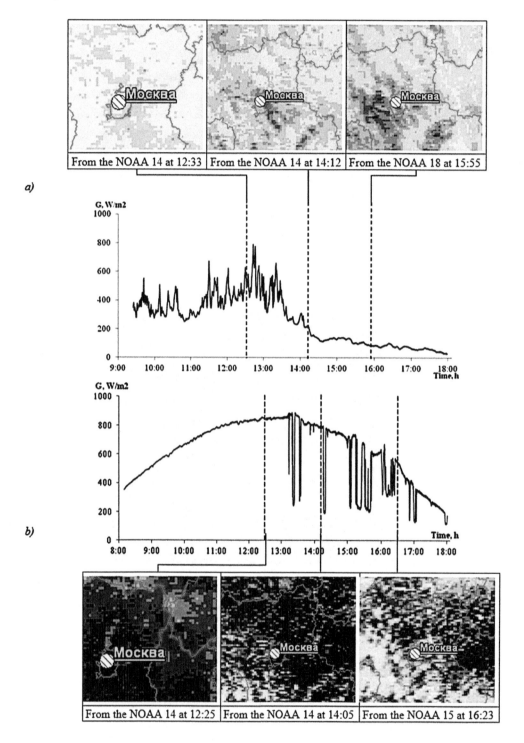

Modern Optimization Algorithms and Applications in Solar Photovoltaic Engineering

Figure 27. Monitoring of the movement of clouds on satellite imagery: 1—the image from the satellite NOAA 14, elevation angle of the satellite 30°; 2—the image from the satellite NOAA 18 elevation angle 25°; 3—the image from the satellite NOAA 14 elevation angle 57°; and 4—the image from the satellite NOAA 18 elevation angle 80

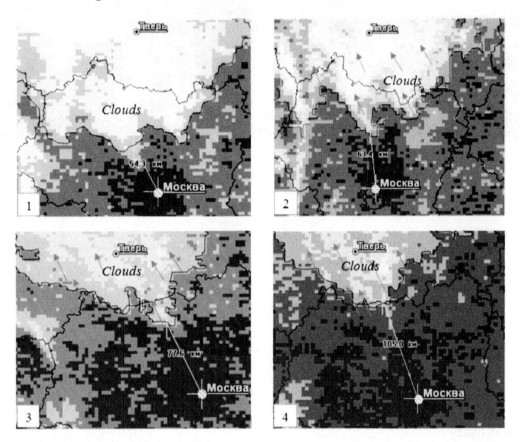

solar power plants. Satellite images of the earth surface give a visual picture about the state of the cloud in the region under study. However, in case of partial cloud, using only satellite images is not able to determine the cloud cover impact on the incident direct solar radiation on specifically selected area or point on the earth's surface. This is due to the fact that the cloud shading may vary depending on the area, height and thickness of clouds, the location and angle of the sun above the horizon. For further information about cloud shadow modeling the incident solar radiation is needed, taking into account the velocity of cloud movement. Using three-dimensional (3D) modeling with aid Google Sketch Up program is considered the best way for simulating sunlight. Figure 28 shows the transformation steps of the satellite image into three-dimensional model. The height of the cloud layers and the temperature of the underlying surface are shown in Figure 28.

After receiving the necessary data on the size and exact location on the cloud space image, a three dimensional model is developed over the map. After that, by enabling the "Shadows" date and time are configured to accurately simulate the arrival of direct solar radiation and shadow clouds are considered in exactly the specified time. Figure 29 shows the result of converting a satellite image from the cloud space image into a three-dimensional model.

Figure 28. The transformation steps of the satellite image into 3D model by using the Google Sketch Up program (satellite image taken in May 16, 2013 at 16:14)

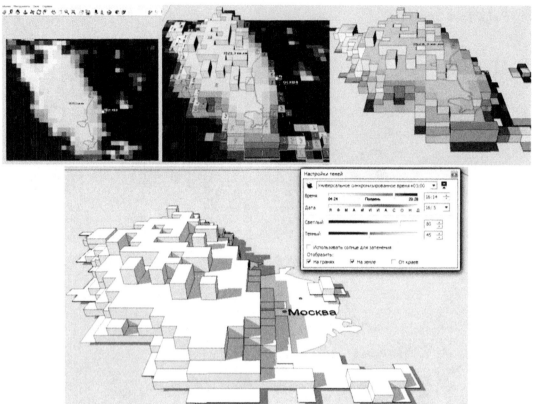

4.4. Summary

An efficient solar monitoring system of meteorological data, photovoltaic solar radiation which enable forecasting sunny and cloudy hours, has been designed and implemented. The new type of PV modules with improved service life time was used because of reduced corrosiveness by using silicone gel laminated. This type of modules is a good candidate for long-term collecting data. By using simple Google Sketch Up program, a three-dimensional model of clouds has been constructed. Also, the height of the cloud layers and the temperature of the underlying surface have been determined for accurate forecasting of sunny or cloudy weather. The described technology and approach is useful not only for research but also for educational purposes because non-expensive components are involved. With the help of the proposed monitoring system, it is possible to study the climatic aspects of solar energy after long-term collecting data for many years. In future PV systems will be aggregating with other renewable systems and GIS systems designed for optimal collecting and distributing electrical power nets according the consumer demand.

Figure 29. The height of the cloud layers and the temperature of the underlying surface: H—height above the Earth's surface; t—temperature

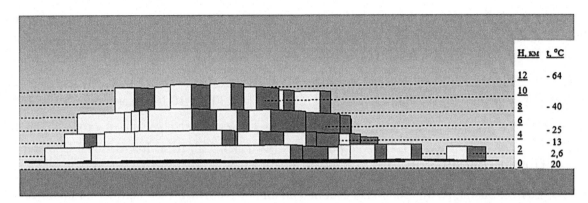

CONCLUSION

The history of human civilization should be considered as the progressive development of new energy sources, their associated conversion technologies and info-communicative infrastructure. Solar energy is the best among the most infinitely available clean sources and photovoltaic conversion is the most appropriate direct conversion solar radiation into electricity.

Great progress in the development of solar cells design and technology allows creating practical PV systems. Optimization of PV systems was analyzed from the point of view of maximum power point tracking technologies (MPPT). The comprehensive comparison of different MPPT techniques for photovoltaic systems was carried out.

The optimization of renewable energy systems demands of combination of energy sources, accumulation and consumption. The case study of elaborated technique and software package allows any user to determine the optimal parameters of the combined systems of autonomous power supply using only renewable energy sources such as solar, wind and water. It was shown that fluctuating and intermittent renewable energy production can be coordinated with the combination of RES, and also the size of the energy demand can be adjusted to the realistic amount of potential renewable sources.

Renewable energy systems using information from GIS systems demands optimization of energy systems taking into account all renewable energy currents at real time regime. The case study of solar monitoring system of meteorological data, photovoltaic solar radiation which enable forecasting sunny and cloudy hours, has been designed and implemented.

It is extraordinary difficulty and exceptional nature of the coming energy transition—but, given the enormous challenges of ushering in a post-fossil world, it may also be because of the possibility of an unprecedented and persistent commitment to a rapid change (Smil, V., 2010).

A general approaching to the challenging task of the implementation of 100 percent renewable energy systems with the next step to integrate RES into existing and future smart energy systems on the large scale were described. A modern concept of smart grid includes a variety of operational and energy measures and systems including smart meters, smart appliances, renewable energy resources, and energy efficiency resources. Electronic power conditioning and control of the production, distribution and consumption infrastructure are important aspects of the smart grid.

REFERENCES

Abouobaida, H., & Cherkaoui, M. (2012, May). Comparative study of maximum power point trackers for fast changing environmental conditions. In *Multimedia Computing and Systems (ICMCS)*, International Conference. doi:10.1109/ICMCS.2012.6320222

Ahmed, J., & Salam, Z. (2014). A Maximum Power Point Tracking (MPPT) for PV system using Cuckoo Search with partial shading capability. *Applied Energy, 119*, 118–130. doi:10.1016/j.apenergy.2013.12.062

Al-Diab, A., & Sourkounis, C. (2010). Variable step size P&O MPPT algorithm for PV systems. In *Optimization of Electrical and Electronic Equipment (OPTIM)*. 12th IEEE International Conference. doi:10.1109/OPTIM.2010.5510441

Altasa, I. H., & Sharaf, A. M. (2008). A novel maximum power fuzzy logic controller for photovoltaic solar energy systems. *Renewable Energy, 33*(3), 388–399. doi:10.1016/j.renene.2007.03.002

Amrouche, B., Belhamel, M., Guessoum, A., 2007. Artificial intelligence based P&O MPPT method for photovoltaic systems. *Revue des Energies Renouvelables ICRESD-07 Tlemcen*, 11–16.

Ashish, P., Nivedita, D., & Ashok, K. M. (2007). Comparison of various MPPT algorithms for cost reduction. *IEEE Transactions on Power Electronics, 22*(2), 698–700.

Ben Salah, C., Chaabenea, M., & Ben Ammara, M. (2008). Multi-criteria fuzzy algorithm for energy management of a domestic photovoltaic panel. *Renewable Energy, 33*(5), 993–1001. doi:10.1016/j.renene.2007.05.036

Benghanema, M., Arabb, A. H., & Mukadamc, K. (1999). Data acquisition system for photovoltaic water pumps. *Renewable Energy, 176*(3), 385–396. doi:10.1016/S0960-1481(98)00126-8

Berrera, M., Dolara, & A., Leva, S. (June, 2009). *Experimental test of seven widely adopted MPPT algorithms*. IEEE Bucharest Power Tech Conference, Bucharest, Romania. doi:10.1109/PTC.2009.5282010

Besheer, A. H., & Adly, M. (2012). *Ant colony system based PI maximum power point tracking for stand-alone photovoltaic system*. IEEE International Conference on Industrial Technology (ICIT). doi:10.1109/ICIT.2012.6210019

Bespalov, D. P. (2011). Atlas oblakov (Cloud Atlas)/Feder. sluzhba po gidrometeorologii i monitoring okruzhayushchey sredy (Rosgidromet), Gl. geofiz. observatoriya im. A.I. Voyeykova; red.: L. K. Surygina. *[Federal Service for Hydrometeorology and Environmental Monitoring (Roshydromet), Ch. Geofiz. Observatory. A.I. Voeikova; Ed.: LK Surygin]. Sankt-Peterburg: D'ART, 2011*. Available at: http://www.dvfu.ru/meteo/book/ AtlasClouds.pdf

Boutasseta, N. (2012). PSO-PI based Control of Photovoltaic Arrays. *International Journal of Computers and Applications, 48*(17), 36–40. doi:10.5120/7444-0557

Cerpinsek, M., Liu, S., & Mernik, L. (2012). A note on teaching learning based optimization algorithm. *Information science, 212*, 79–93. doi:10.1016/j.ins.2012.05.009

Chen, K., Tian, S., Cheng, Y., & Bai, L. (2014). An Improved MPPT Controller for Photovoltaic System under Partial Shading Condition. *IEEE Transaction on Sustainable Energy, 5*(3), 978–985. doi:10.1109/TSTE.2014.2315653

Chen, Y.-T., Lai, Z.-H., & Liang, R.-H. (2014). A novel auto-scaling variable step-size MPPT method for a PV system. *Solar Energy, 102*, 247–256.

Christy, J. S., Raj, M., & Jeyakumar, A. E. (2014). A two stage successive estimation based maximum power point tracking technique for photovoltaic modules. *Solar Energy, 103*, 43–61. doi:10.1016/j.solener.2014.01.042

Chubarova, N. Y. (2008). UV variability in Moscow according to long-term UV measurements and reconstruction model. *Atmospheric Chemistry and Physics Discussion, 8*(8), 893–906. doi:10.5194/acpd-8-893-2008

Coby, S. (2011). Natural resource limitations to terawatt-scale solar cells. *Solar Energy Materials and Solar Cells, 95*(12), 3176–3180. doi:10.1016/j.solmat.2011.06.013

Daraban, S., Petreus, D., & Morel, C. (2013). *A novel global MPPT based on genetic algorithms for photovoltaic systems under the influence of partial shading*. IECON 2013-39th Annual Conference of the IEEE. doi:10.1109/IECON.2013.6699353

Dousoky, G. M. (2010). *On Intelligent Power Electronic Interface for Renewable Energy Systems*. VDM Verlag Publishing House Ltd.

Droege, P. (2009). 100% Renewable Energy Autonomy in Action. Earthscan.

El Sayed, M., (2013). Modeling and simulation of smart maximum power point tracker for photovoltaic system. *Minia J. Eng. Technol., 32* (1).

Eltamaly, A. M. (2010, February). *Modeling of fuzzy logic controller for photovoltaic maximum power point tracker*. Solar Future 2010 Conf. Proc., Istanbul, Turkey.

Eltamaly, A. M., Alolah, A. I., & Abdulghany, M. Y. (2010). *Digital implementation of general purpose fuzzy logic controller for photovoltaic maximum power point tracker*. In Power Electronics Electrical Drives Automation and Motion (SPEEDAM), 2010 International Symposium on Digital Object Identifier. doi:10.1109/SPEEDAM.2010.5542207

Emad, M. A., & Masahito, S. (2010, November). Modified adaptive variable step-size MPPT based-on single current sensor. In TENCON 2010 – IEEE Region 10 Conference.

Emad, M. A., & Masahito, S. (2011, June). *Stability study of variable step size incremental conductance/impedance MPPT for PV systems*. In Power Electronics and ECCE Asia (ICPE & ECCE), 2011 IEEE 8th International Conference.

Energy in Russia. (n.d.). Available at http://en.wikipedia.org/wiki/Energy_in_Russia#Solar_energy

Esram, T., & Chapman, P. L. (2007). Comparison of photovoltaic array maximum power point tracking techniques. *IEEE Transactions on Energy Conversion, 22*(2), 439–449. doi:10.1109/TEC.2006.874230

European Photovoltaic Industry Association (EPIA). (n.d.). Retrieved from http://www.epia.org/

Fangrui, L., Shanxu, D., Fei, L., Bangyin, L., & Yong, K. (2008). A variable step size INC MPPT method for PV systems. *IEEE Transactions on Industrial Electronics*, *55*(7), 2622–2628. doi:10.1109/TIE.2008.920550

Faranda, R., Leva, S., & Maugeri, V. (2008, July). *MPPT techniques for PV systems: energetic and cost comparison*. Power and Energy Society General Meeting – Conversion and Delivery of Electrical Energy in the 21st Century.

Forero, N., Hernández, J., & Gordillo, G. (2006). Development of a monitoring system for a PV solar plant. *Energy Conversion and Management*, *47*(15-16), 2329–2336. doi:10.1016/j.enconman.2005.11.012

GIS for Renewable Energy. (2010). Retrieved from http://www.gisday.com/resources/ebooks/renewable-energy.pdf

Gokmen, N., Karatepe, E., Ugranli, F., & Silvestre, S. (2013). Voltage band based global MPPT controller for photovoltaic systems. *Solar Energy*, *98*, 322–334. doi:10.1016/j.solener.2013.09.025

Goswami, D. Y., & Kreith, F. (2000). *Principles of Solar Engineering* (2nd ed.). Taylor & Francis.

Gounden, N. A., Ann Peter, S., Nallandula, H., & Krithiga, S. (2009). Fuzzy logic controller with MPPT using line-commutated inverter for three-phase grid-connected photovoltaic systems. *Renew. Energy J.*, *34*(3), 909–915. doi:10.1016/j.renene.2008.05.039

Green, M. A. (2006). *Third Generation Photovoltaics*. New York: Advanced Solar Energy Conversion. Springer-Verlag Berlin Heidelberg.

Hammer, A., Heinemann, D., Lorenz, E., & Lückehe, B. (1999). Short-term forecasting of solar radiation: A statistical approach using satellite data. *Solar Energy*, *67*(1–3), 139–150. doi:10.1016/S0038-092X(00)00038-4

Hohm, D. P., & Ropp, M. E. (2000). Comparative study of maximum power point tracking algorithms using an experimental, programmable, maximum power point tracking test bed. In Photovoltaic Specialists Conference, Anchorage, AK. doi:10.1109/PVSC.2000.916230

Ibrahim, H. E., & Houssiny, F. F. (1999, August). *Microcomputer controlled buck regulator for maximum power point tracker for DC pumping system operates from photovoltaic system*. In *IEEE International Fuzzy Systems Conference*. doi:10.1109/FUZZY.1999.793274

Ioan, V. B., & Marcel, I. (2013). *Comparative analysis of the perturb-andobserve and incremental conductance MPPT methods*. In *8th International Symposium on Advanced Topics in Electrical Engineering*, Bucharest, Romania.

Ishaque, K., & Salam, Z. (2013). A Deterministic Particle Swarm Optimization Maximum Power Point Tracker for Photovoltaic System under Partial Shading Condition. *IEEE Transactions on Industrial Electronics*, *60*(8), 3195–3205.

Ishaque, K., Salam, Z., Amjad, M., & Mekhilef, S. (2012). An improved particle swarm optimization (PSO)–based MPPT for PV with reduced steady-state oscillation. *IEEE Transactions on Power Electronics*, *27*(8), 3627–3638. doi:10.1109/TPEL.2012.2185713

Joy Green. (2014). *Solar PV grows exponentially from 2007 to 2014*. Retrieved from http://www.thefuturescentre.org/signals-of-change/3089/solar-pv-grows-exponentially-2007-2014

Karlis, A. D., Kottas, T. L., & Boutalisb, Y. S. (2007). A novel maximum power point tracking method for PV systems using fuzzy cognitive networks (FCN). *Electric Power Systems Research*, 77(3–4), 315–327. doi:10.1016/j.epsr.2006.03.008

Kasyanov, O. V., & Slobodyan, V. A. (n.d.). *Ispolzovaniye kosmicheskikh snimkov dlya kartirovaniya selkhozugodiy*. [Using satellite imagery to map farmland]. Available at: http://www.panorama.kharkov.ua/articles/rarticle.htm?page=15

Kazmi, S., Goto, H., Ichinokura, O., & Guo, H. (2009). An improved and very efficient MPPT controller for PV systems subjected to rapidly varying atmospheric conditions and partial shading. *Power Engineering Conference (AUPEC 2009)*.

Khaehintung, N., Pramotung, K., Tuvirat, B., & Sirisuk, P. (2004). *RISC microcontroller built-in fuzzy logic controller of maximum power point tracking for solar-powered light-flasher applications*. In 30th Annu. Conf. IEEE Ind. Electron. Soc. doi:10.1109/IECON.2004.1432228

Khaehintung, N., Wiangtong, T., & Sirisuk, P. (2006). *FPGA implementation of MPPT using variable step-size P&O algorithm for PV applications*. In International Symposium on Digital Object Identifier, ISCIT. doi:10.1109/ISCIT.2006.340033

Kinattingal, S. (2015). Enhanced Energy Output from a PV System Under Partial Shaded Conditions Through Artificial Bee Colony. *IEEE Transaction on Sustainable Energy*, 6(1), 198–209. doi:10.1109/TSTE.2014.2363521

Kobayashi, K., Takano, I., & Sawada, Y. (2006). A study of a two stage maximum power point tracking control of a photovoltaic system under partially shaded insolation condition. *Solar Energy Materials and Solar Cells*, 90(18-19), 2975–2988. doi:10.1016/j.solmat.2006.06.050

Koutroulis, E., & Kalaitzakis, K. (2003). Development of an integrated data-acquisition system for renewable energy sources systems monitoring. *Renewable Energy*, 28(1), 139–152. doi:10.1016/S0960-1481(01)00197-5

Koutroulis, E., Kalaitzakis, K., & Voulgaris, N. C. (2001). Development of a microcontroller-based photovoltaic maximum power point tracking control system. *IEEE Transactions on Power Electronics*, 16(1), 46–54. doi:10.1109/63.903988

Kulaksız, A. A., & Akkaya, R. (2012). A Genetic Algorithm Optimized ANN–Based MPPT Algorithm for a Stand-Alone PV System with Induction Motor Drive. *Solar Energy*, 86(9), 2366–2375. doi:10.1016/j.solener.2012.05.006

Lei, M., Yaojie, S., Yandan, L., Zhifeng, B., & Liquin, T. (2011). Comprehensive Renewable Energy. In S. Jieqiong (Ed.), *A high performance MPPT control method. Materials for Renewable Energy & Environment (ICMREE), 2011 International Conference*.

Liu, Y., Huang, S.-C., Huang, J.-W., & Liang, W.-C. (2012). A Particle Swarm Optimization–Based Maximum Power Point Tracking Algorithm for PV Systems Operating Under Partially Shaded Conditions. *IEEE Transactions on Energy Conversion, 27*(4), 1027–1035. doi:10.1109/TEC.2012.2219533

Lorenz, E., & Heinemann, D. (2012). Prediction of solar irradiance and photovoltaic power. *Comprehensive Renewable Energy, 1*, 239–292. doi:10.1016/B978-0-08-087872-0.00114-1

Lund, H. (2010). *Renewable Energy Systems. The Choice and Modeling of 100% Renewable Solutions*. Elsevier Inc.

Ma, J. (2013). *Low–Cost Global MPPT Scheme for Photovoltaic Systems under Partially Shaded Conditions. IEEE International Symposium on Circuits and Systems (ISCAS)*.

Mirhassaniet, S. M. (2015). An improved particle swarm optimization based maximum power point-tracking strategy with variable sampling time. *Electrical Power and Energy Systems, 64*, 761–770. doi:10.1016/j.ijepes.2014.07.074

Mukaro, R., & Carelse, X. F. (1999). A microcontroller-based data acquisition system for solar radiation and environmental monitoring. *IEEE Transactions on Instrumentation and Measurement, 48*(6), 1232–1238. doi:10.1109/19.816142

Paraskevadaki, E. V., & Papathanassiou, S. A. (2011). Evaluation of MPP voltage and power of mc-Si PV modules in partial shading conditions. *IEEE Transactions on Energy Conversion, 26*(3), 923–932. doi:10.1109/TEC.2011.2126021

Payne, R. B., Sorenson, M. D., & Kiltz, K. (2005). *The Cuckoos*. Oxford University Press.

Poulek, V., Strebkov, D. S., Persic, I. S., & Libre, M. (2012). Toward 50 years lifetime of PV panels laminated with silicone gel technology. *Solar Energy, 86*(10), 3103–3108. doi:10.1016/j.solener.2012.07.013

Rajesh, R., & Mabel, M. C. (2014). Efficiency analysis of a multi-fuzzy logic controller for the determination of operating points in a PV system. *Solar Energy, 99*, 77–87. doi:10.1016/j.solener.2013.10.036

Rao, R., & Patel, V. (2013). An improved teaching learning based optimization algorithm for solving unconstrained optimization problems. *Scientia Iranica, 20*(3), 710-720.

Rao, R., Savsani, V., & Vakharia, D. (2011). Teaching–learning-based optimization: A novel method for constrained mechanical design optimization problems. *Computer Aided Design, 43*(3), 303–315. doi:10.1016/j.cad.2010.12.015

Real Time Satellite Tracking and Prediction. (n.d.). Available at: www.n2yo.com

Renewables. (2013). *Global Status Report*. Retrieved from http://www.ren21.net/Portals/0/documents/Resources/GSR/2013/GSR2013_lowres.pdf

Reza Reisi, A., Hassan Moradi, M., & Jamasb, S. (2013). Classification and comparison of maximum power point tracking techniques for photovoltaic system: A review. *Renewable & Sustainable Energy Reviews, 19*, 433–443. doi:10.1016/j.rser.2012.11.052

Rezk, H., & El-Sayed, A. H. M. (2013). Sizing of a stand-alone concentrated photovoltaic system in Egyptian site. *International Journal of Electrical Power & Energy Systems*, *45*(1), 325–330. doi:10.1016/j.ijepes.2012.09.001

Rezk, H., & Eltamaly, A. M. (2015). A comprehensive comparison of different MPPT techniques for photovoltaic systems. *Solar Energy*, *112*, 1–11. doi:10.1016/j.solener.2014.11.010

Salam, Z., & Saad, M. S. (2013). Evolutionary based maximum power point tracking technique using differential evolution algorithm. *Energy and Building*, *67*, 245–252. doi:10.1016/j.enbuild.2013.07.085

Sarvi, M., Ahmadi, S., & Abdi, S. (2015). A PSO-based maximum power point tracking for photovoltaic systems under environmental and partially shaded conditions. *Progress in Photovoltaics: Research and Applications*, *23*(2), 201–214. doi:10.1002/pip.2416

Shakhramanyan, M. A., Tyukhov, I. I., & Voshchenkova, N. S. (2009). Space educational technologies. Moscow: Kaluga. (Rus)

Simakin V.V., Tyuhov I.I., Tikhonov A.V. & Smirnov A.V. (2011). A mathematical model to determine the optimal configuration of the combined system of autonomous power supply based on renewable energy sources. *Mechanization and Electrification of Agriculture: A theoretical and scientific journal*. (in Russian).

Smil, V. (2004). World History and Energy. In Encyclopedia of Energy (vol. 6). Elsevier Inc. doi:10.1016/B0-12-176480-X/00025-5

Smil, V. (2008). *Energy in nature and society. General energetics of complex systems*. London: The MIT Press Cambridge.

Smil, V. (2010). *Energy transitions: history, requirements, prospects*. ABC-CLIO, LLC.

Sreekanth, S., & Raglend, I. J. (2012, March). *A comparative and analytical study of various incremental algorithms applied in solar cell*. In International Conference on Computing, Electronics and Electrical Technologies.

Strebkov, D. (n.d.). *Trends in Russian agriculture and rural energy*. International Commission of Agricultural Engineering. Retrieved from http://ecommons.library.cornell.edu/bitstream/1813/10217/1/Russia.PDF

Strebkov, D. S., & Tyukhov, I. I. (2005). Russian Section of the International Solar Energy Society. In The Fifty-Year History of the International Solar Energy Society and its National sections. American Solar Energy Society, Inc.

Sundareswaran, K., Peddapati, S., & Palani, S. (2014). MPPT of PV systems under partial shaded conditions through a colony of flashing fireflies. *IEEE Transactions on Energy Conversion*, *29*(2), 463–472. doi:10.1109/TEC.2014.2298237

Sundareswaran, K., Vignesh kumar, V., & Palani, S. (2015). Application of a combined particle swarm optimization and perturb and observe method for MPPT in PV systems under partial shading conditions. *Renewable Energy*, *75*, 308–317. doi:10.1016/j.renene.2014.09.044

Tajuddin, M. F. N., Arif, M. S., Ayob, S. M., & Salam, Z. (2015). Perturbative methods for maximum power point tracking (MPPT) of photovoltaic (PV) systems: A review. *International Journal of Energy Research*, *39*(9), 1153–1178. doi:10.1002/er.3289

Tey, K. S., Mekhilef, S., Yang, H., & Chuang, M. (2014). A Differential Evolution Based MPPT Method for Photovoltaic Modules under Partial Shading Conditions. *International Journal of Photoenergy*, *2014*, 1–10. doi:10.1155/2014/945906

Trillo-Montero, D., Santiago, I., Luna-Rodriguez, J. J., & Real-Calvo, R. (2014). Development of a software application to evaluate the performance and energy losses of grid-connected photovoltaic systems. *Energy Conversion and Management*, *81*, 144–159. doi:10.1016/j.enconman.2014.02.026

Tyukhov, I., & Mazanov, S. (2009). Opportunities of using satellite information for monitoring solar radiation. *VIESH*, *1*(4), 49–52.

Tyukhov, I., Schakhramanyan, M., Strebkov, D., Mazanov, S., & Vignola, F. (2008). Combined solar PV and earth space monitoring technology for educational and research purposes. *Proceedings of the solar 2008 conference*.

Tyukhov, I., Schakhramanyan, M., Strebkov, D., Simakin, V., & Poulek, V. (2008). PV and GIS Lab for teaching solar energy. *Proceedings of the 23rd European Photovoltaic Solar Energy Conference*.

Tyukhov, I., Schakhramanyana, M., Strebkov, D., Tikhonova, A., & Vignola, F. (2009). Modeling of solar irradiance using satellite images and direct terrestrial measurements with PV modules. *Proc. of SPIE (The International Society for Optical Engineering), Optical Modeling and Measurements for Solar Energy Systems III*. doi:10.1117/12.827821

Tyukhov, I., & Strebkov, D. (2005). Russian Section of the International Solar Energy Society. In The Fifty-Year History of the International Solar Energy Society and its National sections. American Solar Energy Society, Inc.

Tyukhov, I. I. (2012). On development PV technologies directions taking account resource limitations. *Proceedings of Moscow State University of Mechanical Engineering*, *14*(4), 42–46.

Tyukhov, I. I., Raupov, A. H., & Tilov, A. Z. (2012). Global-local monitoring of climate-weather conditions for renewable energy. *Proceedings of the NASA Science Meeting, GOFC-GOLD and NEESPI Workshop and Regional Conference*. Retrieved from http://csfm.marstu.net/publications.html

Veerachary, M., Senjyu, T., & Uezato, K. (2003). Neural-network-based maximum-power-point tracking of coupled-inductor interleaved-boost converter-supplied PV system using fuzzy controller. *IEEE Transactions on Industrial Electronics*, *50*(4), 749–758. doi:10.1109/TIE.2003.814762

Vignola, F., Michalsky, J., & Stoffel, T. (2012). *Solar and Infrared Radiation Measurements (Energy and the Environment)*. CRC Press. doi:10.1201/b12367

Xiao, W., & Dunford, W. G. (2004). *A modified adaptive hill climbing MPPT method for photovoltaic power systems*. In *35th Annual IEEE Power Electronics Specialist Conference*.

Yang, X.-S., & Deb, S. (2009). Cuckoo search via Lévy flights. *Proceedings of World Congress on Nature & Biologically Inspired Computing (NaBIC 2009)*. doi:10.1109/NABIC.2009.5393690

Yang, X.-S., & Deb, S. (2013). Multiobjective cuckoo search for design optimization. *Computers & Operations Research, 40*(6), 1616–1624. doi:10.1016/j.cor.2011.09.026

Zhou, W., Yang, H., & Fang, Z. (2007). A novel model for photovoltaic array performance prediction. *Applied Energy, 84*(12), 1187–1198. doi:10.1016/j.apenergy.2007.04.006

Compilation of References

Abd-Elazim, S. M., & Ali, E. S. (2012). Bacteria foraging optimization algorithm based SVC damping controller design for power system stability enhancement. *International Journal of Electrical Power & Energy Systems*, *43*(1), 933–940. doi:10.1016/j.ijepes.2012.06.048

Abido, M. A. (2002a). Optimal power flow using tabu search algorithm. *Electric Power Components Systems*, *30*(5), 469–483. doi:10.1080/15325000252888425

Abido, M. A. (2002b). Optimal power flow using particle swarm optimization. *International Journal of Electrical Power & Energy Systems*, *24*(7), 563–571. doi:10.1016/S0142-0615(01)00067-9

Abou El Elaa, A. A., & Abido, M. A. (1992). Optimal Operation Strategy for Reactive Power Control Modeling. Simul. Control. *Part A*, *41*(3), 19–40.

Abouobaida, H., & Cherkaoui, M. (2012, May). Comparative study of maximum power point trackers for fast changing environmental conditions. In *Multimedia Computing and Systems (ICMCS)*, International Conference. doi:10.1109/ICMCS.2012.6320222

Abraham, D. R., Mouton, H. D. T., & Akagi, H. (2002). Digital Control of an integrated series active filter and passive rectifier with voltage regulation.*Proceedings of the Power Conversion Conference*, (pp. 68-73).

Abraham, D. R., Mouton, H. D. T., & Akagi, H. (2003). Digital Control of an Integrated Series Active Filter and Diode Rectifier with Voltage regulation. *IEEE Transactions on Industry Applications*, *39*(6), 1–14.

Acha, E., & Ambriz-Perez, H. (2000). A thyristor controlled series compensator model for the power flow solution of practical power networks. *IEEE Transactions on Power Systems*, *15*(1), 58–64. doi:10.1109/59.852101

Adan, I., Gandhi, A., & Harchol-Balter, M. (2013). Server farms with setup costs. Performance Evaluation ScienceDirect. Elsevier.

Ademovic, A., Bisanovic, S., & Hajro, M. (2010). A genetic algorithm solution to the unit commitment problem based on real-coded chromosomes and fuzzy optimization. In *Proc. The 15th IEEE Mediterranean Electrotechnical Conference (MELECON 2010)*. doi:10.1109/MELCON.2010.5476238

Ahmadi, H., Ghasemi, H., Haddadi, A. M., & Lesani, H. (2013). Two approaches to transient stability-constrained optimal power flow. *Electrical Power and Energy Systems*, *47*, 181–192. doi:10.1016/j.ijepes.2012.11.004

Ahmed, J., & Salam, Z. (2014). A Maximum Power Point Tracking (MPPT) for PV system using Cuckoo Search with partial shading capability. *Applied Energy*, *119*, 118–130. doi:10.1016/j.apenergy.2013.12.062

Aho, A. V., Hopcroft, J. E., & Ullman, J. D. (1983). *Data Structures and Algorithms* (1st ed.). Addison Wesley.

Compilation of References

Akagi, H., Kanazawa, Y., & Nabae, A. (1983). Generalized Theory of the Instantaneous Reactive Power in Three-Phase Circuits. *Proc. IPEC Tokyo 93 Int. Conf. Power Electronics*, (pp. 1375-1386).

Akagi, H., Kanazawa, Y., & Nabae, A. (1984). Instantaneous Reactive Power Compensator Comprising Switching Devices without Energy Storage Components. *IEEE Transactions on Industry Applications, IA-20*(3), 625–630. doi:10.1109/TIA.1984.4504460

Al-Diab, A., & Sourkounis, C. (2010). Variable step size P&O MPPT algorithm for PV systems. In *Optimization of Electrical and Electronic Equipment (OPTIM)*. 12th IEEE International Conference. doi:10.1109/OPTIM.2010.5510441

Alekseev, A. V., Novitsky, N. N., Tokarev, V. V., & Shalaginova, Z. I. (2007). *Principles of development and software support of the information-computing environment for computer simulation technique of pipeline and hydraulic systems. In Truboprovodnye sistemy energetic: Metody matematicheskogo modelirovania i optimizatsii* (pp. 221–229). Novosibirsk: Nauka. (in Russian)

Alexanov, A. P. (1995). Distribution of fuel costs of energy supplied by CPs. *Energetik, 1*, 7–8.

Ali J., Fieldhouse J. & Talbot Ch. (2011). Numerical modeling of three-dimensional thermal surface discharges. *Engineering application of computational fluid mechanics, 5* (2), 201-209.

Allaoua, B., & Laoufi, A. (2009). Optimal power flow solution using ant manners for electrical network. *Advances in Electrical and Computer Engineering, 9*(1), 34–40. doi:10.4316/aece.2009.01006

Alsac, O., & Stot, B. (1974). Optimal power flow with steady-state security. *IEEE Transactions on Power Apparatus and Systems, 93*(3), 745–754. doi:10.1109/TPAS.1974.293972

Altasa, I. H., & Sharaf, A. M. (2008). A novel maximum power fuzzy logic controller for photovoltaic solar energy systems. *Renewable Energy, 33*(3), 388–399. doi:10.1016/j.renene.2007.03.002

Ambriz-Perez, H., Acha, E., & Fuerte-Esquivel, C. R. (2006). TCSC-firing angle model for optimal power flow solutions using Newton's method. *International Journal of Electrical Power & Energy Systems, 28*(2), 77–85. doi:10.1016/j.ijepes.2005.10.003

Amrouche, B., Belhamel, M., Guessoum, A., 2007. Artificial intelligence based P&O MPPT method for photovoltaic systems. *Revue des Energies Renouvelables ICRESD-07 Tlemcen*, 11–16.

Ancona, M. A., Bianchi, M., Branchini, L., & Melino, F. (2014). District heating network design and analysis. *Energy Procedia, 45*, 1225–1234. doi:10.1016/j.egypro.2014.01.128

Anita, J. M., Raglend, I. J., & Kothari, D. P. (2012). Solution of unit commitment problem using shuffled frog leaping algorithm. *IOSR Journal of Electrical and Electronics Engineering, 1*(4), 9-26.

Anitha, M., Subramanian, S., & Gnanadass, R. (2007). FDR PSO-Based transient stability constrained optimal power flow solution for deregulated power industry. *Electrical Power Components and Systems, 35*(11), 1619–1632. doi:10.1080/15325000701351641

Aoki, K., Fan, M., & Nishikori, A. (1988). Optimal VAR planning by approximation method for recursive mixed-integer linear programming. *IEEE Transactions on Power Systems, 3*(4), 1741–1747. doi:10.1109/59.192990

Appleyard, J. R. (1978). Optimal design of distribution networks. *The Building Services Engineer, 45*(11), 191-204.

Arakelyan, E. K., Kozhevnikov, N. N., & Kuznetsov, A. M. (2006). Tariffs for electricity and heat. *Teploenergetika, 11*, 60–64.

Aredes, M. (1991). *New Concepts of Power and its Application on Active Filters*. (M.Sc. Thesis). COPPE – Federal University of Rio de Janeiro, Brazil.

Aredes, M., Luís, F. C., Miguel, M., & Jaime, M. (2003). Control Strategies for Series and Shunt Active Filters. In Proceedings of IEEE Bologna Power Tech Conference, (pp. 1122-1126). IEEE.

Arkan, M., & Unsworth, P. J. (1999, October). Stator fault diagnosis in Induction motors using Power decomposition. In *Proceedings of theIEEE Industrial Application Conference 34th Annual Meeting*. doi:10.1109/IAS.1999.805999

Arrow, K., & Debreu, G. (1954). Existence of equilibrium for a competitive economy. *Econometrica*, 22(3), 265–290. doi:10.2307/1907353

Aruldoss, T., Victoire, A., Ebenezer, A., & Jeyakumar. (2005). A simulated annealing based hybrid optimization approach for a fuzzy modeled unit commitment problem. *International Journal of Emerging Electric Power Systems, 3*(1).

Ashish, P., Nivedita, D., & Ashok, K. M. (2007). Comparison of various MPPT algorithms for cost reduction. *IEEE Transactions on Power Electronics, 22*(2), 698–700.

Athay, T., Podmere, R., & Virmani, S. (1979). A practical method for the direct analysis of Transient Stability. *IEEE Transactions on Power Apparatus and Systems, 98*(2), 573–584. doi:10.1109/TPAS.1979.319407

Avik, B., & Chandan, C. (2011). A Shunt Active Power Filter With Enhanced Performance Using ANN-Based Predictive and Adaptive Controllers. *IEEE Transactions on Industrial Electronics, 58*(2), 421–428. doi:10.1109/TIE.2010.2070770

Ayan, K., & Kılıç, U. (2012). Artificial bee colony algorithm solution for optimal reactive power flow. *Applied Soft Computing, 12*(5), 1477–1482. doi:10.1016/j.asoc.2012.01.006

Ayan, K., Kilic, U., & Barakli, B. (2015). Chaotic artificial bee colony algorithm based of security and transient stability constrained optimal power flow. *International Journal of Electrical Power & Energy Systems, 64*, 136–147. doi:10.1016/j.ijepes.2014.07.018

Babak, K., & Abdolreza, R. (2010). Genetic Algorithm Application in Controlling Performance and Power Dissipation of Active Power Filters. *Canadian Journal on Electrical & Electronics Engineering, 1*(1), 15–19.

Balci, H. H., & Valenzuela, J. F. (2004). Scheduling electric power generators using particle swarm optimization combined with the Lagrangian relaxation method. *International Journal of Applied Mathematics and Computer Science, 14*(3), 411–421.

Baldwin, C. J., Dale, K. M., & Dittrich, R. F. (1959). A study of economic shutdown of generating units in daily dispatch. *AIEE Transaction of Power Apparatus and Systems, PAS-78*(4), 1272–1284. doi:10.1109/AIEEPAS.1959.4500539

Barakhtenko, E. A., Sokolov, D. V., & Stennikov, V. A. (2014). New Results in Developing of Algorithms Intended for Parameters Optimization of Heat Supply Systems. In *Proceeding of International Conference Mathematical and Informational Technologies, MIT-2013* (pp. 58-66). Beograd, Serbia.

Barakhtenko, E. A., Oshchepkova, T. B., Sokolov, D. V., & Stennikov, V. A. (2013). New Results in Development of Methods for Optimization of Heat Supply System Parameters and Their Software Implementation. *International Journal of Energy Optimization and Engineering, 2*(4), 80–99. doi:10.4018/ijeoe.2013100105

Barbagallo, D., Di Nitto, E., Dubois, D. J., & Mirandola, R. (2009). A Bio-Inspired Algorithm for Energy Optimization in a Self-organizing Data Center. In *Proceedings of the First international conference on Self-organizing architectures*, (pp. 127-151). Academic Press.

Bard, J. (1998). *Practical Bi-level Optimization*. Dordrecht, Netherlands: Kluwer Academic Publishers; doi:10.1007/978-1-4757-2836-1

Barreau, A., & Moret-Bailly, J. (1977). Prèsentation de deux methods d'optimization de rèseaux de transport d'eau chaude a grande distans. *Entropie, 13*(75), 21–28.

Bartlett, J. (2005). *The art of metaprogramming, Part 1: Introduction to metaprogramming*. Retrieved from http://www-128.ibm.com/developerworks/linux/library/l-metaprog1.html

Basset, M., Winterbone, D., & Pearson, R. (2001). Calculation of Steady Flow Pressure Loss Coefficients for Pipe Junctions. *Proceedings - Institution of Mechanical Engineers*, (215): 861–881.

Basu, M. (2008). Optimal power flow with FACTS devices using differential evolution. *International Journal of Electrical Power & Energy Systems, 30*(2), 150–156. doi:10.1016/j.ijepes.2007.06.011

Beckers, J.-M., & Van Ormelingen, J.-J. (1995). Thermohydrodynamical modelling of a power plant implementation in the Zeebrugge harbor. *Journal of Hydraulic Research, 33*(2), 163–180. doi:10.1080/00221689509498668

Bellman, R. (1957). *Dynamic Programming*. Princeton, NJ: Princeton University Press.

Bellman, R., & Dreyfus, S. (1965). *Applied problems of dynamic programming*. Moscow, USSR: Nauka.

Belyaev, L. S. (2009). *Problems of the electricity market*. Novosibirsk: Nauka.

Ben Salah, C., Chaabenea, M., & Ben Ammara, M. (2008). Multi-criteria fuzzy algorithm for energy management of a domestic photovoltaic panel. *Renewable Energy, 33*(5), 993–1001. doi:10.1016/j.renene.2007.05.036

Benabid, R., Boudour, M., & Abido, M. A. (2009). Optimal location and setting of SVC and TCSC devices using non-dominated sorting particle swarm optimization. *Electric Power Systems Research, 79*(12), 1668–1677. doi:10.1016/j.epsr.2009.07.004

Benghanema, M., Arabb, A. H., & Mukadamc, K. (1999). Data acquisition system for photovoltaic water pumps. *Renewable Energy, 176*(3), 385–396. doi:10.1016/S0960-1481(98)00126-8

Berrera, M., Dolara, & A., Leva, S. (June, 2009). *Experimental test of seven widely adopted MPPT algorithms*. IEEE Bucharest Power Tech Conference, Bucharest, Romania. doi:10.1109/PTC.2009.5282010

Besheer, A. H., & Adly, M. (2012). Ant colony system based PI maximum power point tracking for stand-alone photovoltaic system.*IEEE International Conference on Industrial Technology (ICIT)*. doi:10.1109/ICIT.2012.6210019

Bespalov, D. P. (2011). *Atlas oblakov (Cloud Atlas)/Feder. sluzhba po gidrometeorologii i monitoring okruzhayushchey sredy (Rosgidromet), Gl. geofiz. observatoriya im. A.I. Voyeykova; red.: L. K. Surygina. [Federal Service for Hydrometeorology and Environmental Monitoring (Roshydromet), Ch. Geofiz. Observatory. A.I. Voeikova; Ed.: LK Surygin]. Sankt-Peterburg: D'ART,2011*. Available at: http://www.dvfu.ru/meteo/book/ AtlasClouds.pdf

Bhardwaj, A., Tung, N. S., Shukla, V. K., & Kamboj, V. K. (2012). The important impacts of unit commitment constraints in power system planning. *International Journal of Emerging Trends in Engineering and Development, 5*(2), 301–306.

Bhasaputra, P., & Ongsakul, W. (2002). *Optimal power flow with multi-type of FACTS devices by hybrid TS/SA approach*. IEEE ICIT'02, Bangkok, Thailand. doi: 10.1109/ICIT.2002.1189908

Bhattacharya, A., & Chattopadhyay, P. K. (2010). Hybrid differential evolution with biogeography based optimization for solution of economic load dispatch. *IEEE Transactions on Power Systems, 25*(4), 1955–1964. doi:10.1109/TPWRS.2010.2043270

Bhattacharya, A., & Chattopadhyay, P. K. (2010). Solution of optimal reactive power flow using biogeography-based optimization. *International Journal of Electrical and Electronics Engineering., 4*(8), 580–588.

Bhattacharya, A., & Chattopadhyay, P. K. (2011). Application of biogeography based optimization to solve different optimal power flow problems, IET generation. *Transmission and Distribution, 5,* 72–80.

Bhattacharya, A., & Chattopadhyay, P. K. (2011). Application of biogeography-based optimization to solve different optimal power flows. *IET Gener. Transm. Distrib., 5*(1), 70–80. doi:10.1049/iet-gtd.2010.0237

Bhattacharya, A., & Roy, P. K. (2012). Solution of multi-objective optimal power flow using gravitational search algorithm. *IET Generation Transmission Distribution, 6*(8), 751–763. doi:10.1049/iet-gtd.2011.0593

Bhim, S., Vishal, V., & Jitendra, S. (2007). Neural Network-Based Selective Compensation of Current Quality Problems in Distribution System. *IEEE Transactions on Industrial Electronics, 54*(1), 53–60. doi:10.1109/TIE.2006.888754

Bjelogrlic, M., Calovic, M. S., Babic, B. S., & Ristanovic, P. (1990). Application of Newton's optimal power flow in voltage/reactive power control. *IEEE Transactions on Power Systems, 5*(4), 1447–1454. doi:10.1109/59.99399

Blaze, G., & Marko, C. (2013). Multi-objective powergeneration scheduling: Slovenian power system case study. *Elektrotehniski Vestnik, 80*(5), 222–228.

Bodìk, P., & Griffith, R. (2009). Statistical machine learning makes automatic control practical for internet data centers. In *Proceedings of the 2009 Conference on Hot Topics in Cloud Computing* (HotCloud'09). Academic Press.

Bose Bimal, K. (1990). An adaptive Hysteresis-band current control technique of a voltage fed PWM inverter for machine drive system. *IEEE Transactions on Industrial Electronics, 37*(5), 402–408. doi:10.1109/41.103436

Bouchekara, H. R. E. H., Abido, M. A., & Bouchekara, M. (2014). Optimal Power Flow using Teaching-Learning-Based Optimization Technique. *Electric Power Systems Research, 114,* 49–59. doi:10.1016/j.epsr.2014.03.032

Bouktir, T., Slimani, L., & Belkacemi, M. (2004). A genetic algorithm for solving the optimal power flow problem. *Leonardo Journal of Sciences, 4,* 44–58.

Bousson, K., & Velosa, C. (2015). Robust control and synchronization of chaotic systems with actuator constraints. In P. Vasant (Ed.), Handbook of Research on Artificial Intelligent and Algorithms (Vols. 1–2, pp. 1–43). Hershey, PA: IGI Global. doi:10.4018/978-1-4666-7258-1.ch001

Boutasseta, N. (2012). PSO-PI based Control of Photovoltaic Arrays. *International Journal of Computers and Applications, 48*(17), 36–40. doi:10.5120/7444-0557

Bouyssou, D. (1989). Modeling inaccurate determination, uncertainty, imprecision using multiple criteria. In A. G. Lockett & G. Islei (Eds.), *Improving Decision Making in Organizations, Lecture Notes in Economics and Mathematical Systems.* Springer-Verlag.

Brambilla, M., Cabot, J., & Wimmer, M. (2012). *Model Driven Software Engineering in Practice. Synthesis Lectures on Software Engineering #1.* Morgan & Claypool.

Buck, R., & Schwarzbözl, P. (2009). Solarized gas turbine power systems. *Gas Turbo Technology, 2,* 17–21.

Bull, S. R. (2001). Renewable Energy Today and Tomorrow. *Proceedings of the IEEE, 89*(8), 1216–1226. doi:10.1109/5.940290

Burchet, R. C., Happ, H. H., & Vierath, D. R. (1984). Quadratically convergent optimal power flow. *IEEE Transactions on Power Apparent Systems, PAS-103*(11), 3267–3276. doi:10.1109/TPAS.1984.318568

Compilation of References

Buslenko, N. P., Kalashnikov, V. V., & Kovalenko, I. N. (1983). *Lectures on the theory of complex systems*. Moscow: Sovetskoye radio. (in Russian)

Buslenko, N. P. (1978). *Modeling of complex systems*. Moscow: Nauka. (in Russian)

Cai, H. R., Chung, C. Y., & Wong, K. P. (2008). Application of differential evolution algorithm for transient stability constrained optimal power flow. *IEEE Transactions on Power Systems, 23*(2), 719–728. doi:10.1109/TPWRS.2008.919241

Cai, J., Ma, M., Li, L., Yang, Y., Peng, H., & Wang, X. (2007). Chaotic ant swarm optimization to economic dispatch. *Electric Power Systems Research, 77*(10), 1373–1380. doi:10.1016/j.epsr.2006.10.006

Cai, L. J., & Erlich, I. (2003). Optimal choice and allocation of FACTS devices using genetic algorithms. In *Proceedings on Twelfth Intelligent Systems Application to Power Systems Conference*, (pp. 1–6).

Cai, T. (2015). Application of Soft Computing Techniques for Renewable Energy Network Design and Optimization. In P. Vasant (Ed.), *Handbook of Research on Artificial Intelligence Techniques and Algorithms* (pp. 204–225). Hershey, PA: Information Science Reference; doi:10.4018/978-1-4666-7258-1.ch007

Caizergues, R., & Franenberg, H. (1977). Application des mèthodes d'optimisation des rèseaux de transport d'eau chaude a quelques cas concretes. *Entropie, 13*(75), 29–38.

Caponetto, R., Fortuna, L., Fazzino, S., & Xibilia, M. G. (2003). Chaotic sequences to improve the performance of evolutionary algorithms. *IEEE Transactions on Evolutionary Computation, 7*(3), 289–304. doi:10.1109/TEVC.2003.810069

Castillo-Villar, K. K., & Smith, N. R. (2014). Supply Chain Design Including Quality Considerations: Modeling and Solution Approaches based on Metaheuristics. In P. Vasant (Ed.), *Handbook of Research on Novel Soft Computing Intelligent Algorithms: Theory and Practical Applications* (pp. 102–140). Hershey, PA: Information Science Reference. doi:10.4018/978-1-4666-4450-2.ch004

Cebi, S., Kahraman, C., & Kaya, I. (2012). Soft Computing and Computational Intelligent Techniques in the Evaluation of Emerging Energy Technologies. In P. Vasant, N. Barsoum, & J. Webb (Eds.), Innovation in Power, Control, and Optimization: Emerging Energy Technologies, (pp. 164-197). doi:10.4018/978-1-61350-138-2.ch005

Cerpinsek, M., Liu, S., & Mernik, L. (2012). A note on teaching learning based optimization algorithm. *Information science, 212*, 79–93. doi:10.1016/j.ins.2012.05.009

Chaijarurnudomrung, K., Areerak, K.-N., Areerak, K.-L., & Srikaew, A. (2011). The Controller Design of Three-Phase Controlled Rectifier Using an Adaptive Tabu Search Algorithm. In Proceedings of 8th Electrical Engineering/ Electronics. Computer. Telecommunications and Information Technology (ECTI).

Chalermchaiarbha, S., & Ongsakul, W. (2013). Stochastic Weight Trade-Off Particle Swarm Optimization for Nonconvex Economic Dispatch. *Energy Conversion and Management, 70*, 66–75. doi:10.1016/j.enconman.2013.02.009

Chandrasekaran, K., & Simon, S. P. (2012). Multiobjective unit commitment problem using Cuckoo search Lagrangian method. *International Journal of Engineering Science and Technology, 4*(2), 89–105.

Chan, K. Y., Ling, S. H., Chan, K. W., Iu, H. H. C., & Pong, G. T. Y. (2007), Solving multi-contingency transient stability constrained optimal power flow problems with an improved GA. *IEEE Congress on Evolutionary Computation* (pp. 2901–2908). doi:10.1109/CEC.2007.4424840

Chan, R., Lee, Y., Sudhoff, S., & Zivi, E. (2008). Evolutionary optimization of power electronics based power systems. *IEEE Transactions on Power Electronics, 23*(4), 1908–1917. doi:10.1109/TPEL.2008.925197

Charnes, A., & Cooper, W. (1962). Programming with linear fractional functions. *Naval Research Logistics Quarterly*, *9*(3-4), 181–186. doi:10.1002/nav.3800090303

Chatterjee, A., Ghosal, S. P., & Mukherjee, V. (2012). Solution of combined economic and emission dispatch problems of power systems by an opposition based harmony search algorithm. *International Journal of Electrical Power & Energy Systems*, *39*(1), 9–20. doi:10.1016/j.ijepes.2011.12.004

Chaturvedi, K. T., Pandit, M., & Srivastava, L. (2008). Self-organizing hierarchical particle swarm optimization for non-convex economic dispatch. *IEEE Transactions on Power Systems*, *23*(3), 1079–1087. doi:10.1109/TPWRS.2008.926455

Chebbo, A. M., & Irving, M. R. (1995). Combined active and reactive power dispatch. Part I, Problem formulation and solution algorithm. *IEE Proceedings. Generation, Transmission and Distribution*, *142*(4), 393–400. doi:10.1049/ip-gtd:19951976

Chen, G., He, W., & Liu, J. (2008). Energy-aware server provisioning and load dispatching for connection intensive internet services. In *Proceedings of the 5th USENIX Symposium on Networked Systems Design and Implementation* (NSDI'08), (pp. 337–350). USENIX.

Chen, C. H., & Yeh, S. N. (2006). Particle swarm optimization for economic power dispatch with valve-point effects. In *Proceedings of IEEE PES Transmission and Distribution Conference and Exposition Latin America*. IEEE. doi:10.1109/TDCLA.2006.311397

Cheng, R. T., & Casulli, V. (2004). Modeling a three-dimensional river plume over continental shelf using a 3D unstructured grid model. *The Proceedings of the 8th Inter.Conf. on Estuarine and Coastal Modeling*.

Chen, K., Tian, S., Cheng, Y., & Bai, L. (2014). An Improved MPPT Controller for Photovoltaic System under Partial Shading Condition. *IEEE Transaction on Sustainable Energy*, *5*(3), 978–985. doi:10.1109/TSTE.2014.2315653

Chen, L., Tada, Y., Okamoto, H., Tanabe, R., & Ono, A. (2001). Optimal operation solutions of power systems with transient stability constraints, IEEE Transactions On Circuits and Systems—I: Fundamental Theory. *Appl.*, *48*, 327–339.

Chen, Y.-T., Lai, Z.-H., & Liang, R.-H. (2014). A novel auto-scaling variable step-size MPPT method for a PV system. *Solar Energy*, *102*, 247–256.

Christy, J. S., Raj, M., & Jeyakumar, A. E. (2014). A two stage successive estimation based maximum power point tracking technique for photovoltaic modules. *Solar Energy*, *103*, 43–61. doi:10.1016/j.solener.2014.01.042

Chubarova, N. Y. (2008). UV variability in Moscow according to long-term UV measurements and reconstruction model. *Atmospheric Chemistry and Physics Discussion*, *8*(8), 893–906. doi:10.5194/acpd-8-893-2008

Chung, T. S., & Yun, G. S. (1998). Optimal power flow incorporating FACTS devices and power flow control constraints. *Power System Technology, Proceedings. POWERCON '98. 1998 International Conference on*, (pp. 415 – 419). doi:10.1109/ICPST.1998.728997

Chusanapiputt, S., Nualhong, D., Jantarang, S., & Phoomvuthisarn, S. (2005). *Relative velocity updating in parallel particle swarm optimization based Lagrangian relaxation for large-scale unit commitment problem*. IEEE Region TENCON; doi:10.1109/TENCON.2005.300991

Claudio, M., & Rocco, S. (2003). A rule induction approach to improve Monte Carlo system reliability assessment. *Reliability Engineering & System Safety*, *82*(1), 85–92. doi:10.1016/S0951-8320(03)00137-6

Coby, S. (2011). Natural resource limitations to terawatt-scale solar cells. *Solar Energy Materials and Solar Cells*, *95*(12), 3176–3180. doi:10.1016/j.solmat.2011.06.013

Compilation of References

Conte, A. (2012). *Power consumption of base stations*. Alcatel-Lucent Bell Labs France.

Conway, R. W., Maxwell, W. L., & Miller, L. W. (1967). *Theory of scheduling*. Addison-Wesley.

Corporation, I. (2004). *Serial ATA Staggered Spin-Up*. White Paper.

Courant, R., & Robbins, H. (1941). *What is Mathematics?* Oxford University Press.

Cross, H. (1936). *Analysis of flow in network of conduits or conductors*. Urbana, IL: Eng. Exg. Station of University of Illinois.

Cross, H. (1936). *Analysis of flow in networks of conduits or conductors (Bulletin No 286)*. Urbana, IL: University of Illinois.

Cunha, M., & Sousa, J. (1999). Water distribution network design optimization: Simulated annealing approach. *J. Water Res. Plan. Manage. Div. Soc. Civ. Eng.*, *125*(4), 215–221. doi:10.1061/(ASCE)0733-9496(1999)125:4(215)

Dahidah, M., & Agelidis, V. (2008). Selective harmonic elimination PWM control for cascaded multilevel voltage source converters: A generalized formula. *IEEE Transactions on Power Electronics*, *23*(4), 1620–1630. doi:10.1109/TPEL.2008.925179

Dai, C., Chen, W., Zhu, Y., & Zhang, X. (2009). Reactive power dispatch considering voltage stability with seeker optimization algorithm. *Electric Power Systems Research*, *79*(10), 1462–1471. doi:10.1016/j.epsr.2009.04.020

Dai, C., Chen, W., Zhu, Y., & Zhang, X. (2009). Seeker optimization algorithm for optimal reactive power dispatch. *IEEE Transactions on Power Systems*, *24*(3), 1218–1231. doi:10.1109/TPWRS.2009.2021226

Daño, E. C., & Samonte, E. D. (2005). Public sector intervention in the rice industry in Malaysia. *State intervention in the rice sector in selected countries: Implications for the Philippines*.

Dantzig, G. B. (1955). Linear programming under uncertainty. *Management Science*, *1*(3-4), 197–206. doi:10.1287/mnsc.1.3-4.197

Dantzig, G. B. (1963). *Linear Programming and Extensions*. Princeton, NJ: Princeton University press.

Daraban, S., Petreus, D., & Morel, C. (2013). *A novel global MPPT based on genetic algorithms for photovoltaic systems under the influence of partial shading*. IECON 2013-39th Annual Conference of the IEEE. doi:10.1109/IECON.2013.6699353

Dasgupta, D., & McGregor, D. R. (1993). Short term unit-commitment using genetic algorithms. In *Proceedings of the Fifth International Conference Proceedings, on Tools with Artificial Intelligence* (pp. 240-247).

Datta, D. (2015). Implementation of Fuzzy Random Variable in Imprecise Modeling of Contaminant Migration through a Soil Layer. *International Journal of Scientific & Engineering Research*, *6*(3), 271–281.

Davidson, P. L. (1934). *Methodology for basic calculations on heat networks*. Moscow: Transzheldorizdat. (in Russian)

Dean, J. (2009). *Designs, Lessons and Advice from Building Large Distributed Systems*. Google Fellow.

Deb, N., & Shahidehpour, S. M. (1990). Linear reactive power optimization in large power network using the decomposition approach. *IEEE Transactions on Power Systems*, *5*(2), 393–400.

Deepak, S., & Mariesa, L. C. (2014). An Integrated Active Power Filter–Ultracapacitor Design to Provide Intermittency Smoothing and Reactive Power Support to the Distribution Grid. *IEEE Transactions on Sustainable Energy*, *5*(4), 1116–1125. doi:10.1109/TSTE.2014.2331355

Dehini, R., Bassou, A., & Ferdi, B. (2009). Artificial Neural Networks Application to Improve Shunt Active Power Filter. *International Journal of Computer and Information Engineering, 3*(4), 247–254.

DelCasale, M., Femia, N., Lamberti, P., & Mainardi, V. (2004). Selection of optimal closed-loop controllers for dc–dc voltage regulators based on nominal and tolerance design. *IEEE Transactions on Industrial Electronics, 51*(4), 840–849. doi:10.1109/TIE.2004.831737

Dempe, S. (2002). *Foundations of Bi-level Programming*. Dordrecht, Netherlands: Kluwer Academic Publishers.

Dhillon, J. S., & Kothari, D. P. (2010). *Power system optimization* (2nd ed.). New Delhi, India: Prentice Hall India.

Dieu, V. N., & Ongsakul, W. (2012). Hopfield lagrange network for economic load dispatch. In P. Vasant, B. Barsoum, & J. Webb (Eds.), *Innovation in Power, Control, and Optimization: Emerging Energy Technologies* (pp. 57–94). Hershey, PA: IGI Global. doi:10.4018/978-1-61350-138-2.ch002

Dieu, V. N., & Schegner, P. (2012). An Improved particle swarm optimization for optimal power flow. In P. Vasant (Ed.), *Meta-Heuristics Optimization Algorithms in Engineering, Business, Economics and Finance* (pp. 1–40). Hershey, PA: IGI Global.

Dieu, V. N., Schegner, P., & Ongsakul, W. (2013). Optimization Method for Nonconvex Economic Dispatch. In Lecture Notes. *Electrical Engineering, 239*, 1–27.

Dinh, L. L., Dieu, V. N., & Vasant, P. (2013). Artificial Bee Colony Algorithm for Solving Optimal Power Flow Problem. Research Article. *TheScientificWorldJournal, 2013*, 1–9. doi:10.1155/2013/159040

Dommel, H., & Tinny, W. (1968). Optimal power flow solution. *IEEE Transactions on Power Apparatus and Systems, PAS-87*(10), 1866–1876. doi:10.1109/TPAS.1968.292150

Dorfner, J., & Hamacher, T. (2014). Large-scale district heating network optimization. *IEEE Transactions on Smart Grid, 5*(4), 1884–1891. doi:10.1109/TSG.2013.2295856

Dorigo, M., Maniezzo,, V., &Colorni, A. (1996). Ant system: Optimization by a colony of cooperating agents. *IEEE Trans. Syst., Man, Cybern. B, Cybern., 26*(1), 29–41.

Dorigo, M., & Gambardella, L. (1997). Ant colony system: A cooperative learning approach to the traveling salesman problem. *IEEE Transactions on Evolutionary Computation, 1*(1), 53–66. doi:10.1109/4235.585892

Dorigo, M., & Stützle, T. (2004). *Ant Colony Optimization*. Cambridge, MA: MIT Press. doi:10.1007/b99492

Dousoky, G. M. (2010). *On Intelligent Power Electronic Interface for Renewable Energy Systems*. VDM Verlag Publishing House Ltd.

Droege, P. (2009). 100% Renewable Energy Autonomy in Action. Earthscan.

Du, D. (2009). *Biogeography-based optimization: Synergies with evolutionary strategies, immigration refusal, and Kalman filters*. (M. S. Thesis). Department Electrical and Computer Engineering, Cleveland State University.

Dubois, D., & Prade, H. (2012). Gradualness, uncertainty and bipolarity: Making sense of fuzzy sets. *Fuzzy Sets and Systems, 192*, 3–24. doi:10.1016/j.fss.2010.11.007

Du, D., Simon, D., & Ergezer, M. (2009). Biogeography-based optimization combined with evolutionary strategy and immigration refusal. In *IEEE Proc. International Conference on Systems, Man and Cybernetics* (pp. 997- 1002). doi:10.1109/ICSMC.2009.5346055

Duman, G., & Yorukeren, S. (2012). Optimal power flow using gravitational search algorithm. *Energy Conversion and Management, 59*, 86–95. doi:10.1016/j.enconman.2012.02.024

Eckel, B. (2006). *Thinking in Java* (4th ed.). Prentice-Hall PTR.

El Sayed, M., (2013). Modeling and simulation of smart maximum power point tracker for photovoltaic system. *Minia J. Eng. Technol., 32* (1).

Ela, A. A. A. E., Abido, M. A., & Spea, S. R. (2010). Optimal power flow using differential evolution algorithm. *Electric Power Systems Research, 80*(7), 878–885. doi:10.1016/j.epsr.2009.12.018

Elamvazuthi, I., Ganesan, T., & Vasant, P. (2011). A comparative study of HNN and Hybrid HNN-PSO techniques in the optimization of distributed generation power systems.*Proceedings of the 2011 International Conference on Advanced Computer Science and Information Systems (ICACSIS'11)*.

El-Ela, A. A. A., & El-Sehiemy, R. A-A. (2007). Optimized generation costs using modified particle swarm optimization version. *WSEAS Trans. Power Systems*, 225-232.

El-Habrouk, M., & Darwish, M. K. (2002). A New Control Technique for Active Power Filters Using a Combined Genetic Algorithm/Conventional Analysis. *IEEE Transactions on Industrial Electronics, 49*(1), 58–66. doi:10.1109/41.982249

El-Hawary, M. E. (Ed.). (1996). *Optimal Power Flow: Solution Techniques, Requirement and Challenges, IEEE Tutorial Course*. Piscataway, NJ: IEEE Service Center.

Eltamaly, A. M. (2010, February). *Modeling of fuzzy logic controller for photovoltaic maximum power point tracker*. Solar Future 2010 Conf. Proc., Istanbul, Turkey.

Eltamaly, A. M., Alolah, A. I., & Abdulghany, M. Y. (2010). *Digital implementation of general purpose fuzzy logic controller for photovoltaic maximum power point tracker*. In Power Electronics Electrical Drives Automation and Motion (SPEEDAM), 2010 International Symposium on Digital Object Identifier. doi:10.1109/SPEEDAM.2010.5542207

Eltamaly, A. M., Alolah, A. L., & Hamouda, R. M. (2006). MATLAB simulation of three-phase SCR controller for three-phase induction motor. *Mansoura Engineering Journal, 31*(3), 213–234.

Emad, M. A., & Masahito, S. (2010, November). Modified adaptive variable step-size MPPT based-on single current sensor. In TENCON 2010 – IEEE Region 10 Conference.

Emad, M. A., & Masahito, S. (2011, June). *Stability study of variable step size incremental conductance/impedance MPPT for PV systems*. In Power Electronics and ECCE Asia (ICPE & ECCE), 2011 IEEE 8th International Conference.

Emerson Network Power. (2012). *Energy Logic: Reducing Data Center Energy Consumption by Creating Savings that Cascade Across Systems*. A White Paper from the Experts in Business-Critical Continuity.

Energy in Russia. (n.d.). Available at http://en.wikipedia.org/wiki/Energy_in_Russia#Solar_energy

Esram, T., & Chapman, P. L. (2007). Comparison of photovoltaic array maximum power point tracking techniques. *IEEE Transactions on Energy Conversion, 22*(2), 439–449. doi:10.1109/TEC.2006.874230

European Photovoltaic Industry Association (EPIA). (n.d.). Retrieved from http://www.epia.org/

Fahmi, Z., Samah, B. A., & Abdullah, H. (2013). Paddy Industry and Paddy Farmers Well-being: A Success Recipe for Agriculture Industry in Malaysia. *Asian Social Science, 9*(3), 177. doi:10.5539/ass.v9n3p177

Fang, D. Z., Xiaodong, Y., Jingqiang, S., Shiqiang, Y., & Yao, Z. (2007). An optimal generation rescheduling approach for transient stability enhancement. *IEEE Transactions on Power Systems, 22*(1), 386–394. doi:10.1109/TPWRS.2006.887959

Fangrui, L., Shanxu, D., Fei, L., Bangyin, L., & Yong, K. (2008). A variable step size INC MPPT method for PV systems. *IEEE Transactions on Industrial Electronics*, *55*(7), 2622–2628. doi:10.1109/TIE.2008.920550

Faranda, R., Leva, S., & Maugeri, V. (2008, July). *MPPT techniques for PV systems: energetic and cost comparison.* Power and Energy Society General Meeting – Conversion and Delivery of Electrical Energy in the 21st Century.

Favuzza, S., Graditi, G., Ippolito, M., &Sanseverino, E. (2007). Optimal electrical distribution systems reinforcement planning using gas micro turbines by dynamic ant colony search algorithm. *IEEE Trans. Power Syst*, *22*(2), 580–587.

Fazlollahi, S. (2014). *Decomposition optimization strategy for the design and operation of district energy system.* (PhD Thesis).

Fazlollahi, S., Mandel, P., Becker, G., & Maréchal, F. (2012). Methods for multi-objective investment and operating optimization of complex energy systems. *Energy*, *45*(1), 12–22. doi:10.1016/j.energy.2012.02.046

Feron, R. (1978). Ensembles aleatoires flous. *C.R. Acad. Sci. Paris Ser. A*, *282*, 903–906.

Figueiredo, E. F. (2012). Study Committee A1 (Rotating electrical machines). SC Annual report 2011. *Electra*, *260*, 20–26.

Fletcher, C. A. (1988). Computational Techniques for Fluid Dynamics: Vol. 2. *Special Techniques for Differential Flow Categories*. Berlin: Springer-Verlag.

Ford, L. R., & Fulkerson, D. R. (1962). *Flows in Networks*. Princeton University Press.

Forero, N., Hernández, J., & Gordillo, G. (2006). Development of a monitoring system for a PV solar plant. *Energy Conversion and Management*, *47*(15-16), 2329–2336. doi:10.1016/j.enconman.2005.11.012

Formato, R. (2007). Central Force Optimization: A New Metaheuristic with Applications in Applied Electromagnetics. *Progress in Electromagnetics Research*, *77*, 425–491. doi:10.2528/PIER07082403

Fouad, A. A., & Vittal, V. (1992). *Power System Transient Stability Analysis Using the Transient Energy Function Method.* Englewood Cliffs, NJ: Prentice-Hall.

Fryze, S., Wirk, & Blind, U. (1932). Scheinleistung in elektrischen Strom kainsenmitnicht-sinusfömigenVerlauf von Strom und Spannung. ETZArch. *Elektrotech*, *53*, 596–599.

Fujita, H. T., Yamasaki, T., & Akagi, H. (2000). A Hybrid Active Filter for Damping of Harmonic Resonance in Industrial Power Systems. *IEEE Transactions on Power Electronics*, *15*(2), 2–22. doi:10.1109/63.838093

Gaby, P., Uriel, I., Inessa, A., & Gad, R. (2010). GA for the Resource Sharing and Scheduling Problem. *Global Journal Technology and Optimization*, *1*, 1985–1994.

Gamma, E., Helm, R., Johnson, R., & Vlissides, J. (1995). *Design Patterns: Elements of Reusable Object-Oriented Software*. Addison-Wesley.

Gan, D., Thomas, R. J., & Zimmerman, R. D. (2000). Stability constrained optimal power flow. *IEEE Transactions on Power Systems*, *15*(2), 535–540. doi:10.1109/59.867137

Gandhi, A., Gupta, V., Harchol-Balter, M., & Kozuch, M. A. (2010). Optimality Analysis of Energy-Performance Trade-off for Server Farm Management. Performance Evaluation, ScienceDirect. Elsevier.

Garbai, L., & Molnar, L. (1974). Optimization of urban public utility networks by discrete dynamic programming. Colloquia mathematical society János Bolyai. 12: Progress in operations research (pp. 373-390). Eger, Hungary.

Garbai, L., & Molnar, L. (1974). Optimization of urban public utility networks by discrete dynamic programming. In Colloquia mathematical society János Bolyai. 12: Progress in operations research (pp. 373-390). Eger, Hungary.

Compilation of References

Gavrilova, T. A., & Khoroshevsky, V. F. (2000). *Knowledge bases of intelligent systems*. Saint Petersburg: Piter. (in Russian)

Geng, G., & Jiang, Q. (2012). A Two-Level Parallel Decomposition Approach for Transient Stability Constrained Optimal Power Flow. *IEEE Transactions on Power Systems*, *27*(4), 2063–2073. doi:10.1109/TPWRS.2012.2190111

George, A., & Liu, J. (1981). *Computer Solution of Large Sparse Positive Definite Systems*. Englewood Cliffs, NJ: Prentice Hall.

Ghasemi, A., Valipour, K., & Tohidi, A. (2014). Multi objective optimal reactive power dispatch using a new multi objective strategy. *International Journal of Electrical Power & Energy Systems*, *57*, 318–334. doi:10.1016/j.ijepes.2013.11.049

Ghasemi, M., Ghanbarian, M. M., Ghavidel, S., Rahmani, S., & Moghaddam, E. M. (2014). Modified teaching learning algorithm and double differential evolution algorithm for optimal reactive power dispatch problem: A comparative study. *Information Sciences*, *278*, 231–249. doi:10.1016/j.ins.2014.03.050

Ghasemi, M., Ghavidel, S., Ghanbarian, M. M., & Habibi, A. (2014). A new hybrid algorithm for optimal reactive power dispatch problem with discrete and continuous control variables. *Applied Soft Computing*, *22*, 126–140. doi:10.1016/j.asoc.2014.05.006

Ghasemi, M., Taghizadeh, M., Ghavidel, S., Aghaei, J., & Abbasian, A. (2015). Solving optimal reactive power dispatch problem using a novel teaching–learning-based optimization algorithm. *Engineering Applications of Artificial Intelligence*, *39*, 100–108. doi:10.1016/j.engappai.2014.12.001

GIS for Renewable Energy. (2010). Retrieved from http://www.gisday.com/resources/ebooks/renewable-energy.pdf

Gitelman, L. D., & Ratnikov, B. E. (2008). Energy business: textbook (3rd ed.). Moscow: Publishing House "Delo" ANKh.

Gitel'man, L., Ratnikov, B. (2006). *Energy business*. Moskow: Delo.

Glover, F. (1994). Tabu search for nonlinear and parametric optimization (with links to genetic algorithms). *Discrete Applied Mathematics*, *49*(1-3), 231–255. doi:10.1016/0166-218X(94)90211-9

Gnedenko, B. V., Belyaev, Yu. K., & Soloviev, A. D. (1965). *Mathematical methods in reliability theory*. Moscow: Nauka. (in Russian)

Gokmen, N., Karatepe, E., Ugranli, F., & Silvestre, S. (2013). Voltage band based global MPPT controller for photovoltaic systems. *Solar Energy*, *98*, 322–334. doi:10.1016/j.solener.2013.09.025

Goldberg, D. (1989). *Genetic Algorithms in Search, Optimization and Machine Learning*. Addison-Wesley.

Gomez, J., Oliveira, P., Yusta, J., Villasana, R., & Urdaneta, A. (2004). Ant colony system algorithm for the planning of primary distribution circuits. *IEEE Transactions on Power Systems*, *19*(2), 996–1004. doi:10.1109/TPWRS.2004.825867

Goswami, D. Y., & Kreith, F. (2000). *Principles of Solar Engineering* (2nd ed.). Taylor & Francis.

Gounden, N. A., Ann Peter, S., Nallandula, H., & Krithiga, S. (2009). Fuzzy logic controller with MPPT using line-commutated inverter for three-phase grid-connected photovoltaic systems. *Renew. Energy J.*, *34*(3), 909–915. doi:10.1016/j.renene.2008.05.039

Granada, M., Marcos, B., Rider, J., Mantovani, J. R. S., & Shahidehpour, M. (2012). A decentralized approach for optimal reactive power dispatch using Lagrangian decomposition method. *Electric Power Systems Research*, *89*, 148–156. doi:10.1016/j.epsr.2012.02.015

Granville, S. (1994). Optimal reactive dispatch through interior point methods. *IEEE Transactions on Power Systems*, *9*(1), 136–146. doi:10.1109/59.317548

Green, M. A. (2006). *Third Generation Photovoltaics*. New York: Advanced Solar Energy Conversion. Springer-Verlag Berlin Heidelberg.

Green, R. (1999). The electricity contract market in England and Wales. *The Journal of Industrial Economics*, *47*(1), 107–124. doi:10.1111/1467-6451.00092

Grunwald, D., & Morrey, C. B. III. (2000). Policies for dynamic clock scheduling. In *Proceedings of the 4th Conference on Symposium of Operating System Design and Implementation (OSDI'00)*. OSDI.

Guarino, N. (1997). *Understanding, Building, and Using Ontologies*. Retrieved April 29, 2015, from http://www.academia.edu/516503/Understanding_building_and_using_ontologies

Habiabollahzadeh, H., & Luo, S. A. (1989). Hydrothermal optimal power flow based on combined linear and nonlinear programming methodology. *IEEE Transactions on Power Apparent Systems*, *PWRS-4*(2), 530–537. doi:10.1109/59.193826

Hadidi, R., Mazhari, I., & Kazemi, A. (2007). Simulation Induction Motor with Svc. In *Proceedings of theSecond IEEE Conference on Industrial Electronics and Application*.

Hagh, M., Taghizadeh, H., &Razi, K. (2009). Harmonic minimization in multilevel inverters using modified species-based particle swarm optimization. *IEEE Trans. Power Electron*, *24*(10), 2259–2267.

Haghifam, M., & Manbachi, M. (2011). Reliability and availability modelling of combined heat and power (CHP) systems. *International Journal of Electrical Power & Energy Systems*, *33*(3), 385–393. doi:10.1016/j.ijepes.2010.08.035

Haikarainen, C., Pettersson, F., & Saxén, H. (2014). A model for structural and operational optimization of distributed energy systems. *Applied Thermal Engineering*, *70*(1), 211–218. doi:10.1016/j.applthermaleng.2014.04.049

Haiping, M., Xieyong, R., & Baogen, J. (2010). *Oppositional ant colony. optimization algorithm and its application to fault monitoring*. 2010 29th Chinese Control Conference (CCC).

Halseth, A. (1998). Market power in the Nordic electricity market. *Utilities Policy*, *7*(4), 259–268. doi:10.1016/S0957-1787(99)00003-X

Hammer, A., Heinemann, D., Lorenz, E., & Lückehe, B. (1999). Short-term forecasting of solar radiation: A statistical approach using satellite data. *Solar Energy*, *67*(1-3), 139–150. doi:10.1016/S0038-092X(00)00038-4

Hammons, T., Boyer, J. C., Conners, S. R., Davies, M., Ellis, M., Fraser, M., & Markard, J. et al. (2000). Renewable Energy Alternatives for Developed Countries. *IEEE Transactions on Energy Conversion*, *15*(4), 481–493. doi:10.1109/60.900511

Hao, Y., Fang, Z., & Yanjun, Z. (2014). A Source-Current-Detected Shunt Active Power Filter Control Scheme Based on Vector Resonant Controller. *IEEE Transactions on Industry Applications*, *50*(3), 1953–1965. doi:10.1109/TIA.2013.2289956

Harchol-Balter M. (2010). *Open Problems in Power Management of Data Centers*. IFIP WG7.3 Workshop, University of Namur, Belgium.

Harchol-Balter, M. (2013). *Performance Modeling and Design of Computer Systems*. Cambridge University Press.

Harchol-Balter, M., Gandhi, A., Gupta, V., Raghunathan, R., & Kozuch, M. A. (2012). AutoScale: Dynamic, Robust Capacity Management for Multi-Tier Data Centers. *ACM Transactions on Computer Systems*, *30*(4), 14.

Hassmann, K. (1993). Electric Power Generation. *Proceedings of the IEEE*, *81*(3), 346–354. doi:10.1109/5.241493

Hasuike, T., & Ishii, H. (2009). Robust expectation optimization model using the possibility measure for the fuzzy random programming problem, Applications of Soft Computing: From Theory to Praxis. Advances in Intelligent and Soft Computing, 58, 285 – 294.

Heindrun, C. (1968). Dimensionierung von Wärmeverteilungsnetz nach wirtschaftlichen Gesichtspunkten mit electronischen Rechenanlagen. *Energie (BRD)*, *20*(11), 356–359.

Herrera, F., Lozano, M., & Molina, D. (2005). Continuous scatter search: An analysis of the integration of some combination methods and improvement strategies. *European Journal of Operational Research*, *169*(2), 450–476. doi:10.1016/j.ejor.2004.08.009

Hirsch, M., Pardalos, P., & Resende, M. (2009). Solving systems of nonlinear equations with continuous GRASP. *Nonlinear Analysis Real World Applications*, *10*(4), 2000–2006. doi:10.1016/j.nonrwa.2008.03.006

Hirth, L. (2015). The Optimal Share of Variable Renewables: How the Variability of Wind and Solar Power affects their Welfare-optimal Deployment. *The Energy Journal (Cambridge, Mass.)*, *36*(1), 149–184. doi:10.5547/01956574.36.1.6

Hohm, D. P., & Ropp, M. E. (2000). Comparative study of maximum power point tracking algorithms using an experimental, programmable, maximum power point tracking test bed. In *Photovoltaic Specialists Conference*, Anchorage, AK. doi:10.1109/PVSC.2000.916230

Horstmann, C. S., & Cornell, G. (2004). Core Java™ 2: Vol. II. *Advanced Features* (7th ed.). Prentice Hall PTR.

Huang, C. M., & Huang, Y. C. (2012). Combined differential evolution algorithm and ant system for optimal reactive power dispatch. *Energy Procedia*, *14*, 1238–1243. doi:10.1016/j.egypro.2011.12.1082

Iazeolla, G., & Pieroni, A. (2014). *Power management of server farms*. doi:10.4028/www.scientific.net/AMM.492.453

Iazeolla, G., Pieroni, A., & Scorzini, G. (2013). *Simulation study of server farms power optimization*. RI.01.13 T.R. Software Engineering Lab., University of Roma TorVergata.

Iazeolla, G., Pieroni, A., D'Ambrogio, A., & Gianni, D. (2010). *A distributed approach to the simulation of inherently distributed systems*. Paper presented at the Spring Simulation Multiconference 2010, SpringSim'10. doi:10.1145/1878537.1878675

Iazeolla, G., Pieroni, A., D'Ambrogio, A., & Gianni, D. (2010). *A distributed approach to wireless system simulation*. Paper presented at the 6th Advanced International Conference on Telecommunications, AICT 2010. doi:10.1109/AICT.2010.66

Iazeolla, G., & Pieroni, A. (2014). Energy saving in data processing and communication system. *TheScientificWorldJournal*, *2014*, 1–11. doi:10.1155/2014/452863 PMID:25302326

Iba, K. (1994). Reactive power optimization by genetic algorithm. *IEEE Transactions on Power Systems*, *9*(2), 685–692. doi:10.1109/59.317674

Ibrahim, H. E., & Houssiny, F. F. (1999, August). *Microcomputer controlled buck regulator for maximum power point tracker for DC pumping system operates from photovoltaic system*. In *IEEE International Fuzzy Systems Conference*. doi:10.1109/FUZZY.1999.793274

IEEE. (n.d.). *IEEE 118-bus test system*. Available at: http://www.ee.washington.edu/research/pstca/pf118/pg_tca118bus.htm

Inuiguchi, M., Ichihashi, H., & Tanaka, H. (1990). Fuzzy programming: A survey of recent developments. In R. Slowinski & J. Teghem (Eds.), *Stochastic versus Fuzzy Approaches to multiobjective mathematical programming under uncertainty* (pp. 45–68). Dordrecht: Kluwer Academics. doi:10.1007/978-94-009-2111-5_4

Inuiguchi, M., & Sakawa, M. (1996). Possible and necessary efficiency in possibilistic multiobjective linear programming problems and possible efficiency test. *Fuzzy Sets and Systems*, *78*(2), 231–241. doi:10.1016/0165-0114(95)00169-7

Ioan, V. B., & Marcel, I. (2013). *Comparative analysis of the perturb-andobserve and incremental conductance MPPT methods*. In *8th International Symposium on Advanced Topics in Electrical Engineering*, Bucharest, Romania.

Ionin, A. (1978). Criteria for assessment and calculation of heat network reliability. *Vodosnabzhenie i Sanitarnaya Tekhnika, 12*, 9-10. (in Russian)

Ionin, A. A. (1989). *Reliability of heat network systems*. Moscow: Strojizdat. (in Russian)

Ippolito, L., Cortiglia, A. L., & Petrocelli, M. (2006). Optimal allocation of FACTS devices by using multi-objective optimal power flow and genetic algorithms. *International Journal of Emerging Electric Power Systems, 7*(2). doi:10.2202/1553-779X.1099

Ishaque, K., & Salam, Z. (2013). A Deterministic Particle Swarm Optimization Maximum Power Point Tracker for Photovoltaic System under Partial Shading Condition. *IEEE Transactions on Industrial Electronics, 60*(8), 3195–3205.

Ishaque, K., Salam, Z., Amjad, M., & Mekhilef, S. (2012). An improved particle swarm optimization (PSO)–based MPPT for PV with reduced steady-state oscillation. *IEEE Transactions on Power Electronics, 27*(8), 3627–3638. doi:10.1109/TPEL.2012.2185713

Issakhov A. (2014). Mathematical Modelling of Thermal Process to Aquatic Environment with Different Hydrometeorological Conditions. *The Scientific World Journal*. doi:10.1155/2014/678095

Issakhov, A. (2012a). Parallel algorithm for numerical solution of three-dimensional Poisson equation. *Proceedings of world academy of science, engineering and technolog, 64*, 692-694.

Issakhov, A. (2011). Large eddy simulation of turbulent mixing by using 3D decomposition method. *Journal of Physics: Conference Series, 318*(4), 042051. doi:10.1088/1742-6596/318/4/042051

Issakhov, A. (2012b). Development of parallel algorithm for numerical solution of three-dimensional Poisson equation. *Journal of Communication and Computer, 9*(9), 977–980.

Issakhov, A. (2012c). Mathematical modelling of the influence of thermal power plant to the aquatic environment by using parallel technologies. *AIP Conference Proceedings, 1499*, 15–18. doi:10.1063/1.4768963

Issakhov, A. (2013). Mathematical Modelling of the Influence of Thermal Power Plant on the Aquatic Environment with Different Meteorological Condition by Using Parallel Technologies. *Power, Control and Optimization. Lecture Notes in Electrical Engineering, 239*, 165–179. doi:10.1007/978-3-319-00206-4_11

Jain, J., & Singh, R. (2013). Biogeographic-based optimization algorithm for load dispatch in power system. *International Journal of Emerging Technology and Advanced Engineering, 3*(7), 549-553.

Jalilzadeh, S., & Pirhayati, Y. (2009). An improved genetic algorithm for unit commitment problem with lowest cost. In *Proc. 2009 IEEE International Conference on Intelligent Computing and Intelligent Systems (ICIS 2009)* (pp. 571-575). doi:10.1109/ICICISYS.2009.5357777

Jiang, Y.-H., & Chen, Y.-W. (2009). Neural Network Control Techniques of Hybrid Active Power Filter. *Proceedings of International Conference on Artificial Intelligence and Computational Intelligence*, (pp. 26-30). doi:10.1109/AICI.2009.296

Jian, Q., & Huang, Z. (2008). An enhanced numerical discretization method for transient stability constrained optimal power flow. *IEEE Transactions on Power Systems, 25*(4), 1790–1797. doi:10.1109/TPWRS.2010.2043451

Joksimvoic, G. M., & Penman, J. (2000). Detection of inter turn short circuits in the stator windings of operating motors. In *Proceedings of the 24th International Conference of the IEEE Industrial Electronics Society*. doi:10.1109/41.873216

Jones, D., & Tamiz, M. (2010). *Practical goal programming*. Heidelberg, Germany: Springer New York. doi:10.1007/978-1-4419-5771-9

Jovanovic, M., Turanjanin, V., Bakic, V., Pezo, M., & Vucicevic, B. (2011). Sustainability estimation of energy system options that use gas and renewable resources for domestic hot water production. *Energy*, *36*(4), 2169–2175. doi:10.1016/j.energy.2010.08.042

Joy Green. (2014). *Solar PV grows exponentially from 2007 to 2014*. Retrieved from http://www.thefuturescentre.org/signals-of-change/3089/solar-pv-grows-exponentially-2007-2014

Juang, C., Lu, C., Lo, C., & Wang, C. (2008). Ant colony optimization algorithm for fuzzy controller design and its FPGA implementation. *IEEE Transactions on Industrial Electronics*, *55*(3), 1453–1462. doi:10.1109/TIE.2007.909762

Jubril, A. M. (2012). A nonlinear weights selection in weighted sum for convex multiobjective optimization. *Facta universitatis-series. Mathematics and Informatics*, *27*(3), 357–372.

Julien, A. (1994). An extension to possibilistic linear programming. *Fuzzy Sets and Systems*, *64*(2), 195–206. doi:10.1016/0165-0114(94)90333-6

Kally, E. (1972). Computerized Planning of the Least Cost Water Distribution Network. *Water & Sewage Works*, 121–127.

Kamboj, V. K., & Bath, S. K. (2013). Single area unit commitment using dynamic programming. In *Proceeding of 4th International Conference on Emerging Trends in Engineering and Technology* (IETET- 2013) (pp. 930-936). DOI: 03.AETS.2013.3.260

Kamboj, V. K., & Bath, S. K. (2014). Mathematical formulation of scalar and multiobjective unit commitment problem considering system and physical constraints. In *Proceedings of the 2nd National Conference on Advances in Computing, Communication Network & Electrical Systems* (NCACCNES-2014) (pp. 245-250).

Karaboga, D. (2005). *An idea based on honey bee swarm for numerical optimization, Technical Report-TR06*. Erciyes University, Engineering Faculty, Computer Engineering Department.

Karaboga, D., & Basturk, B. (2007). Artificial bee colony (ABC) optimization algorithm for solving constrained optimization problems, Springer-Verlag, IFSA 2007. *LNAI*, *4529*, 789–798. doi:10.1007/978-3-540-72950-1_77

Karaboga, D., & Basturk, B. (2008). On the performance of artificial bee colony (ABC) algorithm. *Applied Soft Computing*, *8*(1), 687–697. doi:10.1016/j.asoc.2007.05.007

Karlis, A. D., Kottas, T. L., & Boutalisb, Y. S. (2007). A novel maximum power point tracking method for PV systems using fuzzy cognitive networks (FCN). *Electric Power Systems Research*, *77*(3–4), 315–327. doi:10.1016/j.epsr.2006.03.008

Kasyanov, O. V., & Slobodyan, V. A. (n.d.). *Ispolzovaniye kosmicheskikh snimkov dlya kartirovaniya selkhozugodiy*. [Using satellite imagery to map farmland]. Available at: http://www.panorama.kharkov.ua/articles/rarticle.htm?page=15

Kavousi, A., Vahidi, B., Salehi, R., Kazem, M., Farokhnia, N., & Hamid Fathi, S. (2012). Application of the Bee Algorithm for Selective Harmonic Elimination Strategy in Multilevel Inverters. *IEEE Transactions on Power Electronics*, *27*(4), 95–101. doi:10.1109/TPEL.2011.2166124

Kazmi, S., Goto, H., Ichinokura, O., & Guo, H. (2009). An improved and very efficient MPPT controller for PV systems subjected to rapidly varying atmospheric conditions and partial shading. *Power Engineering Conference (AUPEC 2009)*.

Kennedy, J., & Eberhart, R. C. (1995). Particle swarm optimization. In *Proceedings of IEEE International Conference on Neural Networks*. Perth, Australia: IEEE. doi:10.1109/ICNN.1995.488968

Khaehintung, N., Pramotung, K., Tuvirat, B., & Sirisuk, P. (2004). *RISC microcontroller built-in fuzzy logic controller of maximum power point tracking for solar-powered light-flasher applications*. In 30th Annu. Conf. IEEE Ind. Electron. Soc. doi:10.1109/IECON.2004.1432228

Khaehintung, N., Wiangtong, T., & Sirisuk, P. (2006). *FPGA implementation of MPPT using variable step-size P&O algorithm for PV applications*. In International Symposium on Digital Object Identifier, ISCIT. doi:10.1109/ISCIT.2006.340033

Khanh, D. V. K., Vasant, P., Elamvazuthi, I., & Dieu, V. N. (2014). Optimization of thermo-electric coolers using hybrid genetic algorithm and simulated annealing. *Archives of Control Sciences*, *24*(2), 155–176. doi:10.2478/acsc-2014-0010

Kharaim, A. A. (2003). How to calculate the tariffs for heat and electricity generated at CPs without separating the fuel? *Novosti Teplosnabzheniya*, *11*, 3-7.

Khasilev, V. Ya. (1957). Issues about technical and economic calculation of heat networks. *Proektirovaniye gorodskikh teplovykh setei*. M.-L.: Gosenergoizdat. (in Russian)

Khasilev, V. Ya. (1964). Elements of hydraulic circuit theory. *Izv. AN SSSR. Energetika i transport*, (1), 69-88.

Khasilev, V., & Takaishvili, M. (1972). About fundamentals of the technique for calculation and redundancy of heat networks. *Teploenergetika*, *4*, 14–19.

Khazali, A. H., & Kalantar, M. (2011). Optimal reactive power dispatch based on harmony search algorithm. *International Journal of Electrical Power Systems*, *33*(3), 684–692. doi:10.1016/j.ijepes.2010.11.018

Khokhar, B., Parmar, K. P. S., & Dahiya, S. (2012). Application of biogeography-based optimization for economic dispatch problems. *International Journal of Computers and Applications*, *47*(13), 25–30. doi:10.5120/7249-0309

Khorsandi, A., Alimardani, A., Vahidi, B., & Hosseinian, S. H. (2010). Hybrid shuffled frog leaping algorithm and Nelder–Mead simplex search for optimal reactive power dispatch. *IET Generation Transmission Distribution*, *5*(2), 249–256. doi:10.1049/iet-gtd.2010.0256

Khorsandi, A., Hosseinian, S. H., & Ghazanfari, A. (2013). Modified artificial bee colony algorithm based on fuzzy multi-objective technique for optimal power flow problem. *Electric Power Systems Research*, *95*, 206–213. doi:10.1016/j.epsr.2012.09.002

Kim, S., Yoo, G., & Song, J. (1996). A Bifunctional Utility Connected Photovoltaic System with Power Factor Correction and U.P.S. Facility. *Proceedings of the IEEE Conference on Photovoltaic Specialist*, (pp. 1363-1368).

Kinattingal, S. (2015). Enhanced Energy Output from a PV System Under Partial Shaded Conditions Through Artificial Bee Colony. *IEEE Transaction on Sustainable Energy*, *6*(1), 198–209. doi:10.1109/TSTE.2014.2363521

Kirchhoff, G. (1847). Ueber die Auflösung der Gleichungen, auf welche man bei der Untersuchung der linearen Vertheilung galvanische Ströme geführt wird. *Annalen der Physik und Chemie*, *72*(12), 497–508. doi:10.1002/andp.18471481202

Kliman, G. B., Premerlani, W. J., Koegl, R. A., & Hoewelerl, D. (1996). A new approach to on-line turn fault detection in AC motors. In *Proceedings of the IEEE Industrial Application Conference 34th Annual Meeting*. doi:10.1109/IAS.1996.557113

Kobayashi, K., Takano, I., & Sawada, Y. (2006). A study of a two stage maximum power point tracking control of a photovoltaic system under partially shaded insolation condition. *Solar Energy Materials and Solar Cells*, *90*(18-19), 2975–2988. doi:10.1016/j.solmat.2006.06.050

Komatsu, Y. (2002). Application of the Extension pq Theory to a Mains-Coupled Photovoltaic System. *Proceedings of the Power Conversion Conference (PCC)*, (pp. 816-821). doi:10.1109/PCC.2002.997625

Korolyuk, V. S., & Turbin, A. F. (1976). *Semi-Markov processes and their applications*. Kiev, Ukrainian SSR: Naukova Dumka. (in Russian)

Köse, U., & Arslan, A. (2014). Chaotic systems and their recent implementations on improving intelligent systems. In P. Vasant (Ed.), Handbook of Research on Novel Soft Computing Intelligent Algorithms: Theory and Practical Applications (Vols. 1–2, pp. 69–101). Hershey, PA: IGI Global. doi:10.4018/978-1-4666-4450-2.ch003

Koutroulis, E., & Kalaitzakis, K. (2003). Development of an integrated data-acquisition system for renewable energy sources systems monitoring. *Renewable Energy*, *28*(1), 139–152. doi:10.1016/S0960-1481(01)00197-5

Koutroulis, E., Kalaitzakis, K., & Voulgaris, N. C. (2001). Development of a microcontroller-based photovoltaic maximum power point tracking control system. *IEEE Transactions on Power Electronics*, *16*(1), 46–54. doi:10.1109/63.903988

Krost, G., Venayagamoorthy, G. K., & Grant, L. (2008). *Swarm intelligence and evolutionary approaches for reactive power and voltage control*. IEEE Swarm Intelligence Symposium.

Kruse, R., & Meyer, K. D. (1987). *Statistics with vague data*. D. Reidel Publishing Company. doi:10.1007/978-94-009-3943-1

Kulaksız, A. A., & Akkaya, R. (2012). A Genetic Algorithm Optimized ANN–Based MPPT Algorithm for a Stand-Alone PV System with Induction Motor Drive. *Solar Energy*, *86*(9), 2366–2375. doi:10.1016/j.solener.2012.05.006

Kumar, V., & Bath, S. K. (2012, March 27-28). Optimization techniques for unit commitment problem-a review. In *Proceeding of the National Conference at Maharishi Deyanand University, Rohtak* (NCACCNES-2012) (pp. 157.1-157.9).

Kumar, V., & Bath, S. K. (2013). Single area unit commitment problem by modern soft computing techniques. *International Journal of Enhanced Research in Science Technology and Engineering*, *2*(3), 1–6.

Kundra, H., Kaur, S., Verma, A., & Bedi, R. P. S. (2011). PBBO: A new approach for underground water analysis. *International Journal of Computers and Applications*, *30*(3), 37–40. doi:10.5120/3620-5055

Kundur, P. (1994). Power System Stability and Control, McGraw-Hill. Kursat, Y., Ulas, K., Solution of transient stability-constrained optimal power flow using artificial bee colony algorithm. *Turkish Journal of Electrical Engineering & Computer Sciences*, *21*, 360–372.

Kuznetsov, A. M. (2006). Comparison of results of separation of fuel consumption for electricity and for heat supplied by CP by different methods. *Energetik*, *7*, 21.

Kwakernaak, H. (1978). Fuzzy random variables – I. definitions and theorems. *Information Sciences*, *15*(1), 1–29. doi:10.1016/0020-0255(78)90019-1

Kwakernaak, H. (1979). Fuzzy random variables – II. Algorithms and examples for the discreet case. *Information Sciences*, *17*(3), 253–278. doi:10.1016/0020-0255(79)90020-3

Lai, L. L., Ma, J. T., Yokoyama, R., & Zhao, M. (1997). Improved genetic algorithms for optimal power flow under both normal and contingent operation states. *International Journal of Electrical Power Systems*, *19*(5), 287–292. doi:10.1016/S0142-0615(96)00051-8

Larson, R., Hostetler, R., & Edwards, B. H. (2008). *Essential Calculus: Early Transcendental Functions*. Boston: Houghton Mifflin Company.

Laufenberg, M. J., & Pai, M. A. (1998). A new approach to dynamic security assessment using trajectory sensitivities. *IEEE Transactions on Power Systems*, *13*(3), 953–958. doi:10.1109/59.709082

Layden, D., & Jeyasurya, B. (2004). Integrating security constraints in optimal power flow studies. *IEEE Power Engineering Society General Meeting*, *1*, 125-129.

Le, D. A., & Dieu, V. N. (2012). Optimal Reactive Power Dispatch by Pseudo-Gradient Guided Particle Swarm Optimization. In *Proceedings of IPEC,2012 Conference on Power & Energy*, (pp. 7-12). doi:10.1109/ASSCC.2012.6523230

Lee, K. Y., & Park, Y. M. (1995). Optimization method for reactive power planning by using a modified simple genetic algorithm. *IEEE Transactions on Power Systems*, *10*(4), 1843–1850. doi:10.1109/59.476049

Lee, K. Y., Park, Y. M., & Ortiz, J. L. (1984). Optimal real and reactive dispatch. *Electric Power Systems Research*, *7*(3), 201–212. doi:10.1016/0378-7796(84)90004-X

Lee, K., Park, Y., & Ortiz, J. (1985). A united approach to optimal real and reactive power dispatch. *IEEE Transactions on Power Apparatus and Systems*, *104*(5), 1147–1153. doi:10.1109/TPAS.1985.323466

Lee, Y., Wang, W., & Kuo, T. (2008). Soft computing for battery state-of-charge (BSOC) estimation in battery string systems. *IEEE Transactions on Industrial Electronics*, *55*(1), 229–239. doi:10.1109/TIE.2007.896496

Lei, M., Yaojie, S., Yandan, L., Zhifeng, B., & Liquin, T. (2011). Comprehensive Renewable Energy. In S. Jieqiong (Ed.), *A high performance MPPT control method. Materials for Renewable Energy & Environment (ICMREE),2011 International Conference*.

Leobardo, C. B., Sergio H.M., Abraham, C.S., Victor, M., & Cardenas, G. (1998). Single- Phase active power filter and harmonic compensation. Centro Nacional de investigacion y Desarrollo Technologico interior internado Palmira s/n, Cuernavaca, Mor., Mexico, Instituto Technologico y de Estodios Superiores de Monterrey, Campus Ciudad de Mexico Calle del Puente 222. *Esquina Periferico Sur, C.P. D.F. Mexico,* 184-187.

Lesieur, M., Metais, O., & Comte, P. (2005). *Large eddy simulation of turbulence*. New York: Cambridge University Press. doi:10.1017/CBO9780511755507

Levitin, A. (2006). *Introduction to the Design & Analysis of Algorithms* (2nd ed.). Addison Wesley.

Li, M., & Tiaosheng, T. (2001). A gene complementary genetic algorithm for unit commitment. In *Proc. IEEE 5th International Conference on Electrical Machines & Systems* (ICEMS-2001) (Vol. 1, pp. 648-651). IEEE.

Liang, C. H., Chung, C. Y., Wong, K. P., & Dual, X. Z. (2007). Parallel optimal reactive power flow based on cooperative co-evolutionary differential evolution and power system decomposition. *IEEE Transactions on Power Systems*, *22*(1), 249–257. doi:10.1109/TPWRS.2006.887889

Lin, P.-C., & Nureize, A. (2014). Two-Echelon logistic model based on game theory with fuzzy variable. *Recent Advances on Soft Computing and Data Mining. Advances in Intelligent Systems and Computing*, *287*, 325–334. doi:10.1007/978-3-319-07692-8_31

Lin, P.-C., Watada, J., & Wu, B. (2013). Identifying the distribution difference between two populations of fuzzy data based on a nonparametric statistical method. *IEEJ Transactions on Electronics. Information Systems*, *8*(6), 591–598.

Lin, P.-C., Watada, J., & Wu, B. (2014). A parametric assessment approach to solving facility location problems with fuzzy demands. *IEEJ Transactions on Electronics. Information Systems*, *9*(5), 484–493.

Lin, W. M., Cheng, F. S., & Tsay, M. T. (2002). An improved tabu search for economic dispatch with multiple minima. *IEEE Transactions on Power Systems*, *17*(1), 851.

Liserre, M., Dell'Aquila, A., & Blaabjerg, F. (2004). Genetic algorithm-based design of the active damping for an LCL-filter three-phase active rectifier. *IEEE Transactions on Power Electronics*, *19*(1), 76–86. doi:10.1109/TPEL.2003.820540

Lisnianski, A., Elmakias, D., & Hanoch, B. H. (2012). A multi-state Markov model for a short-term reliability analysis of a power generating unit. *Reliability Engineering & System Safety*, *98*(1), 1–6. doi:10.1016/j.ress.2011.10.008

Liu, B., & Liu, Y.-K. (2002). Expected value of fuzzy variable and fuzzy expected value models. *IEEE Transactions on Fuzzy Systems, 10*(4), 445–450. doi:10.1109/TFUZZ.2002.800692

Liu, B., & Liu, Y.-K. (2003). Fuzzy random variable: A scalar expected value operator. *Fuzzy Optimization and Decision Making, 2*(2), 143–160. doi:10.1023/A:1023447217758

Liu, Y., Huang, S.-C., Huang, J.-W., & Liang, W.-C. (2012). A Particle Swarm Optimization–Based Maximum Power Point Tracking Algorithm for PV Systems Operating Under Partially Shaded Conditions. *IEEE Transactions on Energy Conversion, 27*(4), 1027–1035. doi:10.1109/TEC.2012.2219533

Li, Y., Cao, Y., Liu, Z., Liu, Y., & Jiang, Q. (2009). Dynamic optimal reactive power dispatch based on parallel particle swarm optimization algorithm. *Computers & Mathematics with Applications (Oxford, England), 57*(11-12), 1835–1842. doi:10.1016/j.camwa.2008.10.049

Lorenz, E., & Heinemann, D. (2012). Prediction of solar irradiance and photovoltaic power. *Comprehensive Renewable Energy, 1*, 239–292. doi:10.1016/B978-0-08-087872-0.00114-1

Lowe, S. A., Schuepfer, F., & Dunning, D. J. (2009). Case study: Three-dimensional hydrodynamic model of a power plant thermal discharge. *Journal of Hydraulic Engineering, 135*(4), 247–256. doi:10.1061/(ASCE)0733-9429(2009)135:4(247)

Lu, F. C., & Hsu, Y. Y. (1995). Reactive power/voltage control in a distribution substation using dynamic programming. *IEE Proceedings. Generation, Transmission and Distribution, 142*(6), 639–645. doi:10.1049/ip-gtd:19952210

Luis, A. M., Juan, W. D., Rogel, R. W. (1995). A Three- Phase Active Power Filter Operating with Fixed Switching Frequency for Reactive Power and Current Harmonic Compensation. *IEEE Trans. on Ind. Electronics, 42*(4), 14-18.

Lui, Y., & Passino, K. (2002). Biomimicry of Social Foraging Bacteria for Distributed Optimization: Models, Principles, and Emergent Behaviors. *Journal of Optimization Theory and Applications, 115*(3), 603–628. doi:10.1023/A:1021207331209

Lund, H. (2010). *Renewable Energy Systems. The Choice and Modeling of 100% Renewable Solutions*. Elsevier Inc.

Ma, T. T. (2003). Enhancement of power transmission systems by using multiple UPFC on evolutionary programming. IEEE Bologna Power Tech Conference. doi:10.1109/PTC.2003.1304760

Mahadevan, K., & Kannan, P. S. (2010). Comprehensive learning particle swarm optimization for reactive power dispatch. *International Journal of Applied Soft Computing, 10*(2), 641–652. doi:10.1016/j.asoc.2009.08.038

Mahdad, B., Srairia, K., & Bouktir, T. (2008). Optimal power flow for large-scale power system with shunt FACTS using efficient parallel GA. *International Journal of Electrical Power & Energy Systems, 32*, 867–872. doi:10.1109/IECON.2008.4758067

Ma, J. (2013). *Low–Cost Global MPPT Scheme for Photovoltaic Systems under Partially Shaded Conditions.IEEE International Symposium on Circuits and Systems (ISCAS)*.

Ma, J. T., & Lai, L. L. (1996). Evolutionary programming approach to reactive power planning. *IEE Proceedings. Generation, Transmission and Distribution, 143*(4), 365–370. doi:10.1049/ip-gtd:19960296

Majumdar, S., Chakraborty, A. K., Chattopadhyay, P. K., & Bhattacharjee, T. (2012). Modified simulated annealing technique based optimal power flow with FACTS devices. *International Journal of Artificial Intelligence and Soft Computing, 3*(1), 81–93. doi:10.1504/IJAISC.2012.048181

Malafeev, V. A., Smirnov, I. A., Kharaim, A. A., Khrilev, L. S., & Livshits, I. M. (2003). Formation of tariffs at CP in the market environment. *Teploenergetika, 4*, 55–63.

Malesani, L., Rosseto, L., & Tenti. (1986). Active Filter for Reactive Power and Harmonics Compensation. *IEEE – PESC*, (321-330). IEEE.

Mandal, B., & Roy, P. K. (2013). Optimal reactive power dispatch using quasi-oppositional teaching learning based optimization. *International Journal of Electrical Power & Energy Systems*, *53*, 123–134. doi:10.1016/j.ijepes.2013.04.011

Mandal, B., & Roy, P. K. (2014). Multi-objective optimal power flow using quasi-oppositional teaching learning based optimization. *Applied Soft Computing*, *21*, 590–606. doi:10.1016/j.asoc.2014.04.010

Maniya, K. D., & Bhatt, M. G. (2013). A Selection of Optimal Electrical Energy Equipment Using Integrated Multi Criteria Decision Making Methodology. *International Journal of Energy Optimization and Engineering*, *2*(1), 101–116. doi:10.4018/ijeoe.2013010107

Marcu, M., Utu, I., Pana, L., & Orban, M. (2004). Computer simulation of real time identification for induction motor drives. In *Proceedings of the International Conference on Theory and Applications of Mathematics and Informatics*.

Maricar, N. (2003). Photovoltaic Solar Energy Technology Overview for Malaysia Scenario. *Proceedings of the IEEE National Conference on Power and Energy Conference (PECon)*, (pp. 300-305). doi:10.1109/PECON.2003.1437462

Marler, R. T., & Arora, J. S. (2010). Weighted sum method for multi-objective optimization: New insights. *Structural and Multidisciplinary Optimization*, *41*(6), 853–862. doi:10.1007/s00158-009-0460-7

Marouani, I., Guesmi, T., Abdallah, H. H., & Ouali, A. (2009). *Application of a multi-objective evolutionary algorithm for optimal location and parameters of FACTS devices considering the real power loss in transmission lines and voltage deviation buses*. In *6th International Multi-conference on Systems* (pp. 1–6). Djerba: Signals and Devices; doi:10.1109/SSD.2009.4956757

Marsan, M. A. (2011). Energy-Efficient Networking: the views of leaders of international research projects. CCW 2011, Hyannis.

Mathur, R. M., & Basati, R. S. (1981). A thyristor controlled static phase-shifter for AC power transmission. *IEEE Transactions on Power Apparatus and Systems*, *100*(5), 2650–2655. doi:10.1109/TPAS.1981.316780

MATPOWER. (n.d.). Retrieved from http://www.ee.washington.edu/reserach/pstca

Ma, X., El-Keib, A. A., Smith, R. E., & Ma, H. (1995). Genetic algorithm based approach to thermal unit commitment of electric power systems. *Electric Power Systems Research*, *34*(1), 29–36. doi:10.1016/0378-7796(95)00954-G

Maxwell, J. C. (1873). *A treatise of electricity and magnetism* (Vol. 1). Oxford.

Mazur, Yu. Ya. (1986). *Problems of maneuverability in energy development*. Moscow: Nauka. (in Russian)

Melentiev, L. A. (1948). *Cogeneration. Part 2*. Moscow: Publishing House of the USSR Academy of Sciences.

Melentiev, L. A. (1987). *Outlines of Russian energy sector history*. Moscow: Nauka.

Melentiev, L. A. (1993). *Selected works. Scientific fundamentals of cogeneration and energy supply of cities and industrial enterprises*. Moscow: Nauka.

Mellor, S. J., Scott, K., Uhl, A., & Weise, D. (2004). *MDA Distilled*. Addison-Wesley.

Meo, M., & Mellia, M. (2010). *Green Networking. IEEE INFOCOM Workshop on Communications and Control for Sustainable Energy Systems*, Orlando, FL.

Merenkov, A. P. (Ed.). (1992). *Mathematical modeling and optimization of heat, water, oil and gas supply systems*. Novosibirsk: VO "Nauka", Sibirskaya izdatelskaya firma. (in Russian)

Compilation of References

Merenkov, A. P., & Khasilev, V. Ya. (1985). *Theory of hydraulic circuits*. Nauka. (in Russian)

Merenkov, A., & Khasilev, V. (1985). *The theory of hydraulic circuits*. Moscow: Nauka.

Merenkova, N. N., Sennova, E. V., & Stennikov, V. A. (1982). Optimization of schemes and structure of district heating systems.[in Russian]. *Elektronnoye Modelirovaniye, 6*, 76–82.

Mirhassaniet, S. M. (2015). An improved particle swarm optimization based maximum power pointtracking strategy with variable sampling time. *Electrical Power and Energy Systems, 64*, 761–770. doi:10.1016/j.ijepes.2014.07.074

Mladenovic, N., Drazic, M., Kovacevic-Vujcic, V., & Cangalovi, M. (2008). General variable neighborhood search for the continuous optimization. *European Journal of Operational Research, 191*(3), 753–770. doi:10.1016/j.ejor.2006.12.064

Mohamed, Y., & Saadany, E. (2008). Hybrid variable-structure control with evolutionary optimum-tuning algorithm for fast grid-voltage regulation using inverter-based distributed generation. *IEEE Transactions on Power Electronics, 23*(3), 1334–1341. doi:10.1109/TPEL.2008.921106

Mohammed, Q., Parag, K., & Vinod, K. (2014). Artificial-Neural-Network-Based Phase-Locking Scheme for Active Power Filters. *IEEE Transactions on Industrial Informatics, 61*(8), 3857–3866. doi:10.1109/TIE.2013.2284132

Mollazei, S., Farsangi, M. M., Nezamabadi-pour, H., & Lee, K. Y. (2007). Multi-objective optimization of power system performance with TCSC using the MOPSO algorithm. IEEE Power Engineering Society General Meeting. doi:10.1109/PES.2007.385878

Moller, B., Graf, W., & Beer, M. (2003). Safety assessment of structures in view of fuzzy randomness. *Computers & Structures, 81*(15), 1567–1582. doi:10.1016/S0045-7949(03)00147-0

Möller, B., Graf, W., Beer, M., & Sickert, J. U. (2002). Fuzzy randomness-towards a new modeling of uncertainty.*Fifth World Congress on Computational Mechanics*, Vienna, Austria.

Momeh, J. A., Guo, S. X., Oghuobiri, E. C., & Adapa, R. (1994). The quadratic interior point method solving the power system optimization problems. *IEEE Transactions on Power Systems, 9*(3), 1327–1336. doi:10.1109/59.336133

Momoh, J., El-Hawary, M., & Adapa, R. (1999). A review of selected optimal power flow literature to 1993,Parts I and II. *IEEE Transactions on Power Systems, 14*(1), 96–111. doi:10.1109/59.744492

Mo, N., Zou, Z. Y., Chan, K. W., & Pong, T. Y. G. (2007). Transient stability constrained optimal power flow using particle swarm optimization, IET Generation. *Transmission & Distribution, 1*(3), 476–483. doi:10.1049/iet-gtd:20060273

Mondal, D., Chakrabarti, A., & Sengupta, A. (2012). Optimal placement and parameter setting of SVC and TCSC using PSO to mitigate small signal stability problem. *International Journal of Electrical Power & Energy Systems, 42*(1), 334–340. doi:10.1016/j.ijepes.2012.04.017

Moradi, M. H., & Khorasani, P. G. (2008). A new Matlab simulation of induction motor. In *Proceedings of the Australasian Universities Power Engineering Conference*.

Morgan, D., & Goulter, I. (1985). Optimal urban water distribution design. *Water Resources Research, 21*(5), 642–652. doi:10.1029/WR021i005p00642

Mota-Palomino, R., & Quintana, V. H. (1986). Sparse reactive power scheduling by a penalty function linear programming technique. *IEEE Transactions on Power Systems, 1*(3), 31–39. doi:10.1109/TPWRS.1986.4334951

Moulin, H. (1985). *Game theory with examples from mathematical economics*. Moscow: Mir.

Moulin, H. (1991). *Cooperative decision making: Axioms and models*. Moscow: Mir.

Mukaro, R., & Carelse, X. F. (1999). A microcontroller-based data acquisition system for solar radiation and environmental monitoring. *IEEE Transactions on Instrumentation and Measurement*, *48*(6), 1232–1238. doi:10.1109/19.816142

Naess, A., Leira, B., & Batsevych, O. (2009). System reliability analysis by enhanced Monte Carlo simulation. *Structural Safety*, *31*(5), 349–355. doi:10.1016/j.strusafe.2009.02.004

Nekrasov, A. S., & Velikanov, M. A. (1986). Long-term regulation of fuel consumption for heating and ventilation. Progress and prospects. *Energetika*, *46*, 85–98.

Nekrasov, A. S., Velikanov, M. A., & Gorunov, P. V. (1990). *Reliability of fuel supply to power plants: methods and models of studies*. Moscow: Nauka. (in Russian)

Nekrasova, O. A., & Khasilev, V. Ya. (1970). Optimal tree of the pipeline system.[in Russian]. *Ekonomika i Matematicheskiye Metody*, *4*(3), 427–432.

Neumann, J., & Morgenstern, O. (1970). *Theory of game and economic behavior: Tranls. from Engl.* Moscow: Nauka.

Nguyen, T. B., & Pai, M. A. (2003). Dynamic security-constrained rescheduling of power systems using trajectory sensitivities. *IEEE Transactions on Power Systems*, *18*(2), 848–854. doi:10.1109/TPWRS.2003.811002

Niraj, K., Khem, B. T., & Khalid, S. (2013). A Novel Series Active Power Filter Scheme using Adaptive Tabu search Algorithm for Harmonic Compensation. *International Journal of Application or Innovation in Engineering & Management*, *2*(2), 155–159.

Novitskyy, N. (2014). Calculation of the Flow Distribution in Hydraulic Circuits Based oт Their Linearization by Nodal Model of Secants and Chords. *Thermal Engineering*, *60*(14), 1051–1060. doi:10.1134/S004060151314005X

Nureize, A., & Watada, J. (2010). Constructing fuzzy random goal constraints for stochastic goal programming.Integrated Uncertainty Management and Applications, 68, 293–304.

Nureize, A., & Watada, J. (2014). Linear fractional programming for fuzzy random based possibilistic programming problem. *International Journal of Simulation Systems. Science & Technology*, *14*(1), 24–30.

O'Connor, E. (2009). Going Mainstream. *Modern Power Systems*, *11*, 17–18.

Oey, L. Y. (2006). An OGCM with movable land–sea boundaries. *Ocean Modelling*, *13*(2), 176–195. doi:10.1016/j.ocemod.2006.01.001

Ongsakul, W., & Jirapong, P. (2005). Optimal allocation of FACTS devices to enhance total transfer capability using evolutionary programming. In *Proceedings of IEEE Int. Symposium on Circuits and Systems*. doi:10.1109/ISCAS.2005.1465551

Ongsakul, W., & Tantimaporn, T. (2006). Optimal power flow by improved evolutionary programming. *Electric Power Components and Systems*, *34*(1), 79–95. doi:10.1080/15325000691001458

Orero, S. O., & Irving, M. R. (1996).A genetic algorithm for generator scheduling in power systems. *Electric Power Energy Systems*, *18*(1), 19–26. doi:10.1016/0142-0615(94)00017-4

Owen, G. (1971). *Game theory.* Moscow: Mir.

Ozpineci, B., & Tolbert, L. M. (2003). Simulink implementation of induction motor model – A modular approach. In *Proceedings of the International Conference on Electric Machines and Drives*. doi:10.1109/IEMDC.2003.1210317

Ozpineci, B., Tolbert, L., & Chiasson, J. (2005). Harmonic optimization of multilevel converters using genetic algorithms. *IEEE Power Electronics Letters*, *3*(3), 92–95. doi:10.1109/LPEL.2005.856713

Pablo, Y., Juan, M. R., & Carlos, R. C. (2008). Optimal power flow subject to security constraints solved with a particle swarm optimizer. *IEEE Transactions on Power Systems, 23*(1), 33–40. doi:10.1109/TPWRS.2007.913196

Padalko, L. P., & Zaborovsky, A. M. (2006). On the principles of mutually agreed distribution of energy system costs between heat and electric energy. *Energiya i Menedzhment, 1*, 8-12.

Padhy, N. P., Abdel-Moamen, M. A., & Kumar, B. J. (2004). *Optimal location and initial parameter settings of multiple TCSCs for reactive power planning using genetic algorithms*. IEEE Power Engineering Society General Meeting. doi:10.1109/PES.2004.1373013

Pai, M. A. (1989). *Energy Function Analysis for Power System Stability*. Norwell, MA: Kluwer. doi:10.1007/978-1-4613-1635-0

Parag, K., Vinod, K., & Hatem, H. Z. (2015). Optimal Control of Shunt Active Power Filter to Meet IEEE Std. 519 Current Harmonic Constraints Under Non-ideal Supply Condition. *IEEE Transactions on Industrial Electronics, 62*(2), 724–734. doi:10.1109/TIE.2014.2341559

Paraskevadaki, E. V., & Papathanassiou, S. A. (2011). Evaluation of MPP voltage and power of mc-Si PV modules in partial shading conditions. *IEEE Transactions on Energy Conversion, 26*(3), 923–932. doi:10.1109/TEC.2011.2126021

Park, J.-B., Jeong, Y.-W., & Lee, W.-N. (2006). *An improved particle swarm optimization for economic dispatch problems with non-smooth cost functions*. IEEE Power Engineering Society General Meeting.

Park, H. J., Um, J. G., Woo, I., & Kim, J. W. (2012, January). Application of fuzzy set theory to evaluate the probability of failure in rock slopes. *Engineering Geology, 125*, 92–101. doi:10.1016/j.enggeo.2011.11.008

Parpinelli, R., Lopes, H., & Freitas, A. (2002). Data mining with an ant colony optimization algorithm. *IEEE Transactions on Evolutionary Computation, 6*(4), 321–332. doi:10.1109/TEVC.2002.802452

Payne, R. B., Sorenson, M. D., & Kiltz, K. (2005). *The Cuckoos*. Oxford University Press.

Perez-Arriaga, I. J., & Batlle, C. (2012). Impacts of Intermittent Renewables on Electricity Generation System Operation. *Economics of Energy & Environmental Policy, 1*(2), 3–16. doi:10.5547/2160-5890.1.2.1

Pering, T., Burd, T., & Brodersen, R. (1998). The simulation and evaluation of dynamic voltage scaling algorithms. In *Proceedings of the International Symposium on Low Power Electronics and Design* (ISLPED'98) (pp. 76–81). doi:10.1145/280756.280790

Peyret, R., & Taylor, D. Th. (1983). *Computational Methods for Fluid Flow*. New York, Berlin: Springer-Verlag. doi:10.1007/978-3-642-85952-6

Pezzini, P., Gomis-Bellmunt, O., & Sudrià-Andreu, A. (2011). Optimization techniques to improve energy efficiency in power systems. *Renewable & Sustainable Energy Reviews, 15*(4), 2028–2041. doi:10.1016/j.rser.2011.01.009

Phetteplace, G. (1995). *Optimal design of piping systems for district heating. U.S. Army Corps of Engineers*. Cold Regions Research & Engineering Laboratory.

Pissanetzky, S. (1984). *Sparse Matrix Technology*. London: Academic Press Inc.

Pizano-Martinez, A., Fuerte-Esquivel, C. R., & Ruiz-Vega, D. (2011). A new practical approach to global transient stability-constrained optimal power flow. *IEEE Transactions on Power Systems, 26*(3), 1686–1696. doi:10.1109/TPWRS.2010.2095045

Pizano-Martinez, A., Fuerte-Esquivel, C. R., Zamlra-Cardenas, E., & Ruiz-Vega, D. (2014). Selective transient stability-constrained optimal power flow using a SMIE and trajectory sensitivity unified analysis. *Electric Power Systems Research*, *109*, 32–44. doi:10.1016/j.epsr.2013.12.003

Podkovalnikov, S. V., & Khamisov, O. V. (2012). Imperfect electricity markets: Modeling and study of the development of generating capacities, *Izvestiya RAN. Energy*, *2*, 57–76.

Polovko, A. M. (1964). *Fundamentals of reliability theory*. Moscow: Nauka. (in Russian)

Polprasert, J., Ongsakul, W., & Dieu, V. N. (2013). A new improved particle swarm optimization for solving nonconvex economic dispatch problems. *International Journal of Energy Optimization and Engineering*, *2*(1), 66–77. doi:10.4018/ijeoe.2013010105

Popyrin, L. S., Denisov, V. I., & Svetlov, K. S. (1989). On the methods for distributing CP costs. *Elektricheskie Stantsii*, *11*, 20-25.

Poulek, V., Strebkov, D. S., Persic, I. S., & Libre, M. (2012). Toward 50 years lifetime of PV panels laminated with silicone gel technology. *Solar Energy*, *86*(10), 3103–3108. doi:10.1016/j.solener.2012.07.013

Pramod, K., & Alka, M. (2009). Soft Computing Techniques for the Control of an Active Power Filter. *IEEE Transactions on Power Delivery*, *24*(1), 452–461. doi:10.1109/TPWRD.2008.2005881

Preedavichit, P., & Srivastava, S. C. (1997) Optimal reactive power dispatch considering FACTS devices. *Proceedings of the 4th International Conference on Advances in Power System Control, Operation and Management, APSCOM-97*. doi:10.1049/cp:19971906

Puri, M. L., & Ralescu, D. A. (1986). Fuzzy random variables. *Journal of Mathematical Analysis and Applications*, *114*(2), 409–422. doi:10.1016/0022-247X(86)90093-4

Puype, B., Vereecken, W., Colle, D., Pickavet, M., & Demeester, P. (2007). *Power reduction techniques in multilayer traffic engineering*. In Transparent Optical Networks, ICTON '09. 11th International Conference On.

Radha, T., Thanga, R. C., Millie, P. A., & Crina, G. (2010). Optimal gain tuning of PI speed controller in induction motor drives using particle swarm optimization. *Logic Journal of IGPL Advance Access*, *2*(2), 1-14.

Rahnamayan, S., Tizhoosh, H. R., & Salama, M. M. A. (2007). Quasi oppositional differential evolution. In *Proceeding of IEEE Congress on Evolutionary Computation CEC 2007*, (pp. 2229–2236). doi:10.1109/CEC.2007.4424748

Rahnamayan, S., Tizhoosh, H. R., & Salama, M. M. A. (2008). Opposition versus randomness in soft computing techniques. *Applied Soft Computing*, *8*(2), 906–918. doi:10.1016/j.asoc.2007.07.010

Rajan, C. C. A., Mohan, M. R., & Manivannan, K. (2003). Neural based Tabu search method for solving unit commitment problem. *IEE Proceedings. Generation, Transmission and Distribution*, *150*(4), 469–474. doi:10.1049/ip-gtd:20030244

Rajesh, R., & Mabel, M. C. (2014). Efficiency analysis of a multi-fuzzy logic controller for the determination of operating points in a PV system. *Solar Energy*, *99*, 77–87. doi:10.1016/j.solener.2013.10.036

Rakpenthai, C., Premrudeepreechacharn, S., & Uatrongjit, S. (2009). Power system with multi-type FACTS devices states estimation based on predictor–corrector interior point algorithm. *International Journal of Electrical Power & Energy Systems*, *31*(4), 160–166. doi:10.1016/j.ijepes.2008.10.010

Ramakrishna, R. S. (2008). On the scalability of real-coded bayesian optimization algorithm. *IEEE Transactions on Evolutionary Computation*, *12*(3), 307–322. doi:10.1109/TEVC.2007.902856

Compilation of References

Ramesh, S., Kannan, S., & Baskar, S. (2012). An improved generalized differential evolution algorithm for multi-objective reactive power dispatch. *Engineering Optimization, 44*(4), 391–405. doi:10.1080/0305215X.2011.576761

Rang-Mow, J., & Nanming, C. (1995). Application to the fast Newton Raphson economic dispatch and reactive power/voltage dispatch by sensitivity factors to optimal power flow. *IEEE Transactions on Energy Conversion, 10*(2), 293–301. doi:10.1109/60.391895

Rao, R., & Patel, V. (2013). An improved teaching learning based optimization algorithm for solving unconstrained optimization problems. *Scientia Iranica, 20*(3), 710-720.

Rao, R. V., Savsani, V. J., & Vakharia, D. P. (2011). Teaching–learning-based optimization: A novel method for constrained mechanical design optimization problems. *Computer Aided Design, 43*(3), 303–315. doi:10.1016/j.cad.2010.12.015

Rarick, R., Simon, D., Villaseca, F. E., & Vyakaranam, B. (2009). Biogeography-based optimization and the solution of the power flow problem. In *Proceedings of the IEEE Proc. International Conference on Systems, Man and Cybernetics* (pp. 1003-1008). doi:10.1109/ICSMC.2009.5346046

Ratnaweera, A., Halgamuge, S. K., & Watson, H. C. (2004). Self-Organizing Hierarchical Particle Swarm Optimizer with Time-Varying Acceleration Coefficients. *IEEE Transactions on Evolutionary Computation, 8*(3), 240–255. doi:10.1109/TEVC.2004.826071

Razumkhin, B. S. (1986). *Physical models and methods of equilibrium theory in programming and economics*. Moscow: Nauka.

Real Time Satellite Tracking and Prediction. (n.d.). Available at: www.n2yo.com

Reddy, K. S., & Reddy, M. D. (2012). Economic load dispatch using firefly algorithm. *International Journal of Engineering Research and Applications, 2*(4), 2325-2330.

Renewables. (2013). *Global Status Report*. Retrieved from http://www.ren21.net/Portals/0/documents/Resources/GSR/2013/GSR2013_lowres.pdf

Renzi, F. (2007). *Business E^3: Energy, Efficiency, Economy*. IBM Corporation.

Renzo de, D. J. (1982). Wind power: Recent developments. Moscow: Energoatomizdat.

Reza Reisi, A., Hassan Moradi, M., & Jamasb, S. (2013). Classification and comparison of maximum power point tracking techniques for photovoltaic system: A review. *Renewable & Sustainable Energy Reviews, 19*, 433–443. doi:10.1016/j.rser.2012.11.052

Rezk, H., & El-Sayed, A. H. M. (2013). Sizing of a stand-alone concentrated photovoltaic system in Egyptian site. *International Journal of Electrical Power & Energy Systems, 45*(1), 325–330. doi:10.1016/j.ijepes.2012.09.001

Rezk, H., & Eltamaly, A. M. (2015). A comprehensive comparison of different MPPT techniques for photovoltaic systems. *Solar Energy, 112*, 1–11. doi:10.1016/j.solener.2014.11.010

Roache, P. J. (1972). *Computational Fluid Dynamics*. Albuquerque, NM: Hermosa Publications.

Roa-Sepulveda, C. A., & Pavez-Lazo, B. J. (2003). A solution to the optimal power flow using simulated annealing. *International Journal of Electrical Power Systems, 25*(1), 47–57. doi:10.1016/S0142-0615(02)00020-0

Robineau, J., Fazlollahi, S., Fournier, J., Berthalon, A., & Verdier, I. (2014). Multi-objective optimization of the design and operating strategy of a district heating network - application to a case study. In *Proceedings from the 14th International Symposium on District Heating and Cooling* (pp. 79-87). Stockholm, Sweden.

Romero, C. (1991). *Handbook of critical issues in goal programming*. Oxford, UK: Pergamon Press.

Roy, P. K., & Bhui, S. (2013). Multi-objective quasi-oppositional teaching learning based optimization for economic emission load dispatch problem. *International Journal of Electrical Power & Energy Systems*, *53*, 937–948. doi:10.1016/j.ijepes.2013.06.015

Roy, P. K., Ghosal, S. P., & Thakur, S. S. (2010). Biogeography based optimization for multi-constraint optimal power flow with emission and non-smooth cost function. *Expert Systems with Applications*, *37*(12), 8221–8228. doi:10.1016/j.eswa.2010.05.064

Roy, P. K., Ghoshal, S. P., & Thakur, S. S. (2010). Biogeography-based Optimization for economic load dispatch problems. *Electric Power Components and Systems*, *38*(2), 166–181. doi:10.1080/15325000903273379

Roy, P. K., Ghoshal, S. P., & Thakur, S. S. (2010). Multi-objective optimal power flow using biogeography-based optimization. *Electric Power Components and Systems*, *38*(12), 1406–1426. doi:10.1080/15325001003735176

Roy, P. K., Ghoshal, S. P., & Thakur, S. S. (2010). Optimal power flow with TCSC and TCPS modeling using craziness and turbulent crazy particle swarm optimization. *International Journal of Swarm Intelligence Research*, *1*(3), 34–50. doi:10.4018/jsir.2010070103

Roy, P. K., Ghoshal, S. P., & Thakur, S. S. (2011). Optimal reactive power dispatch considering flexible AC transmission system devices using biogeography based optimization. *Electric Power Components and Systems*, *39*(8), 733–750. doi:10.1080/15325008.2010.541410

Roy, P. K., Ghoshal, S. P., & Thakur, S. S. (2012). Optimal VAR control for improvements in voltage profiles and for real power loss minimization using Biogeography Based Optimization. *International Journal of Electrical Power & Energy Systems*, *43*(1), 830–838. doi:10.1016/j.ijepes.2012.05.032

Roy, P. K., Mandal, B., & Bhattacharya, K. (2012). Gravitational search algorithm based optimal reactive power dispatch for voltage stability enhancement, Electric. Power Components and. *Systems*, *40*(9), 956–976.

Roy, P. K., Mandal, B., & Bhattacharya, K. (2012). Gravitational search algorithm based optimal reactive power dispatch for voltage stability enhancement. *Electric Power Components and Systems*, *40*(9), 956–976. doi:10.1080/15325008.2012.675405

Roy, P. K., Paul, C., & Sultana, S. (2014). Oppositional teaching learning based optimization approach for combined heat and power dispatch. *International Journal of Electrical Power & Energy Systems*, *57*, 392–403. doi:10.1016/j.ijepes.2013.12.006

Roy, P. K., Sur, A., & Pradhan, D. (2013). Optimal short-term hydro-thermal scheduling using quasi-oppositional teaching learning based optimization. *Engineering Applications of Artificial Intelligence*.

Rubén, I., & Hirofumi, A. (2005). A 6.6-kV Transformerless Shunt Hybrid Active Filter for Installation on a Power Distribution System. *IEEE Transactions on Power Electronics*, *20*(4), 454–459.

Rudenko, Yu. N., Ushakov, I. A., & Cherkesov, G. N. (1984). Theoretical and methodical aspects of reliability analysis of energy systems. *Energy Reviews: Scientific and Engineering Problems of Energy System Reliability*, *3*, 82–94.

Ruiz-Vega, D., & Pavella, M. (2003). A comprehensive approach to transient stability control. I. Near optimal preventive control. *IEEE Transactions on Power Systems*, *18*(4), 1446–1453. doi:10.1109/TPWRS.2003.818708

Ruminot, P., Morán, L., Aeloiza, E., Enjeti, P., & Dixon, J. (2006). A New Compensation Method for High Current Non-Linear Loads*IEEE International Symposium on Industrial Electronics*, (vol. 2, pp. 1480-1485). doi:10.1109/ISIE.2006.295690

Compilation of References

Saaty, L. (1980). *The analytic hierarchy process: planning, priority setting, resource allocation.* New York: McGraw-Hill.

Sadeghzadeh, S. M., Khazali, A. H., & Zare, S., (2009). *Optimal reactive power dispatch considering TCPAR and UPFC.* EUROCON, St-Petersburg. Doi: 10.1109/EURCON.2009.5167690

Sahuquillo Martínez, S., Pareta, J. S., Otero, B., Ferrari, L., Manzalini, A., & Moiso, C. (2009). Self-Organized Server Farms for Energy Savings. *ICAC-INDST, 9*(June), 16.

Saifullah, K, Dwivedi, B., & Singh, B. (2012). New Optimum Three-Phase Shunt Active Power Filter based on Adaptive Tabu Search and Genetic Algorithm using ANN control in unbalanced and distorted supply conditions. *Elektrika: Journal of Electrical Engineering, 14*(2), 9-14.

Saifullah, K., & Anurag, T. (2012). Comparison of Constant Source Instantaneous Power & Sinusoidal Current Control Strategy for Total Harmonic Reduction for Power Electronic Converters in High Frequency Aircraft System. *International Journal of Advanced Research in Electrical. Electronics and Instrumentation Engineering, l*(5), 314–322.

Saifullah, K., & Anurag, T. (2012). Comparison of Sinusoidal Current Control Strategy & Synchronous Rotating Frame Strategy for Total Harmonic Reduction for Power Electronic Converters in Aircraft System under Different Load Conditions. *International Journal of Advanced Research in Electrical. Electronics and Instrumentation Engineering, l*(5), 305–313.

Saifullah, K., & Anurag, T. (2012). Control and Analysis Of Active Filter Based on Constant Instantaneous Power Control Strategy in High Frequency (400 Hz) Aircraft System with Balanced and Unbalanced Load Conditions. *International Journal of Scientific Computing, 6*(2), 342–345.

Saifullah, K., & Anurag, T. (2013). Simulation of Shunt Active Filters based on Sinusoidal Current Control Strategy in High Frequency Aircraft System. *International Journal of Management. IT and Engineering, 3*(1), 218–230.

Saifullah, K., & Dwivedi, B. (2011). Power Quality Issues. Problems. Standards & their Effects in Industry with Corrective Means. *International Journal of Advances in Engineering & Technology, 1*(2), 1–11.

Sakawa, M. (1993). *Fuzzy sets and interactive multi-objective optimization. Applied Information Technology.* New York: Plenum Press. doi:10.1007/978-1-4899-1633-4

Salam, Z., & Saad, M. S. (2013). Evolutionary based maximum power point tracking technique using differential evolution algorithm. *Energy and Building, 67*, 245–252. doi:10.1016/j.enbuild.2013.07.085

Sangsun, K., & Prasad, N. E. (2002). A New Hybrid Active Power Filter (APF) Topology. *IEEE Transactions on Power Electronics, 17*(1), 23–29.

Saravanan, M., Slochanal, S. M. R., Venkatesh, P., & Abraham, J. P. S. (2007). Application of particle swarm optimization technique for optimal location of FACTS devices considering cost of installation and system loadability. *Electric Power Systems Research, 77*(3-4), 276–283. doi:10.1016/j.epsr.2006.03.006

Sarvi, M., Ahmadi, S., & Abdi, S. (2015). A PSO-based maximum power point tracking for photovoltaic systems under environmental and partially shaded conditions. *Progress in Photovoltaics: Research and Applications, 23*(2), 201–214. doi:10.1002/pip.2416

Sauer, P. W., & Pai, M. A. (1998). *Power System Dynamics and Stability.* Upper Saddle River, NJ: Prentice-Hall.

Savic, D., & Walters, G. (1997). Genetic algorithms for least-cost design of water distribution networks. *J. Water Res. Plan. Manage. Div. Soc. Civ. Eng., 123*(2), 67–77. doi:10.1061/(ASCE)0733-9496(1997)123:2(67)

Sayah, S., & Zehar, K. (2008). Modified differential evolution algorithm for optimal power flow with non-smooth cost functions. *Energy Conversion and Management, 49*(11), 3036–3042. doi:10.1016/j.enconman.2008.06.014

Scheidegger, C., Shah, A., & Simon, D. (2011). Distributed learning with biogeography-based optimization: markov modeling and robot control. In *Proc. 24th International Conference on Industrial Engineering and Other Applications of Applied Intelligent Systems Conference on Modern Approaches in Applied Intelligence* (IEA/ AIE'11) (pp. 203-215).

Semenov, V. (2003). Office of heat supply. *Novosti Teplosnabgenia, 2*(18), 31-39.

Semenov, V. G. (2002). Analysis of possibility for CP operation in the electricity market. *Novosti Teplosnabzheniya, 12*, 45-47.

Sengupta, J. K. (1979). Stochastic goal programming with estimated parameters. *Journal of Economics, 39*(3-4), 225–243. doi:10.1007/BF01283628

Sennova, E. V. (Ed.). (2000). *Reliability of heat supply systems* (Vol. 4). Novosibirsk: Nauka. (in Russian)

Sennova, E. V., & Sidler, V. G. (1987). *Mathematical modeling and optimization of developing heat supply systems*. Novosibirsk: Nauka. (in Russian)

Sennova, E., & Sidler, V. (1987). *Mathematical modeling and optimization of heat supply systems*. Novosibirsk: Nauka. Sib. Otdeleniye.

Shaheen, H. I., Rashed, G. I., & Cheng, S. J. (2011). Optimal location and parameter setting of UPFC for enhancing power system security based on differential evolution algorithm. *International Journal of Electrical Power & Energy Systems, 33*(1), 94–105. doi:10.1016/j.ijepes.2010.06.023

Shakhramanyan, M. A., Tyukhov, I. I., & Voshchenkova, N. S. (2009). Space educational technologies. Moscow: Kaluga. (Rus)

Sharifzadeh, H., & Jazaeri, M. (2009). Optimal reactive power dispatch based on particle swarm optimization considering FACTS devices. *International Conference on Sustainable Power Generation and Supply*. doi:10.1109/SUPERGEN.2009.5347956

Shaw, B., Mukherjee, V., & Ghoshal, S. P. (2012). A novel opposition-based gravitational search algorithm for combined economic and emission dispatch problems of power systems. *International Journal of Electrical Power & Energy Systems, 35*(1), 21–33. doi:10.1016/j.ijepes.2011.08.012

Shi, L., Haoa, J., Zhou, J., & Xu, G. (2004). Ant colony optimization algorithm with random perturbation behavior to the problem of optimal unit commitment with probabilistic spinning reserve determination. *Electric Power Systems Research, 69*(2-3), 295–303. doi:.epsr.2003.10.00810.1016/j

Shifrinson, B. L. (1940). *Main calculation of heat networks*. Gosenergoizdat. (in Russian)

Shifrinson, B.L. (1940). *Main calculation of heat networks*. Gosenergoizdat. (in Russian).

Shifrinson, B. L. (1940). *Basic calculation of heat networks. Theory and computational methods*. Moscow: Gosenergoizdat.

Shiwei, X., Chan, K. W., & Guo, Z. (2014). A novel margin sensitivity based method for transient stability constrained optimal power flow. *Electric Power Systems Research, 108*, 93–102. doi:10.1016/j.epsr.2013.11.002

Siddique, A., Yadava, G. S., & Singh, B. (2005). A Review of stator fault monitoring techniques of induction motors. *IEEE Transactions on Energy Conversion, 20*(1), 215–223. doi:10.1109/TEC.2004.837304

Simakin V.V., Tyuhov I.I., Tikhonov A.V. & Smirnov A.V. (2011). A mathematical model to determine the optimal configuration of the combined system of autonomous power supply based on renewable energy sources. *Mechanization and Electrification of Agriculture: A theoretical and scientific journal*. (in Russian).

Sim, K., & Sun, W. (2003). Ant colony optimization for routing and load balancing: Survey and new directions. *IEEE Transactions on Systems, Man, and Cybernetics. Part A, Systems and Humans, 33*(5), 560–572. doi:10.1109/TSMCA.2003.817391

Simon, D. (2008). Biogeography-based optimization. *IEEE Transactions on Evolutionary Computation, 12*(6), 702–713. doi:10.1109/TEVC.2008.919004

Simon, D., Rarick, R., Ergezer, M., & Du, D. (2011). Analytical and numerical comparisons of biogeography-based optimization and genetic algorithms. *ELSEVIER Journal Information Sciences, 181*(7), 1224–1248. doi:10.1016/j.ins.2010.12.006

Singh, B., & Verma, V. (2003). A New Control Scheme of Series Active Filter for Varying Rectifier Loads. *The Fifth International Conference on Power Electronics and Drive Systems*. doi:10.1109/PEDS.2003.1282900

Sinha, N., Chakrabarti, R., & Chattopadhyay, P. K. (2003). Evolutionary programming techniques for economic load dispatch. *IEEE Transactions on Evolutionary Computation, 7*(1), 83–94. doi:10.1109/TEVC.2002.806788

Sirjani, R., Mohamed, A., & Shareef, H. (2012). optimal allocation of shunt Var compensators in power systems using a novel global harmony search algorithm. *International Journal of Electrical Power & Energy Systems, 43*(1), 562–572. doi:10.1016/j.ijepes.2012.05.068

Sivasubramani, S., & Swarup, K. S. (2012). Multiagent based differential evolution approach to optimal power flow. *Applied Soft Computing, 12*(2), 735–740. doi:10.1016/j.asoc.2011.09.016

Slootweg, J. G., & Kling, W. L. (2003). Is the answer blowing in the wind? *IEEE Power&Energy, 1*(6), 26–33.

Smil, V. (2004). World History and Energy. In Encyclopedia of Energy (vol. 6). Elsevier Inc. doi:10.1016/B0-12-176480-X/00025-5

Smil, V. (2010). *Energy transitions: history, requirements, prospects*. ABC-CLIO, LLC.

Smil, V. (2008). *Energy in nature and society. General energetics of complex systems*. London: The MIT Press Cambridge.

Smirnov, A. V. (1990). *Functional and technological approach to reliability of heat sources in heat supply systems. Heat and energy saving, measurements* (pp. 50–57). Kiev: Institute of Energy Saving Problems of the Ukrainian Academy of Sciences. (in Russian)

Smith, J., Nayak, D. R., & Smith, P. (2014). Wind farms on undegraded peatlands are unlikely to reduce future carbon emission. *Energy Policy, 66*, 585–591. doi:10.1016/j.enpol.2013.10.066

Socha, K., & Dorigo, M. (2008). Ant colony optimization for Continuous domains. *European Journal of Operational Research, 18*(3), 1155–1173. doi:10.1016/j.ejor.2006.06.046

Sokolov, D. V., Stennikov, V. A., Oshchepkova, T. B., & Barakhtenko, Ye. A. (2012). The New Generation of the Software System Used for the Schematic–Parametric Optimization of Multiple Circuit Heat Supply Systems. *Thermal Engineering, 59*(4), 337–343. doi:10.1134/S0040601512040106

Sokolov, E. Ya. (1999). *Combined heat and power production and heat networks*. Moscow: Izdatelstvo MEI.

Sokolov, V. Ya. (1999). *Cogeneration and heat networks*. Moscow, Russia: Izdatelstvo MEI. (in Russian)

Sonmez, Y., Duman, S., Guvenc, U., & Yorukeren, N. (2012). *Optimal power flow incorporating FACTS devices using gravitational search algorithm*. International Symposium on Innovations in Intelligent Systems and Applications (INISTA- 2012). doi:10.1109/INISTA.2012.6246993

Sopapirm, T., Areerak, K.-N., Areerak, K.-L., & Srikaew, A. (2011). The Application of Adaptive Tabu Search Algorithm and Averaging Z`Model to the Optimal Controller Design of Buck Converters. *World Academy of Science. Engineering and Technology, 60*, 477–483.

Sreejith, S., Chandrasekaran, K., & Simon, S. P. (2009). *Application of touring ant colony optimization technique for optimal power flow incorporating thyristor controlled series compensator*. IEEE Region 10 Conference-TENCON 2009. doi:.10.1109/TENCON.2009.5396226

Sreekanth, S., & Raglend, I. J. (2012, March). *A comparative and analytical study of various incremental algorithms applied in solar cell*. In International Conference on Computing, Electronics and Electrical Technologies.

Srinivasan, D., & Chazelas, J. (2004). A priority list-based evolutionary algorithm to solve large scale unit commitment problem. In *Proceedings Int. Conf. on Power System Technology* (pp. 1746–1751). doi:10.1109/ICPST.2004.1460285

Sriyanyong, P., & Song, Y. H. (2005). Unit commitment using particle swarm optimization combined with lagrange relaxation. *IEEE Power Engineering Society General Meeting, 3*, 2752-2759. doi:10.1109/PES.2005.1489390

Stefanov, P. C., & Stank Vic, A. M. (2002). Dynamic phasor in modeling and analysis of unbalanced polyphase ac machines. *IEEE Transactions on Energy Conversion, 17*(1), 107–113. doi:10.1109/60.986446

Stennikov, V. A., & Postnikov, I. V. (2010). Development of methods of reliability analysis heating. *Problemy Energetiki, 5-6*, 28-40. (in Russian)

Stennikov, V. A., & Postnikov, I. V. (2014). Methods for the integrated reliability analysis of heat supply. *Power Technology and Engineering, 6* (47), 446-453.

Stennikov, V. A., & Zharkov, S. V. (2013a). *The gas-burning CPs of the increased fuel efficiency*. First International Forum "Renewable Energy: Towards Raising Energy and Economic Efficiencies" - REENFOR-2013, Moscow.

Stennikov, V. A., & Zharkov, S. V. (2013b). *Problems of WF use in the isolated power systems and possible ways of their solving*. First International Forum "Renewable Energy: Towards Raising Energy and Economic Efficiencies" - REENFOR-2013, Moscow.

Stennikov, V. A., Barakhtenko, E. A., & Sokolov, D. V. (2011). Metaprogramming in the software for solving the problems of heat supply system schematic and parametric optimization. *Programmnaya Inzheneriya, 6*, 31-35. (in Russian)

Stennikov, V. A., Barakhtenko, E. A., & Sokolov, D. V. (2011). Methods of integrated development and reconstruction of heat supply systems on the basis of modern information technologies. *Promyshlennaya Energetika, 4*, 17-22. (in Russian)

Stennikov, V. A., Postnikov, I. V., & Sennova, E. V. (2009). Development of methods for analysis of heat supply reliability. In *Proceedings of the VI International Conference "Mathematical Methods in Reliability Theory. Theory. Methods. Applications"* (pp.459-463). Moscow: Russian Peoples' Friendship University.

Stennikov, V. A., Khamisov, O. V., & Pen`kovskii, A. V. (2011). Optimizing the Heat Market on the Basis of a Two-Level Approach. *Thermal Engineering, 12*(58), 1043–1048. doi:10.1134/S0040601511120111

Stennikov, V. A., Oshchepkova, T. B., & Stennikov, N. V. (2013). Optimal Expansion and Reconstruction of Heat Supply Systems: Methodology and Practice. *International Journal of Energy Optimization and Engineering, 2*(4), 59–79. doi:10.4018/ijeoe.2013100104

Stennikov, V. A., & Postnikov, I. V. (2008). *Integrated assessment of reliability of heat supply to consumers. Pipeline energy systems. Development of theory and methods of mathematical modeling and optimization* (pp. 139–157). Novosibirsk, Russia: Nauka. (in Russian)

Stennikov, V. A., & Postnikov, I. V. (2011). Comprehensive analysis of the heat supply reliability. Izvestya RAN. *Energrtika, 2*, 107–121.

Stennikov, V. A., Sennova, E. V., & Oshchepkova, T. B. (2006). Methods of complex optimization of heat supply systems development. *Izvestiya RAN. Energetika, 3*, 44–54.

Stoft, S. (2006). *Power system economics: Designing markets for electricity*. Moscow: Mir.

Storer, M. W., Greenan, K. M., Miller, E. L., & Voruganti, K. (2008). Pergamum: replacing tape with energy efficient, reliable, disk-based archival storage. FAST'08.

Stott, B. (1979). Power system dynamic response calculations. *Proceedings of the IEEE, 67*(2), 219–241. doi:10.1109/PROC.1979.11233

Strebkov, D. (n.d.). *Trends in Russian agriculture and rural energy*. International Commission of Agricultural Engineering. Retrieved from http://ecommons.library.cornell.edu/bitstream/1813/10217/1/Russia.PDF

Strebkov, D. S., & Tyukhov, I. I. (2005). Russian Section of the International Solar Energy Society. In The Fifty-Year History of the International Solar Energy Society and its National sections. American Solar Energy Society, Inc.

Sultana, S., & Roy, P. K. (2014). Optimal capacitor placement in radial distribution systems using teaching learning based optimization. *International Journal of Electrical Power Systems, 54*, 387–398. doi:10.1016/j.ijepes.2013.07.011

Sumarokov, S. V. (1976). A method of solving the multiextremal network problem. *Ekonomika i Matematicheskiye Metody, 12*(5), 1016-1018. (in Russian)

Sumarokov, S.V. (1976). A method of solving the multiextremal network problem. *Ekonomika i Matematicheskiye Metody, 12*(5), 1016-1018. (in Russian).

Sun, D. I., Ashley, B., Brewer, B., Hughes, A., & Tinney, W. F. (1984). Optimal powerflowby Newton approach. *IEEE Transactions on Power and Apparent Systems, PAS-103*(10), 2864–2875. doi:10.1109/TPAS.1984.318284

Sundareswaran, K., Peddapati, S., & Palani, S. (2014). MPPT of PV systems under partial shaded conditions through a colony of flashing fireflies. *IEEE Transactions on Energy Conversion, 29*(2), 463–472. doi:10.1109/TEC.2014.2298237

Sundareswaran, K., Vignesh kumar, V., & Palani, S. (2015). Application of a combined particle swarm optimization and perturb and observe method for MPPT in PV systems under partial shading conditions. *Renewable Energy, 75*, 308–317. doi:10.1016/j.renene.2014.09.044

Sundar, K. S., & Ravikumar, H. M. (2012). Selection of TCSC location for secured optimal power flow under normal and network contingencies. *International Journal of Electrical Power & Energy Systems, 34*(1), 29–37. doi:10.1016/j.ijepes.2011.09.002

Surekha, P., Archana, N., & Sumathi, S. (2012). Unit commitment and economic load dispatch using self adaptive differential evolution. *WSEAS Transactions on Power Systems, 7*(4), 159-171.

Surmann, H. (1996). Genetic optimization of a fuzzy system for charging batteries. *IEEE Transactions on Industrial Electronics, 43*(5), 541–548. doi:10.1109/41.538611

Suryaman, D., Asvaliantina, V., Nugroho, S., & Kongko, W. (2012). Cooling Water Recirculation Modeling of Cilacap Power Plant. In *Proceeding of the Second International Conference on Port, Coastal, and Offshore Engineering* (2nd ICPCO).

Sutha, S., & Kamaraj, N. (2008). Optimal location of multi type FACTS devices for multiple contingencies using particle swarm optimization. *International Journal of Electrical Systems Science and Engineering, 1*(1), 16–22.

Swain, A. K., & Morris, A. S. (2000). A novel hybrid evolutionary programming method for function optimization. In *Proceedings of 2000 congress on evolutionary computation*, (pp. 699–705). doi:10.1109/CEC.2000.870366

Swarup, K. S. (2006). Swarm intelligence approach to the solution of optimal power flow. Indian Institute of Science.

Swarup, K. S., Yoshimi, M., & Izui, Y. (1994). Genetic algorithm approach to reactive power planning in power systems. In *Proceedings of the 5th Annual Conference of Power and Energy Society*. IEEE. Available online: http://www.ee.washington.edu /research/pstca/

Taher, S. A., & Amooshahi, M. K. (2011). Optimal placement of UPFC in power systems using immune algorithm. *Simulation Modelling Practice and Theory*, *19*(5), 1399–1412. doi:10.1016/j.simpat.2011.03.001

Tajuddin, M. F. N., Arif, M. S., Ayob, S. M., & Salam, Z. (2015). Perturbative methods for maximum power point tracking (MPPT) of photovoltaic (PV) systems: A review. *International Journal of Energy Research*, *39*(9), 1153–1178. doi:10.1002/er.3289

Takaishvili, M. K., & Khasilev, V. Ya. (1972). On the basic concepts of the technique for reliability calculation and reservation of heat networks. *Teploenergetika*, *4*, 14–19.

Tanaka, H., Uejima, S., & Asai, K. (1982). Linear regression analysis with fuzzy model. *IEEE Transactions on Systems. Man and Cybernetics SMC.*, *12*(6), 903–907. doi:10.1109/TSMC.1982.4308925

Tanaka, H., & Watada, J. (1988). Possibilistic linear systems and their application to the linear regression model. *Fuzzy Sets and Systems*, *27*(3), 275–289. doi:10.1016/0165-0114(88)90054-1

Tanaka, H., Watada, J., & Hayashi, I. (1986). Three formulations of fuzzy linear regression analysis. *Transactions of the Society of Instrument and Control Engineers*, *22*(10), 1051–1057. doi:10.9746/sicetr1965.22.1051

Tannehill, J. C., Anderson, D. A., & Pletcher, R. H. (1997). *Computational Fluid Mechanics and Heat Transfer* (2nd ed.). New York: McGraw-Hill.

Tehzeeb, U. H. H., Zafar, R., Mohsin, S. A., & Latee, O. (2012). Reduction in power transmission loss using fully informed particle swarm optimization. *International Journal of Electrical Power & Energy Systems*, *43*(1), 364–368. doi:10.1016/j.ijepes.2012.05.028

Tennekes, H., & Lumley, J. L. (1972). *A first course in turbulence*. The MIT Press.

Terbobri, C. G., Saidon, M. F., & Khanniche, M. S. (2000). Trends of Real Time Controlled Active Power Filters. Power Electronics and variable Speed Drives. Conference publication No. 475 (c) IEE, 410-415. doi:10.1049/cp:20000282

Tewarson, R. P. (1973). *Sparse Matrices (Part of the Mathematics in Science & Engineering series)*. New York: Academic Press Inc.

Tey, K. S., Mekhilef, S., Yang, H., & Chuang, M. (2014). A Differential Evolution Based MPPT Method for Photovoltaic Modules under Partial Shading Conditions. *International Journal of Photoenergy*, *2014*, 1–10. doi:10.1155/2014/945906

Thanushkodi, K., Pandian, S. M. V., & Apragash, R. S. D. (2008). An efficient particle swarm optimization for economic dispatch problems with non-smooth cost functions. *WSEAS Trans. Power Systems*, *4*(3), 257–266.

The IEEE 118-Bus Test System. (n.d.). Available at: http://www.ee.washington.edu/ research/pstca/ pf118/pg_tca118bus.htm

Thukaram, D., & Yesuratnam, G. (2008). Optimal reactive power dispatch in a large power system with AC–DC and FACTS controllers. *IET Generation Transmission and Distribution*, *2*(1), 71–81. doi:10.1049/iet-gtd:20070163

Compilation of References

Titare, L. S., Singh, P., Arya, L. D., & Choube, S. C. (2014). Optimal reactive power rescheduling based on EPSED algorithm to enhance static voltage stability. *International Journal of Electrical Power & Energy Systems, 63*, 588–599. doi:10.1016/j.ijepes.2014.05.078

Tiwari, R. N., Dharmar, S., & Rao, J. R. (1987). Fuzzy goal programming – an additive model. *Fuzzy Sets and Systems, 24*(1), 27–34. doi:10.1016/0165-0114(87)90111-4

Tizhoosh, H. (2005). Opposition-based learning: A new scheme for machine intelligence. In *Proceedings of the International Conference on Computational Intelligence for Modelling Control and Automation (CIMCA-2005)*. doi:10.1109/CIMCA.2005.1631345

Todini, E., & Pilati, S. (1988). A gradient algorithm for the analysis of pipe networks. In *Computer Applications in Water Supply* (Vol. 1, pp. 1–20). London: John Wiley & Sons.

Tolstykh, A. I. (1990). *Compact difference scheme and their applications to fluid dynamics problems. M*. Nauka.

Tolvo, T. (2015). Flue gas condensing and scrubbing: a winning combination. *Modern Power Systems, 35*(3), 18-20.

Tombaz, S., Vastberg, A., & Zander, J. (2011). Energy – and cost-efficient ultra-high-capacity wireless access. IEEE Wireless Communications, 18(5), 18-24.

Trillo-Montero, D., Santiago, I., Luna-Rodriguez, J. J., & Real-Calvo, R. (2014). Development of a software application to evaluate the performance and energy losses of grid-connected photovoltaic systems. *Energy Conversion and Management, 81*, 144–159. doi:10.1016/j.enconman.2014.02.026

Tripathy, M., & Mishra, S. (2007). Bacteria foraging-based solution to optimize both real power loss and voltage stability limit. *IEEE Transactions on Power Systems, 22*(1), 240–248. doi:10.1109/TPWRS.2006.887968

Turan, P., Eren, Ö., Nimet, Y. P., & Gerhard-Wilhelm, W. (2012). Particle Swarm Optimization Approach for Estimation of Energy Demand of Turkey. *Global Journal Technology and Optimization, 3*(1), 2229–8711.

Tu, X., Dessaint, A. L., & Nguyen-Duc, H. (2013). Transient stability constrained optimal power flow using independent dynamic simulation, IET Generation. *Transmission & Distribution, 7*(3), 244–253. doi:10.1049/iet-gtd.2012.0539

Tu, X., Dessaint, L. A., & Kamwa, I. (2013). Fast approach for transient stability constrained optimal power flow based on dynamic reduction method. *IET Gen. Trans. Distr, 8*(7), 1293–1305. doi:10.1049/iet-gtd.2013.0404

Twidell, J., & Weir, A. (1990). *Renewable energy resources: Transl. from the Engl*. Moscow: Energoatomizdat.

Tyukhov, I. I., Raupov, A. H., & Tilov, A. Z. (2012). Global-local monitoring of climate-weather conditions for renewable energy. *Proceedings of the NASA Science Meeting, GOFC-GOLD and NEESPI Workshop and Regional Conference*. Retrieved from http://csfm.marstu.net/publications.html

Tyukhov, I., & Strebkov, D. (2005). Russian Section of the International Solar Energy Society. In The Fifty-Year History of the International Solar Energy Society and its National sections. American Solar Energy Society, Inc.

Tyukhov, I., Schakhramanyan, M., Strebkov, D., Simakin, V., & Poulek, V. (2008). PV and GIS Lab for teaching solar energy. *Proceedings of the 23rd European Photovoltaic Solar Energy Conference*.

Tyukhov, I., Schakhramanyana, M., Strebkov, D., Tikhonova, A., & Vignola, F. (2009). Modeling of solar irradiance using satellite images and direct terrestrial measurements with PV modules. *Proc. of SPIE (The International Society for Optical Engineering), Optical Modeling and Measurements for Solar Energy Systems III*. doi:10.1117/12.827821

Tyukhov, I. I. (2012). On development PV technologies directions taking account resource limitations. *Proceedings of Moscow State University of Mechanical Engineering, 14*(4), 42–46.

Tyukhov, I., & Mazanov, S. (2009). Opportunities of using satellite information for monitoring solar radiation. *VIESH*, *1*(4), 49–52.

Tyukhov, I., Schakhramanyan, M., Strebkov, D., Mazanov, S., & Vignola, F. (2008). Combined solar PV and earth space monitoring technology for educational and research purposes.*Proceedings of the solar 2008 conference.*

U.S. Environmental Protection Agency ENERGY STAR Program. (2007). *Report to Congress on Server and Data Center Energy Efficiency Public Law 109-431*. Author.

Urgaonkar, B., & Chandra, A. (2005). Dynamic provisioning of multi-tier Internet applications.*Proceedings of the 2nd International Conference on Automatic Computing (ICAC'05)*, (pp. 217–228). ICAC.

Ushakov, I. (Ed.). (1985). Reliability of technical systems. Moscow: Radio i svyaz'. (in Russian)

Vaisakh, K., Srinivas, L. R., & Meah, K. (2013). Genetic evolving ant direction particle swarm optimization algorithm for optimal power flow with non-smooth cost functions and statistical analysis. *Applied Soft Computing*, *13*(12), 4579–4593. doi:10.1016/j.asoc.2013.07.002

Valdimarsson, P. (2001). Pipe network diameter optimization by graph theory. In B. Ulanicki (Ed.), Water Software Systems: theory and applications (pp. 1-13). Baldock: Research Studies Press.

Vanderbilt, D., & Louie, S. (1984). A monte-carlo simulated annealing approach to optimization over continuous-variables. *Journal of Computational Physics*, *56*(2), 259–271. doi:10.1016/0021-9991(84)90095-0

Vanitha, M., & Thanushkodi, K. (2012). An effective biogeography based optimization algorithm to solve economic load dispatch problem. *Journal of Computer Science, 8*(9), 1482-1486.

Varadarajan, M., & Swarup, K. S. (2008). Differential evolution approach for optimal reactive power dispatch. *Applied Soft Computing*, *8*(4), 1549–1561. doi:10.1016/j.asoc.2007.12.002

Varadarajan, M., & Swarup, K. S. (2008). Differential evolutionary algorithm for optimal reactive power dispatch. *International Journal of Electrical Power & Energy Systems*, *30*(8), 435–441. doi:10.1016/j.ijepes.2008.03.003

Vasant, P. (2012). Novel Meta-Heuristic Optimization Techniques for Solving Fuzzy Programming Problems. In M. Khan & A. Ansari (Eds.), *Handbook of Research on Industrial Informatics and Manufacturing Intelligence: Innovations and Solutions* (pp. 104–131). Hershey, PA: Information Science Reference; doi:10.4018/978-1-4666-0294-6.ch005

Vasant, P. (2013). Hybrid Optimization Techniques for Industrial Production Planning. In Z. Li & A. Al-Ahmari (Eds.), *Formal Methods in Manufacturing Systems: Recent Advances* (pp. 84–111). Hershey, PA: Engineering Science Reference; doi:10.4018/978-1-4666-4034-4.ch005

Vasant, P. M. (2003). Application of fuzzy linear programming in production planning. *Fuzzy Optimization and Decision Making*, *2*(3), 229–241. doi:10.1023/A:1025094504415

Vasant, P. M. (2013). *Meta-Heuristics Optimization Algorithms in Engineering, Business, Economics and Finans*. Hershey, PA: IGI Global; doi:10.4018/978-1-4666-2086-5

Vasant, P. M. (2014). *Handbook of Research on Novel Solf Computing Intelligent Algorithms: Theory and Practical Applications*. Hershey, PA: IGI Global; doi:10.4018/978-1-4666-4450-2

Vasant, P. M. (2015). *Handbook of Research on Artificial Intelligence Techniques and Algorithms*. Hershey, PA: IGI Global; doi:10.4018/978-1-4666-7258-1

Vasant, P., Barsoum, N., & Webb, J. (2012). *Innovation in Power, Control, and Optimization: Emerging Energy Technolgies*. Hershey, PA: IGI Global; doi:10.4018/978-1-61350-138-2

Compilation of References

Vasant, P., Elamvazuthi, I., & Webb, J. F. (2010). Fuzzy technique for optimization of objective function with uncertain resource variables and technological coefficients. *International Journal of Modeling, Simulation, and Scientific Computing, 1*(3), 349–367. doi:10.1142/S1793962310000225

Vasant, P., Ganesan, T., & Elamvazuthi, I. (2012). Elamvazuthi. Improved Tabu Search Recursive Fuzzy method for crude oil industry. *International Journal of Modeling, Simulation, and Scientific Computing, 3*(1), 20. doi:10.1142/S1793962311500024

Veerachary, M., Senjyu, T., & Uezato, K. (2003). Neural-network-based maximum-power-point tracking of coupled-inductor interleaved-boost converter-supplied PV system using fuzzy controller. *IEEE Transactions on Industrial Electronics, 50*(4), 749–758. doi:10.1109/TIE.2003.814762

Vereecken, W., & VanHeddeghem, W. (2010). Power Consumption in Telecommunication Networks: Overview and Reduction Strategies. *IEEE Communications Magazine, 4*(6).

Verma, A., & Dasgupta, G. (2009). Server workload analysis for power minimization using consolidation. *Proceedings of the 2009 Conference on USENIX Annual Technical Conference (USENIX'09)*. USENIX.

Victoire, T. A. A., & Jeyakumar, A. E. (2005). Unit commitment by a tabu-search-based hybrid optimisation technique. *IEE Proceedings. Generation, Transmission and Distribution, 152*(4), 563–574. doi:10.1049/ip-gtd:20045190

Vignola, F., Michalsky, J., & Stoffel, T. (2012). *Solar and Infrared Radiation Measurements (Energy and the Environment)*. CRC Press. doi:10.1201/b12367

Vijay, B. B., & Prasad, N. E. (1996). An Active line Conditioner to Balance Voltages in a Three-Phase System. *IEEE Trans. on Ind. Application, 32*(2), 14-19.

Vlachogiannis, J. G., & Lee, K. Y. (2008). Quantum-inspired evolutionary algorithm for real and reactive power dispatch. *IEEE Transactions on Power Systems, 23*(4), 1627–1636. doi:10.1109/TPWRS.2008.2004743

Völter, M., Stahl, T., Bettin, J., Haase, A., & Helsen, S. (2006). *Model-Driven Software Development: Technology, Engineering, Management*. Wiley.

Voropai, N. I. (Ed.). (2014). *Reliability of energy systems: problems, models and methods for solving them*. Novosibirsk, Russia: Nauka. (in Russian)

Wai, R., & Tu, C. (2007). Design of total sliding-mode-based genetic algorithm control for hybrid resonant-driven linear piezoelectric ceramic motor. *IEEE Transactions on Power Electronics, 22*(2), 563–575. doi:10.1109/TPEL.2006.889988

Wang, L., Zheng, D. Z., & Tang, F. (2002). An improved evolutionary programming for optimization. In *Proceedings of the Fourth World Congress on Intelligent Control and Automation*. doi:10.1109/WCICA.2002.1021386

Watada, J., & Tanaka, H. (1987). Fuzzy quantification methods. *Proceedings, the 2nd IFSA Congress, at Tokyo*, (pp. 66-69).

Watada, J., & Toyoura, Y. (2002). Formulation of fuzzy switching auto-regression model. *International Journal of Chaos Theory and Applications, 7*(1&2), 67–76.

Watada, J., Wang, S., & Pedrycz, W. (2009). Building confidence interval-based fuzzy random regression model. *IEEE Transactions on Fuzzy Systems, 11*(6), 1273–1283. doi:10.1109/TFUZZ.2009.2028331

Weber, C. I. (2008). *Multi-objective design and optimization of district energy systems including polygeneration energy conversion technologies*. (PhD Thesis).

Wenyin, G., & Zhihua, C. (2013). Differential evolution with ranking-based mutation operators. *IEEE Transaction on Cybernatics, 43*(6), 2066–2081. doi:10.1109/TCYB.2013.2239988 PMID:23757516

Wenzhong, X., & Jiayou, L. (2009). Genetic Algorithm's Optimization for Parameters of Ring-Shaped Heat-Supply Network with Multi-Heat Sources. *Proceedings of the 2009 International Conference on Energy and Environment Technology (ICEET '09)* (Vol. 1, pp. 176-180). doi:10.1109/ICEET.2009.49

Whei-Min, L., & Chih-Ming, H. (2011). A New Elman Neural Network-Based Control Algorithm for Adjustable-Pitch Variable-Speed Wind-Energy Conversion Systems. *IEEE Transactions on Power Electronics, 26*(2), 473–481. doi:10.1109/TPEL.2010.2085454

Wirmond, V. E., Fernandes, T. S. P., & Tortelli, O. L. (2011). TCPST allocation using optimal power flow and genetic algorithms. *International Journal of Electrical Power & Energy Systems, 33*(4), 880–886. doi:10.1016/j.ijepes.2010.11.017

Wood, A. J., & Wollenberg, B. F. (1996). *Power generation operation and control*. New York, NY: John Wiley and Sons.

Wood, D., Reddy, L., & Funk, J. (1993). Modeling Pipe Networks Dominated by Junctions. *Journal of Hydraulic Engineering, 119*(8), 949–958. doi:10.1061/(ASCE)0733-9429(1993)119:8(949)

Woo, S., & Nam, Z. (1991). Semi-Markov reliability analysis of three test/repair policies for standby safety systems in a nuclear power plant. *Reliability Engineering & System Safety, 31*(1), 1–30. doi:10.1016/0951-8320(91)90033-4

Wu, Q. H., & Ma, J. T. (1995). Power system optimal reactive power dispatch using evolutionary programming. *IEEE Transactions on Power Systems, 10*(3), 1243–1249. doi:10.1109/59.466531

Wu, T., Shen, C., Chang, C., & Chiu, J. (2003). 1/spl phi/ 3W Grid-Connection PV Power Inverter with Partial Active Power Filter. *IEEE Transactions on Aerospace and Electronic Systems, 39*(2), 635–646. doi:10.1109/TAES.2003.1207271

Xiao, W., & Dunford, W. G. (2004). *A modified adaptive hill climbing MPPT method for photovoltaic power systems*. In *35th Annual IEEE Power Electronics Specialist Conference*.

Xia, S. W., Zhou, B., Chan, K. W., & Guo, Z. Z. (2015). An improved GSO method for discontinuous non-convex transient stability constrained optimal power flow with complex system model. *International Journal of Electrical Power & Energy Systems, 64*, 483–492. doi:10.1016/j.ijepes.2014.07.051

Xia, X., & Wei, H. (2012). Transient stability constrained optimal power flow based on second-order differential equations. *Procedia Engineering., 29*, 874–878. doi:10.1016/j.proeng.2012.01.057

Xu, Y., Dong, Z. Y., Meng, K., Zhao, J. H., Wong, K. P. (2012). A hybrid method for transient stability-constrained optimal power flow computation. *IEEE Trans. Power Syst., 27*(4), 1769-1777.

Yabuuchi, Y., & Watada, J. (1996). Fuzzy robust regression analysis based on a hyper elliptic function. *Journal of the Operations Research Society of Japan, 39*(4), 512–524.

Yakimov, L. K. (1931). Maximum operating radius of cogeneration-based district heating. *Teplo i Sila, 9*, 8-10. (in Russian)

Yanenko, N. N. (1979). *The Method of Fractional Steps. New York: Springer-Verlag* (J. B. Bunch & D. J. Rose, Eds.). Space Matrix Computations, New York: Academics Press.

Yang, X.-S., & Deb, S. (2009). Cuckoo search via Lévy flights. *Proceedings of World Congress on Nature & Biologically Inspired Computing (NaBIC 2009)*. doi:10.1109/NABIC.2009.5393690

Yang, X.-S., & Deb, S. (2013). Multiobjective cuckoo search for design optimization. *Computers & Operations Research, 40*(6), 1616–1624. doi:10.1016/j.cor.2011.09.026

Yang, Y., & Yu, X. (2007). Cooperative co evolutionary genetic algorithm for digital IIR filter design. *IEEE Transactions on Industrial Electronics, 54*(3), 1311–1318. doi:10.1109/TIE.2007.893063

Yan, W., Yu, J., Yu, D. C., & Bhattarai, K. (2006). A new optimal reactive power flow model in rectangular form and its solution by predictor corrector primal dual interior point method. *IEEE Transactions on Power Systems*, *21*(1), 61–67. doi:10.1109/TPWRS.2005.861978

Yan, X., & Quintana, V. H. (1999). Improving an interior point based OPF by dynamic adjustments of step sizes and tolerances. *IEEE Transactions on Power Systems*, *14*(2), 709–717. doi:10.1109/59.761902

Yan, Z., Xiang, N. D., Zhang, B. M., Wang, S. Y., & Chung, T. S. (1996). A hybrid decoupled approach to optimal power flow. *IEEE Transactions on Power Systems*, *11*(2), 947–954. doi:10.1109/59.496179

Ye, Z., Wu, B., & Zargari, N. (2000). Modelling and simulation of induction motor with mechanical fault based on winding function method. In *Proceedings of the 26th International Conference of the IEEE Industrial Electronics Society*.

Yiqin, Z. (2010). *Optimal reactive power planning based on improved Tabu search algorithm.2010 International Conference on Electrical and Control Engineering*. doi:10.1109/iCECE.2010.961

Yoshida, H., Kawata, K., Fukuyama, Y., Takayama, S., & Nakanishi, Y. (2001). A particle swarm optimization for reactive power and voltage control considering voltage security assessment. *IEEE Transactions on Power Systems*, *15*(4), 1232–1239. doi:10.1109/59.898095

Yuan, Y., Kubokawa, J., & Sasaki, H. (2003). A solution of optimal power flow with multicontingency transient stability constraints. *IEEE Transactions on Power Systems*, *18*(3), 1094–1102. doi:10.1109/TPWRS.2003.814856

Yuryevich, J., & Wong, K. P. (1999). Evolutionary programming based optimal power flow algorithm. *IEEE Transactions on Power Systems*, *14*(4), 1245–1250. doi:10.1109/59.801880

Zadeh, L. A. (1965). Fuzzy sets. *Information and Control*, *8*(3), 338–353. doi:10.1016/S0019-9958(65)90241-X

Zakaryia, M., & Talaq, J. (2011). Economic dispatch by biogeography based optimization method. In *Proceedings of the 2011 International Conference on Signal, Image Processing and Applications with workshop of ICEEA-2011 IPCSIT* (Vol. 21, pp. 161-165).

Zanfirov, A. M. (1933). Technical economic calculation of water and heat networks. *Teplo i Sila*, *11*, 4-10. (in Russian)

Zanfirov, A. M. (1934). On economic calculation of heat networks. *Teplo i Sila*, *12*, 38-39. (in Russian)

Zanfirov, A.M. (1933). Technical economic calculation of water and heat networks. *Teplo i Sila*, *11*, 4-10. (in Russian)

Zanfirov, A.M. (1934). About economic calculation of heat networks. *Teplo i Sila*, (12), 38-39.

Zarate-Minano, R., Cutsem, T. V., Milano, F., & Conejo, A. J. (2010). Securing transient stability using time-domain simulations within an optimal power flow. *IEEE Transactions on Power Systems*, *25*(1), 243–253. doi:10.1109/TPWRS.2009.2030369

Zeleny, M. (1981). The pros and cons of goal programming. *Computers & Operations Research*, *8*(4), 357–359. doi:10.1016/0305-0548(81)90022-8

Zelinka, I., Vasant, P., & Barsoum, N. (2013). Power, Control and Optimization. *Series: Lecture Notes in Electrical Engineering*, *239*, 176.

Zhang, C., Ni, Z., Wu, Z., & Gu, L. (2009). *A novel swarm model with quasi-oppositional particle*. 2009 International Forum on Information Technology and Applications. doi:10.1109/IFITA.2009.525

Zhang, J., Chung, H., Lo, W., Hui, S., & Wu, A. (2001). Implementation of a decoupled optimization technique for design of switching regulators using genetic algorithm. *IEEE Trans. Power Electron*, 752–763.

Zhang, X., Dunn, R. W., & Li, F. (2003). Stability constrained optimal power flow for the balancing market using genetic algorithms. *IEEE Power Engineering Society General Meeting, 2*, 932-937.

Zhang, X., Chen, W., Dai, C., & Cai, W. (2010). Dynamic multi-group self-adaptive differential evolution algorithm for reactive power optimization. *International Journal of Electrical Power & Energy Systems, 32*(5), 351–357. doi:10.1016/j.ijepes.2009.11.009

Zhang, Y. L., Baptista, A. M., & Myers, E. P. (2004). A cross-scale model for 3D baroclinic circulation in estuary-plume-shelf systems. I: Formulation and skill assessment. *Continental Shelf Research, 24*(18), 2187–2214. doi:10.1016/j.csr.2004.07.021

Zhao, Hao, & Wang. (2009). Development strategies for wind power industry in Jiangsu Province, China: Based on the evaluation of resource capacity. *Energy Policy, 37*, 1736–1744.

Zhao, B., Guo, C. X., & Cao, Y. J. (2005). A Multiagent-based particle swarm optimization approach for optimal reactive power dispatch. *IEEE Transactions on Power Systems, 20*(2), 1070–1078. doi:10.1109/TPWRS.2005.846064

Zharkov S. V. (2004). Wind energy use at gas-turbine and steam-turbine plants. *EW, 11*, 58-61.

Zharkov, S. V. (2013a). *Wave power stations as possible export subject of shipyards*. First International Forum "Renewable Energy: Towards Raising Energy and Economic Efficiencies" - REENFOR-2013, Moscow.

Zharkov, S. V. (2013b). *Wind power installation with an inclined axis – the perspective direction of WPIs development for conditions of the RF*. First International Forum "Renewable Energy: Towards Raising Energy and Economic Efficiencies" - REENFOR-2013, Moscow.

Zharkov, S. (2004). Wind use at thermal power plants. *RE-GEN. Wind, 3*, 13–15.

Zharkov, S. V. (2006). Use of wind power plants in power systems. *Энергия: экономика, техника, экология, 10*, 38–39.

Zharkov, S. V. (2009). How to estimate efficiency of energy supply systems. *Gas Turbo Technology, 4*, 7–10.

Zharkov, S. V. (2014). About methods of an assessment of energy supply efficiency and decrease stimulation energy consumption of economy of the RF. *Energetik, 3*, 34–40.

Zhiguo, P., Fang, Z. P., & Suilin, W. (2005). Power Factor Correction Using a Series Active Filter. *IEEE Transactions on Power Electronics, 20*(1), 12–19.

Zhi-Xiang, Z., Keliang, Z., Zheng, W., & Ming, C. (2015). Frequency-Adaptive Fractional-Order Repetitive Control of Shunt Active Power Filters. *IEEE Transactions on Industrial Electronics, 62*(3), 1659–1668. doi:10.1109/TIE.2014.2363442

Zhou, W., Yang, H., & Fang, Z. (2007). A novel model for photovoltaic array performance prediction. *Applied Energy, 84*(12), 1187–1198. doi:10.1016/j.apenergy.2007.04.006

Zhu, J. (2009). *Unit commitment, optimization of power system operation*. Hoboken, NJ: John Wiley & Sons, Inc. Publication. doi:10.1002/9780470466971.ch7

Zhumagulov, B., & Issakhov, A. (2012). Parallel implementation of numerical methods for solving turbulent flows. *Vestnik NEA RK, 1*(43), 12–24.

Zimmermann, H.-J. (1976). Description and optimization of fuzzy systems. *International Journal of General Systems*, *2*(4), 209–215. doi:10.1080/03081077608547470

Zimmermann, H.-J. (1978). Fuzzy programming and linear programming with several objective functions. *Fuzzy Sets and Systems*, *1*(1), 45–55. doi:10.1016/0165-0114(78)90031-3

Zorkaltsev, V. I., & Kolobov, Yu. I. (1984). A simulation model to study reliability of fuel supply to heat generating units. *Trudy Komi filiala Akademii Nauk SSSR* (pp. 33-39). (in Russian)

Zorkaltsev, V., & Ivanova E. (1990). Intensity and synchronism of fluctuations in fuel demand for heating by economic region of the country. Izvestiya AN SSSR. *Energetika i Transport, 6*, 14-22. (in Russian)

Zorkaltsev, V. I., & Ivanova, E. N. (1989). *Analysis of intensity and synchronism of fluctuations in fuel demand for heating*. Syktyvkar: Komi Nauchny Tsentr UrO AN SSSR. (in Russian)

About the Contributors

Pandian Vasant is a senior lecturer at Department of Fundamental and Applied Sciences, Universiti Teknologi PETRONAS in Malaysia. His research interests include Soft Computing, Hybrid Optimization, Holistic Optimization and Applications. He has co-authored research papers and articles in national journals, international journals, conference proceedings, conference paper presentation, and special issues lead guest editor, lead guest editor for book chapters' project, conference abstracts, edited books and book chapters. In the year 2009, P. Vasant was awarded top reviewer for the journal Applied Soft Computing (Elsevier). Currently his Editor-in-Chief of IJCO and IJEOE and Editor of GJTO.

Nikolai I. Voropai is director of the Energy Systems Institute (Siberian Energy Institute until 1997) of the Russian Academy of Science, Irkutsk, Russia. He was born in Belarus in 1943. He graduated from the Leningrad (St. Petersburg) Polytechnic Institute in 1966. N. I. Voropai received his degrees of Candidate of Technical Sciences at the Leningrad Polytechnic Institute in 1974 and Doctor of Technical Sciences at the Siberian Energy Institute in 1990. His research interests include: modeling of power systems, operation and dynamics performance of large power grids; reliability and security of power systems; development of national, international and intercontinental power grids; liberalization of power industry. He is the Chairman of the International Workshop on Reliability Problems of Energy Systems, CIGRE Member, IEEE Fellow Member.

* * *

Nureize Arbaiy is currently with Software Engineering Department at the Faculty of Computer Science and Information Technology, University Tun Hussein Onn Malaysia (UTHM). She received her Dr. of Engineering on 'Satisficing-based Formulation of Fuzzy Random Multi-Criteria Programming Models and Production Applications' from the Graduate School of Information, Production and Systems at Waseda University, Japan. She received her B.S degree from University of Technology Malaysia in 2001 and M.S degree from University Utara of Malaysia in 2004. Her research interests are on multi-criteria decision making, fuzzy logic, expert system, fuzzy regression analysis, and possibilistic theory.

Evgeny Barakhtenko was born on December 19th, 1982 in Russia. He graduated from Irkutsk State Technical University in 2005. Evgeny Barakhtenko received his degree of Candidate of Technical Sciences at Irkutsk State Technical University, Irkutsk, Russia in 2011. He is Senior Researcher at Melentiev Energy Systems Institute of Siberian Branch of the Russian Academy of Sciences, Irkutsk, Russia. Scientific interests area: mathematical models and methods for solving the problems of devel-

About the Contributors

opment and operation control of pipeline systems, study of integrated intelligent energy systems, smart grid technologies, information and intelligence technologies for research into energy pipeline systems, cloud and concurrent computing for energy system research. Evgeny Barakhtenko is the author and the coauthor of more than 40 scientific papers.

Vo Ngoc Dieu received his B.Eng. and M.Eng. degrees in electrical engineering from Ho Chi Minh City University of Technology, Ho Chi Minh city, Vietnam, in 1995 and 2000, respectively and his D.Eng. degree in energy from Asian Institute of Technology (AIT), Pathumthani, Thailand in 2007. He is Research Associate at Energy Field of Study, AIT and lecturer at Department of Power Systems, Faculty of Electrical and Electronic Engineering, Ho Chi Minh City University of Technology, Ho Chi Minh city, Vietnam. His interests are applications of AI in power system optimization, power system operation and control, power system analysis, and power systems under deregulation.

Susanta Dutta was born in 1976 at Bankura, West Bengal, India. He received the BE degree in Electrical Engineering from Dr. B C Roy Engineering College, Durgapur, Burdwan, India in 2004; ME degree from NIT Durgapur, India in 2007. Presently he is working as Assitant Professor in the department of Electrical Engineering, Dr. B. C. Roy Engineering College, Durgapur, India. His field of research interest includes Economic Load Dispatch, Optimal Power flow, FACTS, Automatic Generation Control, Evolutionary computing techniques.

Alibek Issakhov received the BSc, MSc and Ph.D. degree in mathematical and computer modelling specialization in Mechanics and Mathematics faculty from al-Farabi Kazakh National University, Almaty, Kazakhstan, in 2008, 2010 and 2013, respectively. He is a current associated professor in Mathematical and computer modelling department at al-Farabi Kazakh National University. Dr. A. Issakhov has published over 70 peer-reviewed journal and conference paper.

Vikram Kumar Kamboj is presently working as Assistant Professor and Department Coordinator in Electrical Engineering in DAV University, Jalandhar (Punjab), INDIA. He has completed his B.E. (Instrumentation & Control) with Honours from Maharishi Deyanand University, Rohtak and M.Tech (Power System Engineering) with Honours from Rajasthan Technical University, Kota and Ph.D (Electrical Engineering) from Punjab Technical University, Jalandhar. His area of research includes Power System Planning and Optimization, Economic Load Dispatch and Unit Commitment Problem of Electrical Power System.

Saifullah Khalid was born in India. He received his B.E. degree in Electronics & Telecommunication Engineering from SESCOE, Dhule in 2001, M.Tech Degree in Electrical Engineering from MMMEC, Gorakhpur, in 2003 and PhD Degree in 2014 respectively. He worked as a faculty in various engineering colleges from 2003 to 2008. He has been author/co-author of more than 60 papers and 04 books. In 2008, he has joined Airports Authority of India as Air Traffic Control Officer. His current research Interests includes aircraft system, AI Technique and application of optimization algorithms in power quality improvement. He is a member of IEEE, IAENG, IACSIT and the Cleveland engineering society.

Oleg V. Khamisov graduated from mathematical department of Irkutsk State University in 1985. He received the Ph.D. degree in mathematics in 1993 and the Habil. degree in 2011. At present time he is head of mathematical depertment of Energy Systems Institute. His scientific interests are: global optimization, stochastic programming, and operations research. He has about 70 scientific papers published, including one book.

K. Vinoth Kumar received his B.E. degree in Electrical and Electronics Engineering from AnnaUniversity, Chennai, Tamil Nadu, India. He obtained M.Tech in Power Electronics and Drives from VIT University, Vellore, Tamil Nadu, India. Presently he is working as an Assistant Professor in the School of Electrical Science, Karunya Institute of Technology and Sciences (Karunya University), Coimbatore, Tamil Nadu, India. He is pursuing PhD degree in Karunya University, Coimbatore, India. His present research interests are Condition Monitoring of Industrial Drives, Neural Networks and Fuzzy Logic, Special machines, Application of Soft Computing Technique. He has published various papers in international journals and conferences and also published four textbooks. He is a member of IEEE (USA), MISTE and also in International association of Electrical Engineers (IAENG).

R. Saravana Kumar received his B.E. Degree in Electrical and Electronics Engineering from Thiyagarajar College of Engineering, Madurai, and Tamilnadu, India. He obtained M.E., in Power Electronics and Drives from College of Engineering Guindy, Chennai, Tamilnadu, India. He received his PhD degree from the VIT University Vellore, Tamilnadu, India. Presently he is working as a Associate Professor in the School of Electrical Engineering at the VIT University, Vellore. He has published technical and research paper in three international journals and 10 national and 12 international conferences. He has written four books in the area of Special machines and control, Power electronics with Matlab/Simulink, Fuzzy logic and Neural Network, Basic Electrical and Electronics Engineering .His present research interests are Condition Monitoring of Industrial Drives, Real Time control of Electrical systems using LabVIEW &Matlab /dSPACE, Modeling and simulation of Electrical system, Multilevel Inverter, Matrix converter, Special machines, Application of Soft computing Technique, Power Quality.

S. Suresh Kumar received his B.E. degree in Electrical and Electronics Engineering from Bharathiar University, Coimbatore, Tamil Nadu, India. He has obtained M.E. from Bharathiar University, Coimbatore, Tamil Nadu, India. He has received doctoral degree from Bharathiar University, Coimbatore, Tamil Nadu, India in 2007. Presently he is working as a Professor and Head of the Department for Electronics and Communication Engineering in Dr.N.G.P.Institute of Technology, Coimbatore, Tamil Nadu, India. He is having 18 years of teaching experience from PSG College of technology. His present research interests are Electrical Machines and Power Quality. He has already published 107 papers in international journals and international conferences. He is a member of IEEE (USA), ASE, ISCA, MCSI, and MISTE and also in International association of Electrical Engineers.

Pei-Chun Lin received B.Sc. degree in the department of mathematics from National Kaohsiung Normal University, Kaohsiung, Taiwan and received M.Sc. degree in the department of mathematical sciences from National Chengchi University, Taipei, Taiwan. Moreover, she received Ph.D degree from Graduate School of Information, Production and Systems (IPS), Waseda University, Fukuoka, Japan. She is currently an adjunct researcher in the Graduate school of IPS, Waseda University. Her research interests include soft computing, intelligent knowledge, fuzzy statistic analysis and management engineering.

About the Contributors

Debashis Nandi received his B.E. degree in Electronics and Communication Engineering from R. E. College, Durgapur (University of Burdwan), India, in 1994 and M. Tech. Degree from Burdwan University on Microwave Engineering in 1997. He received his Ph.D degree from IIT, Kharagpur, India on Medical Imaging Technology in 2012. His area of research includes Computer security and cryptography, Secure chaotic communication, Image processing, Medical imaging, Video coding optimization etc. He has published more than 25 research papers in national and international journals and two patents. He is an Associate Professor in the Department of Information Technology, National Institute of Technology, Durgapur, India.

Weerakorn Ongsakul obtained B.Eng. (Electrical Eng.) in 1988 from Chulalongkorn University, Thailand; M.S. and Ph.D. (Electrical Eng.) from Texas A&M University, USA in 1991 and 1994, respectively. He is currently an Associate Professor of Energy and former Dean of School of Environment, Resources and Development, Asian Institute of Technology. His research interests are in power system operation, artificial intelligence applications to power systems, smart grid and renewable energy. He has conducted projects sponsored by Sida, EC-ASEAN Energy Facility/ACE and EU-Thailand Economic Co-operation Small Project Facility, and projects sponsored by Energy Conservation and Promotion Fund and Electricity Generating Authority of Thailand (EGAT), Provincial Electricity Authority (PEA) with a combined funding of US$3.0 million. Based on his research work, he has published more than 180 international refereed journal and conference proceedings paper. He served as an Energy Specialist, Energy Standing Committee, Senate of Thailand during 2008-2011. He served as a consultant of Asian Development Bank Institute (ADBI) in 2011-2012. He has been serving as a Secretary General of the Greater Mekong Subregion Academic and Research Network (GMSARN) since 2006. He co-authored one book entitled Artificial Intelligence in Power System Optimization, published by CRC Press/Taylor & Francis in March 2013.

Tamara Oshchepkova was born on September, 13th, 1943 in Russia. She graduated from Kazan State University in 1965. Tamara Oshchepkova received her degree of Candidate of Technical Sciences at Sobolev Institute of Mathematics of the Siberian Branch of the Russian Academy of Sciences, Novosibirsk, Russia in 1983. She is senior researcher at the Energy Systems Institute of the Siberian Branch of the Russian Academy of Sciences, Irkutsk, Russia. Her research interests include: methodical bases and software for solving the problems of optimal development and reconstruction of pipeline systems. Dr. Oshchepkova is the coauthor of several monographs and more than 100 scientific papers.

Sourav Paul was born in 1987 at Dhanbad, Jharkhand, India. He completed his B-Tech in Electrical Engineering from Dr. B. C. Roy Engineering College, Durgapur, West Bengal, India in 2009; M-Tech in Power Systems from Dr. B. C. Roy Engineering College, Durgapur, West Bengal, India in 2012. Presently he is working as Assistant Professor of Electrical Engineering at Dr. B. C. Roy Engineering College, Durgapur, West Bengal, India. His field of interest includes Power System Stabilizer, Automatic Generation Control and Soft computing Techniques.

Andrey V. Penkovskii was born on April 7th, 1982 in Russia. He graduated from Irkutsk State Technical University in 2004. He is Researcher at Melentiev Energy Systems Institute of Siberian Branch of the Russian Academy of Sciences, Irkutsk, Russia. His research interests include: development of methods of mathematical modeling and optimization heat supply system in conditions of the market. Penkovskii is the author and the coauthor of more than 37 scientific papers.

Ivan V. Postnikov was born on January 27th, 1982 in the Irkutsk region of Russia. He is graduated from Irkutsk State Technical University in 2004. Ivan is the heat-power engineer and Senior Researcher at Melentiev Energy Systems Institute of Siberian Branch of the Russian Academy of Sciences. In 2013 defended his thesis on "Methodological support of a comprehensive research of heating reliability". His research interests include: researches in the field of reliability of a heat supply, working out of methods and mathematical models for analysis and synthesis of heat supply systems. Ivan is the co-author of a several monographs and more than 20 scientific papers.

Provas Kumar Roy was born in 1973 at Mejia, Bankura, West Bengal, India. He Received the B.E Degree in Electrical Engineering from R. E. College, Durgapur, Burdwan, India in 1997; M.E Degree in Electrical Machine from Jadavpur University, Kolkata, India in 2001 and Ph.D from NIT Durgapur in 2011. Presently he is working as Associate Professor in the department of Electrical Engineering, Jalpaiguri Government Engineering College, Jalpaiguri, West Bengal, India. His field of research interest includes Economic Load Dispatch, Optimal Power flow, FACTS, Unit Commitment, Automatic Generation Control, Power System Stabilizer, Radial Distribution System, State Estimation and Evolutionary computing techniques.

A. Immanuel Selvakumar received his doctoral degree from Anna University, Chennai, Tamil Nadu, India. Presently he is working as a Professor and Head of the department for Electrical and Electronics Engineering in Karunya Institute of Technology and Sciences (Karunya University), Coimbatore, Tamil Nadu, India. He is having 15 years of teaching experience. His present research interests are Power Quality. He has published several papers in international journals and international conferences. He is a member of IEEE (USA).

Dmitry Sokolov was born in December 7th, 1984 in Irkutsk. He defended PhD thesis in 2013. Dmitry is the senior researcher in Department of Pipeline Energy Systems in Energy Systems Institute. His scientific interests: methods of mathematical modeling and optimization of heat supply systems, development of high-speed algorithms, application of concurrent computing in power engineering, modern software development methodology. Dmitry Sokolov is the co-author of more than 30 scientific papers.

Nikolay Stennikov was born on August 25th, 1983 in Russia. He graduated from Irkutsk State Technical University in 2005. Nikolay Stennikov received his degree of Candidate of Technical Sciences at Melentiev Energy Systems Institute of Siberian Branch of the Russian Academy of Sciences, Irkutsk, Russia in 2009. He is Leading engineer at Melentiev Energy Systems Institute of Siberian Branch of the Russian Academy of Sciences. His research interests include: optimization of structure and functioning of heat supply systems, methods and models for the organization of work of heatsources on the general thermal networks. Nikolay Stennikov is the author and the coauthor of more than 20 scientific papers. ore than 100 scientific papers.

About the Contributors

Valery Stennikov was born on July 9th, 1954 in Russia, PhD, Professor, Doctor of Science. He is the Deputy director of Melentiev Energy Systems Institute of Siberian Branch of the Russian Academy of Sciences, Irkutsk, Russia. Valery Stennikov received his degree of Candidate of Technical Sciences at Energy Systems Institute (Irkutsk) in 1985 and Doctor of Technical Sciences at Melentiev Energy Systems Institute of Siberian Branch of the Russian Academy of Sciences (Irkutsk) in 2002. His research interests include: The methodology, mathematical models and methods for a development substantiation heat supply systems taking into account reliability and controllability requirements; basic tendencies and laws of development of heat supply system, power effective technologies and the equipment; energy saving; methods and algorithms of calculation of economically well-founded differentiated tariffs for thermal energy. Professor Stennikov is the author and co-author of more than 270 scientific papers.

Igor I. Tyukhov is currently Deputy Chair Holder of the UNESCO Chair "Renewable energy sources and rural electrification" at the All-Russian Research Institute for Electrification of Agriculture (VIESH) from 1997 up to now. He has been affiliated with the Moscow Power Engineering Institute for more than 35 years teaching various physics disciplines (from general physics to solid state physics, photovoltaics and semiconductor lasers) and conducting research work on solar energy, solar concentrators, optical metrology, semiconductor physics and technology, solar cells, renewable energy cooperating with the All-Russian Electrotechnical Research Institute. He worked as invited Prof. at the George Mason University (1999-2000), the Oregon State University (2002-2003), and the Oregon Technological Institute (2003). He is Assoc. Editor of Solar Energy Journal. I. Tyukhov is visiting Prof. at Mari El State University and teaches students at the Russian State Agrarian University - Moscow Timiryazev Agricultural Academy (part time). He is the author of more than 300 papers, patents, and book chapters.

Junzo Watada received the B.Sc. and M.Sc. degrees in electrical engineering from Osaka City University, and the Ph.D. degree from Osaka Prefecture University, Osaka, Japan. Currently, he is a Professor of management engineering, knowledge engineering, and soft computing at the Graduate School of Information, Production and Systems, Waseda University, Fukuoka, Japan. His research interests include soft computing, tracking system, knowledge engineering, and management engineering.

Sergey V. Zharkov was born on April 17, 1963 in Russia, PhD. He graduated from Irkutsk State Technical University in 1986. Sergey Zharkov received his degree of Candidate of Technical Sciences at Energy Systems Institute (Irkutsk) in 1996. He is senior researcher at Melentiev Energy Systems Institute of Siberian Branch of the Russian Academy of Sciences, Irkutsk, Russia. His research interests include: technologies of renewable energy sources, thermal power plants and energy supply systems. He has about 70 scientific papers, including 10 invention patents.

Index

A

Adaptive Tabu search algorithm 276, 278, 284, 286, 288-290, 300
Aquatic Environment 227, 243
Artificial bee colony 149-150, 165, 282, 326, 328-329, 335, 341, 347, 352, 402

B

base station 211
Bilevel Program 73
Biogeography Based Optimization (BBO) 149, 178, 244-245, 251, 326-327, 335, 352, 360, 362, 366-367, 369, 377

C

Chaotic Weight Factor 177, 183
Coefficient 5-10, 19, 31-32, 35, 66, 81, 94, 116, 127-129, 132, 134-140, 145, 193, 231, 409, 411, 417-418, 428
Conflict of Reserve 126

D

Decomposition Reliability Analysis 116
Discreteness 49, 52, 57, 78-79
Dynamic Programming 29, 57, 76, 78-79, 82-84, 89, 101, 245

E

Economic Load Dispatch (ELD) 149, 251, 361, 369
Efficiency of Energy Supply Systems 1-2, 11
Energy 1-2, 4, 6-9, 11-12, 15-22, 27-29, 40, 43, 52, 54-57, 59-60, 65-71, 73, 78-79, 82, 85, 102-110, 113, 115-116, 119, 149, 208-209, 211-215, 220, 223-224, 228-229, 240, 243, 276, 281, 283, 295, 297, 309, 391-394, 403, 413-421, 423-425, 427-428, 430, 436-437
energy saving 55, 103, 208-209, 211-212, 214-215, 220, 223-224
Evolutionary Algorithm 148, 150, 154, 160, 245, 252, 269, 328
Evolutionary optimization 140, 169, 244-245, 326

F

Failure Rate of Element 126
Fault Detection 324
Flexible AC transmission system (FACTS) 327
Fuel Saving 1-2, 6, 8, 12
Fuzzy Decision 127, 129, 135-136, 139, 145
Fuzzy Random Regression 127, 129-138, 140, 146
Fuzzy Random Variable 129-132, 146

G

genetic algorithm 150, 178, 245, 279, 282, 286-287, 297, 299, 327, 340
Goal Programming 129, 134-136, 138, 140, 146
Graph Rossander 73
Greenhouse Gas Emissions 2

H

Harmonics 276-284, 288, 292-293, 295, 324
Heat Energy Market 54-55, 59, 73
Heat Network 19, 27-30, 32, 34-36, 39, 42-43, 47-48, 54-55, 57, 61, 65-68, 70, 79, 84, 96-97, 102-103, 106-107, 115, 117-123, 126
Heat Source 13, 31-32, 38, 41, 64-65, 68, 70, 78, 105-107, 109-111, 113-114, 117-120, 228
Heat Supply Complex 107, 110
Heat Supply System 26-28, 40-41, 44-49, 52, 54-58, 60, 62, 68, 73, 76, 80, 86, 92, 96, 101-103, 107, 113
Hybrid power filter 276, 284, 291-293
Hydraulic Circuit Theory 57, 115
Hydrometeorological Conditions 227, 240, 243

I

Information 30, 44, 58, 89, 101, 103-107, 110, 116, 119, 127-129, 134, 139, 262, 282, 287, 335, 337-338, 340, 362-363, 369, 372, 390-391, 403, 407, 419, 424-425, 428, 431-432, 435, 437

L

Level of Heat Supply Reliability 123, 126

M

Markov Random Process 113, 119
Mathematical Formulation 151, 337, 360, 377
Mathematical Model 31, 39, 54, 58, 71, 127-128, 140, 157-158, 227-228, 238, 241, 243, 362, 398, 417-419
Mathematical Modeling 52, 57-58, 77, 94, 102, 230-231
Matlab 157, 185, 255, 276, 282, 284, 287, 290-291, 295, 309, 311, 315, 324, 346, 367, 413-414
Metaprogramming 92-94, 101
Method of Multi-Loop Optimization 29, 43, 52
Methods for Assessing Economic 1
Migration 245, 251, 269, 337-338, 360, 362-364
Model-Driven Engineering 90, 101
Multi-Loop Optimization Method 29, 76, 83, 101
Multi-Objective Evaluation 140, 146
Mutation 150, 245, 251, 269, 337-338, 340-341, 360, 362-366, 408, 413

N

Navier–Stokes Equations 233, 243
Nodal Reliability Indices 106, 109, 115, 126

O

Oligapoliya 73
Ontology 91-95, 101
opposition-based learning 246, 252
Optimal power flow 147-148, 151, 165, 177-178, 251
optimal reactive power dispatch 149, 178, 244, 326-327, 352

P

Parallel Computing 243
parallel technologies 227

Parameters 6, 8, 11, 16, 26-27, 29-31, 39-40, 42-45, 49, 52-53, 57, 59, 76-80, 82-83, 85-86, 88, 90, 94, 101, 110, 116, 118, 128, 130, 135, 137-138, 149-150, 159, 218, 230, 234, 241, 246, 255, 262, 281, 283-284, 287-288, 290, 293-294, 308-310, 322, 324, 327, 339, 342-343, 347, 352, 402, 413-415, 418, 420-421, 423-425, 428, 431, 437
Particle Swarm Optimization 148-150, 177-178, 181, 194, 245, 279, 282, 297, 327, 339, 402
Passive filter 277-278, 284-286, 288, 294
PDT Technique 324
Pipeline 26, 29, 32-33, 35, 37-38, 42, 53, 57, 66, 73, 77-79, 81-84, 88, 90, 94, 97, 101, 126
Post-Emergency Conditions (in the Heat Network) 126
Pseudo-Gradient Search 177-178, 181-182

Q

QoS 208-209, 211-212, 218, 224

R

Real coded genetic algorithm 340
Reliability 27, 38, 44-46, 48, 53, 55, 57, 77, 102-124, 126, 283, 309, 371, 373, 414
Renewable Energy 1-2, 12, 17, 79, 281, 390, 392, 413-415, 417, 419-421, 423-424, 437
Reservoir – Cooler 243
Resource and Environmental Efficiency 1-2
Restoration Rate of Element 126

S

Scheme 13, 15-16, 28, 30, 38-42, 46-48, 56, 60, 62, 68, 77, 81, 86-89, 91, 93, 95-96, 105, 108-109, 115-118, 140, 227, 234, 282, 284, 291, 297, 300, 309, 338
Series active filter 276, 279, 284-286, 288
Shunt active power filter 276, 278, 282, 284, 294, 297
Sinusoidal Fryze voltage control 276, 284, 288
Solar radiation 104, 392-393, 403, 424-426, 428-437
Space monitoring system 426
stratified medium 227
Structure 5, 11, 27, 30, 40, 42-43, 46, 53, 57, 62, 73, 76, 85, 103-104, 106-108, 113, 118-119, 132, 211, 390, 396, 402, 413-415, 423, 426
symmetrical component 309-311

T

Teaching learning based optimization 150, 154, 245
Theory of Hydraulic Circuits 40, 54, 73, 101, 126
Thermal Power Plant 227-228, 243
Thyristor controlled phase shifter (TCPS) 327, 332
Thyristor controlled series compensator (TCSC) 327, 331
Time Reserve of Consumer 126
Total System Efficiency 10
transient stability 147-149, 154, 158-159, 165, 167, 170
Transmission loss 166, 249, 256, 258-259, 261, 263-264, 327-328, 331, 333-334, 342, 347-349, 351-353, 361-362

U

Uncertainties 127-129, 140, 146
Unit Commitment Problem (UCP) 372

V

Voltage Deviation 178, 185, 189, 193-194, 327-328, 333, 342, 347-353
Voltage Stability Index 177-178, 185, 189, 193-194

W

Wind Power 2, 12, 15, 18, 414

Become an IRMA Member

Members of the **Information Resources Management Association (IRMA)** understand the importance of community within their field of study. The Information Resources Management Association is an ideal venue through which professionals, students, and academicians can convene and share the latest industry innovations and scholarly research that is changing the field of information science and technology. Become a member today and enjoy the benefits of membership as well as the opportunity to collaborate and network with fellow experts in the field.

IRMA Membership Benefits:

- **One FREE Journal Subscription**
- **30% Off Additional Journal Subscriptions**
- **20% Off Book Purchases**
- Updates on the latest events and research on Information Resources Management through the IRMA-L listserv.
- Updates on new open access and downloadable content added to Research IRM.
- A copy of the Information Technology Management Newsletter twice a year.
- A certificate of membership.

IRMA Membership $195

Scan code to visit irma-international.org and begin by selecting your free journal subscription.

Membership is good for one full year.

www.irma-international.org

CPSIA information can be obtained at www.ICGtesting.com
Printed in the USA
BVOW07*1816150116

432671BV00011B/197/P